Climate Change and Environmental Sustainability-Volume 1

Climate Change and Environmental Sustainability-Volume 1

Editors

Bao-Jie He
Ayyoob Sharifi
Chi Feng
Jun Yang

MDPI • Basel • Beijing • Wuhan • Barcelona • Belgrade • Manchester • Tokyo • Cluj • Tianjin

Editors
Bao-Jie He
Chongqing University
China

Chi Feng
Chongqing University
China

Jun Yang
Northeastern University
China

Ayyoob Sharifi
Hiroshima University
Japan

Editorial Office
MDPI
St. Alban-Anlage 66
4052 Basel, Switzerland

This is a reprint of articles from the Topics published online in the open access journal *Atmosphere* (ISSN 2073-4433), *Buildings* (ISSN 2075-5309), *Land* (ISSN 2073-445X), *Remote Sensing* (ISSN 2072-4292), *Sustainability* (ISSN 2071-1050) (available at: https://www.mdpi.com/topics/Climate_Environmental).

For citation purposes, cite each article independently as indicated on the article page online and as indicated below:

LastName, A.A.; LastName, B.B.; LastName, C.C. Article Title. *Journal Name* **Year**, *Volume Number*, Page Range.

ISBN 978-3-0365-2672-0 (Hbk)
ISBN 978-3-0365-2673-7 (PDF)

190 k words

Contents

About the Editors vii

Preface to "Climate Change and Environmental Sustainability-Volume 1" ix

Haiyan Liu, Kangning Xiong, Yanghua Yu, Tingling Li, Yao Qing, Zhifu Wang and Shihao Zhang
A Review of Forest Ecosystem Vulnerability and Resilience: Implications for the Rocky Desertification Control
Reprinted from: *Sustainability* **2021**, *13*, 11849, doi:10.3390/su132111849 1

Nate Kauffman and Kristina Hill
Climate Change, Adaptation Planning and Institutional Integration: A Literature Review and Framework
Reprinted from: *Sustainability* **2021**, *13*, 10708, doi:10.3390/su131910708 17

Meiling Zhang, Stephen Nazieh, Teddy Nkrumah and Xingyu Wang
Simulating Grassland Carbon Dynamics in Gansu for the Past Fifty (50) Years (1968–2018) Using the Century Model
Reprinted from: *Sustainability* **2021**, *13*, 9434, doi:10.3390/su13169434 45

Keke Yu, Le Wang, Lipeng Liu, Enguo Sheng, Xingxing Liu and Jianghu Lan
Temperature Variations and Possible Forcing Mechanisms over the Past 300 Years Recorded at Lake Chaonaqiu in the Western Loess Plateau
Reprinted from: *Sustainability* **2021**, *13*, 11376, doi:10.3390/su132011376 65

Jing Peng, Li Dan, Jinming Feng, Kairan Ying, Xiba Tang and Fuqiang Yang
Absolute Contribution of the Non-Uniform Spatial Distribution of Atmospheric CO_2 to Net Primary Production through CO_2-Radiative Forcing
Reprinted from: *Sustainability* **2021**, *13*, 10897, doi:10.3390/su131910897 81

Alex Taylor, Maarten Wynants, Linus Munishi, Claire Kelly, Kelvin Mtei, Francis Mkilema, Patrick Ndakidemi, Mona Nasseri, Alice Kalnins, Aloyce Patrick, David Gilvear and William Blake
Building Climate Change Adaptation and Resilience through Soil Organic Carbon Restoration in Sub-Saharan Rural Communities: Challenges and Opportunities
Reprinted from: *Sustainability* **2021**, *13*, 10966, doi:10.3390/su131910966 99

Jiejie Han, Xi Zhao, Hao Zhang and Yu Liu
Analyzing the Spatial Heterogeneity of the Built Environment and Its Impact on the Urban Thermal Environment—Case Study of Downtown Shanghai
Reprinted from: *Sustainability* **2021**, *13*, 11302, doi:10.3390/su132011302 121

Harold L. W. Chisale, Paxie W. Chirwa, Folaranmi D. Babalola and Samuel O. M. Manda
Perceived Effects of Climate Change and Extreme Weather Events on Forests and Forest-Based Livelihoods in Malawi
Reprinted from: *Sustainability* **2021**, *13*, 11748, doi:10.3390/su132111748 139

Ha Pham and Marc Saner
A Systematic Literature Review of Inclusive Climate Change Adaption
Reprinted from: *Sustainability* **2021**, *13*, 10617, doi:10.3390/su131910617 155

Ion V. Ion and Antoaneta Ene
Evaluation of Greenhouse Gas Emissions from Reservoirs: A Review
Reprinted from: *Sustainability* **2021**, *13*, 11621, doi:10.3390/su132111621 . . . . **173**

Ie Zheng Goh and Nitanan Koshy Matthew
Residents' Willingness to Pay for a Carbon Tax
Reprinted from: *Sustainability* **2021**, *13*, 10118, doi:10.3390/su131810118 . . . . **187**

Anton Galich, Simon Nieland, Barbara Lenz and Jan Blechschmidt
How Would We Cycle Today If We Had the Weather of Tomorrow? An Analysis of the Impact
of Climate Change on Bicycle Traffic
Reprinted from: *Sustainability* **2021**, *13*, 10254, doi:10.3390/su131810254 . . . . **213**

Shiran Victoria Shen
Integrating Political Science into Climate Modeling: An Example of Internalizing the Costs of
Climate-Induced Violence in the Optimal Management of the Climate
Reprinted from: *Sustainability* **2021**, *13*, 10587, doi:10.3390/su131910587 . . . . **235**

Karla G. Cedano, Tiare Robles-Bonilla, Oscar S. Santillán and Manuel Martínez
Assessing Energy Poverty in Urban Regions of Mexico: The Role of Thermal Comfort and
Bioclimatic Context
Reprinted from: *Sustainability* **2021**, *13*, 10646, doi:10.3390/su131910646 . . . . **259**

**Anni Arumsari Fitriany, Piotr J. Flatau, Khoirunurrofik Khoirunurrofik and Nelly Florida
Riama**
Assessment on the Use of Meteorological and Social Media Information for Forest Fire Detection
and Prediction in Riau, Indonesia
Reprinted from: *Sustainability* **2021**, *13*, 11188, doi:10.3390/su132011188 . . . . **273**

About the Editors

Bao-Jie He is a research professor of urban climate and built environment at the School of Architecture and Urban Planning, Chongqing University, China. Prior to Chongqing University, Bao-Jie He was a PhD researcher at the Faculty of Built Environment, University of New South Wales, Australia. Bao-Jie works for Cool Cities and Communities and Net Zero Carbon Built Environment. Bao-Jie has strong academic capability, with about 80 peer-reviewed papers published in high-ranking journals and oral presentations given at reputable conferences. Bao-Jie acts as Topic Editor-in-Chief, Leading Guest Editor, Associate Editor, Editorial Board Member, Conference Chair, Sessional Chair and on the Scientific Committee of a variety of international journals and conferences. Dr. He received the Green Talents Award (Germany) in 2021 and the National Scholarship for Outstanding Self-Funded Foreign Students (China) in 2019. Dr. He was ranked as one of the top 100,000 global scientists (both single-year and career top 2%) by Mendeley, 2021.

Ayyoob Sharifi is with the Graduate School of Humanities and Social Sciences, Hiroshima University. He also has a cross-appointment at the Graduate School of Advanced Science and Engineering. Ayyoob's research is mainly at the interface of urbanism and climate change mitigation and adaptation. He actively contributes to global change research programs, such as Future Earth, and is currently serving as a lead author for the Sixth Assessment Report (AR6) of the Intergovernmental Panel on Climate Change (IPCC). Before joining Hiroshima University, he was the Executive Director of the Global Carbon Project (GCP)—a Future Earth core project—leading the urban flagship activity of the project, which is focused on conducting cutting-edge research to support climate change mitigation and adaptation in cities.

Chi Feng received his joint PhD training in South China University of Technology (China) and KU Leuven (Belgium). He is now a research professor in the School of Architecture and Urban Planning, Chongqing University (China), and is leading a research group of more than 10 members. His research topics cover the coupled heat and moisture transfer in porous building materials as well as the hygrothermal performance of building envelopes and built environments. He has led eight international and national research projects, including a China–Europe round robin campaign on material property determination (nine countries participated), the National Natural Science Foundation of China and the National Key R&D Program of China. He has published more than 60 peer-reviewed journal/conference papers at home and abroad. He has been drafting two Chinese standards and participating in another nine international/national ones.

Jun Yang is working at the Urban Climate and Human Settlements Lab, Northeastern University (Shenyang China). His research expertise involves urban climate zones, urban ecology, urban human settlements and sustainability. As PI or Co-PI, he has been involved in 50 research projects, receiving a total of 15 million in RMB from EGOV.CN (e.g., NFC, MOST and MOE) since 2002. He has authored and co-authored more than 160 papers and book chapters and published more than 50 English papers as well as more than 110 Chinese papers in academic journals. He is now an Associate Editor of *SN Social Sciences* and the *International Journal of Environmental Science and Technology*, on the Editorial Board of *PLOS One*, *PLOS Climate* and *Frontiers in Built Environment*, the Lead Guest Editor of *Complexity* and a Guest Editor of the *IEEE Journal of Selected Topics in Applied Earth Observations and Remote Sensing*.

Preface to "Climate Change and Environmental Sustainability-Volume 1"

The Earth's climate is changing; the global average temperature is estimated to already be about 1.1 °C above pre-industrial levels. Indeed, we are now living in conditions of a climate emergency. Climate change leads to many adverse events, such as extreme heat, flooding, bushfire, drought, and many other associated economic and social consequences. Further warming is projected to occur in the coming decades, and climate-induced impacts may exceed the capacity of society to cope and adaptive in a 1.5 °C or 2 °C world. Therefore, urgent actions should be taken to address climate change and avoid irreversible environmental damages.

Climate change is interrelated with many other challenges such as urbanisation, population increase and economic growth. For instance, cities are now the main settlements of human being and are major sources of greenhouse gas emissions that are key contributors to climate change. Moreover, rapid and unregulated urbanisation in some contexts further causes urban problems such as environmental pollution, traffic congestion, urban flooding and heat island intensification. In the absence of well-designed measures, increasing urbanisation trends in the next two–three decades are likely to further aggravate such problems. Overall, climate change and many other challenges have deteriorated the sustainable development of the world.

The United Nations proposed the Sustainable Development Goals in 2015. Goal 13, Climate Action, emphasises the need for urgent action to combat climate change and its impacts in order to enhance sustainability. To achieve this, there is a need to develop a holistic framework that considers mitigation—the decarbonisation of society—to address the challenge of climate change from the root, and adaptation—an immediate action—to increase the resilience of and protect society from climate-induced hazards. The framework prioritises the transformation of the traditional methods of environmental modifications in various fields, including transportation, industry, building, energy generation, agriculture, land use and forestry, towards sustainable ones to limit greenhouse gas emissions. The framework also highlights the significance of sustainable environmental planning and design for adaptation in order to reduce climate-induced threats and risks. Moreover, it encourages the involvement and participation of all stakeholders to accelerate climate change mitigation and adaptation progress by developing sound climate-related governance systems.

The framework also calls for the support and engagement of all societal stakeholders. To support the achievement and implementation of the framework, this book focuses on climate change and environmental sustainability by covering four key aspects, including climate change mitigation and adaptation, sustainable urban–rural planning and design, decarbonisation of the built environment in addition to climate-related governance and challenges. Climate change mitigation and adaptation covers topics of greenhouse gas emissions and measurement, climate-related disasters and reduction, risk and vulnerability assessment and visualisation, impacts of climate change on health and well-being, ecosystem services and carbon sequestration, sustainable transport and climate change mitigation and adaptation, sustainable building and construction, industry decarbonisation and economic growth, renewable and clean energy potential and implementation in addition to environmental, economic and social benefits of climate change mitigation.

Sustainable urban–rural planning and design deals with questions of climate change and regional economic development, territorial spatial planning and carbon neutrality, urban overheating mitigation and adaptation, water-sensitive urban design, smart development for urban habitats, sustainable land use and planning, low-carbon cities and communities, wind-sensitive urban planning and design, nature-based solutions, urban morphology and environmental performance in addition to innovative technologies, models, methods and tools for spatial planning. Decarbonisation of the built environment addresses issues of climate-related impacts on the built environment, the health and well-being of occupants, demands on energy, materials and water, assessment methods, systems and tools, sustainable energy, materials and water systems, energy-efficient design technologies and appliances, smart technology and sustainable operation, the uptake and integration of clean energy, innovative materials for carbon reduction and environmental regulation, building demolition and material recycling and reusing in addition to sustainable building retrofitting and assessment. Climate-related governance and challenges concerns problems of targets, pathways and roadmaps towards carbon neutrality, pathways for climate resilience and future sustainability, challenges, opportunities and solutions for climate resilience, the development and challenges climate change governance coalitions (networks), co-benefits and synergies between adaptation and mitigation measures, conflicts and trade-offs between adaptation and mitigation measures, mapping, accounting and trading carbon emissions, governance models, policies, regulations and programs, financing urban climate change mitigation, education, policy and advocacy of climate change mitigation and adaptation in addition to the impacts and lessons of COVID-19 and similar crises.

Overall, this book aims to introduce innovative systems, ideas, pathways, solutions, strategies, technologies, pilot cases and exemplars that are relevant to measuring and assessing the impact of climate change, mitigation and adaptation strategies and techniques in addition to public participation and governance. The outcomes of this book are expected to support decision makers and stakeholders to address climate change and promote environmental sustainability. Lastly, this book aims to provide support for the implementation of the United Nations Sustainable Development Goals and carbon neutrality in efforts aimed at achieving a more resilient, liveable and sustainable future.

Climate change has been widely recognised as a major challenge to the world, with significant environmental, economic and social consequences. Given this, addressing climate change is an urgent and profound task of society, a complex and difficult mission of several generations. To address the challenge of climate change, there is a need to develop a holistic climate change mitigation and adaptation framework that can cover as many climate-related topics as possible and connect as many stakeholders as possible across the globe. This book is an important one, bringing together key climate-related topics, including climate-induced impact assessment, environmental vulnerability and resilience assessment, greenhouse gas emission dynamics and sequestration, climate change mitigation and adaptation strategies in addition to climate-related governance. Results reported in this book are conducive to a better understanding of the climate emergency, climate-related impacts and the solutions. We expect the book to benefit decision makers, practitioners and researchers in different fields such as climate modelling and prediction, forest ecosystems, land management, urban planning and design, urban governance in addition to institutional operation.

Prof. Bao-Jie He acknowledges Project NO. 2021CDJQY-004, supported by the Fundamental Research Funds for the Central Universities. We appreciate the assistance from Mr. Lifeng Xiong, Mr. Wei Wang, Ms. Xueke Chen and Ms. Anxian Chen at the School of Architecture and Urban Planning, Chongqing University, China.

Bao-Jie He, Ayyoob Sharifi , Chi Feng, Jun Yang
Editors

Review

A Review of Forest Ecosystem Vulnerability and Resilience: Implications for the Rocky Desertification Control

Haiyan Liu, Kangning Xiong *, Yanghua Yu, Tingling Li, Yao Qing, Zhifu Wang and Shihao Zhang

School of Karst Science, State Engineering Technology Institute for Karst Desertification Control, Guizhou Normal University, Guiyang 550001, China; lhy1679454421@163.com (H.L.); yuyanghua2003@163.com (Y.Y.); tinglingli2021@163.com (T.L.); qin333yao444@163.com (Y.Q.); wzfkst@163.com (Z.W.); sh_zhang1993@163.com (S.Z.)
* Correspondence: xiongkn@163.com

Abstract: With a changing climate and socio-economic development, ecological problems are increasingly serious, research on ecosystem vulnerability and ecological resilience has become a hot topic of study for various institutions. Forests, the "lungs of the earth", have also been damaged to varying degrees. In recent years, scholars have conducted numerous studies on the vulnerability and resilience of forest ecosystems, but there is a lack of a systematic elaboration of them. The results of a statistical analysis of 217 related documents show: (1) the number of studies published rises wave upon wave in time series, which indicates that this area of study is still at the stage of rising; (2) the research content is concentrated in four dimensions—ecosystem vulnerability assessment, ecosystem vulnerability model prediction, ecological resilience, and management strategies—among which the ecosystem vulnerability assessment research content mainly discusses the evaluation methods and models; (3) the research areas are mainly concentrated in China and the United States, with different degrees of distribution in European countries; and (4) the research institutions are mainly the educational institutions and forestry bureaus in various countries. In addition, this paper also reveals the frontier theory of forest ecosystem vulnerability and resilience research from three aspects—theoretical research, index system, and technical methods—puts forward the problems of current research, and suggests that a universally applicable framework for forest ecosystem vulnerability and resilience research should be built in the future, and theoretical research should be strengthened to comprehensively understand the characteristics of forest ecosystems so that sustainable management strategies can be proposed according to local conditions.

Keywords: ecosystem vulnerability; ecological resilience; forest; review

Citation: Liu, H.; Xiong, K.; Yu, Y.; Li, T.; Qing, Y.; Wang, Z.; Zhang, S. A Review of Forest Ecosystem Vulnerability and Resilience: Implications for the Rocky Desertification Control. *Sustainability* **2021**, *13*, 11849. https://doi.org/10.3390/su132111849

Academic Editors: Baojie He, Ayyoob Sharifi, Chi Feng and Jun Yang

Received: 10 October 2021
Accepted: 25 October 2021
Published: 27 October 2021

1. Introduction

Forests, which play a key role in the global carbon cycle, are an important part of the global biosphere [1]. As the mainstay of terrestrial ecosystems and an important renewable resource, forests provide a wide range of ecosystem services to humans [2], not only providing timber and forest products, but also rich species and genetic diversity, which can regulate climate, prevent soil erosion, and suppress wind and sand damage, etc., which plays an extremely significant role in human survival and development [3]. The value created by global ecosystems is estimated to be around USD 33 trillion per year [4]. However, under the influence of natural factors and human disturbances, including climate change, drought, storms, insect invasion, fire, and deforestation [5–10], forest ecosystems have been damaged to varying degrees. For example, selective human logging has reduced vegetation canopy cover and posed a significant threat to soil erosion, then leading to a reduction in the ecological resilience of forests, which in turn presents greater vulnerability [11]. In 2015, Article 5 of the United Nations Framework Convention on Climate Change (UNFCCC) Paris Agreement highlighted the central role of forest ecosystems in achieving the goal of limiting temperature rise objectives and encouraged Parties to manage and protect forest

1

ecosystems, recognizing the importance of forest ecosystems in terms of their potential to mitigate climate change and achieve non-carbon benefits [12]. Therefore, it is scientifically important to capture the vulnerability and improve the resistance and resilience of the forest ecosystem.

The concept of vulnerability was first used in the social sciences and is now more commonly applied in ecological research [13]. The ecosystem vulnerability was first introduced to research by the American scholar Clements in 1905 [14]. Later, Niu [15] redefined Ecotone as the "interface" between two or more material, energy, structural, and functional systems in an ecosystem, and the spatial domain of the "transition zone" that extends outwards around the interface, which is called the ecosystem vulnerability zone. At the beginning of the 21st century, the IPCC third assessment report defined vulnerability in the context of climate change as "The degree to which a natural or social system is vulnerable or incapable of coping with the adverse effects of climate change as a function of the characteristics, magnitude and rate of change of a system's climate and its sensitivity and adaptive capacity" [16] and Turner et al. [17] defined vulnerability as a function of the exposure, sensitivity, and adaptive composition of a system under stress. Most current research follows the IPCC definition and considers that the basic components of vulnerability research include the assessment of system change, the evaluation of the sensitivity of the system to respond to change, the estimation of the potential impact of change on the system, and the evaluation of the system's adaptability to change and its possible impacts.

In the mid-nineteenth century, along with the development of western industry, 'resilience' was mainly used in mechanics to describe the ability of metals to recover after deformation by external forces [18]. It was not until the 1970s that Canadian scholar Holling [19] first introduced the concept of resilience to the field of ecology; in the 1980s and 1990s, ecological resilience was considered as the ability of complex systems to absorb external shocks and ensure the continuation of the original functions and structures of the system, thus stimulating thoughts on how to make systems more resilient and stronger by allowing them to absorb greater external shocks [20]; subsequently, Gunderson and Holling [21] proposed the Panarchy model of ecosystem evolutionary dynamics and a multi-scale nested adaptive cycle model with core concepts describing the adaptive evolution of complex systems, a time–space-conscious hierarchical order, and adaptive cycles. Liao [22] argues that ecological resilience does not need to take into account changes in the state of the system, but should indicate the system's ability to survive; based on previous research, George et al. [23] argues that ecological resilience is the system's ability to adapt and recover in the face of external disturbances, and that its maintenance is a key objective of ecological restoration; a further conceptualization sees ecological resilience as emphasizing the need for systems to reach new and diverse states of equilibrium with the ability to adapt and change to transform to these different states [24].

The study of forest ecosystem vulnerability and ecological resilience has important implications for revegetation and sustainable forest management. Forest ecosystems are inherently resilient, and many species and ecosystems have already adapted to historically changing climatic conditions, but the scale and rate of future change may exceed the natural adaptive capacity of forest species and ecosystems [25]. Including the current global warming has caused many changes in forests [26], such as droughts that have increased forest wood mortality [27], and these changes may be exacerbated by responses to human activities, such as the introduction of exotic pests, habitat destruction, and fires [28], resulting in forest ecosystem vulnerability. Ecosystem vulnerability studies are an important tool for capturing the state and spatial distribution characteristics of regional habitat vulnerability [29] however, studies of vulnerability vary by region and by genesis, limiting the general applicability of the concept of vulnerability and comparisons between different domains. In addition, ecological resilience is a concept closely related to ecosystem vulnerability, which is often quantified as the recovery time of a system after a disturbance [30]. However, current research lacks a general framework for linking vulnerability and resilience [31].

Additionally, although resilience is currently widely considered in forests, it has not been widely implemented in forest research and management [32]. Vulnerability studies can therefore be conducted to identify risks arising from future changes and to identify key vulnerable resources; resilience studies can address the uncertain future of forestry by quantifying the resilience of systems to disturbance, thus providing guidance for enhancing the ecological services of forests.

This study provides a comprehensive analysis of existing research on forest ecosystem vulnerability and ecological resilience, and classifies the literature according to annual distribution, research content, research areas, and institutions. It summarizes the current domestic and international research progress and main achievements related to ecosystem vulnerability and ecological resilience, proposes nine key scientific questions to provide directions for future research on forest ecosystem vulnerability and ecological resilience, and provides a theoretical basis for the current enhancement of forest ecosystem functions under the rocky desertification control in karst areas.

2. Acquisition and Argumentation of Literatures

In order to understand the current state of research on the vulnerability and resilience of forest ecosystems. In this paper, the English-language literature is from the "Web of Science" database, which has extensive and multidisciplinary bibliographic data on cutting-edge scientific publications, and this database is commonly used for academic research and bibliometric analysis in the field [33,34]; the Chinese literature is from "China National Knowledge Infrastructure". The aim of the literature search was to focus on the research theme of vulnerability and resilience of forest ecosystems, with the terms "vulnerability" and "resilience" used to express theoretical knowledge related to the ecological field. Therefore, the first search was conducted using "topic" as the search term and "forest" as the first search word; among the results, "Ecological/Ecosystem vulnerability" and "Ecological/Ecosystem resilience" were used as the search words for the second search; the search time range was the maximum time range of the database.

Through an initial review of article titles, abstracts, and keywords, the screened articles focused on forest ecosystems and associated systems (e.g., tree species and forest-dependent communities) and explicitly examined concepts, methods, and models of vulnerability and resilience; we also accepted studies that provided methods for assessing vulnerability and resilience of non-specific ecosystems, as they apply to forests as well. Then, the full text was further screened to eliminate duplicates, and 217 articles were finally selected for review, including 61 in Chinese journals and 156 in English journals. Among them included 197 journal articles, 13 master's theses, 5 doctoral theses, and 2 conference articles. Microsoft Excel 2016 was used to analyze the trend of the number of publications and, at the same time, the statistical analysis function that comes with the Web of Science and CNKI systems was used to statistically analyze the time, country, and institution of publications collected. This paper systematically analyses the literature on forest ecosystem vulnerability and resilience; summarizes main research results and landmark achievements in three aspects, including basic theory, indicator systems, and technical methods; and proposes key scientific questions based on existing research, in order to provide a scientific basis for future research on forest ecosystem vulnerability and resilience, promote forest ecosystem conservation and sustainability, and ultimately achieve harmonious development between humans and nature.

2.1. Annual Distribution of the Literature

As shown in Figure 1, domestic and international research on forest ecosystem vulnerability and resilience can be roughly divided into three stages. The first stage was from 1996 to 2008, with a relatively small fluctuation in the curve and no more than five publications per year, as it was an embryonic stage; at the second stage (2008–2014), the number began to grow slowly and the curve showed a relatively large increase; in the third stage (2014–2021), the curve rose rapidly and the number of publications showed a substantial increase, reaching the highest level in 2020. Overall, English literature follows much the same trend as the overall literature, while the Chinese literature is smaller.

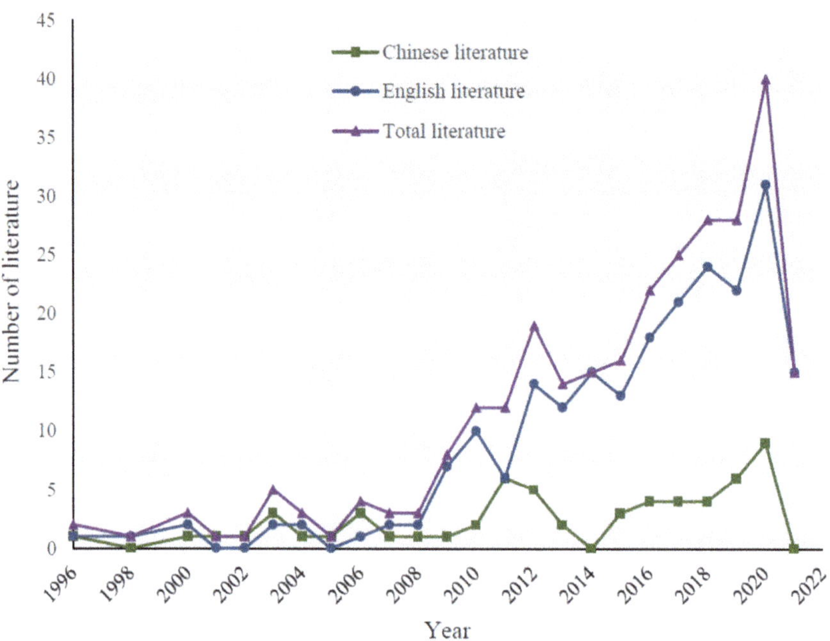

Figure 1. Trends in the literature related to ecosystem vulnerability and ecological resilience research of forest ecosystem by year until 2021.

2.2. Content Distribution of the Literature

The content of forest ecosystem vulnerability and resilience research is shown in Figure 2, which divides all literature into ecosystem vulnerability assessment, modelling prediction of ecosystem vulnerability, ecological resilience, management strategies, and other types, according to their research content. The ecosystem vulnerability assessment literature accounted for 36.74% and its modelling prediction category accounted for 9.77%. Ecosystem vulnerability research includes theoretical studies, indicator system construction, evaluation methods, and modelling prediction, among which evaluation methods and indicator system construction are the mainstream, while modelling prediction is an emerging element of ecosystem vulnerability research in recent years [6,35]. Ecological resilience research accounts for 34.42%, with studies mainly aiming to enhance system resistance and reduce system vulnerability [36,37]. Management strategies accounted for 12.56% and other types of literature accounted for 6.51%. Overall, there is a predominance of research on forest ecosystem vulnerability and resilience, but it is mostly independent and less theoretical research.

2.3. Country Distribution of the Literature

The literature regions are organized as shown in Figure 3. Due to spatial constraints, only regions with more than three articles are listed in this article. Studies on the vulnerability and resilience of forest ecosystems are dominated by China and the United States, accounting for 25% and 18% of the overall studies, respectively. China has conducted studies to varying degrees in the northwest desert belt, the northeast forest–grass interlacing belt and the southwest karst–karst fragile belt, followed by Canada, Germany, Spain, Australia, Brazil, and France, with Canada accounting for 7%, Germany accounted for 6%. The rest of the countries, including the UK, India, and Switzerland, also have a small number of studies. Overall, most of the study areas are located in temperate, tropical, and subtropical regions. Additionally, they are concentrated in Asia (China), North America (USA, Canada, etc.), and Europe (Germany, Spain, etc.). There are fewer studies of vulnerable areas of karst rocky desertification.

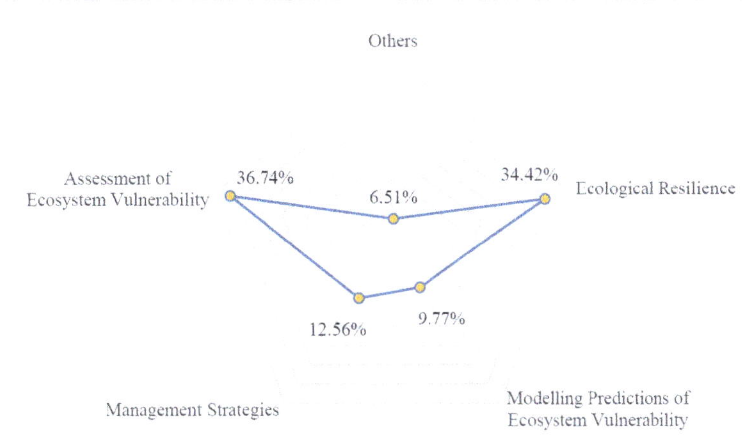

Figure 2. Literature by content.

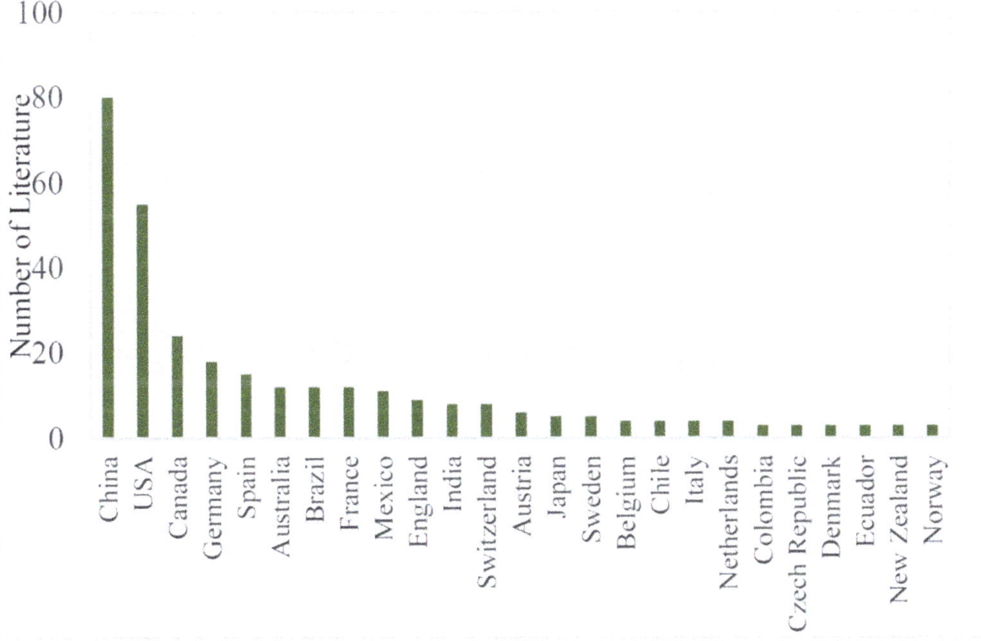

Figure 3. Literature by country.

2.4. Institution Distribution of the Literature

A count of all the literature reveals a wide range of research units. Due to the length of the chart, only units with three or more articles were counted, with a total of 118 articles and 23 units (Figure 4). Among them, the United States Forest Service has the largest number of publications, with a cumulative total of 13, and Beijing Forestry University has 11, followed by the United States Geological Survey, Technical University of Munich, University of Minnesota System, University of Wisconsin Madison, University of Vermont, and University of Nacional Autonoma de Mexico. All the research units, except in China, are mostly institutions of higher education or forestry in developed countries, such as the United States of America and Germany.

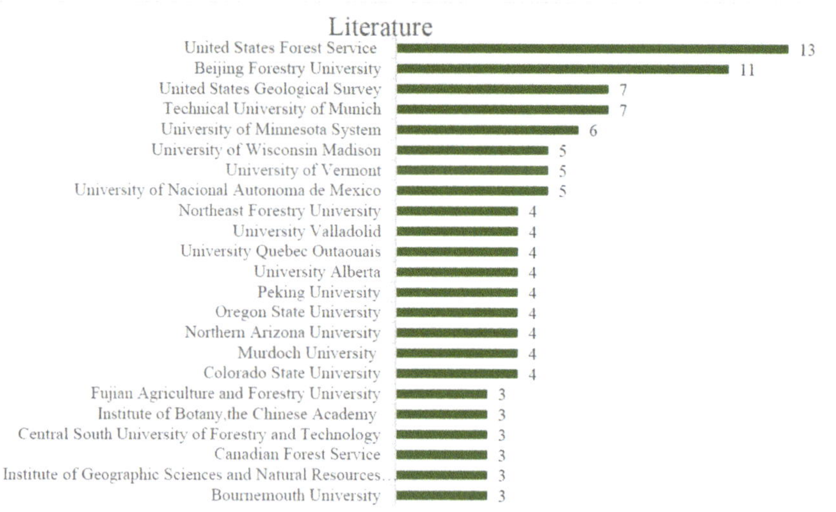

Figure 4. Literature by institution.

3. Main Progress and Landmark Achievements

3.1. Theoretical Research

3.1.1. The Study of Ecosystem Vulnerability Theory Consists Mainly of the Conceptual Study of Vulnerability, the Analysis of Its Causes, and the Classification of Vulnerability Types

Following the introduction of the concept of ecotone by Clements in 1905 [14], the concept of vulnerability was first introduced and elaborated by Timmerman in the early 1980s in the field of geography, where he argued that vulnerability was an indication of the degree of response within a system to disturbances from hazardous events both within and outside the system [38], followed by a large number of studies which have been conducted on the meaning of ecosystem vulnerability. The study of ecosystem vulnerability covers a wide range of fields and subjects, and the meaning of ecosystem vulnerability varies from one subject to another. The vulnerability of forest ecosystems refers to the combination of natural and anthropogenic factors that cause the structure and function of forest ecosystems to shift in the opposite direction from their original stable state or direction of succession in a certain spatial and temporal context, resulting in the destruction of their basic structure and the weakening of their stability and resilience [39]. For example, forest fire vulnerability is also defined as the potential loss from fire, including impacts on property, people, and environmental stability [40]. Additionally, for drought-induced forest vulnerability, Mildrexler et al. [41] defined its ecosystem vulnerability based on the processes of water and energy exchange due to drought and high temperatures. The vulnerability of forests in karstic desertification areas is reflected in the low vegetation cover, the predominance of secondary forest vegetation, the poor water-holding capacity of the soil itself, the looseness of the soil, and the homogeneous mix of species [42]. Different ecologically vulnerable areas have different causes of ecosystem vulnerability, early studies on forest ecosystem vulnerability were mainly based on global climate change [43] and natural disasters [44], but as research progressed, it began to expand to ecosystem vulnerability caused by a combination of anthropogenic and natural disturbances, and the research field was gradually extended from the initial natural ecosystem to the field of research has gradually extended from the initial study of natural ecosystems to the integrated study of social and natural systems.

3.1.2. Ecologists Generally Accept That the Concept of Ecological Resilience Includes the Concept of "Engineering and Ecological Resilience", but There Is No Consensus on How to Define or Apply Resilience in a Forestry Context

There has been considerable debate on the concept of ecological resilience, including "engineering, ecological and socio-ecological resilience" [45]. However, ecologists generally agree on two definitions of resilience: one focuses on the return to its equilibrium state after a disturbance, usually measured by the rate of recovery, which is often called "engineered resilience" [46]. The second focuses on the ability of a system to maintain its current state, defined as the extent to which a system can tolerate or resist disturbances without or before rearranging into another functionally and structurally different state, also known as "ecological resilience" [19,21]. Hodgson et al. [47] argue that both types of resilience are important in most ecological studies and that they should be considered together. For ecological resilience in forest ecosystems, Thompson et al. [48] consider it to be the ability of trees to resist disturbance and to recover to their original state after disturbance. The study of tree ecological resilience is a necessary element in the study of forest dynamics in the context of frequent times of extreme disturbance [49], for example, drought can make a difference to the growth recovery process of trees by affecting their ecological resilience [50]. The relevance of existing resilience concepts to forest management has also been reviewed [51], but to date there is no consensus on how to define or apply resilience in the context of forestry.

3.2. Indicator System

3.2.1. Ecosystem Vulnerability Assessment Indicator System

The establishment of an indicator system is the basis for ecosystem vulnerability and resilience assessment, and the construction of the indicator system is mainly reflected in three dimensions: firstly, in view of the regional differences in the formation of ecosystem vulnerability, the dominant factors leading to regional environmental vulnerability are clarified in the light of regional characteristics, and a single-type indicator system is constructed for regional vulnerability. For example, Yu and Lu [52] selected erosion-related water erosion factors and wind erosion factors to establish an indicator system for vulnerability assessment of the alpine region of the Qinghai–Tibet Plateau; the single-type indicator system is simple in structure, targeted, and has a strong regional dimension, and can identify key factors that lead to regional environmental vulnerability according to regional characteristics.

The second is to combine natural–social–economic aspects, taking into account both the intrinsic structure and functional characteristics of the system, as well as external disturbances, to build a comprehensive indicator system. Currently, scholars are using the "stress-state-response (PSR)" [53], which is based on the idea of cause and effect, and the "exposure-sensitivity-adaptability (VSD)" [54], which is based on the idea of "traitability". These conceptual models have been used to develop indicator systems, such as the work by Cinco-Castro and Herrera-Silveira [55], which constructs an indicator system for mangrove vulnerability in Mexico based on exposure, sensitivity, and adaptability. Comprehensive evaluation indicator systems contain a wide range of indicators, but are less operational due to the limitations of data availability and correlation between indicators [56].

The third is to predict the ecosystem vulnerability of systems in a changing future by establishing a simulation-based indicator system in the context of ecosystem vulnerability due to climate change or other changes. This includes climate change modelling, ecosystem modeling, socio-economic development modeling, and land use change modeling [57]. For example, Turkes et al. [58] developed a multi-factor spatial model of forest ecosystem vulnerability to climate change to assess the ecosystem vulnerability of forests and future scenarios of forests under climate change. Modelling forecasting is one of the more popular areas of vulnerability research in recent years.

3.2.2. Ecological Resilience Evaluation Indicator System

Since the end of the 20th century, many scholars have explored the factors of ecological resilience, but in general, research on this topic is still lacking, and a consensus concept and model for ecological resilience has not yet been obtained. Therefore, the indicators used for each resilience concept differ in two main ways. First, based on engineering and ecological resilience, the indicators used for assessment reflect state-based indicators, such as forest biodiversity, which is the most used [59,60], as the ability of an ecosystem to remain in an environment that remains unchanged over time related to how quickly it transitions to another environment. Furthermore, Folke et al. [61] show that biodiversity is one of the factors influencing this degree of slowness, in addition to forest structure and function [62,63]. The second is based on socio-ecological resilience, mainly stress and response-based indicators, such as population and economy [64]. Overall, indicator systems based on engineering and ecological resilience are the most widely used.

3.2.3. Ecosystem Vulnerability-Ecological Resilience Indicator Systems

There is currently no universal ecosystem vulnerability–resilience indicator system, with the IPCC in its Fifth Assessment Report proposing climate vulnerability as a function of exposure, sensitivity to climate change, and resilience or system adaptive capacity [65]. Angeler et al. [66] also showed that assessing system vulnerability by quantifying ecological resilience can help address the uncertainty in predicting ecosystem responses to environmental changes across ecosystems. Additionally, the difficulty of quantifying resilience has been a common challenge encountered by scholars. With the advent of Earth Observation Systems (EOS), satellite remote sensing can quantify resilience in response to disturbances [67,68], but this approach has certain limitations, such as limited spatial resolution, inapplicability to smaller geographical units, and the quantification of ecological resilience for different causes of disturbance does not universally applicable. Overall, further research is needed on the ecosystem vulnerability–resilience indicator system in the future.

3.3. Technical Method

3.3.1. Due to the Different Systems of Vulnerability Evaluation Indicators and Data Characteristics, There Are Additionally Many Types of Evaluation Methods, and the Study Needs to Select a Responsive Method for Evaluation According to the Study Scale and the Study Population

Vulnerability evaluation methods mainly include the AHP-fuzzy integrated evaluation method [69], landscape ecology method [70], grey cluster analysis method [71], principal component analysis method [72], set-pair analysis [73], integrated index method [74], and matter element extension method [75]. Among them, the principal component analysis method and the composite index method are widely used, with the former being suitable for quantitative vulnerability evaluation with more complete data and the latter being more widely applicable, mainly by normalizing the data, establishing weights, and then weighting and summing to obtain the vulnerability index. As research progresses, evaluation methods become increasingly diverse and complex. Therefore, when carrying out research, we should select suitable methods for evaluation according to different regional characteristics and research objects to make the results more scientific and credible.

3.3.2. Ecological Resilience Characterizes the Resistance and Resilience of a System, so Research Methods Need to Be Temporally and Spatially Specific, Which Additionally Means That Simulation Models Must Follow Principles of Applicability and Innovation

Assessing the impacts of environmental change is particularly challenging in forest ecosystems, which are long-lived and often persistent in their response to change [76]. Holling proposed two measures of resilience, the total area of the attraction domain and the height of the lowest point of the attraction basin above the equilibrium point [19]. Carpenter et al. [77] argue that ecological resilience cannot be measured directly and must be estimated by means of resilience surrogates, namely indirect proxies that are derived from theory; in addition, there are rapid assessment methods [78] and threshold analysis methods [76]. However, all these methods and models have certain limitations, and the lack of a suitable model and input explanatory variables can hinder the assessment of true vegetation resilience in complex ecosystems [79]. Therefore, simulation models must follow the principles of applicability and innovation. For example, Duveneck and Scheller [80] considered resistance as a change in species composition over time and used two forest landscape simulation models to simulate forest resilience under three future climate change scenarios and four management regimes, respectively. Additionally, for forest ecosystems to consider resistance and resilience in time and space, Cantarello et al. [81] provided a new quantitative approach using a spatially explicit model of forest dynamics focusing on the three components of resilience, namely resistance, recovery, and net change. New ideas are offered for forest resilience research.

3.3.3. As Ecosystem Vulnerability and Resilience Research Shifts from Mathematical and Theoretical Modelling to "3S" Spatial Analysis, i.e., RS, GIS, and GPS, "3S" Techniques Are Being Used Extensively in the Field of Ecology and the Environment

Vulnerability and resilience research involving multiple domains has a complex structure of research methods and indicator systems. Therefore, the analysis techniques have also changed from the previous mathematical model analysis to "3S" spatial analysis [82]. "3S" technologies have been widely used in the field of spatial information research, and their application in the field of ecological environment is also developing rapidly. Using "3S" technologies to obtain, process, and dynamically analyze resource and environmental information has become an important trend in the study of fragile ecosystems, and is widely used in ecosystem vulnerability assessment [83]. Sahana and Ganaie [84] analyzed landscape vulnerability in the Rudraprayag region of India based on GIS technology to explore the sensitivity of forests to fire; Hart et al. [85] combined GIS techniques and field trial data to assess the resistance and resilience of eight vegetation species to fire.

4. Key Scientific Issues to Be Solved

An extensive survey of the literature shows that significant progress has been made in research on the vulnerability and resilience of forest ecosystems. In general, much of the research has still focused on assessing the vulnerability, response, and future impacts of ecosystems or species to climate change or natural disasters. For example, Perie and de Blois [86] assessed the habitat suitability of tree species and predicted the suitability conditions of species as a function of future climate change, demonstrating that, despite overall regional habitat conditions, the study also demonstrated that different tree species have different adaptive characteristics despite the overall habitat conditions in the region. In contrast, research on forest resilience focuses on assessing the capacity of forest ecosystem composition, structure, and function to both resist and recover from disturbance, and quantifying this capacity through modeling. Seidl et al. [87], for example, propose an approach to disturbance ecology that uses the concept of resilience to quantify and address forest ecosystem services, providing a theoretical basis for the management of constantly disturbed forests. It is worth mentioning that most of the current studies on ecosystem vulnerability and resilience in karst fragile areas are regional large-scale or water resource vulnerability studies, and fewer studies have been conducted on forest ecosystems. In this regard, this paper will raise several scientific questions as follows:

In view of the problem of less theoretical research. theoretical research on ecosystem vulnerability was mainly at the early stage of qualitative exploration. As ecosystem vulnerability research has matured, theoretical research has become scarcer, and scholars at home and abroad are more likely to conduct ecosystem vulnerability simulation and evaluation of the system, transforming from the previous theoretical research on concepts, causes, and laws to empirical research. Especially with the development of remote sensing technology, technology has become an important data source for ecosystem vulnerability studies. Subsequently, its empirical studies have tended to become more widespread, and in general the development of vulnerability theory studies lags behind the studies of its methodological system. Although ecologically fragile zones have innate ecosystem vulnerability characteristics, with the continuous development of social and economic development, the impact of human interference and other factors on ecologically fragile areas is becoming more and more significant, and the ecological fragility characteristics shown are no longer caused by pure imbalance in the development of the ecological environment itself, but are the result of the superposition of natural and human influences. Therefore, the study of the theoretical system of ecologically fragile areas covers a wide range of contents and should be strengthened in the future under the combined effect of natural and social conditions.

In view of the problem of the lack of unified theoretical norms and evaluation models due to the wide range of evaluation indicators, the complexity and variability of the causes of ecosystem vulnerability have led to a wide range of evaluation indicators and a lack of unified theoretical norms and conceptual models. For forest ecosystems, the causes of vulnerability include internal system factors (community structure and function, etc.), environmental disturbance factors (climate and natural disasters, etc.), and socio-economic (population, income, and livestock situation, etc.). Given the difficulty of quantifying evaluation indicators and the different scales of study, the indicators chosen differ. Some researchers have assessed the vulnerability of forests in different distribution zones by modelling the distribution area of forests and the number of limiting species in order to assess the vulnerability of global forest ecological zones to future climate change [88]. At the same time, some researchers, in order to study the impact of agricultural expansion on forest cover in human-dominated tropical landscapes, have also selected indicators of exposure to cropland expansion, sensitivity, and forest capacity to respond to assess forest vulnerability [89]. Therefore, the indicator system must be selected according to the causes of forest ecosystem vulnerability and the scale of the study. Of course, theoretical research also needs to be strengthened in the future to select appropriate indicators on this basis, and establish a complete, integrated, scientific assessment system of forest ecosystems involving various scales and objects.

In view of the problem of uncertainty in the quantitative evaluation and analysis for ecosystem vulnerability, there are many methods to analyze ecosystem vulnerability, but different climatic factors have different impacts on different ecological factors, and these ecological factors also interact with each other, which together have negative impacts on the ecosystem [90]. Therefore, when conducting a quantitative evaluation of ecosystem vulnerability, a comprehensive analysis of the ecosystem vulnerability characteristics of the study area should be carried out, and the analytical methods and models should be selected in accordance with the principles of scientificity and rationality, to improve the accuracy of the evaluation results.

The current ecosystem vulnerability is limited to large-scale research and lacks clarification from the perspective of small to medium scale and single ecosystem. Due to the booming development of 3S technology, ecosystem vulnerability is mostly limited to large scale studies based on its ability to quickly acquire spatial data [91,92], and there is a lack of clarification of ecosystem vulnerability from the perspective of small and medium scales or single ecosystems issues. Therefore, ecosystem vulnerability studies should be carried out more on small and medium scales, considering the combination of field monitoring and "3S" technology to obtain experimental data, and to explore in depth the processes, characteristics, and mechanisms of fragile ecosystem degradation, to conduct vulnerability assessment. This allows managers to tailor system management strategies to local and situational contexts.

In view of the difficulty of identifying the concept of forest ecological resilience, despite the increasing research on resilience in forest ecosystems, forest resilience is still a vague concept and no resilience concept has been defined or applied in a forestry context to date [30]. Therefore, in the future, we should select a suitable framework of resilience concepts and evaluation indicators based on a recognition of how resilience approaches operate in forest management practice and a clear understanding of how to make resilience-related forest management decisions.

The problem of quantifying indicators in resilience evaluation is the biggest difficulty in current research, and exploring the establishment of more accurate mathematical models and the normalization of indicators is a pressing challenge in ecological resilience evaluation research. The required time span and spatial extent make it difficult to reason experimentally about the resilience of forest ecosystems, making simulation models an important tool in research [93]. As research on forest ecological resilience continues, simulation models need to be innovated; for example, Liu et al. [94] used the intensive Landsat time series model to establish forest resilience indicators in order to cope with the maximum magnitude of disturbance to the forest, as a way to quantitatively assess forest resilience. Alternatively, we can quantify forest resilience by selecting appropriate variables. For example, forest ecosystem productivity is a good indicator of ecosystem health, and some forest resilience studies have used normalized tree-ring width index as a state variable [95], while others have used normalized vegetation index, enhanced vegetation indices, and leaf area indices [96–98], all of which are parameters for estimating ecosystem productivity. In conclusion, when conducting forest resilience assessment, it is important to consider the characteristics of the study population and regional features in selecting indicators and building models to improve the rationality and accuracy of the results.

In view of the current lack of a universal framework linking ecosystem vulnerability and resilience, overall, ecological resilience is still mostly focused on studies of social–ecological systems such as urban resilience [99], and ecological resilience based on natural ecosystems is less common and mostly independent. Promoting the organic coupling and coordination of ecosystem vulnerability and resilience helps to develop forest conservation strategies and improve forest ecosystem services. Based on the previous discussion, it is known that ecosystem vulnerability characterizes the state, and indicators are easily accessible, while ecological resilience characterizes the process and indicators are difficult to quantify. Therefore, in order to fully grasp the characteristics of forest ecosystems and ensure sustainable development of ecosystems, we can use 3S modelling techniques or a combination of 3S technology and field monitoring to establish a quantitative model linking the vulnerability and resilience of forest ecosystems, considering the selection of indicators from natural, disturbance, and social factors.

Addressing the weak practical application of ecosystem vulnerability and resilience research, ecosystem vulnerability studies can identify key problems in the system, while ecological resilience studies can clarify the resistance and resilience of the system, so that forest managers can apply precise policies to the forest and achieve sustainable management. Ecosystem-based adaptation is seen as having the potential to integrate sustainable management, conservation, and restoration of ecosystems into adaptation to climate change [100]. Therefore, forests in ecologically fragile areas require standardized management and assessment systems that focus on applying theory and practice to maintain ecosystem balance in three ways: firstly, based on research into the vulnerability of forest ecosystems, management policies can be developed to target key vulnerable resources and improve system resilience. Secondly, based on research on habitat suitability of tree species, cultivate plantation forests or introduce new tree species into forest communities to increase forest cover, biodiversity, and system productivity, thereby reducing forest vulnerability and increasing resistance to sustainable development. Thirdly, policies should be formulated to reduce anthropogenic disturbances, such as forest closure and reforestation, to improve the resilience of the system.

Addressing the low level of research on the vulnerability and resilience of forest ecosystems in the karst areas, karst rocky desertification represents a relatively unique type of desert in the world [101], and the problem of fragile ecological environment and social development in karst mountain area is an international issue [102]. In response, Xiong et al. proposed a series of vegetation restoration measures for rocky desertification control, including mountain closure [103] and cultivation of ecological and economic forests [104]. However, most of the forests in the region are still vulnerable and less resilient. The reasons are mainly the fragility of the karst itself and the impact of human activities. Therefore, when conducting vulnerability and resilience analyses, the indicators selected should focus on environmental and human disturbance factors. For large scales, 3S technology is used to build a vegetation distribution–anthropogenic activity model, while placing resilience variables (e.g., productivity) in the model to achieve a combination of vulnerability and resilience. For small scales, 3S technology is considered to be combined with field monitoring; community structure and function indicators are added to environmental and anthropogenic factors to accurately identify vulnerable resources in forest ecosystems. In addition, studies can be carried out to assess the vulnerability and sensitivity of species from the perspective of habitat suitability, just as some scholars have used the tree-ring width data to simulate the resistance of the species to drought [9], habitat suitability studies on tree species are particularly important in rocky desertification areas due to the dichotomous structure above and below ground [105] and severe soil erosion. Based on the above studies, we can implement sustainable management policies for karst forests to improve the homogeneous community structure and weak resilience, thus enhancing ecosystem services.

5. Conclusions and Future Research

In this paper, we conducted a systematic literature review by analyzing 217 papers retrieved from CNKI and Web of Science. The following conclusions were drawn: (1) research on the vulnerability and resilience of forest ecosystems is rapidly increasing; (2) among the studies on ecosystem vulnerability, ecological resilience, management strategies, and other aspects of forest, ecosystem vulnerability analysis is the most common, accounting for 46.17%, of which vulnerability modeling predictions based on future changes accounts for 9.77%, and is an emerging area for future research; (3) research on forest ecosystem vulnerability and resilience is more widely studied in China and the United States, accounting for 25% and 18% of the total, respectively; and (4) most of the research units on forest ecosystem vulnerability and resilience are universities or forestry institutions in developed countries, such as the USA and Germany, with the exception of China.

Following the quantitative analysis, the article also proposes nine key scientific questions to be addressed in response to the current state of research at home and abroad, providing directions for continued in-depth research in the future.

The key questions for future research on the vulnerability and resilience of forest ecosystems can be summarized in the following areas: ecosystem vulnerability and resilience of forests, how to establish a framework for a universally applicable vulnerability indicator system, how to develop models to quantify the resistance and resilience of forest ecosystems, how to link vulnerability and resilience for a comprehensive analysis of forest ecosystems, and how to accurately study forest vulnerability and resilience in the context of rocky desertification control. These are the issues and challenges in urgent need for future research that need to be addressed in the future.

It should be noted here that this paper may be subject to potential uncertainties in the quantitative data of the search results due to the limitations of the search engine, but it has little impact on the subsequent exploration of theoretical advances in forest ecosystem vulnerability and resilience research.

Author Contributions: K.X. and Y.Y. conceptualized the framework, acquired the funding, and supervised the overall project; H.L., T.L., Y.Q., Z.W. and S.Z. collected data; H.L. analyzed data, and wrote the manuscript; K.X. and Y.Y. provided modification comments; and K.X. and H.L. reviewed the final manuscript. All authors have read and agreed to the published version of the manuscript.

Funding: Funding for the project was supported by the World Top Discipline Program of Guizhou Province (No. 125 2019 Qianjiao Keyan Fa), the Key Science and Technology Program of Guizhou Province (No. 5411 2017 Qiankehe Pingtai Rencai) and the China Overseas Expertise Introduction Program for Discipline Innovation (D17016).

Institutional Review Board Statement: Not applicable.

Informed Consent Statement: Not applicable.

Data Availability Statement: Data are contained within the article.

Conflicts of Interest: The authors declare no conflict of interest.

References

1. Boonstra, R.; Andreassen, H.P.; Boutin, S.; Hušek, J.; Ims, R.A.; Krebs, C.J.; Skarpe, C.; Wabakken, P. Why do the boreal forest ecosystems of northwestern Europe differ from those of western North America? *Bioscience* **2016**, *66*, 722–734. [CrossRef]
2. Millennium Ecosystem Assessment Board. *Ecosystems and Human Well-Being*; Island Press: Washington, DC, USA, 2005.
3. Miyawaki, A. A vegetation-ecological study on regeneration and restoration of vegetation. *Veg. Sci. News* **2001**, *6*, 55–62.
4. Costanza, R.; d'Arge, R.; de Groot, R.; Farber, S.; Grasso, M.; Hannon, B.; Limburg, K.; Naeem, S.; O'Neill, R.V.; Paruelo, J.; et al. The value of the world's ecosystem services and natural capital. *Nature* **1997**, *387*, 253–260. [CrossRef]
5. Zemp, D.C.; Schleussner, C.-F.; Barbosa, H.M.J.; Rammig, A. Deforestation effects on Amazon forest resilience. *Geophys. Res. Lett.* **2007**, *44*, 6182–6190. [CrossRef]
6. Ozcan, O.; Musaoglu, N.; Tyrkeş, M. Assessing vulnerability of a forest ecosystem to climate change and variability in the western Mediterranean sub-region of Turkey. *J. For. Res.* **2018**, *29*, 709–725. [CrossRef]
7. Akinola, O.V.; Adegoke, J. Assessment of forest fire vulnerability zones in Missouri, United States of America. *Int. J. Sustain. Dev. World Ecol.* **2019**, *26*, 251–257. [CrossRef]
8. Zimmerman, J.K.; Willig, M.R.; Hernandez-Delgado, E.A. Resistance, resilience, and vulnerability of social-ecological systems to hurricanes in Puerto Rico. *Ecosphere* **2020**, *11*, e03159. [CrossRef]
9. Fang, O.; Zhang, Q.B.; Vitasse, Y.; Zweifel, R.; Cherubini, P. The frequency and severity of past droughts shape the drought sensitivity of juniper trees on the Tibetan plateau. *For. Ecol. Manag.* **2021**, *486*, 118968. [CrossRef]
10. Kneeshaw, D.D.; Sturtevant, B.R.; DeGrandpé, L.; Doblas-Miranda, E.; James, P.M.A.; Tardif, D.; Burton, P.J. The Vision of Managing for Pest-Resistant Landscapes: Realistic or Utopic? *Curr. For. Rep.* **2021**, *7*, 97 113.
11. Truman, C.C.; Strickland, T.C.; Potter, T.L.; Franklin, D.H.; Bosch, D.D.; Bednarz, C.W. Variable rainfall intensity and tillage effects on runoff, sediment, and carbon losses from a loamy sand under simulated rainfall. *J. Environ. Qual.* **2007**, *36*, 1495–1502. [CrossRef]
12. Guidi, C.; Di Matteo, G.; Grego, S. An overview of proven Climate Change Vulnerability Assessment tools for forests and forest-dependent communities across the globe: A literature analysis. *J. For. Res.* **2018**, *29*, 1167–1175. [CrossRef]
13. De Lange, H.J.; Sala, S.; Vighi, M.; Faber, J.H. Ecological vulnerability in risk assessment—A review and perspectives. *Sci. Total Environ.* **2010**, *408*, 3871–3879. [CrossRef] [PubMed]
14. Clements, F.E. *Research Methods in Ecology*; University Publishing Company: Lincoln, NE, USA, 1905.
15. Niu, W.Y. The discriminatory index with regard to the weakness, overlapness, and breadth of ecotone. *Acta Ecol. Sin.* **1989**, *9*, 97–105.
16. IPCC. *Climate Change 2001: Impacts, Adaptation, and Vulnerability*; Cambridge University Press: Cambridge, UK; New York, NY, USA, 2001.
17. Turner, B.L.; Kasperson, R.E.; Matson, P.A.; McCarthy, J.J.; Corell, R.W.; Christensen, L.; Eckley, N.; Kasperson, J.X.; Luers, A.; Martello, M.L.; et al. A framework for vulnerability analysis in sustainability science. *Proc. Natl. Acad. Sci. USA* **2003**, *100*, 8074–8079. [CrossRef]
18. Shao, Y.W.; Xu, J. Understanding urban resilience: A conceptual analysis based on integrated international literature review. *Urban Plan. Int.* **2015**, *30*, 48–54.
19. Holling, C.S. Resilience and Stability of Ecological Systems. *Annu. Rev. Ecol. Syst.* **1973**, *4*, 1–23. [CrossRef]
20. Ouyang, H.; Ye, Q. A review on the evolution of resilient city theory: Concept, context and tendency. *City Plan. Rev.* **2016**, *40*, 34–42.
21. Gunderson, L.H.; Holling, C.S. (Eds.) *Panarchy: Understanding Transformations in Human and Natural Systems*; Island Press: Washington, DC, USA, 2002.
22. Liao, K.H. A Theory on Urban Resilience to Floods-A Basis for Alternative Planning Practices. *Ecol. Soc.* **2012**, *17*, 48. [CrossRef]

23. Gann, G.D.; McDonald, T.; Walder, B.; Aronson, J.; Nelson, C.R.; Jonson, J.; Hallett, J.G.; Eisenberg, C.; Guariguata, M.R.; Liu, J.; et al. International Principles and Standards for the Practice of Ecological Restoration. Second edition. *Restor. Ecol.* **2019**, *27*, S1–S46. [CrossRef]

24. Ma, X.Y. *Spatial Analysis, Evaluation and Optimization of the Yellow River Beach Area in Xinxiang City from the Perspective of Resilient Cities*; Beijing Forestry University: Beijing, China, 2019.

25. Keenan, R.J. Climate change impacts and adaptation in forest management: A review. *Ann. For. Sci.* **2015**, *72*, 145–167. [CrossRef]

26. Lucier, A.; Ayres, M.; Karnosky, D.; Thompson, I.; Loehle, C.; Percy, K.; Sohngen, B. Forest responses and vulnerabilities to recent climate change. *IUFRO World Ser.* **2009**, *22*, 29–52.

27. De La Serrana, R.G.; Vilagrosa, A.; Alloza, J.A. Pine mortality in southeast Spain after an extreme dry and warm year: Interactions among drought stress, carbohydrates and bark beetle attack. *Trees* **2015**, *29*, 1791–1804. [CrossRef]

28. Bernier, P.; Schoene, D. *Adapting Forests and Their Management to Climate Change: An Overiew*; Information Service of FAO: Rome, Italy, 2009; Volume 60, pp. 5–11.

29. Hou, K.; Tao, W.; Wang, L.; Li, X.X. Study on hierarchical transformation mechanisms of regional ecological vulnerability and its applicability. *Ecol. Indic.* **2020**, *114*, 106343. [CrossRef]

30. Nikinmaa, L.; Lindner, M.; Cantarello, E.; Jump, A.S.; Seidl, R.; Winkel, G.; Muys, B. Reviewing the use of resilience concepts in forest sciences. *Curr. For. Rep.* **2020**, *6*, 61–80. [CrossRef]

31. Lecina-Diaz, J.; Martinez-Vilalta, J.; Alvarez, A.; Banque, M.; Birkmann, J.; Feldmeyer, D.; Vayredm, J.; Retana, J. Characterizing forest vulnerability and risk to climate-change hazards. *Front. Ecol. Environ.* **2021**, *19*, 126–133. [CrossRef]

32. Reyer, C.P.O.; Brouwers, N.; Rammig, A.; Brook, B.W.; Epila, J.; Grant, R.F.; Holmgren, M.; Langerwisch, F.; Leuzinger, S.; Lucht, W.; et al. Forest resilience and tipping points at different spatio-temporal scales: Approaches and challenges. *J. Ecol.* **2015**, *103*, 5–15. [CrossRef]

33. Weißhuhn, P.; Müller, F.; Wiggering, H. Ecosystem vulnerability review: Proposal of an interdisciplinary ecosystem assessment approach. *Environ. Manag.* **2018**, *61*, 904–915. [CrossRef]

34. Rana, I.A. Disaster and climate change resilience: A bibliometric analysis. *Int. J. Disaster Risk Reduct.* **2020**, *50*, 101839. [CrossRef]

35. Kumar, M.; Kalra, N.; Singh, H.; Sharma, S.; Singh Rawat, P.; Singh, R.K.; Gupta, A.K.; Jumar, P.; Ravindranath, N.H. Indicator-based vulnerability assessment of forest ecosystem in the Indian Western Himalayas: An analytical hierarchy process integrated approach. *Ecol. Indic.* **2021**, *125*, 107568. [CrossRef]

36. Johnstone, J.F.; Allen, C.D.; Franklin, J.F.; Frelich, L.E.; Harvey, B.J.; Higuera, P.E.; Mack, M.C.; Meentemeyer, R.K.; Metz, M.R.; Perry, G.L.W.; et al. Changing disturbance regimes, ecological memory, and forest resilience. *Front. Ecol. Environ.* **2016**, *14*, 369–378. [CrossRef]

37. Ibarra, J.T.; Cockle, K.; Altamirano, T.; van der Hoek, Y.; Simard, S.W.; Bonacic, C.; Martin, K. Nurturing resilient forest biodiversity: Nest webs as complex adaptive systems. *Ecol. Soc.* **2020**, *25*, 27. [CrossRef]

38. Timmerman, P. *Vulnerability, Resilience and the Collapse of Society: A Review of Models and Possible Climatic Applications*; Institute for Environmental Studies, University of Toronto: Toronto, ON, Canada, 1981.

39. Gong, W.; Hu, T.X.; Gong, Y.B. A brief discussion on the formation characteristics of fragile forest ecosystems and their management measures. *Sichuan For. Explor. Des.* **2004**, *3*, 5–10.

40. Schelhaas, M.J.; Hengeveld, G.; Moriondo, M.; Reinds, G.J.; Kundzewicz ter Maat, H.W.; Bindi, M. Assessing risk and adaptation options to fires and windstorms in European forestry. *Mitig. Adapt. Strateg. Glob. Chang.* **2010**, *15*, 681–701. [CrossRef]

41. Mildrexler, D.J.; Yang, Z.; Cohen, W.; Bell, D. A forest vulnerability index based on drought and high temperatures. *Remote Sens. Environ.* **2016**, *173*, 314–325. [CrossRef]

42. Peng, Q.; Lin, C.H.; He, T.F. The development of study on the characteristics of soil and water loss and its conservation in Guizhou karst mountain area. *Guizhou Sci.* **2006**, *24*, 66–70.

43. Ali, A. Climate change impacts and adaptation assessment in Bangladesh. *Clim. Res.* **1999**, *12*, 109–116. [CrossRef]

44. Chuluundorj, O. *A Multi-Level Study of Vulnerability of Mongolian Pastoralists to Natural Hazards and Its Consequences on Individual and Household Well-Being*; University of Colorado: Denver, CO, USA, 2006.

45. Bone, C.; Moseley, C.; Vinyeta, K.; Bixler, R.P. Employing resilience in the United States forest service. *Land Use Policy* **2016**, *52*, 430–438. [CrossRef]

46. Pimm, S.L. The complexity and stability of ecosystems. *Nature* **1984**, *307*, 321–326. [CrossRef]

47. Hodgson, D.; McDonald, J.L.; Hosken, D.J. What do you mean, 'resilient'? *Trends Ecol. Evol.* **2015**, *30*, 503–506. [CrossRef] [PubMed]

48. Thompson, I.; Mackey, B.; McNulty, S.; Mosseler, A. *Forest Resilience, Biodiversity and Climate Change: A Synthesis of the Biodiversity/Resilience/Stability Relationship in Forest Ecosystems*; Technical Series no. 43; Secretariat of the Convention on Biological Diversity: Montreal, QC, Canada, 2009; pp. 1–67.

49. Law, B.E.; Waring, R.H. Carbon implications of current and future effects of drought, fire and management on Pacific Northwest forests. *For. Ecol. Manag.* **2015**, *355*, 4–14. [CrossRef]

50. Huang, M.; Wang, X.; Keenan, T.F.; Piao, S. Drought timing influences the legacy of tree growth recovery. *Glob. Chang. Biol.* **2018**, *24*, 3546–3559. [CrossRef]

51. Brown, E.D.; Williams, B.K. Resilience and resource management. *Environ. Manag.* **2018**, *56*, 1416–1427. [CrossRef] [PubMed]

52. Yu, B.H.; Lu, C.H. Assessment of ecological vulnerability on the Tibetan plateau. *Geogr. Res.* **2011**, *30*, 2289–2295.

53. Berger, A.R.; Hodge, R.A. Natural change in the environment: A challenge to the pressure-state-response concept. *Soc. Indic. Res.* **1998**, *44*, 255–265. [CrossRef]
54. Polsky, C.; Neff, R.; Yarnal, B. Building comparable global change vulnerability assessments:The vulnerability scoping diagram. *Glob. Environ. Chang. Hum. Policy Dimens.* **2007**, *17*, 472–485. [CrossRef]
55. Cinco-Castro, S.; Herrera-Silveira, J. Vulnerability of mangrove ecosystems to climate change effects: The case of the Yucatan Peninsula. *Ocean Coast. Manag.* **2020**, *192*, 105196. [CrossRef]
56. Wang, J.Y.; Zhao, G.X.; Wang, X.F.; Wang, L.H.; Liu, M.S.; Liu, T. On the vulnerability of China's ecological environment and its assessment. *Shandong Agric. Sci.* **2004**, *2*, 9–11.
57. Xu, G.C.; Kang, M.Y.; He, L.N.; Li, Y.F.; Chen, Y.R. Advance in research on ecological vulnerability. *Acta Ecol. Sin.* **2009**, *29*, 2578–2588.
58. Turkes, M.; Musaoglu, N.; Ozcan, O. Assessing the vulnerability of a forest ecosystem to climate change and variability in the western Mediterranean sub-region of Turkey: Future Evaluation. *J. For. Res.* **2018**, *29*, 1177–1186. [CrossRef]
59. Derroire, G.; Balvanera, P.; Castellanos-Castro, C.; Decocq, G.; Kennard, D.K.; Lebrija-Trejos, E.; Leiva, J.A.; Oden, P.-C.; Powers, J.S.; Rico-Gray, V.; et al. Resilience of tropical dry forests–a meta-analysis of changes in species diversity and composition during secondary succession. *Oikos* **2016**, *125*, 1386–1397. [CrossRef]
60. Lennox, G.D.; Gardner, T.A.; Thomson, J.R.; Ferreira, J.; Berenguer, E.; Lees, A.C.; Nally, R.M.; Aragao, L.E.O.C.; Ferraz, S.F.B.; Louzada, J.; et al. Second rate or a second chance? Assessing biomass and biodiversity recovery in regenerating Amazonian forests. *Glob. Chang. Biol.* **2018**, *24*, 5680–5694. [CrossRef] [PubMed]
61. Folke, C.; Carpenter, S.; Walker, B.; Scheffer, M. Regime shifts, resilience, and biodiversity in ecosystem management. *Annu. Rev. Ecol. Evol. Syst.* **2004**, *35*, 557–581. [CrossRef]
62. Craven, D.; Filotas, E.; Angers, V.A.; Messier, C. Evaluating resilience of tree communities in fragmented landscapes: Linking functional response diversity with landscape connectivity. *Divers. Distrib.* **2016**, *22*, 505–518. [CrossRef]
63. López-Jiménez, L.N.; Duran-Garcia, R.; Dupuy-Rada, J.M. Recovery of structure, diversity and composition of a tropical semievergreen forest in Yucatan, Mexico. *Madera Bosques* **2019**, *25*, e2511587.
64. Rivera-Monroy, V.H.; Danielson, T.M.; Castañeda-Moya, E.; Marx, B.D.; Travieso, R.; Zhao, X.; Gaiser, E.E.; Farfan, L.M. Long-term demography and stem productivity of Everglades mangrove forests (Florida, USA): Resistance to hurricane disturbance. *For. Ecol. Manag.* **2019**, *440*, 79–91. [CrossRef]
65. IPCC. *Climate Change 2014: Impacts, Adaptation, and Vulnerability*; Cambridge University Press: Cambridge, UK, 2014.
66. Angeler, D.G.; Baho, D.L.; Allen, C.R.; Johnson, R.K. Linking degradation status with ecosystem vulnerability to environmental change. *Oecologia* **2015**, *178*, 899–913. [CrossRef] [PubMed]
67. Cui, X.; Gibbes, C.; Southworth, J.; Waylen, P. Using remote sensing to quantify vegetation change and ecological resilience in a semi-arid system. *Land* **2013**, *2*, 108–130. [CrossRef]
68. Dwomoh, F.K.; Wimberly, M.C. Fire regimes and forest resilience: Alternative vegetation states in the West African tropics. *Landsc. Ecol.* **2017**, *32*, 1849–1865. [CrossRef]
69. Li, J.D.; Xu, D. Evaluation on exploitation potential of forest health tourism based on AHP and fuzzy synthetic evaluation-A case study of Liaodong mountain area. *Chin. J. Agric. Resour. Reg. Plan.* **2018**, *39*, 135–142.
70. Giglio, E. Landscape ecological method to study agriculture vegetation: Some examples from the po valley. *Ann. Bot.* **2016**, *6*, 95–110.
71. Delgado, A.; Vidal, J.; Castro, J.; Felix, J.; Saenz, J. Assessment of Surface Water Quality on the Upper Watershed of Huallaga River, in Peru, using Grey Systems and Shannon Entropy. *Int. J. Adv. Comput. Sci. Appl.* **2020**, *11*, 437–444. [CrossRef]
72. Xenarios, S.; Nemes, A.; Sarker, G.W.; Nagothu, U.S. Assessing vulnerability to climate change: Are communities in flood-prone areas in Bangladesh more vulnerable than those in drought-prone areas? *Water Resour. Rural Dev.* **2016**, *7*, 1–19. [CrossRef]
73. Li, B.; Yang, Z.; Su, F. Vulnerability measurement of Chinese marine economic system based on set pair analysis. *Sci. Geogr. Sin.* **2016**, *36*, 47–54.
74. Wirehn, L.; Danielsson, A.; Neset, T.-S.S. Assessment of composite index methods for agricultural vulnerability to climate change. *J. Environ. Manag.* **2015**, *156*, 70–80. [CrossRef]
75. Huang, Z.; Wan, J. Vulnerability assessment of safe operation of utility tunnel based on improved matter-element extension method. *Constr. Sci. Technol.* **2020**, *8*, 78–81.
76. Standish, R.J.; Hobbs, R.J.; Mayfield, M.M.; Bestelmeyer, B.; Suding, K.N.; Battaglia, L.; Eviner, V.T.; Hawkes, C.V.; Temperton, V.M.; Harris, V.A.C.J. Resilience in ecology: Abstraction, distraction, or where the action is? *Biol. Conserv.* **2014**, *177*, 43–51. [CrossRef]
77. Carpenter, S.R.; Westley, F.; Turner, M.G. Surrogates for resilience of social–ecological systems. *Ecosystems* **2005**, *8*, 941–944. [CrossRef]
78. Nemec, K.T.; Chan, J.; Hoffman, C.; Spanbauer, T.L.; Hamm, J.A.; Allen, C.R.; Hefley, T.; Pan, D.; Shrestha, P. Assessing resilience in stressed watersheds. *Ecol. Soc.* **2014**, *19*, 34. [CrossRef]
79. Behera, M.D.; Murthy, M.S.R.; Das, P.; Sharma, E. Modelling forest resilience in Hindu Kush Himalaya using geoinformation. *J. Earth Syst. Sci.* **2018**, *127*, 95. [CrossRef]
80. Duveneck, M.J.; Scheller, R.M. Measuring and managing resistance and resilience under climate change in northern Great Lake forests (USA). *Landsc. Ecol.* **2016**, *31*, 669–686. [CrossRef]

81. Cantarello, E.; Newton, A.C.; Martin, P.A.; Evans, P.M.; Gosal, A.; Lucash, M.S. Quantifying resilience of multiple ecosystem services and biodiversity in a temperate forest landscape. *Ecol. Evol.* **2017**, *7*, 9661–9675. [CrossRef] [PubMed]
82. Zhang, X.L.; Yu, W.B.; Cai, H.S.; Guo, X.M. Review of the evaluation methods of regional eco-environmental vulnerability. *Acta Ecol. Sin.* **2018**, *38*, 5970–5981.
83. Fang, C.; Wang, Y.; Fang, J. A comprehensive assessment of urban vulnerability and its spatial differentiation in China. *J. Geogr. Sci.* **2016**, *26*, 153–170. [CrossRef]
84. Sahana, M.; Ganaie, T.A. GIS-based landscape vulnerability assessment to forest fire susceptibility of Rudraprayag district, Uttarakhand, India. *Environ. Earth Sci.* **2017**, *76*, 676. [CrossRef]
85. Hart, S.J.; Henkelman, J.; McLoughlin, P.D.; Nielsen, S.E.; Truchon-Savard, A.; Johnstone, J.F. Examining forest resilience to changing fire frequency in a fire-prone region of boreal forest. *Glob. Chang. Biol.* **2019**, *25*, 869–884. [CrossRef]
86. Perie, C.; de Blois, S. Dominant forest tree species are potentially vulnerable to climate change over large portions of their range even at high latitudes. *Peer J.* **2016**, *4*, e2218. [CrossRef]
87. Seidl, R.; Spies, T.A.; Peterson, D.L.; Stephens, S.L.; Hicke, J.A. Searching for resilience: Addressing the impacts of changing disturbance regimes on forest ecosystem services. *J. Appl. Ecol.* **2016**, *53*, 120–129. [CrossRef]
88. Wang, C.J.; Zhang, Z.X.; Wan, J.Z. Vulnerability of global forest ecoregions to future climate change. *Glob. Ecol. Conserv.* **2019**, *20*, e00760. [CrossRef]
89. Bourgoin, C.; Oszwald, J.; Bourgoin, J.; Gond, V.; Blanc, L.; Dessard, H.; Van Phan, T.; Sist, P.; Laderach, P.; Reymondin, L. Assessing the ecological vulnerability of forest landscape to agricultural frontier expansion in the Central Highlands of Vietnam. *Int. J. Appl. Earth Obs. Geoinf.* **2020**, *84*, 101958. [CrossRef]
90. Prentice, I.C.; Cramer, W.; Harrison, S.P.; Leemans, R.; Monserud, R.A.; Solomon, A.M. A global biome model based on plant physiology and dominance, soil properties and climate. *J. Biogeogr.* **1992**, *19*, 117–134. [CrossRef]
91. Guo, B.; Zang, W.; Luo, W. Spatial-temporal shifts of ecological vulnerability of Karst Mountain ecosystem-impacts of global change and anthropogenic interference. *Sci. Total Environ.* **2020**, *741*, 140256. [CrossRef]
92. Hu, W.J.; Chen, B.; Arnupap, P.; Zhang, D.; Liu, X.M.; Gu, H.F.; Ajcharaporn, P.; Chao, B.X.; Zheng, X.Q. Ecological vulnerability assessment of coral reef: A case study of Si Chang island, Thailand. *Chin. J. Ecol.* **2020**, *39*, 979–989.
93. Albrich, K.; Rammer, W.; Turner, M.G.; Ratajczak, Z.; Brazinuas, K.H.; Hansen, W.D.; Seidl, R. Simulating forest resilience: A review. *Glob. Ecol. Biogeogr.* **2020**, *29*, 2082–2096. [CrossRef] [PubMed]
94. Liu, M.; Liu, X.; Wu, L.; Tang, Y.; Li, Y.; Zhang, Y.; Ye, L.; Zhang, B. Establishing forest resilience indicators in the hilly red soil region of southern China from vegetation greenness and landscape metrics using dense Landsat time series. *Ecol. Indic.* **2021**, *121*, 106985. [CrossRef]
95. Gao, S.; Liu, R.; Zhou, T.; Fang, W.; Yi, C.; Lu, R.; Luo, H. Dynamic responses of tree-ring growth to multiple dimensions of drought. *Glob. Chang. Biol.* **2018**, *24*, 5380–5390. [CrossRef]
96. Di Mauro, B.; Fava, F.; Busetto, L.; Crosta, G.F.; Colombo, R. Post-fire resilience in the Alpine region estimated from MODIS satellite multispectral data. *Int. J. Appl. Earth Obs. Geoinf.* **2014**, *32*, 163–172. [CrossRef]
97. Spasojevic, M.J.; Bahlai, C.A.; Bradley, B.A.; Butterfield, B.J.; Tuanmu, M.N.; Sistla, S.; Suding, K.N. Scaling up the diversity–resilience relationship with trait databases and remote sensing data: The recovery of productivity after wildfire. *Glob. Chang. Biol.* **2016**, *22*, 1421–1432. [CrossRef]
98. Danielson, T.M.; Rivera-Monroy, V.H.; Castañeda-Moya, E.; Briceño, H.; Travieso, R.; Marx, B.D.; Farfán, L.M. Assessment of Everglades mangrove forest resilience: Implications for above-ground net primary productivity and carbon dynamics. *For. Ecol. Manag.* **2017**, *404*, 115–125. [CrossRef]
99. Wu, X.; Zhang, J.; Geng, X.; Wang, T.; Wang, K.; Liu, S. Increasing green infrastructure-based ecological resilience in urban systems: A perspective from locating ecological and disturbance sources in a resource-based city. *Sustain. Cities Soc.* **2020**, *61*, 102354. [CrossRef]
100. IUCN. *Ecosystem-Based Adaptation: An Approach for Building Resilience and Reducing Risk for Local Communities and Ecosystems. Submission to the Chair of the AWG-LCA with respect to the Shared Vision and Enhanced Action on Adaptation*; International Union for the Conservation of Nature: Grand, Switzerland, 2008.
101. Yang, M.D. On the fragility of karst environment. *Yunnan Geogr. Environ. Res.* **1990**, *2*, 21–29.
102. Ouyang, Z.Y. On the comprehensive management, development and breaking away from poorness of the ecologically fragile karst area in south-west China. *World Sci.-Tech. R. D.* **1998**, *2*, 3–5.
103. Xiong, K.N.; Li, P.; Zhou, Z.F. *A Typical Study on Remote Sensing-GIS of Karst Desertification in Guizhou Province*; Geological Publishing House: Beijing, China, 2002; pp. 128–150.
104. Xiong, K.N.; Mei, Z.M.; Peng, X.W.; Lan, A.J. Integrated management of the karst rock desertification areas—A case study of Huangjiang karst gorge. *Guizhou For. Sci. Technol.* **2006**, *1*, 5–8.
105. Yang, M.D. On the geographical structure of karst landscapes and their environmental effects. *Guizhou Soc. Environ. Sci.* **1998**, *11*, e03159.

 *sustainability*

Article

Climate Change, Adaptation Planning and Institutional Integration: A Literature Review and Framework

Nate Kauffman * and Kristina Hill *

College of Environmental Design, University of California, Berkeley, CA 94720, USA
* Correspondence: natekauffman@berkeley.edu (N.K.); kzhill@berkeley.edu (K.H.)

Abstract: The scale and scope of climate change has triggered widespread acknowledgement of the need to adapt to it. Out of recent work attempting to understand, define, and contribute to the family of concepts related to adaptation efforts, considerable contributions and research have emerged. Yet, the field of climate adaptation constantly grapples with complex ideas whose relational interplay is not always clear. Similarly, understanding how applied climate change adaptation efforts unfold through planning processes that are embedded in broader institutional settings can be difficult to apprehend. We present a review of important theory, themes, and terms evident in the literature of spatial planning and climate change adaptation to integrate them and synthesize a conceptual framework illustrating their dynamic interplay. This leads to consideration of how institutions, urban governance, and the practice of planning are involved, and evolving, in shaping climate adaptation efforts. While examining the practice of adaptation planning is useful in framing how core climate change concepts are related, the role of institutional processes in shaping and defining these concepts—and adaptation planning itself—remains complex. Our framework presents a useful tool for approaching and improving an understanding of the interactive relationships of central climate change adaptation concepts, with implications for future work focused on change within the domains of planning and institutions addressing challenges in the climate change era.

Keywords: climate change; climate change adaptation; spatial planning; institutions; sustainability; resilience; uncertainty; vulnerability; adaptive capacity; urban governance

Citation: Kauffman, N.; Hill, K. Climate Change, Adaptation Planning and Institutional Integration: A Literature Review and Framework. *Sustainability* **2021**, *13*, 10708. https://doi.org/10.3390/su131910708

Academic Editor: Baojie He

Received: 1 August 2021
Accepted: 23 September 2021
Published: 27 September 2021

Publisher's Note: MDPI stays neutral with regard to jurisdictional claims in published maps and institutional affiliations.

1. Introduction

The environmental severity and enormity of climate change is coming into sharper focus, as are considerations of crucial and complex impacts on society and daunting demands of the requisite efforts to adapt to it [1]. Climate Change Adaptation (CCA) is understood as a challenge ensnaring numerous actors across multiple societal sectors, acting as a nexus of overlapping concerns and connections [2]. Significant increases in literature concerned with climate change adaptation is evident, with commensurate scholarship dedicated to exploring key concepts in the field [3–5]. Hurdles to effectively engaging with climate adaptation concepts run the gamut: from the inaccessibility of scientific "jargon" [6] to the need to synthesize research and identify areas lacking attention [7,8]. Disentangling the roles and relationships between modes of preparing adaptive responses to climate change (planning) and the social patterns that govern these practices (institutions) reveal more areas of confusion and needed consideration, especially for examining how these practices and patterns may themselves adapt or be adapted [4,9,10]. While conceptual frameworks used to streamline and simplify complex ideas are common, frameworks constructed for the purpose of clarifying key concepts in the field of climate change adaptation planning are lacking.

Planning is a concept with wide and diverse meaning across numerous scales and disciplines [11]. While climate impacts on the atmosphere and oceans of earth are increasingly severe (and entail their own planning considerations), we are concerned here with

17

spatial planning, which frames the landscape as a crucial, dynamic medium—a geographic template—upon and within which effects of climate change will be experienced most acutely by humans [12]. Spatial planning uses diverse scientific methods and information to shape decisions about how features of the landscape are designed, constructed, and managed. Berkes and Folke [13] sought to formalize the concept of *social-ecological systems* (SES) as linked human and natural systems that somehow "fit" together [14]; and a framework for "match[ing] the dynamics of *institutions* with the dynamics of *ecosystems* for mutual social-ecological *resilience* and improved performance." While earlier work on the concept was undertaken by Ratzlaff (1970) and later Cherkasskii (1988) reflects that the SES initialization is also used to denote '*socio-ecological*' or '*socioecological*' systems, Berkes and Folke sought to avoid a modifier (socio-) that would imply a subordinate role of the social features of SES (Colding, 2019). Nonetheless, they remain largely interchangeable in the literature. The concept's presence in publications across numerous subject areas has exploded in the 21st century [15], perhaps reflecting or coinciding with increasing interest in the climate crisis and the human role and responses to it. SESs are useful here as a way of examining human interactions with and within the geographic template, and determining how technical and scientific knowledge about SESs are used to inform action in order to shape it and its future states: the essence of spatial planning [16,17]. Planning decisions about shaping SESs are implicitly ethical because they may generate opportunities and challenges for future generations [18].

Because climate change is characterized by significant and potentially increasing uncertainty, decision-making processes are encountering complexity in planning adaptation efforts to address these "(super)wicked" problems [9,19–22]. This is especially true in urban regions complicated by the concentration, entanglement, exposure, and diversity of citizens, resources, assets, and the systems for their management evident there, as well as the numerous, multileveled and/or polycentric governance structures employed as administrative actors [23,24]. Urban areas are complex geographies, where deep and complicated histories, cultures, and institutions generate important questions about the social aspects of power, resources, and environmental health, safety, and justice [25,26]. For these reasons, while we do not rigorously analyze or compare issues arising from various scales of consideration that spatial planning constantly confronts (local vs. national; site-based vs. regional), we examine central ideas and themes related to CCA that are especially evident in densely populated, developed areas. Extensive research on the role and function of multi-level governance (MLG) is evident in CCA circles, as are discussions of various traditions, processes, and planning cultures across nations and regions of the globe (including recent work by Ishtiaque (2021) and DiGreggorio (2019) useful for deeper examination of multilevel governance dynamics.) Most of the discussion within this article is derived from—and applies most directly to—developed nations and western planning traditions whose similarities and features lend toward the generalization and synthesis useful in the construction of the proposed framework.

Meadows' [27,28] landmark 1972 study, *Limits to Growth*, was recently assessed to examine the "fit" between projections of troubling development trends modeled a half-century ago—and their potential implications for countless (and planetary) SESs. Specifically, the "Business as Usual" description of a scenario describing unsustainable development practices (in this instance, particularly as a function of pollution increases including atmospheric greenhouse gas concentrations) appears to be playing out today, potentially portending calamitous impacts for society by or before midcentury [29]. Given that countless planning endeavors have unfolded for decades within the context of a finite planet articulated by *Limits to Growth*, major questions emerge about what planning is fundamentally *for*, how it functions (or can fail), and how it is positioned to operate in the climate change era.

Moreover, insofar as planning is understood as a *practice* utilized for governing the use of resources and space, the *institutions*—rules, norms, customs, and conventions—that simultaneously overarch and undergird planning are crucial to consider, and perhaps the fundamental relationship *between* planning and institutions most of all [30]. This frames the basic question at the center of this review: how is climate change driving transformation of the human systems that must confront it? What are the prominent and salient concepts that characterize this confrontation, and how are they related—to one another and to the planning and institutional domains grappling with climate change? This literature review draws upon important concepts and themes from these fields and areas of interest, as well as synthesizes and integrates prominent concepts into a broadly applicable framework to further research and consideration of the relationships between these fields and ideas. We demonstrate that core concerns stemming from climate change studies are commonplace and of increasing relevance in planning and institutional domains, and that logical links between them can be articulated to illustrate relationships framing notable conceptual and thematic intersections and interactions; these, in turn, work to clarify areas of emphasis, key linkages, and important "blind spots" that persist in CCA research.

This article is structured as follows: Section 2 describes the review approach, and briefly situates spatial planning within a historical and theoretical context that frames consideration of important concepts in the climate adaptation literature. Section 3 integrates these into a Climate Change Adaptation Planning (CCAP) schema, and we describe its key phases. Section 4 examines how, in turn, the practice of adaptation planning is related to theory about adaptation features of interest. Synthesis and integration of these features produces a conceptual framework that exhibits the 'nested' and covalent relationships and dynamics therein, which is followed by an examination of the role of institutions in these dynamics. We close with a brief discussion and conclusion examining insights and further questions framed by the work.

2. Climate Change Adaptation Planning: Prologue, Practice, Paradigm

Our research is focused around a literature review that examines prominent themes related across several domains of interest to CCA: spatial planning, climate change, and institutions. Comparing ideas and terminology of importance across diverse fields and phenomena involving various sociocultural dynamics is complex for a variety of reasons [31]. This is especially true when theories of *change* in social patterns are involved because framing and contextualizing historical trends inevitably entails consideration of broad themes [32]. Our review considered highly-cited articles in the domains of interest to assemble a network of conceptual and empirical articles and studies engaging concepts with broad prominence in CCA research. This formed the basis of an approach articulated by Paré, as geared towards "identifying, describing, and transforming [important] concepts, constructs and relationships ... [to build a] higher order of theoretical structure" [33]. In turn, this approach was used as a theoretical and narrative basis for constructing a *conceptual framework*. This is a common goal and outcome of research linking interdisciplinary bodies of knowledge to explore associated phenomena by articulating "key factors, constructs, or variables" to describe logical relationships among them that correspond to the main tenets of the research [34,35]. Accompanying the narrative review, the framework is used to consider relevant issues in the institutional domain, as well as for framing a discussion about persistent challenges, emergent insights, and potential applications.

2.1. A Very Brief History of Modern Spatial Planning

Landscape architecture arose as a formal design discipline in the 19th century based partially on the increasing recognition of connections between environmental and social health, out of which the sub-discipline of *landscape planning* emerged [36]. Landscape design and planning's interests in large-scale (watershed, regional) geographies and dynamic environmental and human (system, network) processes led to a broader rationale for incorporating ecological considerations into multi-scalar spatial planning [12]. In the postwar era, *ecological planning* entered common parlance, further shaped by the concerns of the modern environmental movement's discontent with harmful effects of unbridled development [37,38]. One of the overarching themes in ecological views of spatial planning is the concept of the *suitability* of landscapes: how their inherent and potential qualities predispose them to various uses by humans [12].

Modern perspectives focusing on Sustainable Development (SD) emerged in the late 20th century largely to address the obvious tensions between intensifying resource management practices and future prosperity [28]. Goals to achieve SD have become key concerns in the climate change era, especially in urban areas of high development intensity [39,40]. The means by which these goals are achieved—the "pathways" taken to reach them—inherently entail *strategic* planning approaches because limited resources force choices that entail tradeoffs [19,41–43]. The scope and scale of climate change is coming into sharper focus in the 21st century, as are its implications for significant change and uncertainty over time [20,22,44,45].

The failure of society to curb GHG emissions through climate *mitigation* has increased the need for climate *adaptation*, emerging as a central concern of spatial planners across the globe; with some anticipating a paradigm shift in the fields of spatial planning concerned with adaptation to more effectively address it [46–48]. Challenges especially evident for spatial planning in the climate era emerge when administrative units delineated in space (as municipal boundaries, borders, zones, etc.) do not adequately address or 'fit' well with the climate phenomena that defy socio-politically conceived and articulated 'lines on the (proverbial) map' [49,50]. Indeed, as the landscape itself is modified by climate change, increased flexibility will surely be required of the very planning processes meant to effectively manage it.

2.2. Climate Change Adaptation: Central Concepts

To situate the practice of spatial planning within CCA efforts and the diversity of interactions that SESs in the climate change era will confront, we summarize several core concepts important in climate adaptation work. These ideas serve to populate our conceptual framework in the next section, which, in turn, displays their relational and dynamic qualities within an integrated theoretical construct.

2.2.1. Sustainability

The harvesting, commodification, distribution, (re)uses, and disposal of resources is a ubiquitous human activity [51]. This is especially true in (and for the provision of) urban areas, where intense turnover and concentration of stocks occurs, recognition of which has given rise to studies of urban ecology and metabolism [52–55]. These processes also entail significant energy "footprints", and numerous environmental impacts, including pollution, result from them [56]. The concept of sustainability may be understood to mean the maintenance of some (economic, social, environmental) entity, process, and/or outcome over time, framed in the environmental context of SESs [13,57,58]. Thus, while resource management remains a central consideration of sustainability in general (and SD specifically), it is also understood as a concept with applications in broader social realms [14].

Resource scarcity (and competition) resulting from unsustainable management practices carries equity implications, both across extant socioeconomic classes and for future generations who may be disadvantaged or disenfranchised by prior resource usage [59–61]. Because planning is a core component of development, SD is frequently invoked as a concept to guide both the means and ends of planning-for-sustainability, a topic of increasing importance in an era of rising environmental concern, uncertainty, and flux [62–64]. Some authors have argued that SES are the logical analytical unit for SD research, with others asserting that they contain inherently interrelated concepts with special relevance to adaptation, or the quality of adaptability [16,65].

2.2.2. Adaptation and Adaptive Capacity

Influential scholarship concerning fundamentals about adaptation is extensive. For the purposes of CCA, it entails altering or adjusting systems and behavior to "alleviate adverse impacts of change or take advantage of new opportunities" through anticipation or response to climate change impacts [66]. Adaptation can be differentiated based on *who* is involved in adjustment, *what* prompts this adjustment, and *how* it is undertaken [67,68]. Together, they "manifest" *adaptive capacity*, through a variety of institutional and social mechanisms (Ibid.). While non-human (eco)systems may also be said to display CCA behavior (and possess adaptive capacity), we are concerned primarily with the active inception and application of human efforts to "influence the direction of change" in SESs affected by climate change [69–71]. Pelling [72] articulates transformation of SESs as a pathway along which adaptation may play out, arguing that adaptation may trigger fundamental changes that decouple systems from more linear modes of progression.

Efforts to manifest adaptive capacity may "backfire", potentially increasing vulnerability [73]. This is known as maladaptation [74,75]. Maladaptive outcomes bear the double burden of generally worsening conditions (reducing resilience or increasing vulnerability) at the implied mutual exclusion of building adaptive capacity due to resource limits [76]. While noting various viewpoints and definitions, Gallopín [77] describes adaptive capacity in SES generally as the capability to cope with environmental change combined with the ability to improve in relation to it. Eakin [78] argues that there are *generic* (development-focused) and *specific* (climate impact-focused) domains of adaptive capacity, and that pursuit of one may exclude, subordinate, or otherwise reduce the other. Whereas adaptation actions might be understood in intuitive ways as relating to adaptive capacity (a quality), these interact in the context of additional qualities—namely vulnerability and resilience—which define SESs in important ways.

2.2.3. Vulnerability and Risk

Vulnerability concerns adverse impacts that occur due to a state's "susceptibility to harm" resulting from potentially complex interplays of exposure and sensitivity to stresses; and it is amplified by a lack of adaptive capacity [68,79,80]. When harmful, these stresses take the form of hazards representing threats to systems, events that "realize" hazards in significant ways by causing damage are *disasters*, and those stemming from or involving natural phenomena are natural disasters [65,81–83]. *Risk* essentially describes the condition and degree(s) of being vulnerable (based on exposure, sensitivity, and capacity) to hazards [84]; and risks shape and define adaptive capacity itself [85]. Risks are generally thought to be, in some sense, quantifiable, i.e., capable of being rendered in terms of probabilities describing the likelihood of outcomes [86–89]. The concentration of people, resources, and systems in urban spaces implies increased exposure, and additional risk based on the location of urban assets (in coastal areas, for example) may arise [26,41]. Risk operates in and across various societal domains: it should be considered in social and economic terms in addition to physical ones, including their interactions [90].

2.2.4. Resilience and Robustness

Systems exposed to risk and experiencing vulnerability may cope with it by drawing upon internal resources, whose realization may reduce impacts. Since Holling's [91] pioneering work in studying ecosystems' capacity to withstand and rebound from states of disturbance,—to "absorb" and "persist"—resilience has become something of a darling within adaptation circles; prompting some to caution that its over-invocation might dilute its meaning [92]. Resilience is of particular importance in the context of climate change because it represents a desirable quality of interacting designed and natural systems, and their relationship to risk and vulnerability [93,94].

Systems that are resilient possess features, including flexibility and diversity, redundancy and modularity, and safe failure characteristics [43]. These work to reduce risk from disasters, which manifests in various types that include interacting, interconnected, compound, and cascading risks [84]. The UN's [95] adoption of frameworks for identifying and evaluating these risks speaks to the centrality of disaster risk reduction (DRR) in adaptation and resilience concerns and approaches. If resilience is seen as flexibility in the face of disturbance, *robustness* might be understood as the capability to resist and withstand it [16]. According to this view, resilient and/or robust systems maintain their core structure despite disturbance, enough so as to avoid becoming vulnerable to the point of significant structural deformation or collapse [96].

2.2.5. Uncertainty

Planning is a process of anticipating, preparing for, and influencing future states of affairs. Uncertainty is a critically important epistemic situation that is inherent to planning because these 'affairs' of future states are influenced by numerous processes that engender and shape events, eventualities, and exigencies [21,97]. This is the meta-context of planning: the temporal dimension within which all socioecological systems play out. Uncertainty intrinsically implies what is unknown and/or unknowable [98]. It is a matter of degree; hence, "levels" of uncertainty exist [99]. Uncertainty is generally understood to increase as more distant futures are considered; and uncertainty may reflect, or be considered as a function of, complexity [88,100].

As planning is intended to inform decision-making, it must ultimately confront uncertainty in that context, influencing the selection of options for coping with or managing it in acceptable ways [101–104]. In this sense, uncertainty actually produces the need to make decisions [105]. These decisions address, but can also *producel*, uncertainty; environmental uncertainty (uncertainty *for* planning) and process uncertainty (uncertainty *from* planning) may also exist, emerge, and interact [88,106]. Christensen's [102] elegant rendering of planning problems hinges on two related processes and their relationship with uncertainty: identifying what to do (a goal) and determining how to do it (through technology), effectively invoking the "ends and means" dyad familiar across all disciplines of planning. The capacity to *learn* new information that changes how uncertainty is *characterized* (and, therefore, changes degrees of *belief*) is a fundamentally adaptive ability [107].

The sheer scale and scope of potential impacts that CCA seeks to address entail significant uncertainty about how and when they will play out, thus shaping the 'menu of options' for responding to them [100,108,109]. Uncertainty might be epistemic (stemming from a lack of knowledge), aleatory (due to intrinsic stochasticity), or both, and it can produce delays in decision-making [104]. A striking example of how the very conceptualization of uncertainty is evolving in the climate change era concerns the asserted "death" of stationarity [110]. Stationarity refers to the statistical concept that environmental fluctuations are bounded inside a value range that is stable (or stationary) over meaningfully long time scales, an assumption that undergirds countless modeling approaches in environmental science and engineering [111,112]. Whether or not reports of stationarity's death have been greatly exaggerated, uncertainty is certainly growing, in actuality and/or as a topic of interest and importance [20].

2.3. Planning: Practice, Policy and Governance
2.3.1. Why Plan(ning)?

The practice of planning is the professionalized implementation of planning efforts, processes shaped by and based on the application of planning theories [88,113]. In exploring what the ultimate purpose of planning is, institutional perspectives have positioned it as operating, in effect, as a mode of *governing societal actions* through processes of "regulation, coordination and control" [114], while others have extended this view to ideally incorporate progressive values linked to social justice and democracy more broadly [115,116]. Generally speaking, planning is practiced in order to use knowledge to shape and implement action by informing decision-making. While noting a multitude of theoretical approaches to spatial planning, Morphet [117] acknowledges planning's inherent power as a redistributive social force, with implications for how power itself is mediated. For our purposes, planning occurs through governmentally-sanctioned processes that concern access to goods and services deemed socially beneficial, and which maintain or enhance public health, safety, and welfare within a particular place; these provisions are often simplified as public "good(s)" [118].

2.3.2. Planning's Mandate: Service to the Public Good(s)

Defining what, exactly, constitutes the public good—much less deciding how to go about achieving, maintaining, or enhancing it—is well-recognized as complex, contentious, and dynamic, involving many diverse stakeholders across multiple levels of society [119–121]. Accordingly, Kunzmann [122] identifies the planning process as one preferably led by the public sector. Numerous climate effects are expected to disproportionately impact (by definition) vulnerable communities, and greater concern for the wellbeing and livelihoods impacted by the products of the adaptation process are, thus, linked closely to planning [123]. Erikson and Brown [124] and Ribot [125] articulate challenges for planning associated with sustainability, resilience, and vulnerability related to uncertainty and complexity in the climate era. Transformative adaptation resulting from effective planning ideally reinforces the legitimacy of the social contract underlying public consent that is granted to planning authorities, ostensibly in their efforts to protect and expand the public good [126].

Planning is understood on basic terms to be a collaborative process that must address what Myers and Kitsuse [127] identified as one of planning's "twin hazards", disagreement (the other being uncertainty), which is confronted through a number of different techniques for conflict resolution in planning, including communication, collaboration, mediation, dialogue, discussion, deliberation, and debate [128–133]. Innes [134] offers an examination of consensus-building as a crucial process for approaching various planning and policy-based disagreements. These serve to discover and define that of which the public good(s) actually consist, and doing so is where the practice of planning partially derives its validity [135]. Owing to numerous factors emerging from climate impacts on the public sector, planning is being deeply reexamined in the context of climate change [41,88,136].

2.3.3. So . . . What Is the Plan?

A plan involves articulating and orienting towards a *vision* for the future—what some human geographers refer to as environmental imaginaries. These frame discourses for structuring the relationship of human processes within places, based on societal imperatives and aspirations amounting to the "virtualities" of future states of affairs [137]. This articulation, in the context of producing the "instrument" of a plan, might involve constructing a declarative set of goals, while orienting towards them identifies steps, stages, or strategies for their realization, though both should embody flexibility to changing circumstances, thus possibly entailing "menus" of scenarios that could be encountered [120,138]. This serves to "situate" the future within an as-yet unrealized (imaginary) SES, towards which the plan is intended to guide decision-making [139,140]. Strategic plans are generally flexible, longer-term, and less fine-grained than more near-term and discrete project plans, owing partially to greater uncertainty existing in "further off" futures [141].

Plan-making may be challenged as a function of numerous horizontal (sector and actor-related) and vertical (multi-level governance-related) connections and the legal, regulatory, and institutional standards at play [142–144]. Plans themselves must define and address the community they are intended to serve; and adopted plans represent, to some acceptable degree, the resolution of various disputes and tensions that arise based on the interests of various stakeholders involved, as well as how they may have constructed their own visions for the future [21,145]. From an adaptation standpoint, this principle also applies to plans that could impact broader communities, so that adaptation actions undertaken within or for one community do not unduly disadvantage another [45]. Resolving these overlaps, tensions, and tradeoffs is, therefore, part of mediating the planning process that shapes and, subsequently, manifests in the scope and strategy of a given adaptation plan [146].

3. 'Sketching' Climate Change Adaptation Planning: Important Features of Interest

The considerations and theories outlined in the last section illustrate features of planning that are useful in apprehending the fast-emerging practices (and problems) involved in Climate Change Adaptation Planning (CCAP). In this section, we illustrate a conceptual schematic (schema), describing the interplay of notable, generalized features of CCAP (Figure 1). Walker [147] describes a *thinking* (planning) and *implementation* (action) phase in adaptive theory applied to policy, to which we add a third phase related to the ongoing assessment of applied work: adaptive management [148]. These echo Peter Hall's [149] trifurcated policy paradigm: overall goal-setting (planning), techniques or instruments (actions), and their "calibration" (management).

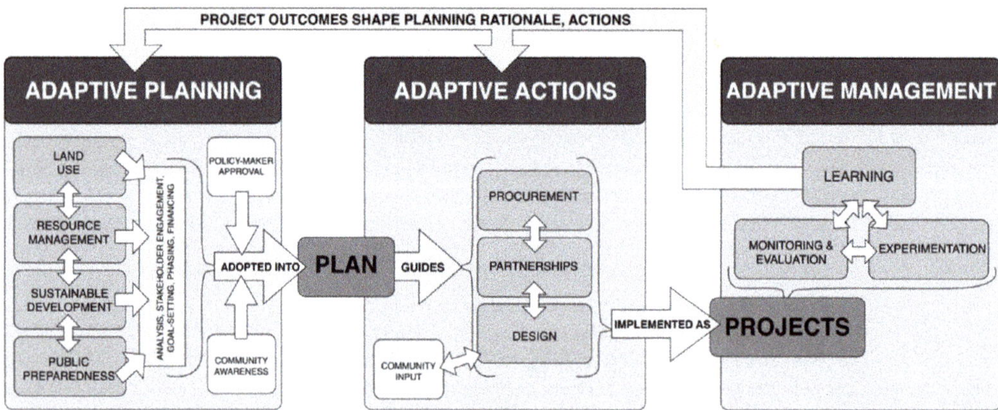

Figure 1. A Climate Change Adaptation Planning (CCAP) Schema. In the *Adaptive Planning* phase, prominent planning concerns are addressed to produce a plan; Implementation based on guidance from plans yields *Adaptive Actions* in the forms of projects; these, in turn become subject to *Adaptive Management* practices for improving upstream and scaled-up efforts.

3.1. Adaptive Planning

Aspects of the planning process are inherently *anticipatory* in nature, wherein complex public policy decision-making occurs in the context of preparing for uncertain future states, thereby naturally engendering adaptive approaches [46]. As a feature of adaptive governance, adaptive planning naturally entails complexities owing to the diversity of actors and actions involved, especially in urban areas [23,150,151]. Anticipatory and planned adaptation within this phase prepare for (instead of react to) future states of affairs, in theory reducing vulnerability and costs [152–154]. Adaptive planning entails stakeholder engagement that takes many forms, but the familiar top-down/bottom-up heuristic is useful in that planners operationalize the interactions of political decision makers in governance (top) and a broader public (bottom), though this group can be defined in various fashions [155,156]. Corfee [145], citing Mitchell [157] and Cash [158], identify requirements for science-policy assessments that inform and influence planning to be deemed publicly acceptable, namely that they be credible, legitimate, and salient.

Plans emerge as products of governance that identify steps for realizing goals in accordance with rules observed by the actor-networks involved, and they gain approval and adoption by passage through the "sluices of democratic and constitutional procedures" [159,160]. Adaptive planning ideally embraces *learning* processes concerned with the structure and effects of the overarching institutional contexts as a useful principle for improving outcomes [159,161,162]. Adaptation plans may include financing components or supplementary plans for funding implementation [163,164]. "Evolutionary" processes in institutional and governance systems, in which processes of reframing and transforming learning occur, are understood as critical for adaptive and equitable systems and are conceptually well-oriented toward adaptation, in general [24,129,165,166]. Limitations in validity assessment and/or forecasting methods may serve to constrain the adaptive planning applications to some extent, though climate change's overall uncertainty implies that flexible, adaptive approaches to planning for it are logical [9,20,167,168].

3.2. Adaptive Actions

We borrow from Aylett's [2] description of adaptive governance as relying on distinct adaptation *planning* and *action* processes, thus echoing Ostrom's [169] notion of the *action situation*. We use the term *adaptive actions* essentially to describe the inception of *projects*. Adaptation projects in urban areas might entail activities involving construction, such as urban greening to reduce heat island effects, improved shoreline defenses as approaches to coastal zone management, integration of "green" stormwater networks to mitigate upland flooding, and the regional management of "upstream" watersheds, and many municipal infrastructure systems represent adaptation imperatives and opportunities in some fashion [48,170–172]. Yet, adaptive actions might also include community initiatives involving outreach, education, and participation without resulting in changes to the physical environment [173]. Thus, broad CCA interest categories in applied adaptation include land use planning (for reclamation, restoration, preservation, conservation aims, for example), natural resource management regimes (concerning water, for example), sustainable development projects (for housing, infrastructure, and public amenities), and community engagement initiatives (for educational or preparedness purposes) [24,69,120,174–179].

Large, complex, or costly adaptive actions that exceed the capacity of public policy and governance institutions often necessitate NGO and private sector involvement, in which planners operate at the "boundary" between the public and private entities [180–182]. Public-Private Partnerships (PPP) describe arrangements in which collaborative, mutually-beneficial relationships are assembled; they are common in urban and municipal settings and a subject of interest in sustainable development circles, with noted promise for adaptation, despite their inherent complexities [123,183–186]. Procurement processes and partnerships are generally intended to alleviate capacity constraints of government. These arrangements can distribute risk and integrate diverse skills and resources into projects involving infrastructure, DRR, urban development, and, increasingly, adaptation projects (and which may entail some or all of the aforementioned project goals and concerns), though these arrangements in the context of CCA are still relatively novel [186,187].

3.3. Adaptive Management

CCA inherently acknowledges that traditional, linear project implementation "pipelines" for realizing plans may be of limited value in an era characterized by increasing uncertainty and complexity [150]. While ancient in practice, recent interest in sustainable resource use, conservation, and ecosystem management have popularized the concept of *adaptive management* [91,188–190]. Other authors have stressed the ties of adaptive management to system resilience and flexibility [191]. Drawing on work from Allen [150] and his work with Garmestani [148], Chaffin [44] defines adaptive management as "implementation of management actions as experiments, followed by monitoring, evaluation and adjustment". Because of the prominence of nature-based solutions and green infrastructure in applied adaptation projects, numerous concerns of adaptive management are relevant to CCAP [192]. Adaptive management applies flexible strategies that take into account emergent opportunities and are generally intended as modes of increasing knowledge, thereby arguably building adaptive capacity and aiding adaptive governance [20,176].

Numerous approaches to understanding change in SESs exist, though central interest in investigating causal processes are especially relevant to planning, a notion termed by Dewey [193] as "experimental knowing". Despite its experimental and flexible nature, adaptive management's potential to induce change (in broader practice and approaches) may be limited by institutional settings where change is itself is problematized or opposed [152]. The experimental underpinning of adaptive management may be useful for learning and information sharing across scales, theoretically aiding in expanding resourcefulness and responsiveness, thereby increasing adaptive capacity [25,43]. The potential for *specific* adaptive actions (in the form of demonstration projects, for example) to broadly inform others might create synergies for syntheses of learning, testing, and adjustment across other sectors and policy realms [152]. Experiments also may be *efficient* in the sense that small scales (and costs) may generate knowledge that is useful at broader scales, though experimentation itself—especially in large (landscape), complex (urban), and dynamic (climate-related) contexts—presents numerous challenges [148,194]. While "scaling up" projects for broader regional application remains complex and daunting [195–197], Hallegatte's [20] identification of the desirable "low regret" quality of adaptation strategies and projects represents obvious conceptual correspondence with experimentation.

Adaptive management also presents opportunities to improve the planning process by incorporating enhanced social inclusiveness, including the dissemination and sharing of information [198,199]. Monitoring that produces data useful for policy consideration is subject to a "reuptake mechanism", whereby conditions observed in adaptation actions may then inform improved planning practices of future or concurrent ones [145]. Fankhauser [45] asserts that adaptation potential is predicated on having "room" to change behavior. By providing the public, planners, and policymakers with real-time, real-world feedback that illustrates how selected adaptive actions are functioning, the "room" for adaptation may become better-parameterized through the reduction of uncertainty (especially relevant in the climate change era) provided by experimental observations. The "feedback loops" inherent to adaptive management suggest that CCAP is, thus, better conceived as looped processes, which are common in conceptualizations of SESs [129,161,200].

4. Zooming Out: CCAP in Broader Context

Partially owing to the varied and multi-scale concerns and methods of practice, the literature exploring what CCAP is and how it operates contains no shortage of concepts and terminology for intellectualizing relevant ideas, themes, theories, and describing a diversity of applied work. While it is beyond the scope of this article and our study to exhaustively compare and square the myriad notions and constructs put forth to describe CCA, we offer a summary of important and interesting concepts, which we synthesize in this section. We then construct a conceptual, graphic framework (Figure 2) that strives to integrate these concepts into a holistic logic, offering a mode of rendering the important ideas and their relationships in a conceptual "space" that captures essential ideas of how important features and forces of CCA interact.

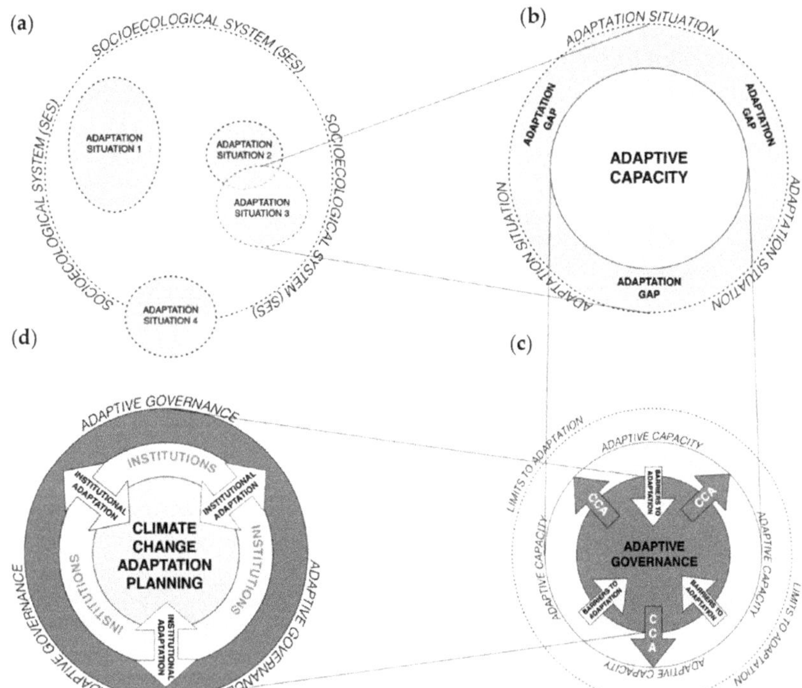

Figure 2. A framework displaying the 'nested' and 'coupled' nature of concepts and interactions of importance in climate change adaptation literature. Arrows denote force directionality, indicating how efforts or concepts "push or pull" towards and/or against other conceptual features or 'spaces'. A 'glossary' of key terms from the framework above. A theoretical discussion follows. (**a**) The *socioecological system* (SES) forms the basic conceptual unit of consideration for framing the adaptation situation. Numerous and interacting adaptation situations may exist within a given SES, or overlap, or "spill" into others. *Adaptation Situations* are characterized by features of the SES, including those in *sociotechnical* (human-based) and *biophysical* (natural setting and context-based) domains, which interact. Phenomena in the biophysical domain engender sociotechnical efforts to establish or expand ("realize") adaptative capacity. (**b**) *Adaptive capacity* is generated by sociotechnical efforts to adapt to biophysical features of the adaptation situation. In general, it is realized by building resilience and reducing vulnerability. An *adaptation gap* exists in the portion of the adaptation situation that lies beyond the adaptive capacity realized within it. It represents the amount of unrealized adaptive capacity. (**c**) *Adaptive governance* describes sociotechnical efforts in shaping the adaptation situation: when effective, adaptive governance increases adaptive capacity, thereby, ideally, shrinking the adaptation gap. Maladaptive (ineffective or counter-productive) efforts reduce adaptive capacity. *Barriers to adaptation* are produced, encountered, and addressed by the sociotechnical and biophysical domains, and in their interactions. Barriers constrain and shrink adaptive capacity, often by hindering adaptive governance or exceeding its reach. *Limits to adaptation* describe the extents of possible adaptation efforts, beyond which increasing adaptive capacity is (actually or considered) infeasible or impossible. Limits may be unknown. (**d**) *Adaptive governance* employs formal practices (*planning*) as modes of realizing efforts, and it is shaped by broader characteristic cultural features and processes (*institutions*). Its efficacy is the sum of institutional and planning efforts performed in the interest of CCA. *Integrated adaptation* refers to the coordination and feedback between adaptation planning-based practices and institutional processes of adaptive change that.

4.1. Conceptualizing Climate Change Adaptation: Framework Features and Forces

4.1.1. Context: Defining Social-Ecological Systems

Pioneering work by Berkes and Folke [13,58] to articulate the interactive dimensions and interplay between humans and their environments introduced the keystone concept of *social-ecological systems* (SES), based partly on work regarding the systematic nature of aspects of the human-nature interaction illustrated by concepts, including vulnerability, resilience, and sustainability [17,65,201]. These insights became key components of numerous interpretive framework approaches to understanding socioecological interdependencies [13]. Of particular importance to planners is that SESs are inherently *spatially contextualized*. That is, because of the entanglement of particular and countless effects of some given environmental situation on sociotechnical (human) systems (and vice versa), they are understood as being in some way at play within a spatially distinct or discernible setting. However, this quality is also, by implication, malleable; and its definition or delimiting is based partially on the interest and perspective of the individual(s) considering or using it as a construct for understanding, planning and managing actions to intentionally alter SESs—the basis of adaptation [200].

4.1.2. Problem: Emergence of Adaptation Situations

Insofar as SESs contain or capture the dynamics between human drives to utilize natural resources and systems, dilemmas stemming from these drives and the capacity of the environment to accommodate them emerge constantly [202,203]. This produces phenomena in which the social and ecologic system aspects relate (or are situated with respect) to one another, generally impelling tensions regarding resources and governance, thus engendering situations in which, according to Ostrom [169], actions may be taken to address or resolve them—generating the concept of the *action situation* [204–206].

The magnitude of climate change on earth's biogeophysical systems has compelled some authors to refine Ostrom's original notion to define *adaptation situations* as a particular form of action situation [118]. Citing previous work, Bisaro and Hinkel [207] describe the adaptation situation as one involving "one or more actors interacting within a common biophysical and institutional environment in which outcomes are altered through climate change". This implies that social features of the situation may be interested in adapting *to* climate change, as well as that, regardless of their interest or efforts, outcomes will be shaped by biophysical effects *of* climate change, which resonates with other scholarship describing the centrality of human endeavors to shape the adaptation situation [208–210].

4.1.3. Manifesting Adaptive Capacity: Adaptive Governance

The sociotechnical (human) features of SESs address the adaptation situation by making decisions about taking actions. These actions amount to Smit and Wandel's [68] description of adaptation(s) as the "manifestation of *adaptive capacity*". The dominant means by which adaptive capacity is manifested by the sociotechnical entities of an SES is through *adaptive governance*, in large part because of the scale at which governmentally-organized action can operate, [24,211,212]. Chaffin [44], in reviewing adaptive governance and synthesizing the perspectives of others, describes adaptive governance as emerging from the search for "modes of managing uncertainty and complexity in SESs". Adaptive governance might be understood as the exercised portion of adaptive capacity—the part that "people use" [213]. Accordingly, depending on how and *when* adaptive capacity is used, it is dynamic over time, unfolding across scales in "coupled cycles of change" [68,214]. While we examine adaptive governance through the lens of climate change, concepts from theories of evolutionary governance may also be useful to consider and apply.

Though adaptive capacity is doubtless considered a desirable quality to possess, the particular and various ways in which adaptive governance is conceived and practiced may give rise to effects that tend to reduce or constrain adaptive capacity; or outcomes that are maladaptive [136,215,216]. Likewise, while adaptive capacity may reflect or express component qualities of the adaptation situation, including vulnerability, resilience, and sustainability, understanding how adaptive capacity is designed or generated (or not) remains complex [16,214]. Carter [41], drawing upon work by Rosenzweig [217], after Mehrotra [218], positions adaptive capacity in relation to vulnerability and hazards, the interactions of all three in essence serving to define risk. In this view, a system's adaptive capacity serves as a kind of counterweight against its vulnerability. While capacity intuitively refers to the *amount* of something (of which one might possess more or less), governance is not the only *source* of adaptive capacity, which can be possessed or provided by non-human features of an adaptation situation, or through non-governance-mediated human actions [43,162]. We focus on adaptive governance because of its centrality to CCAP.

4.1.4. Aspirations: The Adaptation Gap

Lying between the optimal and actual adaptive capacity characterized within a given adaptation situation is a *"gap"*, wherein the potential actions and outcomes of becoming optimally or fully adapted have not (yet) been realized. Moser and Eckstrom [215], echoing Burton [219], note this as a form of adaptation deficit. In describing the analytical methodology of gap analysis for assessing climate hazards, Chen [220] defines the adaptation gap as a "difference between existing adaptation efforts and adaptation need". The United Nations' recently published Adaptation Gap Report focuses on nature-based solutions in conceptualizing and further defining the adaptation gap, though previous volumes with different emphases all include the adaptation gap as a centralizing theme [221]. Numerous complications arise from attempts to quantify subjective, complex, and dynamic features of an adaptation situation that, in theory, define the adaptation gap, including the potential "unknowability" of what, precisely, the gap actually entails and includes [98,222]. Nonetheless, the concept of the adaptation gap is intuitive and useful in the same sense that adaptive capacity is, the former describing an amount of adaptation work *to* be done, and the latter describing the work that *has been* done (thereby establishing existing capacity) or *can be* done as a function of this work. If adaptive governance and other adaptation-oriented sociotechnical efforts are understood as seeking to build adaptive capacity, what forces and phenomena serve to constrain or diminish it?

4.1.5. Challenges: Barriers and Limits to Adaptation

A subject of broad interest is *barriers* to adaptation. Moser and Eckstrom [200] define these as "impediments that can stop, delay, or divert the adaptation process", specifying that they may be surmounted through "concerted effort, creative management, change of thinking, prioritization, and related shifts in resources, land uses, institutions, etc.". Work from Anderies [16], Ostrom [223], and Adger [224] helps situate this concept within the SES literature which, by extension, we project and integrate as features of adaptation situations [207]. Some authors have invoked the notion of adaptation "obstacles", which we consider essentially analogous to barriers [145]. Barriers arise at different stages and levels of adaptation; and they may emerge because of features of governance itself—potentially influencing exactly how adaptive such governance can claim to be—and, by extension, defining its degree of adaptive capacity [69,152,200]. Importantly, Bisaro [225], questioning the utility of the concept, points out that barriers that are easily identified might mask larger, structural, and *institutional* forces that produce the effect(s) of barriers without presenting obvious modes of addressing them.

A common phenomena that arises from and promulgates barriers to adaptation (thus, in theory, reducing adaptive capacity) is *path dependency*, which occurs when institutions or organizations "fail to effectively adapt established practices to face changing circumstances", a pattern of behavior observed across numerous sectors and organizational endeavors, though maladaptive outcomes are a common effect—with obvious and sector-specific implications for CCA, especially in urban settings [2,226,227]. From an economic perspective, situations in which inferior practices perpetuated by path dependency may serve to "lock-in" inefficient (or maladaptive) behaviors and outcomes [228]. Citing Pierson [229] and Wilson [70], among others, Fischer [69] notes path dependency as a kind of *inertia* that results when future actions are shaped in profound or pernicious ways by previous ones. Path dependency, in this sense, is of particular importance for CCAP because of planning's stepwise, cyclical, discursive, and constantly-unfolding nature; the ubiquity of decision-making points and processes therein; diverse sets of actors taking part in the process(es); and the variety of "embedded" cultural features and forces that steer and constrain them [186,230–232].

Whereas the notion of barriers (and obstacles) naturally conjures ideas about surmounting them, *limits* to adaptation refer to bounds that describe "level(s) of adaptive capacity ... that cannot be surpassed", potentially defining the boundary between acceptable and intolerable risks, and which might require transformative change to avoid [85,233]. Barnett [226] distinguishes between "hard" limits that are essentially defined by the environment and "soft" ones that are socially determined and, thus, theoretically malleable. Indeed, Eisenhauer [73], in defining these limits as "factors that prevent adaptation from succeeding", points out that they have been articulated as both objectively identifiable (as in the case of certain biotic and economic examples) and, from a more constructivist perspective, presenting as difficult-to-define endogenous effects emerging from societies' "goals, values, risk perceptions and actions". Limits are perhaps also worth considering as "blended" between hard and soft characterizations because sociopolitical conceptualizations of limits emerge in response to environmental ones, which may then be redefined by human intervention. In general, limits define the extent to which adaptive capacity *could* be realized—apart from how effectively barriers *are* overcome in the practice of adaptive governance (to increase adaptive capacity). Again, this resonates with Adger's [224] view that limits are situational thresholds beyond which "adaptation actions fail to protect things stakeholders care about", which we understand to include non-physical "things", such as social cohesion, morale, trust in institutions, etc.

4.2. CCAP: Integrating Institutional Adaptation

4.2.1. The Role of Institutions

Gupta [4] elegantly renders institutions as "social patterns", while a more expansive view, according to Oberlack [74], citing several others, articulates institutions as "rules and procedures that structure action situations within which individual and collective decision-making [is affected to] constrain, enable and incentivize actions; link individual actions, events and outcomes; distribute authority and power; define reciprocal rights and duties; and shape beliefs, motivations and social learning" [24,169,234,235]. These may be formal or informal [159]. Vatn [236] describes the invisible or even unselfconsciously natural instantiation of institutions in behavior as conventions that are observed, referencing work by Crawford and Ostrom [237], to compose a "grammar" of institutions and their functions. Institutions might be understood as self-reinforcing "regularities": patterns of behavior evident in networks of social actors who "tacitly create [them] to solve a wide variety of recurrent problems" [238]. Yet, despite regularities and recurrences, institutions are not static; they "distribute obligations and entitlements to resources as well as the power to change such obligations and entitlements" [239]. Though they may be nonmaterial (informal), institutions reify actual, tangible outcomes.

Institutional analyses focused on resources (components or products of the environment) and how the notion of property (which entails ownership, often of the landscape itself) factors into their management, is a well-established field of institutional interest, and planning has been articulated as a mode of "bundling the rights" of ownership associated with property in this sense [240]. From an economic perspective, the linkages between humans and their environment are mediated by countless rules that shape and reinforce beliefs and values, but these are dynamic and responsive [241]. Where public policy is concerned, this dynamic quality of institutions has important implications because the question of how power and influence is distributed within society—including this critical capacity to alter existing situations and arrangements—is of enormous importance in the climate change era [74], insofar as planning efforts are understood as being shaped by larger cultural and institutional forces, and these may fail to present obvious, accessible, and discrete decision-making processes themselves [172,225].

4.2.2. Institutions and Change

In theorizing about the evolutionary nature of governance, Van Assche [166] positions institutions as being designed for change, even postulating that the essence of democracy lies in the "rules of self-transformation; *rules to change the rules*". As institutions occupy important features of SESs and spatial discourse generally, they are tightly linked with conceptions of the environmental imaginary [242], entailing consideration of the distribution and access to power and influence involved in its realization, recalling Bromley's [239] obligations and entitlements [243]. In other words, institutions structure what is possible based partially on how society mediates the tensions arising from multitudes (citizens, actors) shaping and sharing something more unified (the environment) [244]. Institutions influence aspirations (for a more healthy and just environment, for example), even while subject to inertia (perpetuating the status quo), and the outright resistance to change, termed the *precautionary principle*, which is important in situations involving uncertainty [245–247]. Similar to the concept of path dependency in organizational endeavors, institutional inertia and "lock in" may occur when regimes and patterns of behavior become ossified due to various factors [229,241]. Institutions within or across SESs may constrain or delimit the actions of organizations by conformation and homogenization, producing institutional isomorphism [248], which may be induced by coercive, mimetic, or normative means [142]. Storbjörk and Hedrén [172] describe clashing cultures, knowledge claims, and cross-sectoral integration problems as several notable barriers to institutional change.

While approaches to determining how institutions resist change (in inertial, oppositional, and isomorphic ways) are evident, factors that instigate change within and across institutions are complex to identify, perhaps owing to requisite "concatenations" of underlying mechanisms [249,250]. Hodgson [251] identified two dominant institutional modes: agent-*sensitive* and agent-in*sensitive*, the latter describing an institution in which significant change affected by institution-shaping actors (agents) is unlikely or difficult. Individuals, organizations, and governance structures that cut across public and private sectors constantly respond to environmental change (thereby engendering change); thus, environmental change does not occur in an "institutional vacuum" [249,252]. Influential individuals (leaders) [253], sociopolitical mobilization [254], and/or catalytic or vivid events [255] that impose or focus urgency upon some situation may induce institutional change by creating, though other factors have been identified as important "drivers" precipitating change dynamics [10,249,256]. Aggregating these behavioral changes across scales and social structures—and mediating or coordinating them through planning mechanisms—, in turn, changes the institutional environment itself, in theory providing conditions for institutional adaptation [117]. Planning that attempts to engage these institutional change dynamics confronts a duality in that institutions are both behavior patterns "out in the world" (actions) and internal ones "in the head" (thoughts and feelings), which obviously presents complexities to planners attempting to derive institutional origins [257,258]. All of these qualities speak to the difficulty in clearly formalizing or mapping institutional dynamics, made especially complex in a situation in which the underlying environmental context is also in a state of flux.

4.2.3. Institutions, Climate Adaptation, Planning

Smit and Wandel [68] note that adaptive capacity may be increased through improvements in *technology* and/or institutions, while Rodima-Taylor [123] echoes Koppel's [259] position that technological innovation is *induced* by institutional change. Christensen [102] considers technology in the context of planning to be the "knowledge of how to do something"—literally, the *means*. Our CCAP schema illustrates that these means might be expanded by integrating adaptive principles into planning that make it more "nimble" (thus, resistant to path-dependence). Yet, how these qualities relate to an institutional adaptation discourse remains complex, in part owing to the need to disentangle the functions and mechanics of institutions themselves [10,260,261]. In developing a framework for assessing institutional adaptive capacity, Gupta [4] identifies two core characteristics: one essentially describing their inherent, extant qualities; and the second relating to the degree to which they "allow or encourage" their own (institutional) change, essentially describing adaptability itself. The *rate* of change, or timing, also matters: disparities between non-institutional changes that occur within SESs and that at which institutions are fundamentally *able to affect change* may lead to missed opportunities, including from a lack of timely collaboration and cooperation [4,215,226].

Roggero [208] explores how one aspect of institutional change is positioned with respect to CCA in his iteration of Hagedorn's [234] notion of *integrative* institutions (that address climate-related *interdependencies*) versus *segregative* ones (that focus only on climate-impacted *resources* under their effective purview). Institutional complexity itself may work against institutional change or adaptation simply as a function of the increased "work" required to do so in complex networks, though structured learning processes may be useful [24,155,262]. Informal, 'behind-the-scenes' "shadow" processes may be important factors for inducing institutional change [175], in addition to the identification and inception of "additional or adjusted institutional design propositions" to address climate uncertainties and complexities [161].

A critical question for CCAP and its role in building adaptive capacity seems to concern the scope of its *influence* and *intentions*, particularly in relationship to institutional forces that define, delimit, and direct them, as well as how these may differ or mesh with planning practices and processes as traditionally understood. For example, failures to adapt may be due to issues of governance more so than the planned, technical implementation of applied adaptation efforts, reflecting complexity inherent to multi-level governance [24,263,264]. Patterson's [10] work investigating dimensions and possible drivers of institutional adaptation in urban governance reveals that, in formal terms, "planning" is limited in its role: for example, it is not the job of planners to cultivate charismatic leaders, nor to foment community pressure (much less political disruptions), even though these may occur partially as a function of adaptation planning. The lack of real or perceived alignment of institutions with climate change adaptation risks the governance processes for achieving it being less adaptive and/or less strategic than optimal: a condition describing or producing institutional "voids" [256].

5. Discussion

5.1. Central Insights

As explored and illustrated in this review, planning and institutional domains are being challenged or are changing because of the emergence, intensity, and importance of climate change within policy and governance spheres. The core goal of this review is to explore complicated topics across several domains and, based on thematic and conceptual linkages prominent in the literature, to construct an integrative perspective to increase clarity in comprehension of complex and *related* topics relevant to CCA. Several insights based on this work are notable. First, important concepts of climate change literature have been increasingly encountered and integrated into spatial planning practices, which have led to distinct *forms* of planning. Our CCAP schema demonstrates how, for example, uncertainty is being addressed not only as an increasing "fact of life" for planners to manage but one that can be understood and approached opportunistically and as a force driving innovation and learning processes that increase adaptive capacity. In other words, the emphasis and engagement with climate change issues is leading to adaptation in the *practice* of planning itself.

Second, prominent and complex concepts of interest evident in climate change literature can be organized into a holistic construct that displays important tenets of the research, and displayed in such a fashion as to clarify their interplay, as through the proposed framework. These interplay may take the form of *positional* properties of features within a framework that group or separate concepts, nest or embed them in one another, or imply some connective linkage(s) or couplings. They can also be rendered in *mechanistic* terms, where dynamics of some feature of interest logically or implicitly affect others, thus illustrating *causal* relationships. These are of particular importance in adaptation work in a similar fashion to features of our CCAP schema, in that, fundamentally, *being adaptive* entails processes of feedbacks and responses in *systems*. Because our framework's foundational feature (within and through which other features interact) are SESs, we can intuitively grasp this systematic structure and behavior. The framework, in this regard, is useful in two primary ways: it organizes and simplifies information; and it provides its own *logic* that is both emergent (arising from themes and ideas in the literature examined) and can be utilized, altered, or critiqued by practitioners for case-specific or applied work, or as a basis for expansion or alteration through introducing additional or different theoretical components.

Finally, as a function of the deeply complex, subtle, and dynamic nature of institutions (including merely identifying or agreeing upon them), we display the limits of the framework; prompt consideration of how planning and institutions are, in theory and reality, bound together; and provide context for considering relevant connections or patterns as theses domains unfold and interact through CCA endeavors. For example, we discuss that organizational path dependency and institutional lock-in both serve to reduce adaptive capacity, while the modes of surmounting these barriers to adaptation are nonetheless domain-distinct, in terms of the means for assessing, addressing, or ameliorating them. Likewise, planning and institutions must be understood in a temporal context in important ways: planning because its legitimacy and efficacy depend on the results of its implementation and "follow through"; and institutions because their social utility, acceptance and adherence are derived, at least partially, by way of their durability. The examination of key features of the climate change era, namely uncertainty and change itself, present vexing questions and prompt provocative, perhaps even subversive, perspectives from which to consider the practice of planning and its institutional context.

5.2. Adoption, Application, Adaptation of the Framework

This article seeks to articulate the ways in which important concepts relevant to climate adaptation might be more clearly differentiated and understood in their relational dynamics, partially through illustrating schema that can be adapted to various actual situations or case studies, and linking these with prominent themes and patterns from our literature review. An overarching challenge in CCA, planning, and institutional change (especially) is measuring or quantifying the magnitude or effects of concepts that, to some extent, resist or defy efforts to do so. Certain aspects of SESs are, after all, based on informal, constantly-changing, and nonmaterial qualities with which it is, nonetheless, important to grapple. Our "schematizing" of concepts in ways that can be visualized, to some extent, might provide interesting opportunities for researchers seeking to understand how individuals (within or across organizations, levels of government, and/or demographic groups) comprehend, or (literally) "picture", some of these concepts.

Future use of the framework along these lines might take the form of research employing templates that are used to gauge (for example) how different groups render adaptive capacity inside an adaptation situation, define magnitudes of effects for various barriers; order hierarchies of adaptation planning issues, "connect" causal influences or tensions between features and how they are situated relative to others, or articulate the "distance(s)" they imagine limits lie from adaptive capacity. Clearly, these exercises would yield abstractions: sketches or diagrams, that stand in for more nuanced work. Yet, these might reveal insights and/or patterns valuable to managers seeking to understand institutional or organizational dynamics, public sentiment, or differences across divisions, or even the age or career seniority of individuals. While not the focus of this article, social science methods applied to constructing impressions and understanding of how various groups apprehend the concepts explored here—and their relationships to each other—may be illuminating. A consistent theme of this research seems to be that what people believe is *possible* (and the institutional ramifications therein) is strongly linked with problem definition and framing, with obvious impacts on decision-making and commensurately dramatic implications for CCAP.

5.3. Critical Considerations and Questions

One of the appeals of institutions that are not only adaptive but well-integrated into CCAP is that their influence and capacity to "structure ... political decision-making ... [and] shape practices and behaviors" is understood as being vital for the success of large-scale, strategic efforts necessary in complex urban settings [10,23,137]. In this context, the utility of local knowledge and local institutions has been emphasized as a driver of adaptive capacity but also as *processes*, not merely *information or rules* (content) [68,265,266]. In one sense, planning is a practice of more than instrumentalizing content; it inherently represents engagement with ongoing processes. Yet, precisely because planning entities (individuals, agencies, departments, divisions, authorities) are empowered *by and within* overarching institutional milieus, questions emerge about planning as a force for transformational, fundamental change in the ongoing adaptation quest, which some see as amounting to the proposition of a paradigm shift for planning itself [47]. In other words, can planning "unlock" institutions from nonadaptive tendencies, and to what degree?

We have examined the relationships between these concepts and their underlying theories to situate planning in a critical light, insofar as we question its agency and the scope of its traditionally-conceived responsibilities. Planning, in the face of massive environmental change and uncertainty, may *itself* obscure the clarity of future visions and complicate the steps for manifesting them, in no small part due to institutional inertia *and* dynamics. That is, uncertainties rooted in the institutional domain may amplify overall situational uncertainty and complicate planning processes attempting to address it. Dovers [267] points out that even constructing an understanding of the limits to adaptation is fraught in part because of the institutional dimension, whose sheer complexity grows with the scale considered [268]. With climate change altering resource regimes and shaping the public good(s) of citizens linked through institutional behavior and (ideally) aligned through adaptation planning practices, questions about how common-pool resources and common-pool institutions can or should shape planning's role in allocating entitlements and obligations emerge [60,263,269–271]. This, in turn (and in ways beyond the scope of this article), ensnares any number of private sector considerations and the need to, among other things, understand how planning and institutions are positioned to address or adapt to markets relevant in adaptation [66,143].

6. Conclusions

Our review examined important concepts related to the CCA plight by examining the theoretical and applied linkages between the practice of spatial planning and role of institutions in the governance of adaptation, with an emphasis on issues and dynamics broadly relevant in urban regions. Through this process, we sought to illustrate and situate prominent themes and concepts in climate adaptation work that connect to engage planning and institutional dynamics, as well as their effects on SESs, which Berkes and Folke originally termed the "linkages between ecosystems and institutions" [13]. Epstein expanded on this concept and considered the differentiation between social and ecological systems as reconciled by "fitting" them together through institutions themselves; in doing so, this revealed strengths and limitations of the institutional *couplings* of these systems [14]. Planning, as we have discussed, represents a mode of instrumentalizing adaptive governance largely in the interest of increasing adaptive capacity; and, in the climate era, our schema demonstrates how planning employs various techniques to do so in the context of uncertainty and change, in fact, by embracing it and approaching it opportunistically. Likewise, our framework illustrates the nested and linked—or *coupled*—mechanics of planning to larger concepts and displays how their interconnections might be understood. For their part, institutions, while playing important roles in shaping and constraining planning and defining various aspects of SESs, remain difficult to fully comprehend and describe when the same considerations of uncertainty and change characterize the (conceptual) landscape in which they are realized.

In his treatise articulating the global, intergenerational ethical and moral implications of climate change, Stephen Gardiner [272] identifies *institutional inadequacy* as a key characteristic; one that, for various reasons, cannot simply be overcome by better governance. This article situates adaptation planning as a critical link between governance and institutions: in the case of the former, as a "downstream" tool for facilitating policy decision-making; in the latter, by triggering feedback from features of the SESs that have "upstream" implications for the "rules of the game" themselves, which define and constrain what futures are considered possible or desirable [273]. Planning, as a field seeking to integrate science and knowledge into decision-making, is surely constrained in its capacity to do so by various political and institutional arrangements and realities, though Roggero [209] asserts that organizing knowledge in *"institutionally meaningful ways* can advance . . . understanding of the link between institutions and adaptation". What precisely constitutes institutional meaningfulness in the context of climate change remains complex, dynamic, and, surely, case-specific, to some degree.

Insofar as we consider institutions to be collectivized social patterns of behavior that are "rendered durable" over time by routine and habits, the task for planning to break from reinforced tendencies that reduce adaptive capacity seems pressing [251,258]. These reflections position planning in a crucial position that prompts consideration about the nature or characterization of planning entities themselves: are they primarily *agents* within Hodgson's [251] reckoning (to whom institutions may be sensitive/responsive in terms of change), or merely a *means* by which those agents interact? If they fall into the former category (or if they are understood to be both), the question of intent emerges: is it the role and responsibility of planning to actively, aggressively attempt to alter—*or even do away with*—institutions in light of the knowledge planning inevitably encounters and frames? If so, which institutions? In what circumstances, to what degree, why, and—critically—*how*? While this last question involves what Dover [267] calls the *practicalities* of institutional change, the challenge for adaptation planning in the 21st century may be as much about principles as practicalities.

Author Contributions: Writing—original draft, N.K.; Writing—review & editing, K.H. All authors have read and agreed to the published version of the manuscript.

Funding: This research received no external funding.

Conflicts of Interest: The authors declare no conflict of interest.

References

1. Nordgren, J.; Stults, M.; Meerow, S. Supporting Local Climate Change Adaptation: Where We Are and Where We Need to Go. *Environ. Sci. Pol.* **2016**, *66*, 344–352. [CrossRef]
2. Aylett, A. Institutionalizing the Urban Governance of Climate Change Adaptation: Results of an International Survey. *Urban Clim.* **2015**, *14*, 4–16. [CrossRef]
3. Einecker, R.; Kirby, A. Climate Change: A Bibliometric Study of Adaptation, Mitigation and Resilience. *Sustainability* **2020**, *12*, 6935. [CrossRef]
4. Gupta, J.; Termeer, C.; Klostermann, J.; Meijerink, S.; van den Brink, M.; Jong, P.; Nooteboom, S.; Bergsma, E. The Adaptive Capacity Wheel: A Method to Assess the Inherent Characteristics of Institutions to Enable the Adaptive Capacity of Society. *Environ. Sci. Policy* **2010**, *13*, 459–471. [CrossRef]
5. Wang, Z.; Zhao, Y.; Wang, B. A Bibliometric Analysis of Climate Change Adaptation Based on Massive Research Literature Data. *J. Clean. Prod.* **2018**, *199*, 1072–1082. [CrossRef]
6. Tribbia, J.; Moser, S.C. More than Information: What Coastal Managers Need to Plan for Climate Change. *Environ. Sci. Policy* **2008**, *11*, 315–328. [CrossRef]
7. Berrang-Ford, L.; Pearce, T.; Ford, J.D. Systematic Review Approaches for Climate Change Adaptation Research. *Reg. Environ. Chang.* **2015**, *15*, 755–769. [CrossRef]

8.   Ford, J.D.; Pearce, T. What We Know, Do Not Know, and Need to Know about Climate Change Vulnerability in the Western Canadian Arctic: A Systematic Literature Review. *Environ. Res. Lett.* **2010**, *5*, 014008. [CrossRef]
9.   Giordano, T. Adaptive Planning for Climate Resilient Long-Lived Infrastructures. *Utilities Policy* **2012**, *23*, 80–89. [CrossRef]
10.   Patterson, J.J. More than Planning: Diversity and Drivers of Institutional Adaptation under Climate Change in 96 Major Cities. *Glob. Environ. Chang.* **2021**, *68*, 102279. [CrossRef]
11.   Lawrence, D.P. Planning Theories and Environmental Impact Assessment. *Environ. Impact Assess. Rev.* **2000**, *20*, 607–625. [CrossRef]
12.   Ndubisi, F. *Ecological Planning: A historical and Comparative Synthesis*; Johns Hopkins University Press: Baltimore, MD, USA, 2002.
13.   Fikret, B.; Carl, F. *Linking Social and Ecological Systems: Management Practices and Social Mechanisms for Building Resilience*; Cambridge University Press: Cambridge, UK, 1998.
14.   Epstein, G.; Pittman, J.; Alexander, S.M.; Berdej, S.; Dyck, T.; Kreitmair, U.; Rathwell, K.J.; Villamayor-Tomas, S.; Vogt, J.; Armitage, D. Institutional Fit and the Sustainability of Social–Ecological Systems. *Curr. Opin. Environ. Sustain.* **2015**, *14*, 34–40. [CrossRef]
15.   Colding, J.; Barthel, S. Exploring the Social-Ecological Systems Discourse 20 Years Later. *Ecol. Soc.* **2019**, *24*, 2. [CrossRef]
16.   Anderies, J.M.; Janssen, M.A.; Ostrom, E. A Framework to Analyze the Robustness of Social-Ecological Systems from an Institutional Perspective. *E S* **2004**, *9*, art18. [CrossRef]
17.   Gallopin, G.C.; Gutman, P.; Maletta, H. Global Impoverishment, Sustainable Development and the Environment: A Conceptual Approach. *Int. Soc. Sci. J.* **1989**, *121*, 375–397.
18.   Aldo, L.; Schwartz, C.W. *A Sand County Almanac: With Other Essays on Conservation from Round River*; Oxford University Press: New York, NY, USA, 1949.
19.   Albrechts, L. Strategic (Spatial) Planning Reexamined. *Environ. Plan. B Plan. Des.* **2004**, *31*, 743–758. [CrossRef]
20.   Hallegatte, S. Strategies to Adapt to an Uncertain Climate Change. *Glob. Environ. Chang.* **2009**, *19*, 240–247. [CrossRef]
21.   Levin, K.; Cashore, B.; Bernstein, S.; Auld, G. Overcoming the Tragedy of Super Wicked Problems: Constraining Our Future Selves to Ameliorate Global Climate Change. *Policy Sci.* **2012**, *45*, 123–152. [CrossRef]
22.   Toimil, A.; Losada, I.J.; Nicholls, R.J.; Dalrymple, R.A.; Stive, M.J.F. Addressing the Challenges of Climate Change Risks and Adaptation in Coastal Areas: A Review. *Coast. Eng.* **2020**, *156*, 103611. [CrossRef]
23.   Castán Broto, V. Urban Governance and the Politics of Climate Change. *World Dev.* **2017**, *93*, 1–15. [CrossRef]
24.   Pahl-Wostl, C. A Conceptual Framework for Analysing Adaptive Capacity and Multi-Level Learning Processes in Resource Governance Regimes. *Glob. Environ. Chang.* **2009**, *19*, 354–365. [CrossRef]
25.   Bulkeley, H.; Castán Broto, V. Government by Experiment? Global Cities and the Governing of Climate Change. *Trans. Inst. Br. Geogr.* **2013**, *38*, 361–375. [CrossRef]
26.   Rasmussen, D.J.; Oppenheimer, M.; Kopp, R.E.; Shwom, R. The Political Complexity of Coastal Flood Risk Reduction: Lessons for Climate Adaptation Public Works in the U.S. *Earth's Future* **2021**, *9*, 1–19. [CrossRef]
27.   Meadows, D.H. *The Limits to Growth: A Report for the Club of Rome's Project on the Predicament of Mankind*; Earth Island: London, UK, 1972; ISBN 0-85644-008-6.
28.   Meadows, D.H.; Randers, J.; Meadows, D.L. *The Limits to Growth: The 30-Year Update*; Chelsea Green Pub.: Hartford, VT, USA, 2004; ISBN 1-931498-51-2.
29.   Herrington, G. Update to Limits to Growth: Comparing the World3 Model with Empirical Data. *J. Ind. Ecol.* **2021**, *25*, 614–626. [CrossRef]
30.   Gualini, E. *Planning and the Intelligence of Institutions: Interactive Approaches to Territorial Policy-Making between Institutional Design and Institution-Building*, 1st ed.; Routledge: London, UK, 2001. [CrossRef]
31.   Ritchie, J.; Lewis, J.; Nicholls, C.M.; Ormston, R. *Qualitative Research Practice: A Guide for Social Science Students and Researchers*, 2nd ed.; Sage: London, UK, 2014.
32.   Webster, J.; Watson, R.T. Analyzing the Past to Prepare for the Future: Writing a Literature Review. *Manag. Inf. Syst. Q.* **2002**, *26*, xiii–xxiii.
33.   Paré, G.; Trudel, M.-C.; Jaana, M.; Kitsiou, S. Synthesizing Information Systems Knowledge: A Typology of Literature Reviews. *Inform. Manag.* **2015**, *52*, 183–199. [CrossRef]
34.   Jabareen, Y. Building a Conceptual Framework: Philosophy, Definitions, and Procedure. *Int. J. Qual. Methods* **2009**, *8*, 49–62. [CrossRef]
35.   Miles, M.B.; Huberman, A.M. *Qualitative Data Analysis: An Expanded Source Book*, 2nd ed.; Sage: Newbury Park, CA, USA, 1994.
36.   Hill, K. *Landscape Architecture and Environmental Planning*; Oxford University Press: Oxford, UK, 2018. [CrossRef]
37.   McHarg, I.L. *Design with Nature*; Natural History Press: Garden City, NY, USA, 1969.
38.   Swaffield, S.R. *Theory in Landscape Architecture: A Reader*; Penn Studies in Landscape Architecture; University of Pennsylvania Press: Philadelphia, PA, USA, 2002; ISBN 978-0-8122-1821-3.
39.   Lélé, S.M. Sustainable Development: A Critical Review. *World Dev.* **1991**, *19*, 607–621. [CrossRef]
40.   Sauvé, S.; Bernard, S.; Sloan, P. Environmental Sciences, Sustainable Development and Circular Economy: Alternative Concepts for Trans-Disciplinary Research. *Environ. Dev.* **2016**, *17*, 48–56. [CrossRef]
41.   Carter, J.G.; Cavan, G.; Connelly, A.; Guy, S.; Handley, J.; Kazmierczak, A. Climate Change and the City: Building Capacity for Urban Adaptation. *Prog. Plan.* **2015**, *95*, 1–66. [CrossRef]
42.   Rondinelli, D.A. Public Planning and Political Strategy. *Long Range Plan.* **1976**, *9*, 75–82. [CrossRef]

43. Tyler, S.; Moench, M. A Framework for Urban Climate Resilience. *Clim. Dev.* **2012**, *4*, 311–326. [CrossRef]
44. Chaffin, B.C.; Gosnell, H.; Cosens, B.A. A Decade of Adaptive Governance Scholarship: Synthesis and Future Directions. *E S* **2014**, *19*, art56. [CrossRef]
45. Fankhauser, S.; Smith, J.B.; Tol, R.S.J. Weathering Climate Change: Some Simple Rules to Guide Adaptation Decisions. *Ecol. Econ.* **1999**, *30*, 67–78. [CrossRef]
46. Birchall, S.J.; MacDonald, S.; Slater, T. Anticipatory Planning: Finding Balance in Climate Change Adaptation Governance. *Urban Clim.* **2021**, *37*, 100859. [CrossRef]
47. Hill, K. Climate Change: Implications for the Assumptions, Goals and Methods of Urban Environmental Planning. *UP* **2016**, *1*, 103–113. [CrossRef]
48. Lawrence, J.; Bell, R.; Blackett, P.; Stephens, S.; Allan, S. National Guidance for Adapting to Coastal Hazards and Sea-Level Rise: Anticipating Change, When and How to Change Pathway. *Environ. Sci. Policy* **2018**, *82*, 100–107. [CrossRef]
49. Hannah, L. A Global Conservation System for Climate-Change Adaptation. *Conserv. Biol.* **2010**, *24*, 70–77. [CrossRef] [PubMed]
50. Wilder, M.; Scott, C.A.; Pablos, N.P.; Varady, R.G.; Garfin, G.M.; McEvoy, J. Adapting Across Boundaries: Climate Change, Social Learning, and Resilience in the U.S.–Mexico Border Region. *Ann. Assoc. Am. Geogr.* **2010**, *100*, 917–928. [CrossRef]
51. Graedel, T.E.; Allenby, B.R. *Industrial Ecology and Sustainable Engineering*; Prentice Hall: Upper Saddle River, NJ, USA, 2010; ISBN 978-0-13-600806-4.
52. Céspedes Restrepo, J.D.; Morales-Pinzón, T. Urban Metabolism and Sustainability: Precedents, Genesis and Research Perspectives. *Resour. Conserv. Recycl.* **2018**, *131*, 216–224. [CrossRef]
53. Ioppolo, G.; Heijungs, R.; Cucurachi, S.; Salomone, R.; Kleijn, R. Urban metabolism: Many open questions for future answers. In *Pathways to Environmental Sustainability: Methodologies and Experiences*; Springer: Cham, Switzerland, 2013; p. 23. [CrossRef]
54. Jianguo, W. Urban Ecology and Sustainability: The State-of-the-Science and Future Directions: Actionable Urban Ecology in China and the World: Integrating Ecology and Planning for Sustainable Cities. *Landsc. Urban Plan.* **2014**, *125*, 209.
55. Kennedy, C.; Pincetl, S.; Bunje, P. The Study of Urban Metabolism and Its Applications to Urban Planning and Design. *Environ. Pollut.* **2011**, *159*, 1965–1973. [CrossRef]
56. Tarr, J.A. *The Search for the Ultimate Sink. [Electronic Resource]: Urban Pollution in Historical Perspective*; Technology and the Environment; University of Akron Press: Akron, OH, USA, 1996; Project MUSE: Baltimore, MD, USA, 2015; ISBN 978-1-935603-38-2.
57. Basiago, A.D. *Economic, Social, and Environmental Sustainability in Development Theory and Urban Plan-Ning Practice: The Environmentalist*; Kluwer Academic Publishers: Boston, MA, USA, 1999.
58. Berkes, F.; Colding, J.; Folke, C. (Eds.) *Navigating Social-Ecological Systems: Building Resilience for Complexity and Change*; Cambridge University Press: Cambridge, UK, 2003.
59. Baccini, P.; Brunner, P.H. *Metabolism of the Anthroposphere: Analysis, Evaluation, Design*; MIT Press: Cambridge, MA, USA, 2012; ISBN 978-0-262-30132-9.
60. Dipierri, A.A.; Zikos, D. The Role of Common-Pool Resources' Institutional Robustness in a Collective Action Dilemma under Environmental Variations. *Sustainability* **2020**, *12*, 10526. [CrossRef]
61. Stoddart, H.; Schneeberger, K.; Dodds, F.; Shaw, A.; Bottero, M.; Cornforth, J.; White, R. *A Pocket Guide to Sustainable Development Governance*; Commonwealth Secretariat; Stakeholder Forum: London, UK, 2011.
62. Gopalakrishnan, V.; Bakshi, B.R. Including nature in engineering decisions for sustainability. In *Encyclopedia of Sustainable Technologies*; Elsevier: Amsterdam, The Netherlands, 2017; pp. 107–116, ISBN 978-0-12-804792-7.
63. Lemons, J.; Westra, L.; Goodland, R. (Eds.) *Ecological Sustainability and Integrity: Concepts and Approaches*; Environmental Science and Technology Library; Springer: Dordrecht, The Netherlands, 1998; Volume 13, ISBN 978-90-481-4980-3.
64. Wheeler, S.M. *Planning for Sustainability: Creating Livable, Equitable, and Ecological Communities*; Routledge: London, UK, 2004; p. 289.
65. Young, O.R.; Berkhout, F.; Gallopin, G.C.; Janssen, M.A.; Ostrom, E.; van der Leeuw, S. The Globalization of Socio-Ecological Systems: An Agenda for Scientific Research. *Glob. Environ. Chang.* **2006**, *16*, 304–316. [CrossRef]
66. Neil Adger, W.; Arnell, N.W.; Tompkins, E.L. Successful Adaptation to Climate Change across Scales. *Glob. Environ. Chang.* **2005**, *15*, 77–86. [CrossRef]
67. Fischer, A.P. Pathways of Adaptation to External Stressors in Coastal Natural-Resource-Dependent Communities: Implications for Climate Change. *World Dev.* **2018**, *108*, 235–248. [CrossRef]
68. Smit, B.; Wandel, J. Adaptation, Adaptive Capacity and Vulnerability. *Glob. Environ. Chang.* **2006**, *16*, 282–292. [CrossRef]
69. Wilson, G.A. Community Resilience, Globalization, and Transitional Pathways of Decision-Making. *Geoforum* **2012**, *43*, 1218–1231. [CrossRef]
70. Fazey, I.; Wise, R.M.; Lyon, C.; Câmpeanu, C.; Moug, P.; Davies, T.E. Past and Future Adaptation Pathways. *Clim. Dev.* **2016**, *8*, 26–44. [CrossRef]
71. Pelling, M.; O'Brien, K.; Matyas, D. Adaptation and Transformation. *Clim. Chang.* **2015**, *133*, 113–127. [CrossRef]
72. Eisenhauer, D.C. Climate change; Adaptation. In *International Encyclopedia of Human Geography*; Elsevier: Amsterdam, The Netherlands, 2020; pp. 281–291, ISBN 978-0-08-102296-2.
73. Oberlack, C. Diagnosing Institutional Barriers and Opportunities for Adaptation to Climate Change. *Mitig. Adapt. Strateg. Glob. Chang.* **2017**, *22*, 805–838. [CrossRef]

74. Scott, C.A.; Shrestha, P.P.; Lutz-Ley, A.N. The Re-Adaptation Challenge: Limits and Opportunities of Existing Infrastructure and Institutions in Adaptive Water Governance. *Curr. Opin. Environ. Sustain.* **2020**, *44*, 104–112. [CrossRef]
75. Kondo, T.; Lizarralde, G. Maladaptation, Fragmentation, and Other Secondary Effects of Centralized Post-Disaster Urban Planning: The Case of the 2011 "Cascading" Disaster in Japan. *Int. J. Disaster Risk Reduct.* **2021**, *58*, 102219. [CrossRef]
76. Gallopín, G.C. Linkages between Vulnerability, Resilience, and Adaptive Capacity. *Glob. Environ. Chang.* **2006**, *16*, 293–303. [CrossRef]
77. Eakin, H.C.; Lemos, M.C.; Nelson, D.R. Differentiating Capacities as a Means to Sustainable Climate Change Adaptation. *Glob. Environ. Chang.* **2014**, *27*, 1–8. [CrossRef]
78. Mccarthy, J.; Canziani, O.; Leary, N.; Dokken, D.; White, K. *Climate Change 2001: Impacts, Adaptation, and Vulnerability. Contribution of Working Group II to the Fourth Assessment Report of the Intergovernmental Panel on Climate Change*; Cambridge University Press: Cambridge, UK, 2001; Volume 19.
79. Adger, W.N. Vulnerability. *Glob. Environ. Chang.* **2006**, *16*, 268–281. [CrossRef]
80. Revi, A.; Satterthwaite, D.; Aragón-Durand, F.; Corfee-Morlot, J.; Kiunsi, R.B.R.; Pelling, M.; Roberts, D.; Solecki, W.; Gajjar, S.P.; Sverdlik, A. Towards Transformative Adaptation in Cities: The IPCC's Fifth Assessment. *Environ. Urban.* **2014**, *26*, 11–28. [CrossRef]
81. Oliver-Smith, A. Anthropological research on hazards and disasters. *Annu. Rev. Anthropol.* **1996**, *25*, 303–328. [CrossRef]
82. Alexander, D.E. *Natural Disasters*; Kluwer Academic Publishers: Boston, MA, USA, 1993.
83. Pescaroli, G.; Alexander, D. Understanding Compound, Interconnected, Interacting, and Cascading Risks: A Holistic Framework: A Holistic Framework for Understanding Complex Risks. *Risk Anal.* **2018**, *38*, 2245–2257. [CrossRef]
84. Dow, K.; Berkhout, F.; Preston, B.L.; Klein, R.J.T.; Midgley, G.; Shaw, M.R. Limits to Adaptation. *Nat. Clim. Chang.* **2013**, *3*, 305–307. [CrossRef]
85. Mack, R. *Planning on Uncertainty: Decision Making in Business and Government Administration*; Wiley Interscience: New York, NY, USA, 1971.
86. Haimes, Y.Y. *Risk Modeling, Assessment, and Management*; Wiley: Hoboken, NJ, USA, 2004.
87. Abbott, J. Understanding and Managing the Unknown: The Nature of Uncertainty in Planning. *J. Plan. Educ. Res.* **2005**, *24*, 237–251. [CrossRef]
88. Van Der Heijden, K. *Scenarios: The Art of Strategic Conversation*; John Wiley: Chichester, UK, 1996.
89. Martins, A.N.; Fayazi, M.; Kikano, F.; Hobeica, L. *Enhancing Disaster Preparedness: From Humanitarian Architecture to Community Resilience*, 1st ed.; Elsevier: Amsterdam, The Netherlands, 2020.
90. Holling, C.S. Resilience and Stability of Ecological Systems. *Annu. Rev. Ecol. Syst.* **1973**, *4*, 1–23. [CrossRef]
91. Rose, A. Economic Resilience to Natural and Man-Made Disasters: Multidisciplinary Origins and Contextual Dimensions. *Environ. Hazards* **2007**, *7*, 383–398. [CrossRef]
92. Twigg, J. *Characteristics of a Disaster-Resilient Community: A Guidance Note*; DFID DRR Interagency Coordination Group: London, UK, 2007.
93. UNISDR. *How to Make Cities More Resilient: A Handbook for Local Government Leaders*; United Nations: Geneva, Switzerland, 2012.
94. UNISDR. *Sendai Framework for Disaster Risk Reduction*; UNISDR: Geneva, Switzerland, 2015.
95. Holling, C.S.; Meffe, G.K. Command and Control and the Pathology of Natural Resource Management. *Conserv. Biol.* **1996**, *10*, 328–337. [CrossRef]
96. Lipshitz, R.; Strauss, O. Coping with Uncertainty: A Naturalistic Decision-Making Analysis. *Organ. Behav. Hum. Decis. Process.* **1997**, *69*, 149–163. [CrossRef]
97. Chow, C.C.; Sarin, R.K. Known, Unknown, and Unknowable Uncertainties. *Theory Decis.* **2002**, *52*, 127–138. [CrossRef]
98. Van der Heijden, J. Studying Urban Climate Governance: Where to Begin, What to Look for, and How to Make a Meaningful Contribution to Scholarship and Practice. *Earth Syst. Gov.* **2019**, *1*, 100005. [CrossRef]
99. Rauws, W. Embracing Uncertainty Without Abandoning Planning: Exploring an Adaptive Planning Approach for Guiding Urban Transformations. *disP Plan. Rev.* **2017**, *53*, 32–45. [CrossRef]
100. Fred, E.; Trist, E. The Causal Texture of Organisational Environments. *Hum. Relat.* **1965**, *18*, 21–32.
101. Christensen, K.S. Coping with Uncertainty in Planning. *J. Am. Plan. Assoc.* **1985**, *51*, 63–73. [CrossRef]
102. Fischhoff, B.; Davis, A.L. Communicating Scientific Uncertainty. *Proc. Natl. Acad. Sci. USA* **2014**, *111*, 13664–13671. [CrossRef] [PubMed]
103. Van der Bles, A.M.; van der Linden, S.; Freeman, A.L.J.; Mitchell, J.; Galvao, A.B.; Zaval, L.; Spiegelhalter, D.J. Communicating Uncertainty about Facts, Numbers and Science. *R. Soc. Open Sci.* **2019**, *6*, 181870. [CrossRef]
104. Shackle, G.L.S. *Decision Order and Time in Human Affairs*; Cambridge University Press: Cambridge, UK, 1969.
105. Gruber, J. *Coordinating Growth Management through Consensus-Building: Incentives and the Generation of Social, Intellectual, and Political Capital*; Institute of Urban and Regional Development (IURD), University of California, Berkeley: Berkeley, CA, USA, 1994.
106. Oppenheimer, M.; O'Neill, B.C.; Webster, M. Negative Learning. *Clim. Chang.* **2008**, *89*, 155–172. [CrossRef]
107. McInerney, D.; Lempert, R.; Keller, K. What Are Robust Strategies in the Face of Uncertain Climate Threshold Responses?: Robust Climate Strategies. *Clim. Chang.* **2012**, *112*, 547–568. [CrossRef]

108. Reeder, T.; Ranger, N. *How Do You Adapt in an Uncertain World? Lessons from the Thames Estuary 2100 Project*; World Resources Report: Washington, DC, USA, 2010; p. 16.
109. Milly, P.C.D.; Betancourt, J.; Falkenmark, M.; Hirsch, R.M.; Kundzewicz, Z.W.; Lettenmaier, D.P.; Stouffer, R.J. Stationarity Is Dead: Whither Water Management? *Science* **2008**, *319*, 573–574. [CrossRef]
110. Stedinger, J.R.; Griffis, V.W. Getting From Here to Where? Flood Frequency Analysis and Climate1: Getting From Here to Where? Flood Frequency Analysis and Climate. *JAWRA J. Am. Water Resour. Assoc.* **2011**, *47*, 506–513. [CrossRef]
111. Stroup, L.J. Adaptation of U.S. Water Management to Climate and Environmental Change. *Prof. Geogr.* **2011**, *63*, 414–428. [CrossRef]
112. Cartwright, T.J. Problems, Solutions and Strategies: A Contribution to the Theory and Practice of Planning. *J. Am. Inst. Plan.* **1973**, *39*, 179–187. [CrossRef]
113. Pierre, N.J. Models of Urban Governance: The Institutional Dimension of Urban Politics. *Urban Aff. Rev.* **1999**, *34*, 372–396. [CrossRef]
114. Healey, P. *Planning: Shaping Places in Fragmented Societies*; Macmillan: Bassingstoke, UK, 1998.
115. Alexander, E. Dilemmas in Evaluating Planning, or Back to Basics: What Is Planning For? *Plan. Theory Pract.* **2009**, *10*, 233–244. [CrossRef]
116. Morphet, J. *Effective Practice in Spatial Planning*; The RTPI Library Series; Routledge: New York, NY, USA, 2011; ISBN 978-0-415-49281-2.
117. Reyes Plata, J.A. Urban planning, urban design, and the creation of public goods. In *Sustainable Cities and Communities*; Leal Filho, W., Marisa Azul, A., Brandli, L., Gökçin Özuyar, P., Wall, T., Eds.; Encyclopedia of the UN Sustainable Development Goals; Springer International Publishing: Cham, Switzerland, 2020; pp. 899–905, ISBN 978-3-319-95716-6.
118. Bolan, R.S. Emerging Views of Planning. *J. Am. Inst. Plan.* **1967**, *33*, 233–245. [CrossRef]
119. Faludi, A. The Performance of Spatial Planning. *Plan. Pract. Res.* **2000**, *15*, 299–318. [CrossRef]
120. Forester, J. Critical Theory and Planning Practice. *J. Am. Plan. Assoc.* **1980**, *46*, 275–286. [CrossRef]
121. Kunzmann, K. Strategic spatial development through information and communication. In *The Revival of Strategic Spatial Planning*; Salet, W., Faludi, A., Eds.; Royal Netherlands Academy of Arts and Sciences: Amsterdam, The Netherlands, 2000; pp. 259–265.
122. Rodima-Taylor, D.; Olwig, M.F.; Chhetri, N. Adaptation as Innovation, Innovation as Adaptation: An Institutional Approach to Climate Change. *Appl. Geogr.* **2012**, *33*, 107–111. [CrossRef]
123. Eriksen, S.; Brown, K. Sustainable Adaptation to Climate Change. *Clim. Dev.* **2011**, *3*, 3e6. [CrossRef]
124. Ribot, J.C. Vulnerability does not aall from the sky: Toward multi-scale propoor climate policy. In *Social Dimensions of Climate Change: Equity and Vulnerability in a Warming World*; Mearns, R., Norton, A., Eds.; The World Bank: Washington, DC, USA, 2010; pp. 47–74.
125. Pelling, M. *Adaptation to Climate Change: From Resilience to Transformation*; Routledge: London, UK, 2011.
126. Dowell, M.; Kitsuse, A. Constructing the Future in Planning: A Survey of Theories and Tools. *J. Plan. Educ. Res.* **2000**, *19*, 221–231.
127. Leach, W.; Sabatier, P. Facilitators, coordinators, and outcomes. In *The Promise and Performance of Environmental Conflict Resolution*; O'Leary, R., Bingham, L.B., Eds.; Resources for the Future: Washington, DC, USA, 2003; pp. 148–171.
128. Ostrom, E. *Governing the Commons: The Evolution of Institutions for Collective Action*; Cambridge University Press: Cambridge, MA, USA, 1990.
129. Roberts, N. Public Deliberation: An Alternative Approach to Crafting Policy and Setting Direction. *Public Admin. Rev.* **1997**, *57*, 124–132. [CrossRef]
130. Roberts, N. Keeping Public Officials Accountable through Dialogue: Resolving the Accountability Paradox. *Public Admin. Rev.* **2002**, *62*, 658–669. [CrossRef]
131. Ryan, C.M. Leadership in Collaborative Policy Making: An Analysis of Agency Roles in Regulatory Negotiation. *Policy Sci.* **2001**, *34*, 221–245. [CrossRef]
132. Moore, C.W. *The Mediation Process: Practical Strategies for Resolving Conflict*; Jossey-Bass: San Francisco, CA, USA, 1987.
133. Innes, J.E. Consensus Building: Clarifications for the Critics. *Plan. Theory* **2004**, *3*, 5–20. [CrossRef]
134. Susskind, L.; McKearnan, S.; Thomas-Larmer, J. (Eds.) *The Consensus Building Handbook: A Comprehensive Guide to Reaching Agreement*; Sage: Thousand Oaks, CA, USA, 1999.
135. Macintosh, A. Coastal Climate Hazards and Urban Planning: How Planning Responses Can Lead to Maladaptation. *Mitig. Adapt. Strateg. Glob. Chang.* **2013**, *18*, 1035–1055. [CrossRef]
136. Bulkeley, H.; Betsill, M.M. Revisiting the Urban Politics of Climate Change. *Environ. Polit.* **2013**, *22*, 136–154. [CrossRef]
137. Peterson, G.D.; Cumming, G.S.; Carpenter, S.R. Scenario Planning: A Tool for Conservation in an Uncertain World. *Conserv. Biol.* **2003**, *17*, 358–366. [CrossRef]
138. Albrechts, L. Bridge the Gap: From Spatial Planning to Strategic Projects. *Eur. Plan. Stud.* **2006**, *14*, 1487–1500. [CrossRef]
139. Soden, R.; Kauffman, N. Infrastructuring the imaginary: How sea-level rise comes to matter in the San Francisco Bay Area. In Proceedings of the 2019 CHI Conference on Human Factors in Computing Systems, Glasgow, UK, 4–9 May 2019; ACM: Glasgow, UK, 2019; pp. 1–11.
140. Balducci, A.; Boelens, L.; Hillier, J.; Nyseth, T.; Wilkinson, C. Introduction: Strategic Spatial Planning in Uncertainty: Theory and Exploratory Practice. *Town Plan. Rev.* **2011**, *82*, 481–501. [CrossRef]

141. Daddi, T.; Bleischwitz, R.; Todaro, N.M.; Gusmerotti, N.M.; De Giacomo, M.R. The Influence of Institutional Pressures on Climate Mitigation and Adaptation Strategies. *J. Clean. Prod.* **2020**, *244*, 118879. [CrossRef]
142. Hughes, S. A Meta-Analysis of Urban Climate Change Adaptation Planning in the U.S. *Urban Clim.* **2015**, *14*, 17–29. [CrossRef]
143. Nalau, J.; Torabi, E.; Edwards, N.; Howes, M.; Morgan, E. A Critical Exploration of Adaptation Heuristics. *Clim. Risk Manag.* **2021**, *32*, 100292. [CrossRef]
144. Corfee-Morlot, J.; Cochran, I.; Hallegatte, S.; Teasdale, P.-J. Multilevel Risk Governance and Urban Adaptation Policy. *Clim. Chang.* **2011**, *104*, 169–197. [CrossRef]
145. Turkelboom, F.; Leone, M.; Jacobs, S.; Kelemen, E.; García-Llorente, M.; Baró, F.; Termansen, M.; Barton, D.N.; Berry, P.; Stange, E.; et al. When We Cannot Have It All: Ecosystem Services Trade-Offs in the Context of Spatial Planning. *Ecosyst. Serv.* **2018**, *29*, 566–578. [CrossRef]
146. Walker, W.E.; Rahman, S.A.; Cave, J. Adaptive Policies, Policy Analysis, and Policy-Making. *Eur. J. Oper. Res.* **2001**, *128*, 282–289. [CrossRef]
147. Allen, C.R.; Garmestani, A.S. (Eds.) *Adaptive Management of Social-Ecological Systems*; Springer: Dordrecht, The Netherlands, 2015; ISBN 978-94-017-9681-1.
148. Hall, P.A. Policy Paradigms, Social Learning, and the State: The Case of Economic Policymaking in Britain. *Comp. Polit.* **1993**, *25*, 275–296. [CrossRef]
149. Allen, C.R.; Fontaine, J.J.; Pope, K.L.; Garmestani, A.S. Adaptive Management for a Turbulent Future. *J. Environ. Manag.* **2011**, *92*, 1339–1345. [CrossRef] [PubMed]
150. Folke, C.; Hahn, T.; Olsson, P.; Norberg, J. Adaptive governance of social-ecological systems. *Annu. Rev. Environ. Resour.* **2005**, *30*, 441–473. [CrossRef]
151. Burley, J.G.; McAllister, R.R.J.; Collins, K.A.; Lovelock, C.E. Integration, Synthesis and Climate Change Adaptation: A Narrative Based on Coastal Wetlands at the Regional Scale. *Reg. Environ. Chang.* **2012**, *12*, 581–593. [CrossRef]
152. Tol, R.S.J.; Klein, R.J.T.; Nicholls, R.J. Towards Successful Adaptation to Sea-Level Rise along Europe's Coasts. *J. Coast. Res.* **2008**, *242*, 432–442. [CrossRef]
153. Klein, R.J.T.; Tol, R.S.J. *Adaptation to Climate Change: Options and Technologies. An Overview Paper*; Technical Paper FCCC/TP/1997/3; United Nations Framework Convention on Climate Change Secretariat: Bonn, Germany, 1997; 36p.
154. Urwin, K.; Jordan, A. Does Public Policy Support or Undermine Climate Change Adaptation? Exploring Policy Interplay across Different Scales of Governance. *Glob. Environ. Chang.* **2008**, *18*, 180–191. [CrossRef]
155. Sabatier, P.A. Top-down and Bottom-up Approaches to Implementation Research: A Critical Analysis and Suggested Synthesis. *J. Public Policy* **1986**, *6*, 21–48. [CrossRef]
156. Mitchell, R.B.; Clark, W.C.; Cash, D.W.; Dickson, N. (Eds.) *Global Environmental Assessments: Information, Institutions, and Influence*; MIT: Cambridge, UK, 2006.
157. Cash, D.W.; Clark, W.C.; Alcock, F.; Dickson, N.M.; Eckley, N.; Guston, D.H.; Jaeger, J.; Mitchell, R.B. Knowledge Systems for Sustainable Development. *Proc. Natl. Acad. Sci. USA* **2003**, *100*, 8086–8091. [CrossRef]
158. Schroeder, H.; Kobayashi, Y. Climate change governance: Responding to an existential crisis. In *The Impacts of Climate Change*; Elsevier: Amsterdam, The Netherlands, 2021; pp. 479–489, ISBN 978-0-12-822373-4.
159. Habermas, J. *Between Facts and Norms: Contributions to a Discourse Theory of Law and Democracy*; MIT: Cambridge, UK, 1998.
160. Huntjens, P.; Lebel, L.; Pahl-Wostl, C.; Camkin, J.; Schulze, R.; Kranz, N. Institutional Design Propositions for the Governance of Adaptation to Climate Change in the Water Sector. *Glob. Environ. Chang.* **2012**, *22*, 67–81. [CrossRef]
161. Torabi, E.; Dedekorkut-Howes, A.; Howes, M. Adapting or Maladapting: Building Resilience to Climate-Related Disasters in Coastal Cities. *Cities* **2018**, *72*, 295–309. [CrossRef]
162. Barrett, S. Local Level Climate Justice? Adaptation Finance and Vulnerability Reduction. *Glob. Environ. Chang.* **2013**, *23*, 1819–1829. [CrossRef]
163. Moser, S.C.; Ekstrom, J.A.; Kim, J.; Heitsch, S. Adaptation Finance Archetypes: Local Governments' Persistent Challenges of Funding Adaptation to Climate Change and Ways to Overcome Them. *E S* **2019**, *24*, art28. [CrossRef]
164. Geels, F.W. Technological Transitions as Evolutionary Reconfiguration Processes: A Multi-Level Perspective and a Case-Study. *Res. Policy* **2002**, *31*, 1257–1274. [CrossRef]
165. Van Assche, K.; Beunen, R.; Duineveld, M. *Evolutionary Governance Theory*; Springer Briefs in Economics; Springer International Publishing: Cham, Switzerland, 2014; ISBN 978-3-319-00983-4.
166. Kwakkel, J.; van Der Pas, J.W.G.M. Evaluation of Infrastructure Planning Approaches: An Analogy with Medicine. *Futures* **2011**, *43*, 934–946. [CrossRef]
167. Goodwin, P.; Wright, G. The Limits of Forecasting Methods in Anticipating Rare Events. *Technol. Forecast. Soc. Chang.* **2010**, *77*, 355–368. [CrossRef]
168. Ostrom, E. Doing institutional analysis digging deeper than markets and hierarchies. In *Handbook of New Institutional Economics*; Menard, C., Shirley, M.M., Eds.; Springer: Boston, MA, USA, 2005; pp. 819–848, ISBN 978-0-387-25092-2.
169. Chaffin, B.C.; Shuster, W.D.; Garmestani, A.S.; Furio, B.; Albro, S.L.; Gardiner, M.; Spring, M.; Green, O.O. A Tale of Two Rain Gardens: Barriers and Bridges to Adaptive Management of Urban Stormwater in Cleveland, Ohio. *J. Environ. Manag.* **2016**, *183*, 431–441. [CrossRef] [PubMed]

170. Andersson, E.; Barthel, S.; Borgström, S.; Colding, J.; Elmqvist, T.; Folke, C.; Gren, Å. Reconnecting Cities to the Biosphere: Stewardship of Green Infrastructure and Urban Ecosystem Services. *Ambio* **2014**, *43*, 445. [CrossRef] [PubMed]
171. Storbjörk, S.; Hedrén, J. Institutional Capacity-Building for Targeting Sea-Level Rise in the Climate Adaptation of Swedish Coastal Zone Management. Lessons from Coastby. *Ocean Coast. Manag.* **2011**, *54*, 265–273. [CrossRef]
172. Allen, K.M. Community-Based Disaster Preparedness and Climate Adaptation: Local Capacity-Building in the Philippines: Community-Based Disaster Preparedness and Climate Adaptation. *Disasters* **2006**, *30*, 81–101. [CrossRef]
173. Albrechts, L. More of the Same Is Not Enough! How Could Strategic Spatial Planning Be Instrumental in Dealing with the Challenges Ahead? *Environ. Plan. B* **2010**, *37*, 1115–1127. [CrossRef]
174. Leck, H. What Lies beneath: Understanding the Invisible Aspects of Municipal Climate Change Governance. *Curr. Opin. Environ. Sustain.* **2015**, *13*, 61–67. [CrossRef]
175. Main, K.L.; Mazereeuw, M.; Masoud, F.; Lu, J.; Barve, A.; Ojha, M.; Krishna, C. Climate action zones: A clustering methodology for resilient spatial planning in climate uncertainty. In *Enhancing Disaster Preparedness*; Elsevier: Amsterdam, The Netherlands, 2021; pp. 241–258, ISBN 978-0-12-819078-4.
176. Nalau, J.; Preston, B.L.; Maloney, M.C. Is Adaptation a Local Responsibility? *Environ. Sci. Policy* **2015**, *48*, 89–98. [CrossRef]
177. Vogel, B.; Henstra, D. Studying Local Climate Adaptation: A Heuristic Research Framework for Comparative Policy Analysis. *Glob. Environ. Chang.* **2015**, *31*, 110–120. [CrossRef]
178. Bicknell, J.; Dodman, D.; Satterthwaite, D. (Eds.) Conclusions: Local development and adaptation. In *Adapting Cities to Climate Change: Understanding and Addressing the Development Challenges*; Earthscan: London, UK, 2009; pp. 359–383.
179. Bierbaum, R.; Smith, J.B.; Lee, A.; Blair, M.; Carter, L.; Chapin, F.S.; Fleming, P.; Ruffo, S.; Stults, M.; McNeeley, S.; et al. A Comprehensive Review of Climate Adaptation in the United States: More than before, but Less than Needed. *Mitig. Adapt. Strateg. Glob. Chang.* **2013**, *18*, 361–406. [CrossRef]
180. Warsen, R.; Nederhand, J.; Klijn, E.H.; Grotenbreg, S.; Koppenjan, J. What Makes Public-Private Partnerships Work? Survey Research into the Outcomes and the Quality of Cooperation in PPPs. *Public Manag. Rev.* **2018**, *20*, 1165–1185. [CrossRef]
181. Guston, D.H. Boundary Organizations in Environmental Policy and Science: An Introduction. *Sci. Technol. Hum. Values* **2001**, *26*, 339–408. [CrossRef]
182. Leck, H.; Simon, D. Fostering Multiscalar Collaboration and Co-Operation for Effective Governance of Climate Change Adaptation. *Urban Stud.* **2013**, *50*, 1221–1238. [CrossRef]
183. Glasbergen, P. Setting the scene: The partnership paradign in the making. In *Partnerships, Governance and Sustainable Development: Reflections on Theory and Practice*; Glasbergen, P., Biermann, F., Mol, A., Eds.; Elgar Publishing: Cheltenham, UK, 2007; p. 314.
184. Agrawal, A. Local institutions and adaptation to climate change. In *Social Dimensions of Climate Change: Equity and Vulnerability in the Warming World*; Mearns, R., Norton, A., Eds.; The World Bank: Washington, DC, USA, 2010; pp. 173–198.
185. Harman, B.P.; Taylor, B.M.; Lane, M.B. Urban Partnerships and Climate Adaptation: Challenges and Opportunities. *Curr. Opin. Environ. Sustain.* **2015**, *12*, 74–79. [CrossRef]
186. Bauer, A.; Steurer, R. Innovation in Climate Adaptation Policy: Are Regional Partnerships Catalysts or Talking Shops? *Environ. Polit.* **2014**, *23*, 818–838. [CrossRef]
187. Buck, L.E.; Geisler, C.C.; Schelhas, J.; Wollenberg, E. *Biological Diversity: Balancing Interests through Adaptive Collaborative Management*; CRC Press: Boca Raton, FL, USA, 2001.
188. Walters, C.J. *Adaptive Management of Renewable Resources*; McGraw Hill: New York, NY, USA, 1986.
189. Williams, B.K. Adaptive Management of Natural Resourcesframework and Issues. *J. Environ. Manag.* **2011**, *92*, 1346–1353. [CrossRef] [PubMed]
190. Gunderson, L. Resilience, Flexibility and Adaptive Management—Antidotes for Spurious Certitude? *Conserv. Ecol.* **1999**, *3*, 7. [CrossRef]
191. Demuzere, M.; Orru, K.; Heidrich, O.; Olazabal, E.; Geneletti, D.; Orru, H.; Bhave, A.G.; Mittal, N.; Feliu, E.; Faehnle, M. Mitigating and Adapting to Climate Change: Multi-Functional and Multi-Scale Assessment of Green Urban Infrastructure. *J. Environ. Manag.* **2014**, *146*, 107–115. [CrossRef]
192. Dewey, J. The Quest for Certainty: A Study of the Relation of Knowledge and Action. *J. Philos.* **1929**, *27*, 14–25.
193. Walters, C.J.; Holling, C.S. Large-Scale Experimentation and Learning by Doing. *Ecology* **1990**, *71*, 2060–2068. [CrossRef]
194. Allan, C.; Curtis, A. Nipped in the Bud: Why Regional Scale Adaptive Management Is Not Blooming. *Environ. Manag.* **2005**, *36*, 414–425. [CrossRef]
195. Garmestani, A.S.; Allen, C.R.; Cabezas, H. Panarchy, Adaptive Management and Governance: Policy Options for Building Resilience. *Nebraska Law Review.* **2008**, *87*, 1036.
196. Lee, K.N. Appraising adaptive management. In *Biological Diversity*; CRC Press: Boca Raton, FL, USA, 2021; p. 22.
197. Buijs, A.E.; Mattijssen, T.J.; Van der Jagt, A.P.; Ambrose-Oji, B.; Andersson, E.; Elands, B.H.; Steen Møller, M. Active Citizenship for Urban Green Infrastructure: Fostering the Diversity and Dynamics of Citizen Contributions through Mosaic Governance. *Curr. Opin. Environ. Sustain.* **2016**, *22*, 1–6. [CrossRef]
198. Stringer, L.C.; Dougill, A.J.; Fraser, E.; Hubacek, K.; Prell, C.; Reed, M.S. Unpacking "Participation" in the Adaptive Management of Social—Ecological Systems: A Critical Review. *E S* **2006**, *11*, art39. [CrossRef]
199. Moser, S.C.; Ekstrom, J.A. A Framework to Diagnose Barriers to Climate Change Adaptation. *Proc. Natl. Acad. Sci. USA* **2010**, *107*, 22026–22031. [CrossRef] [PubMed]

200. Füssel, H.-M.; Klein, R.J.T. Climate Change Vulnerability Assessments: An Evolution of Conceptual Thinking. *Clim. Chang.* **2006**, *75*, 301–329. [CrossRef]
201. Hardin, G. The Tragedy of the Commons. *Science* **1968**, *162*, 1243–1248. [CrossRef]
202. Anderies, J.M. Economic Development, Demographics, and Renewable Resources: A Dynamical Systems Approach. *Environ. Dev. Econ.* **2003**, *8*, 219–246. [CrossRef]
203. Andersson, D.; Bratsberg, S.; Ringsmuth, A.K.; de Wijn, A.S. Dynamics of Collective Action to Conserve a Large Common-Pool Resource. *Sci. Rep.* **2021**, *11*, 9208. [CrossRef] [PubMed]
204. Marshall, G.R. Transaction Costs, Collective Action and Adaptation in Managing Complex Social–Ecological Systems. *Ecol. Econ.* **2013**, *88*, 185–194. [CrossRef]
205. Obeng-Odoom, F. The Meaning, Prospects, and Future of the Commons: Revisiting the Legacies of Elinor Ostrom and Henry George: The Meaning, Prospects, and Future of the Commons. *Am. J. Econ. Sociol.* **2016**, *75*, 372–414. [CrossRef]
206. Hinkel, J.; Bisaro, A. A Review and Classification of Analytical Methods for Climate Change Adaptation: Analytical Methods for Climate Change Adaptation. *Wiley Interdiscip. Rev. Clim. Chang.* **2015**, *6*, 171–188. [CrossRef]
207. Roggero, M. Adapting Institutions: Exploring Climate Adaptation through Institutional Economics and Set Relations. *Ecol. Econ.* **2015**, *118*, 114–122. [CrossRef]
208. Roggero, M.; Bisaro, A.; Villamayor-Tomas, S. Institutions in the Climate Adaptation Literature: A Systematic Literature Review through the Lens of the Institutional Analysis and Development Framework. *J. Inst. Econ.* **2018**, *14*, 423–448. [CrossRef]
209. Eakin, H. Institutional Change, Climate Risk, and Rural Vulnerability: Cases from Central Mexico. *World Dev.* **2005**, *33*, 1923–1938. [CrossRef]
210. Adger, W.N.; Huq, S.; Brown, K.; Conway, D.; Hulme, M. Adaptation to Climate Change in the Developing World. *Prog. Dev. Stud.* **2003**, *3*, 179–195. [CrossRef]
211. Pelling, M.; High, C. Understanding Adaptation: What Can Social Capital Offer Assessments of Adaptive Capacity? *Glob. Environ. Chang.* **2005**, *15*, 308–319. [CrossRef]
212. Wamsler, C.; Brink, E. Interfacing Citizens' and Institutions' Practice and Responsibilities for Climate Change Adaptation. *Urban Clim.* **2014**, *7*, 64–91. [CrossRef]
213. Gunderson, L.; Holling, C.S. (Eds.) *Panarchy: Understanding Transformations in Human and Natural Systems*; Island Press: Chicago, IL, USA, 2002.
214. Ekstrom, J.A.; Moser, S.C. Identifying and Overcoming Barriers in Urban Climate Adaptation: Case Study Findings from the San Francisco Bay Area, California, USA. *Urban Clim.* **2014**, *9*, 54–74. [CrossRef]
215. Juhola, S.; Glaas, E.; Linnér, B.-O.; Neset, T.-S. Redefining Maladaptation. *Environ. Sci. Policy* **2016**, *55*, 135–140. [CrossRef]
216. Rosenzweig, C.; Solecki, W.D.; Hammer, S.A.; Mehrohtra, S. (Eds.) *Climate Change and Cities First Assessment Report of the Urban Climate Change Research Network*; Cambridge University Press: Cambridge, UK, 2011.
217. Mehrotra, S.; Natenzon, C.; Omojola, A.; Folorunsho, R.; Gilbride, J.; Rosenzweig, C. Framework for city climate risk assess. In Proceedings of the Fifth Urban Research Symposium, Marseille, France, 28–30 June 2009; World Bank: Washington, DC, USA, 2009.
218. Burton, I. *Climate Change and the Adaptation Deficit. Earthscan Reader on Adaptation to Climate Change*; Schipper, E.L.F., Burton, I., Eds.; Earthscan: Sterling, VA, USA, 2009; pp. 89–95.
219. Chen, C.; Doherty, M.; Coffee, J.; Wong, T.; Hellmann, J. Measuring the Adaptation Gap: A Framework for Evaluating Climate Hazards and Opportunities in Urban Areas. *Environ. Sci. Policy* **2016**, *66*, 403–419. [CrossRef]
220. United Nations Environment Programme. *Adaptation Gap Report*; UNEP DTU Partnership (UDP): Nairobi, Kenya, 2020.
221. Davoudi, S.; Mehmood, A.; Brooks, E. *The London Climate Change Adaptation Strategy: Gap Analysis*; Newcastle University: Newcastle, UK, 2011.
222. Ostrom, E. A Diagnostic Approach for Going beyond Panaceas. *Proc. Natl. Acad. Sci. USA* **2007**, *104*, 15181–15187. [CrossRef] [PubMed]
223. Adger, W.N.; Dessai, S.; Goulden, M.; Hulme, M.; Lorenzoni, I.; Nelson, D.R.; Naess, L.O.; Wolf, J.; Wreford, A. Are There Social Limits to Adaptation to Climate Change? *Clim. Chang.* **2009**, *93*, 335–354. [CrossRef]
224. Bisaro, A.; Roggero, M.; Villamayor-Tomas, S. Analysis: Institutional Analysis in Climate Change Adaptation Research: A Systematic Literature Review. *Ecol. Econ.* **2018**, *151*, 34–43. [CrossRef]
225. Barnett, J.; Evans, L.S.; Gross, C.; Kiem, A.S.; Kingsford, R.T.; Palutikof, J.P.; Pickering, C.M.; Smithers, S.G. From Barriers to Limits to Climate Change Adaptation: Path Dependency and the Speed of Change. *E S* **2015**, *20*, art5. [CrossRef]
226. Healey, P. Transforming Governance: Challenges of Institutional Adaptation and a New Politics of Space. *Eur. Plan. Stud.* **2006**, *14*, 299–319. [CrossRef]
227. Arthur, W.B. *Increasing Returns and Path Dependence in the Economy*; University of Michigan Press: Ann Arbor, MI, USA, 1994.
228. Pierson, P. Increasing Returns, Path Dependence, and the Study of Politics. *Am. Polit. Sci. Rev.* **2000**, *94*, 251–267. [CrossRef]
229. Tilly, C. *Big Structures, Large Processes, Huge Comparisons*; Russel Sage: New York, NY, USA, 1984.
230. Sanyal, B. Planning as Anticipation of Resistance. *Plan. Theory* **2005**, *4*, 225–245. [CrossRef]
231. Booth, P. Culture, Planning and Path Dependence: Some Reflections on the Problems of Comparison. *Town Plan. Rev.* **2011**, *82*, 13–28. [CrossRef]

232. Klein, R.J.T.; Midgley, G.F.; Preston, B.L.; Alam, M.; Berkhout, F.G.H.; Dow, K.; Shaw, M.R. Adaptation opportunities, constraints, and limits. In *Climate Change 2014: Impacts, Adaptation, and Vulnerability. Part A: Global and Sectoral Aspects. Contribution of Working Group II to the Fifth Assessment Report of the Intergovernmental Panel on Climate Change*; Cambridge University Press: Cambridge, NY, USA, 2014; pp. 899–943.
233. Hagedorn, K. Particular Requirements for Institutional Analysis in Nature-Related Sectors. *Eur. Rev. Agric. Econ.* **2008**, *35*, 357–384. [CrossRef]
234. Paavola, J. Institutions and Environmental Governance: A Reconceptualization. *Ecol. Econ.* **2007**, *63*, 93–103. [CrossRef]
235. Vatn, A. *Institutions and the Environment*; Edward Elgar Publishing: Cheltenham, UK, 2006.
236. Crawford, S.E.S.; Ostrom, E. A Grammar of Institutions. *Am. Polit. Sci. Rev.* **1995**, *89*, 582–600. [CrossRef]
237. Schotter, A. *The Economic Theory of Social Institutions*; Cambridge University Press: Cambridge, UK, 2008.
238. Basili, M.; Franzini, M.; Vercelli, A. *Environment, Inequality and Collective Action*; Routledge: London, UK; New York, NY, USA, 2006; ISBN 978-0-415-34234-6.
239. Sorensen, A. Institutions and Urban Space: Land, Infrastructure, and Governance in the Production of Urban Property. *Plan. Theory Pract.* **2018**, *19*, 21–38. [CrossRef]
240. Young, O.R. Land Use, Environmental Change, and Sustainable Development. *Environ. Chang.* **2011**, *5*, 66–85.
241. Knight, J.; Douglass, N. Explaining Economic Change: The Interplay Be- Tween Cognition and Institutions. *Leg. Theory* **1977**, *3*, 211–226. [CrossRef]
242. Milkoreit, M. Imaginary Politics: Climate Change and Making the Future. *Elem. Sci. Anthropocene* **2017**, *5*, 62. [CrossRef]
243. Ekers, M.; Loftus, A. The Power of Water: Developing Dialogues between Foucault and Gramsci. *Environ. Plan. D Soc. Space* **2008**, *26*, 698–718.
244. Swyngedouw, E. The Antinomies of the Postpolitical City: In Search of a Democratic Politics of Environmental Production. *Int. J. Urban Reg. Res.* **2009**, *33*, 601–620. [CrossRef]
245. Chhetri, N.; Easterling, W.E.; Terando, A.; Mearns, L. Modeling Path Dependence in Agricultural Adaptation to Climate Variability and Change. *Ann. Assoc. Am. Geogr.* **2010**, *100*, 894–907. [CrossRef]
246. Gollier, C.; Treich, N. Decision-Making under Scientific Uncertainty: The Economics of the Precautionary Principle. *J. Risk Uncertain.* **2003**, *27*, 77–103. [CrossRef]
247. Lempert, R.J.; Collins, M.T. Managing the Risk of Uncertain Thresholds Responses: Comparison of Robust, Optimum, and Precautionary Approaches. *Risk Anal.* **2007**, *27*, 1009–1026. [PubMed]
248. Scott, W. Institutions and Organizations. *Leadersh. Organ. Dev. J.* **2003**, *24*, 469–470. [CrossRef]
249. Smets, M.; Morris, T.; Greenwood, R. From Practice to Field: A Multilevel Model of Practice-Driven Institutional Change. *AMJ* **2012**, *55*, 877–904. [CrossRef]
250. Tilly, C. Mechanisms in Political Processes. *Annu. Rev. Polit. Sci.* **2001**, *4*, 21–41. [CrossRef]
251. Hodgson, G.M. What Are Institutions? *J. Econ. Issues* **2006**, *40*, 1–25. [CrossRef]
252. Agrawal, A. *The Role of Local Institutions in Adaptation to Climate Change*; The World Bank: Washington, DC, USA, 2008.
253. Mimura, N.; Pulwarty, R.S.; Duc, D.M.; Elshinnawy, I.; Redsteer, M.H.; Huang, H.Q.; Nkem, J.N.; Sanchez Rodriguez, R.A. Chapter 15: Adaptation planning and implementation. In *Climate Change 2014: Impacts, Adaptation, and Vulnerability. Part A: Global and Sectoral Aspects. Contribution of Working Group II to the Fifth Assessment Report of the Intergovernmental Panel on Climate Change*; Field, C.B., Barros, V.R., Dokken, D.J., Mach, K.J., Mastrandrea, M.D., Bilir, T.E., Chatterjee, M., Ebi, K.L., Estrada, Y.O., Genova, R.C., et al., Eds.; Cambridge University Press: Cambridge, UK; New York, NY, USA, 2014.
254. Keskitalo, E.C.H. (Ed.) *Developing Adaptation Policy and Practice in Europe: Multi-Level Governance of Climate Change*; Springer: Dordrecht, The Netherlands, 2010; ISBN 978-90-481-9324-0.
255. Bazerman, M. Climate Change as a Predictable Surprise. *Clim. Chang.* **2005**, *77*, 179–193. [CrossRef]
256. Biesbroek, G.R.; Termeer, C.J.A.M.; Kabat, P.; Klostermann, J.E.M. Institutional covernance barriers for the development and implementation of climate adaptation strategies. In Proceedings of the International Human Dimensions Programme (IHDP) Conference 'Earth System Governance: People, Places, and the Planet', Amsterdam, The Netherlands, 2–4 December 2009.
257. Hodgson, G.M.; Knudsen, T. Why We Need a Generalized Darwinism, and Why Generalized Darwinism Is Not Enough. *J. Econ. Behav. Organ.* **2006**, *61*, 1–19. [CrossRef]
258. MacKinnon, D.; Cumbers, A.; Pike, A.; Birch, K.; McMaster, R. Evolution in Economic Geography: Institutions, Political Economy, and Adaptation. *Econ. Geogr.* **2009**, *85*, 129–150. [CrossRef]
259. Koppel, B.M. (Ed.) Induced innovation theory, agricultural research, and Asia's green revolution: A reappraisal. In *Induced Innovation Theory and International Agricultural Development: A Reassessment*; John Hopkins University Press: Baltimore, MD, USA, 1995; pp. 56–72.
260. Petersen-Rockney, M.; Baur, P.; Guzman, A.; Bender, S.F.; Calo, A.; Castillo, F.; De Master, K.; Dumont, A.; Esquivel, K.; Kremen, C.; et al. Narrow and Brittle or Broad and Nimble? Comparing Adaptive Capacity in Simplifying and Diversifying Farming Systems. *Front. Sustain. Food Syst.* **2021**, *5*, 564900. [CrossRef]
261. Voigt, S. How (Not) to Measure Institutions. *J. Inst. Econ.* **2013**, *9*, 1–26. [CrossRef]
262. Lubell, M.; Robins, G.; Wang, P. Network Structure and Institutional Complexity in an Ecology of Water Management Games. *E S* **2014**, *19*, art23. [CrossRef]
263. Armitage, D. Governance and the Commons in a Multi-Level World. *Int. J. Commons* **2008**, *2*, 7–32. [CrossRef]

264. Huitema, D.; Adger, W.N.; Berkhout, F.; Massey, E.; Mazmanian, D.; Munaretto, S.; Plummer, R.; Termeer, C.C.J.A.M. The Governance of Adaptation: Choices, Reasons, and Effects. Introduction to the Special Feature. *E S* **2016**, *21*, art37. [CrossRef]
265. Berkes, F. Indigenous Ways of Knowing and the Study of Environmental Change. *J. R. Soc. N. Z.* **2009**, *39*, 151–156. [CrossRef]
266. Naess, L.O. The Role of Local Knowledge in Adaptation to Climate Change: Role of Local Knowledge in Adaptation. *WIREs Clim. Chang.* **2013**, *4*, 99–106. [CrossRef]
267. Dovers, S.R.; Hezri, A.A. Institutions and Policy Processes: The Means to the Ends of Adaptation. *WIREs Clim. Chang.* **2010**, *1*, 212–231. [CrossRef]
268. Ostrom, E. Nested Externalities and Polycentric Institutions: Must We Wait for Global Solutions to Climate Change before Taking Actions at Other Scales? *Econ. Theory* **2012**, *49*, 353–369. [CrossRef]
269. Bromley, D.W. Searching for Sustainability: The Poverty of Spontaneous Order. *Ecol. Econ.* **1998**, *24*, 231–240. [CrossRef]
270. Van Klingeren, F.; de Graaf, N.D. Heterogeneity, Trust and Common-Pool Resource Management. *J. Environ. Stud. Sci.* **2021**, *11*, 37–64. [CrossRef]
271. Wilson, J. *Scientific Uncertainty, Complex Systems, and the Design of Common-Pool Institutions*; National Resource Council: Washington, DC, USA, 2002; p. 24.
272. Gardiner, S.M. A Perfect Moral Storm: Climate Change, Intergenerational Ethics and the Problem of Moral Corruption. *Environ. Values* **2006**, *15*, 397–413. [CrossRef]
273. Greif, A.; Kingston, C. Institutions: Rules or equilibria? In *Political Economy of Institutions, Democracy and Voting*; Schofield, N., Caballero, G., Eds.; Springer: Berlin/Heidelberg, Germany, 2011; pp. 13–43, ISBN 978-3-642-19518-1.

 *sustainability*

Article

Simulating Grassland Carbon Dynamics in Gansu for the Past Fifty (50) Years (1968–2018) Using the Century Model

Meiling Zhang *, Stephen Nazieh, Teddy Nkrumah and Xingyu Wang

Applied Statistics, College of Science, Gansu Agricultural University, Lanzhou 730070, China; nazerodavinci@gmail.com (S.N.); teddynkrumah2002@yahoo.com (T.N.); 18893711885@139.com (X.W.)
* Correspondence: zhangml@gsau.edu.cn

Citation: Zhang, M.; Nazieh, S.; Nkrumah, T.; Wang, X. Simulating Grassland Carbon Dynamics in Gansu for the Past Fifty (50) Years (1968–2018) Using the Century Model. *Sustainability* **2021**, *13*, 9434. https://doi.org/10.3390/su13169434

Academic Editors: Baojie He, Ayyoob Sharifi, Chi Feng and Jun Yang

Received: 21 July 2021
Accepted: 15 August 2021
Published: 23 August 2021

Publisher's Note: MDPI stays neutral with regard to jurisdictional claims in published maps and institutional affiliations.

Abstract: China is one of the countries most impacted by desertification, with Gansu Province in the northwest being one of the most affected areas. Efforts have been made in recent decades to restore the natural vegetation, while also producing food. This has implications for the soil carbon sequestration and, as a result, the country's carbon budget. Studies of carbon (C) dynamics in this region would help to understand the effect of management practices on soil organic carbon (SOC) as well as aboveground biomass (ABVG), and to aid informed decision-making and policy implementation to alleviate the rate of global warming. It would also help to understand the region's contribution to the national C inventory of China. The CENTURY model, a process-based model that is capable of simulating C dynamics over a long period, has not been calibrated to suit Gansu Province, despite being an effective model for soil C estimation. Using the soil and grassland maps of Gansu, together with weather, soil, and reliable historical data on management practices in the province, we calibrated the CENTURY model for the province's grasslands. The calibrated model was then used to simulate the C dynamics between 1968 and 2018. The results show that the model is capable of simulating C with significant accuracy. Our measured and observed SOC density (SOCD) and ABVG had correlation coefficients of 0.76 and 0.50, respectively, at $p < 0.01$. Precipitation correlated with SOCD and ABVG with correlation coefficients of 0.57 and 0.89, respectively, at $p < 0.01$. The total SOC storage (SOCS) was 436.098×10^6 t C (approximately 0.4356% of the national average) and the average SOCD was 15.75 t C/ha. There was a high ABVG in the southeast and it decreased towards the northwest. The same phenomenon was observed in the spatial distribution of SOCD. Among the soils studied, Hostosols had the highest SOC sequestration rate (25.6 t C/ha) with Gypsisols having the least (7.8 t C/ha). Between 1968 and 2018, the soil carbon stock gradually increased, with the southeast experiencing the greatest increase.

Keywords: century model; soil organic carbon; gypsisols; hostosols; grasslands

1. Introduction

The key finding at the United Nations 48th session of the Intergovernmental Panel on Climate Change (IPCC) was that meeting a 1.5 °C (2.7 F) target was possible, but would require significant emission reductions [1]. This confirms the inevitable phenomenon of climate change [2]. The main drivers of global warming include industrial production and burning of vast amounts of fossil fuels. This has led to problems like the melting of polar glaciers, leading to an increase in sea levels and a decrease in freshwater resources and crop production [3,4]. This has a direct impact on global food security and related problems. The reduction of global warming has therefore become a concern, not only for professionals in global warming, but also for researchers in a range of fields.

Research suggests that the soil contains an enormous amount of C, so any activities that release this C storage may affect the atmospheric CO_2 concentration. An effective way to alleviate climate warming is the rational use and management of agricultural soils [5,6]. Wise use of land can greatly reduce atmospheric CO_2 [7]. SOC losses may, however, be

45

huge when mismanaged [8], as shown by Sornpoon et al. [9]. Soil water dynamics can also affect CO_2 emissions and absorption rates [10], which is a direct consequence of soil management.

Studies of SOC have been conducted at different scales with diverse methods with varying results. From the literature, the global SOCS ranges between 1195×10^9 and 2946×10^9 [11–13]. Research shows that the SOCS in China's soils is about 100.118×10^9 t C [14]. However, there is a wide spatial variation in the SOC across the country. For instance, Luo et al. estimated 4.1×10^8 t C in the Guangdong Province [15], whereas Han, B., Wang, X.K., and Ou, Y.Z.Y. estimated 1.27×10^9 t C in Northeast China [16]. In the arid regions of Western China, the SOCS was estimated to be 122.66×10^8 t C [17] and 116.128×10^7 t C in the Sanjiang Plain [18].

Simulating C sequestration dynamics is a complex process and can be extremely costly. Therefore, reliable, efficient, and cost-effective methods that encompass complex chemical, physical, and biological soil processes are required to monitor the impact of land use and soil management on SOC dynamics. In addition, long-term experiments are valuable for studying the temporal dynamics of SOC as affected by land use and soil management. The CENTURY model has been used by several researchers to estimate SOC storage, with certain success and accuracy [19,20]. For instance, Althoff et al., 2018 [21], estimated an SOCS of 21.0 Mg ha^{-1}, which was strikingly similar to the average field values observed (20.2 Mg ha^{-1}). In addition, for the Shandong province of China, Tang et al. [22] estimated the SOC dynamics using the Century model with a coefficient of determination (R^2) of 0.722 and a mean error (ME) of 0.37 at $p < 0.01$. This gives evidence that the Century model promises an effective way of estimating SOC dynamics under a variety of conditions.

However, it has not been adapted to the grasslands of Gansu Province in China. In this study, we seek to calibrate and validate the Century model to adapt to the SOC of the grasslands of Gansu, as influenced by grazing and land management practices. We also seek to determine the contribution of the grasslands of Gansu to the Chinese national C inventory by simulating the SOC and ABVG dynamics of Gansu grasslands from 1968 to 2018, and to make recommendations to policy-makers as well as the developers of the CENTURY model. This, we believe, would bring enormous benefits to studies of carbon sequestration in the province, and would contribute to national and global SOC monitoring.

2. Materials and Methods

2.1. Study Location

Located in the inland of Northwestern China, Gansu Province is between longitudes $92°13'$ E and $108°46'$ E and latitudes $32°31'$ N and $42°57'$ N. The terrain is largely mountainous, with an elevation ranging between 556 m and 5722 m above sea level. It decreases from high mountains in the southeast to low lands in the northwest. With a total area of 453,700 km^2 [23], the province accounts for approximately 4.436% of China's total land surface area. Figure 1 shows Gansu province with the validation points indicated.

The climate is mainly semi-arid arid continental [24], with warm to hot summers and cold to very cold winters. Temperatures are generally high with varying degrees among areas. The mean annual temperature and precipitation range between 273.16 and 287.26 K and 36.60 to 734.90 mm, respectively. The province has a rich variety of soils, supporting a large range of vegetation. There are at least 16 major soil groups [25], dominated by Gypsisols, Cambisols, and Leptosols (Table 1).

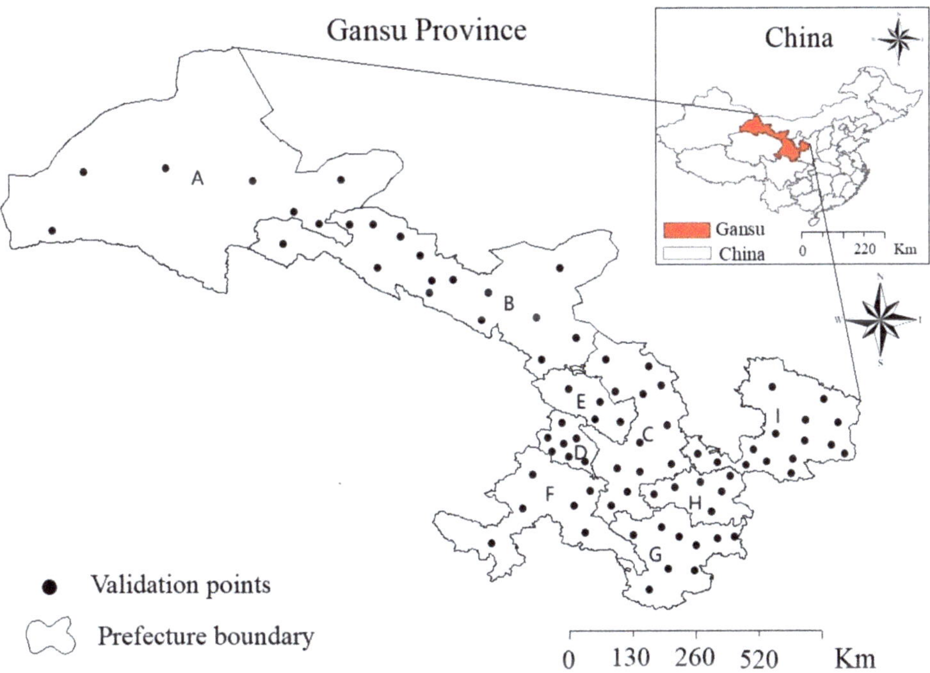

Figure 1. Gansu Province, China, indicating study sites (validation points). (**A**) Jiayuguan, Aksai Kazakh, Anxi, Dunhuang, Jinta, Subei Mongol, Suzhou, and Yumen; (**B**) Jinchuan, Yongchang, Heshui, Minqin, Tianzhu Tibetan, Wuwei, Gaotai, Linze, Minle, Shandan, Sunan Yugur, and Zhangye; (**C**) Baiyin, Huining, Jingtai, Jingyuan, Pingchuan, Dingxi, Lintao, Longxi, Min, Tongwei, Weiyuan, and Zhang; (**D**) Gaolan Liangzhou, and Yongdeng Yuzhong; (**E**) Dongxiang, Guanghe, Hezheng (Linxia), Jishishan Bonan, Kangle, Linxia Shi, Linxia, and Yongjing; (**F**) Zhuoni, Lintan, Luqu, Maqu, Diebu, Xiahe, and Zhugqu; (**G**) Cheng, Dangchang, Hui, Kang, Liangdang, Li, Wen, Wudu, and Xihe; (**H**) Gangu, Qin'an, Qingcheng, Tianshui, Wushan, and Zhangjiachuan Hui; (**I**) Chongxin, Huating, Jingchuan, Jingning, Lingtai, Pingliang, Zhuanglang, Hezheng (Qingyang), Huachi, Huan, Ning, Qingyang, Xifeng, Zhengning, and Zhengyuan.

Table 1. Distribution of major soil groups in Gansu Province, China [25–27].

Soil Group	Area (Km²)	Percentage (%)	BD (gcm⁻³)	Silt (%)	Sand (%)	Clay (%)	pH
					Physiochemical Properties		
Gypsisols	105,236.7	23.19	0.25	35	25	40	7.0–9.0
Cambisols	94,781.1	20.89	1.12	30	58.8	11.2	5.7–7.0
Leptosols	81,108.7	17.88	1.34	28	57.9	14.1	6.8–8.0
Luvisols	29,081.7	6.41	1.27	30.2	49.5	20.3	7.0–8.0
Arenosols	27,751.2	6.12	1.5–1.7	13.04	68.96	18	7.0–8.0
Calcisols	27,379.5	6.03	1.25	28.0	53.0	19.0	7.0–8.5
Kastanozems	15,859	3.5	1.13	27.3	51.2	21.5	7.0–8.5
Chernozems	14,469.6	3.19	1.30	30.9	50.5	18.6	6.5–7.5
Solonchaks	14,204.7	3.13	1.16	34	49.07	16.0	>8.3
Anthrosols	13,391.7	2.95	1.20	25	58.6	20	4.0–4.5
Fluvisols	8438.1	1.86	1.28	16	68.75	24	6.8–7.5
Regosols	8086.2	1.78	1.26	5.75	67.25	26.25	7.1–8.5
Phaeozems	7444	1.64	1.25	28	63.25	19	5.0–7.0
Gleysols	3668.7	0.81	1.27	30	49.5	20	6.0–8.0
Greyzems	2137	0.47	1.32	26	53.86	20	6.75–7.9
Histosols	680.7	0.15	1.32	26	53.86	20.14	7.8
Total	453,719	100					

Only a few of the soil groups, such as Regosols and Phaeozems, can be found throughout the province. However, the majority are more localized. For example, Solonchaks and Gypsisols are found only in the northwest, whereas Luvisols and Histosols are found only in the southeast. Kastanozems are located in the central part of the province. Calcisols, Arenosols, and Anthrosols are mostly localized in the northwest and central part of the province. Chernonozems are mainly in the southeast, with very little coverage in the central part of the province. Histosols, with the highest concentration of SOC, are localized in the Gannan alpine region. Leptosols are found mainly in the mountain ranges bordering Qinghai and Gansu. Cambilsols are the most prominent soil type in the southeast, apart from the Gannan alpine region and the shrub lands.

As of 2010, the dominant vegetation type was grassland, covering approximately 31% of the total land surface [28]. This does not include desert areas and grasslands with minor coverage. For this study, we selected 13 out of the 19 grassland types in the province. These are dominated by temperate typical, alpine meadow, stipa desert steppe, and subalpine deciduous broadleaf, as shown in Table 2.

Table 2. Distribution of grassland types in Gansu Province, China.

Grassland Type	Area (Km^2)	Percentage (%)
Temperate Typical	52,581.6	37.2
Alpine Meadows	21,715.2	15.4
Stipa Desert Steppe	16,123.2	11.4
Subalpine Deciduous Broadleaf Shrubs	15,945.7	11.3
Typical Meadows	9006.1	6.4
Temperate Deciduous	6138.5	4.3
Alpine Sparse	5376.4	3.8
Halophyte	4774.9	3.4
Temperate Meadows	3346.5	2.4
Subalpine Hard-leaf Evergreen Broadleaf Shrubs	2781.6	2
Artemisia Ordosica	2059.2	1.5
Tropical and Subtropical Evergreen Broadleaf Shrubs	1404.1	1
Total	141,253	100

2.2. Data

2.2.1. Meteorological Data

Meteorological data were downloaded from www.worldclim.org (accessed on 1 July 2020) [29] in raster format at a 2.5 min (~21 km^2) spatial resolution. The variables include monthly average minimum and maximum temperatures (°C) and monthly total precipitation (mm) covering the period from 1961 to 2018. These data were downscaled from CRU-TS-4.03 by the Climatic Research Unit, University of East Anglia [30]. Data on climatic regions were downloaded from koeppen-geiger.vu-wien.ac.at [31] in vector format. The climate vector file was based on the Koppen–Geiger Climate Classification [24]. It was updated based on values provided by the Climate Research Unit (CRU) of the University of East Anglia.

2.2.2. Terrestrial Ecoregions

We obtained the terrestrial eco-regions data from the Database for Ecosystems and Ecosystem Services Zoning of China in vector format (http://www.ecosystem.csdb.cn/ accessed on 1 August 2020) [28]. The dataset includes the spatial distribution information of the national first, second, and third level ecosystem classification in 2000, 2005, and 2010, respectively. The dataset contains the spatial distribution patterns of forests, grasslands, shrubs, farmland, towns, deserts, bare land, and other ecosystems [32].

2.2.3. Soil Data

The soil data used in this study was developed by the Food and Agriculture Organization (FAO) of the United Nations and the International Institute for Applied Systems Analysis (IIASA). It was obtained from the Harmonized World Soil Database Version 1.2 (HWSD V1.2) [27] in vector format. The Soil Map of China is based on data from the office for the Second National Soil Survey of China (1995), and is distributed by the Nanjing Institute of Soil Science (Shi et al., 2004) [33]. Data on soil properties were obtained from version 3.6 of the digitized Soil Map of the World (DSMW) [25], prepared by the Food and Agriculture Organization of the United Nations. The variables in the dataset include the bulk density, percent sand, silt, clay, and the pH of the top and subsoil of each sol type. Valuable information on some soil properties was obtained from R. Kodešová et al., 2011 [34].

2.2.4. Validation Data

To validate our model, we obtained Global Aboveground and Belowground Biomass Carbon Density Maps for the Year 2010 from the FAO [35], which are available at the Oak Ridge National Laboratory (ORNL) Distributed Active Archive Center (DAAC) in GeoTIFF format. It provides temporally consistent and harmonized global maps of the above- and below-ground biomass carbon density (Mg C/ha) for 2010 at a 300-m spatial resolution.

2.3. Method

The climate, ecoregion, and soil maps were intersected using ArcGIS, creating a vector file of the study area. This gave rise to 81 sites, each of which had unique characteristics of soil, vegetation, and climate. This was used as a boundary map to calculate the relevant statistics unique to each site using the zonal statistics tool in ArcGIS. To obtain the required input data for the CENTURY model, the mean minimum and maximum temperatures, the mean, standard deviation, and skewness of the precipitation for each site were calculated using the zonal statistics tool of ArcGIS. Figure 2 illustrates the following process.

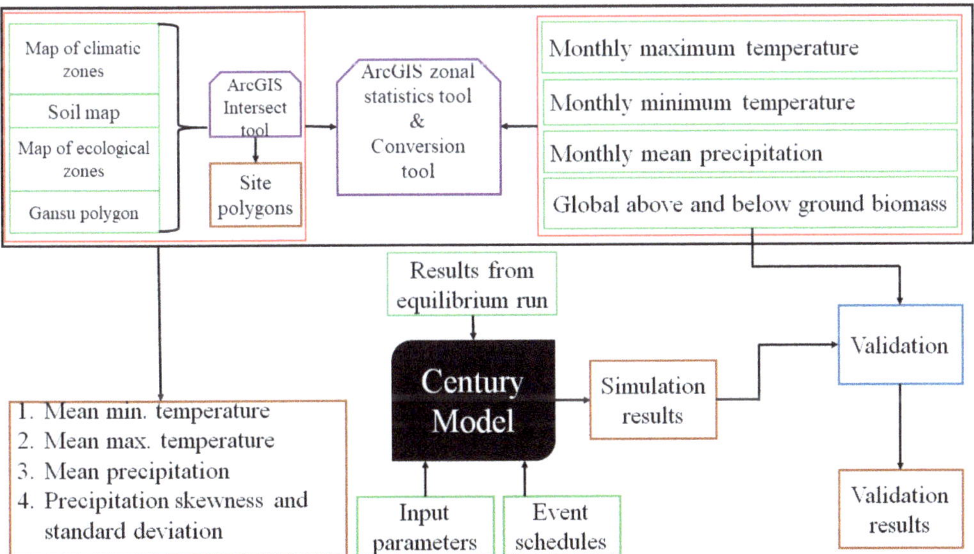

Figure 2. Flowchart of the model calibration, parameterization, and simulation.

- The CENTURY Model

The Century model, a general FORTRAN model of the plant−soil ecosystem that has been used to represent carbon and nutrient dynamics for different types of ecosystems [36], comes with several sub-models. In this study, the grass/crop sub-model of CENTURY 4.0 was used. It was developed by the National Renewable Energy Laboratory (NREL) to provide a tool for ecosystem analysis and to test the consistency of data as well as evaluate the effects of changes in management and climate on ecosystems. It is freely available at nrel.colostate.edu [36,37]. The model is capable of integrating the effects of climate and soil variables and agricultural management practices to simulate carbon, nitrogen, and water dynamics in the soil−plant system [36].

- Model Parameterization

A proper adjustment of the parameters for the underlying environmental context is essential for a near-accurate output of the CENTURY model, as one of the objectives of this study was to calibrate and validate the CENTURY model for the analysis of SOCdynamics for a fifty-year period (1968 to 2018) in Gansu Province, China. The CENTURY model has a number default parameterized. However, each of our study sites has specific climate, soil, vegetation, and management practices. Therefore, some parameters were adjusted for all 81 sites to reflect these site-specific characteristics. We discuss below the model's initialization, calibration, and validation processes.

- Model initialization

The model was initialized in three stages, as follows: In Stage 1, we entered the mean monthly precipitation, standard deviation, and skewness of precipitation, as well as the mean minimum and maximum temperatures for 50 years. In stage two, we entered the site control parameters, which included pH, bulk density, silt, clay, and sand content (Table 1). In stage 3, each of the 81 sites was parameterized using crop parameters, and the organic matter initial values obtained from the equilibrium simulation were entered for each of the 81 sites. Two sites were selected; one from the alpine regions and the other from the low-lying arid region, and the model was calibrated to match their observed output. The equilibrium simulation was run for 2968 years for each of the calibration sites to stabilize the output variable SOMTC (total soil organic carbon). This was done using the CENTURY default grass/crop parameters.

- Model calibration

This is an essential step as it aims at fine-tuning the internal parameters of the model to improve the correlation between the simulated and measured values. The calibration was done by iteratively adjusting the internal parameters until the SOMTC output closely represented the measured soil C, as suggested by E.S.O. Bortolon et al. [38]. The potential aboveground monthly production (g/m^2) was altered at each iteration, until the SOMTC matched the measured soil C stocks with a considerable accuracy. The new parameters were then used to parameterize the other sites. In addition to these parameters, site-specific parameters such as pH were also changed according to the soil type, as shown in Table 1.

- Event Scheduling

To accurately estimate the SOC dynamics in the province, we accounted for historical land use and soil management practices. We developed nine land-use and soil management scenarios to schedule the model using the EVENT100 program of the century model, based on historical sources and surveys. Details are shown in Table 3.

Table 3. Land use/management practice scenarios per zone.

Scenario	Zone	Land Use/Management Practice
1	A	Intensive grazing, ploughing, and no-till after
2	B	Intensive grazing and land till
3	C	Row—cultivator and moderate erosion
4	D	Ploughing and moderate grazing,
5	E	Cultivator and medium grazing
6	F	Hay harvest and no till
7	G	Low grazing and ploughing
8	H	Moderate grazing with no till
9	I	Land till and high erosion

The soil management scenarios were maintained as closely as possible to the available historical information. All 81 sites were put into one of the nine (9) groups (A to I) according to the management practices.

- Equilibrium run

As stated early on, two calibration sites were selected; one from the arid region in the northwest and the other from the alpine region in the south. The output, SOMTC, is the total soil organic carbon at the end of the equilibrium simulation period. The initial soil carbon is usually derived from an initial simulation, which usually covers several thousands of years [39]. We ran the equilibrium simulation for 2968 years to obtain the initial carbon pool for the two calibration sites.

- Running the Model

At this stage, all the sites have been calibrated and parameterized and the schedule files made ready. The SOMTC values obtained from the equilibrium run were used as initial inputs. A batch file (.bat) was created to run the model on all the sites without the need to repeat the process for each site. Along with the batch file was an accompanying list of variables in text .txt format. From this stage, the .lis files, which contain the output data, were obtained. These were then converted to Microsoft Excel files for subsequent analyses. For each of the 81 sites, we obtained the monthly SOC and aboveground biomass for 50 years.

- Model validation

The following output variables were used to assess the performance of the model: aboveground C in plant biomass (ABVG), and soil organic carbon. As suggested by Bockstaller and Girardin, 2003 [39], and Gomes and Varriale, 2004 [40], the model was validated by comparing the outputs of the model with a set of independently measured data. SOC data from the FAO in 2010 were used to validate the model. We calculated the coefficient of determination, R^2 between each pair of variables as follows:

$$R^2 = \left| \frac{\sum_{i=1}^{n}\left(S_i - \overline{S}\right)\left(O_i - \overline{O}\right)}{\sqrt{\sum_{i=1}^{n}\left(S_i - \overline{S}\right)^2\left(O_i - \overline{O}\right)^2}} \right|^2 \tag{1}$$

R^2 is a statistical measure that represents the proportion of the variance for a dependent variable. The closer R^2 is to 1, the better the fitting effect, and the more realistic the fitting function.

We also calculated the mean absolute percentage error, MAPE, as follows:

$$MAPE = \frac{1}{n}\sum_{i=1}^{n}\left|\frac{S_i - P_i}{S_i}\right| \tag{2}$$

where S_i is the SOC simulated value of the *i*th site validation point, O_i is the observed *SOC* value at the *i*th site validation point, $\overline{O}$ is the average of the observed *SOC* values of the dry farmland validation point from the period 1968 to 2018, and $\bar{S}$ is the average value of the simulated *SOC* value of the site validation point from 1969 to 2018. P_i is the predicted value at the *i*th site.

To assess how well the sample represents the population, we calculated the standard error (*SE*) at each site as follows:

$$SE = \frac{\sigma}{\sqrt{n}} \tag{3}$$

where σ is the standard deviation of the samples and n is the number of samples

$$SOC(Y) = SOCD(Y) * area(Y) \tag{4}$$

3. Results

3.1. Results from the Equilibrium Run

Table 4 shows the results of the equilibrium simulation for the two calibration sites.

Table 4. Results from the equilibrium run for the two calibration sites.

Calibration Site	SOMTC (t C)
Arid region	1680.7
Alpine region	2028.07

These values are important for initializing the model for the actual simulation, because they are used as the amount of soil organic matter at the beginning of the actual simulation period. During the equilibrium run, we assumed that 21,968 years ago, the arid region was all grass without desert, but experienced little rain and intensive grazing; a recipe for desertification. In the southeast, however, we assumed ample rainfall and moderate grazing. This led to a higher SOMTC in the calibration site to the southeast than in the northwest.

3.2. Results from Actual Simulation

We computed the SOCD and total SOC in each site and their respective averages. We found that at 0-20cm, Gansu's grasslands stored $436.098 \times 106 \pm 87.25 \times 104$ t C of SOC. The SOC density (SOCD) ranged between 11.09 t C/ha and 33.08 t C/ha, and averaged 15.75 t C/ha across the province. We multiplied the SOCD of each site by their corresponding areas and the average, standard deviation, and minimum and maximum found. The detailed results are presented in Table 5.

Table 5. Summary of the results.

	Aboveground Biomass Density (g/m²)	SOCS (t C)	SOCD (t C/ha)
Mean	2.95×10^2	436.098×10^6	15.75
Standard Deviation	1.15×10^2	87.25×10^4	5.67
Minimum	7.20	12.44	5.23
Maximum	1207.56	26.13	30.68

The standard deviations illustrate the great ecological variation among the various ecological zones in the province, namely, arid, alpine, and shrub lands. The lowest ABVG was recorded in the northwest and Qingyang between 1979 and 1989, and the highest was recorded in the southeast between 2001 and 2018. The highest SOCS was estimated in the alpine region, whereas the highest SOCD was recorded in the subalpine region between 2001 and 2018.

3.2.1. Point Validation

The average ABVG and SOC values across all of the sites were calculated and then compared with the 2010 aboveground and SOC data from FAO in a correlation analysis. Table 6 shows the correlation between mean precipitation, simulated and observed aboveground biomass, and the simulated and observed SOC.

Table 6. Correlation analysis between SOC Obs. (observed SOCD (g C/ha)), SOCD Sim. (simulated SOCD (g C/ha)), ABVG Sim. (simulated ABVG (g C/m^2)), ABVG Obs. (observed ABVG (g C/m^2)), and MMP (monthly mean precipitation (mm)) at 81 sites in Gansu Province in 2010.

	MMP	ABVG Sim.	ABVG Obs.	SOCD Sim.
MMP				
ABVG Sim.	0.57			
ABVG Obs.	0.67	0.50		
SOCD Sim.	0.89	0.58	0.82	
SOCD Obs.	0.71	0.64	0.56	0.76

All of the correlations were significantly positive at $\alpha < 0.01$. For instance, there is a strong correlation between our simulated and observed SOCD ($R^2 = 0.76$, $p < 0.01$), which was even stronger between precipitation and simulated SOCD ($R^2 = 0.89$, $p < 0.01$). This indicates the strong dependence of SOC and the aboveground biomass on precipitation. Overall, our simulated values fit well with the observed values.

To further validate our model, we calculated two error metrics, namely the standard error (SE) and the mean absolute percentage error ($MAPE$). $MAPE$ is a measure (in percentage) of the accuracy of a forecast system. SE assesses how well the sample represents the population. For instance, from Table 7, the MAPE between the observed and simulated SOCD is 0.0824, indicating that the simulated and observed values differ by 8%. The between the simulated and observed aboveground biomass is 0.1411, indicating they differ by 14%. All measures were highly significant at $p < 0.01$.

Table 7. Correlation and error metrics of point validation.

		Mean Precipitation (mm)	Aboveground C Observed (t C/ha)	Aboveground C Simulated (t C/ha)	SOCD Observed (t C/ha)
Aboveground C Observed (t C/ha)	*p*-value	***			
	SE	0.083			
Aboveground C Simulated (t C/ha)	*p*-value	***	***		
	SE	0.093	0.098		
	MAPE	—	0.1411783		
SOCD Observed (t C/ha)	*p*-value	***	***	***	
	SE	0.079	0.093	0.087	
SOCD Simulated (t C/ha)	*p*-value	***	***	***	***
	SE	0.052	0.064	0.092	0.073
	MAPE	—	—	—-	0.0824219

Note: *** indicates that the *p*-value is highly significant.

We further developed linear models between the observed and simulated $SOCD$ and $ABVG$ as shown below:

$$Y_{ABVGsim} = 30.934 + 0.00695 X_{ABVGobs} \tag{5}$$

$$Y_{SOCsim} = 238.917 + 0.316 X_{SOCobs} \tag{6}$$

where $Y_{ABVGsim}$ and Y_{SOCsim} are the simulated aboveground biomass and $SOCD$, respectively, and $X_{ABVGobs}$ and X_{SOCobs} are the observed aboveground biomass and $SOCD$, respectively. The intercepts and coefficients are highly significant ($p \leq 0.01$).

3.2.2. Spatial and Temporal Distribution of SOC and ABVG

Table 8 shows the summary statistics of the output variables for the various zones:

Table 8. Zonal distribution of Gansu grassland carbon stock and vital statistics.

Zone	No. of Sites	Soil Organic Carbon Density (SOCD) t C/ha				SOC Storage (SOCS) ($\times 10^6$ t C)				Aboveground Biomass (ABVG) (g cm^{-2})			
		Min	Max	Av	Std	Min	Max	Av	Std	Min	Max	Av	Std
A	8	5.23	12.67	8.67	1.89	11.61	19.73	16.66	6.14	7.2	250.34	80.69	72.14
B	12	7.52	15.55	10.4	2.03	15.96	34.91	30.44	5.04	90.89	234.56	125.24	34.33
C	12	9.38	17.45	13.4	2.44	30.46	35.05	34.86	1.25	300.5	790.78	500.25	149.22
D	4	10.45	20.23	15.53	3.01	25.76	45.42	45.46	5.71	275.34	1004.01	800.3	199.89
E	8	13.44	23.78	19.21	2.01	49.5	53.39	51.87	1.08	359.3	1575.05	900.45	384.18
F	7	16.55	30.68	23.17	4.21	95.7	112.5	89.62	4.15	464.34	1207.56	790.5	209.29
G	9	15.67	29.56	22.8	3.85	83.29	96.89	75.46	3.79	87.05	400.58	284.23	104.19
H	6	11.25	21.56	16.68	2.98	42.91	48.41	48.82	1.64	314.6	689.45	505.4	124.22
I	15	14.3	22.75	19.13	2.27	46.72	51.36	42.91	1.28	45.5	945.05	300.9	248.73

We grouped all the 81 study sites into nine zones (A to I) (Figure 1). Each zone has a specific number of sites. In each zone, we computed the average and standard deviation of SOC storage (SOCS) and ABVG. Zones F and G, both located in the southeast, have the most accumulated SOC. However, zone A in the northwest arid region had the least accumulated SOC. We also computed the average aboveground biomass and SOCD at intervals of at least 10 years for each site. We then interpolated the values to obtain a spatiotemporal map of the province. Figure 3 illustrates the SOCD dynamics in Gansu between 1968 and 2018 for 50 years.

Higher values of SOC density were estimated in the southeast throughout the simulation period. However, lower values of SOC were recorded in arid regions to the northwest. Generally, SOCD decreased from the southeast to the northwest. The average SOCD between 1979 and 1989 was lower than that of the previous decade, and increased afterward until 2018. Between 1968 and 2018, the average SOCD increased by 38% in parts of the southeast, whereas the increase in most parts of the northeast was approximately 9%. This is mainly due to the intensive land management simulated to fight desertification. In our model, we assumed intensive tilling to convert deserts into farmlands. This led to a tremendous release of SOC from the soil. However, after the 1980s, we simulated a no-till management practice and this led to a significant accumulation of SOC. In the alpine regions, SOC has always been on an increasing trend, as we simulated no-till land management practices throughout the period. Consequently, the soil was undisturbed and therefore accumulated SOC progressively, with the exception of Qingyang (zone I), where SOCD decreased from 1968 to 2000 and increased from 2000 to 2018. As a result, most of the SOC is concentrated in the southeast, especially in zones F and G (Figure 3).

We also estimated the amount of aboveground biomass for the study period. Figure 4 shows the spatial and temporal distribution of aboveground biomass between 1968 and 2018.

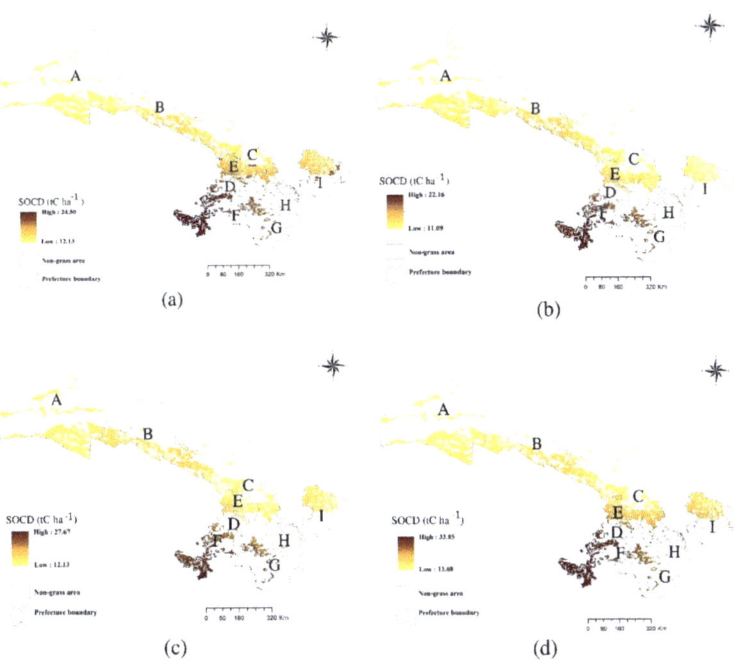

Figure 3. Spatial and temporal distribution of SOC of the grasslands of Gansu from 1968 to 2018: (**a**) 1968 to 1978, (**b**) 1979 to 1989, (**c**) 1990 to 2000, and (**d**) 2001 to 2018.

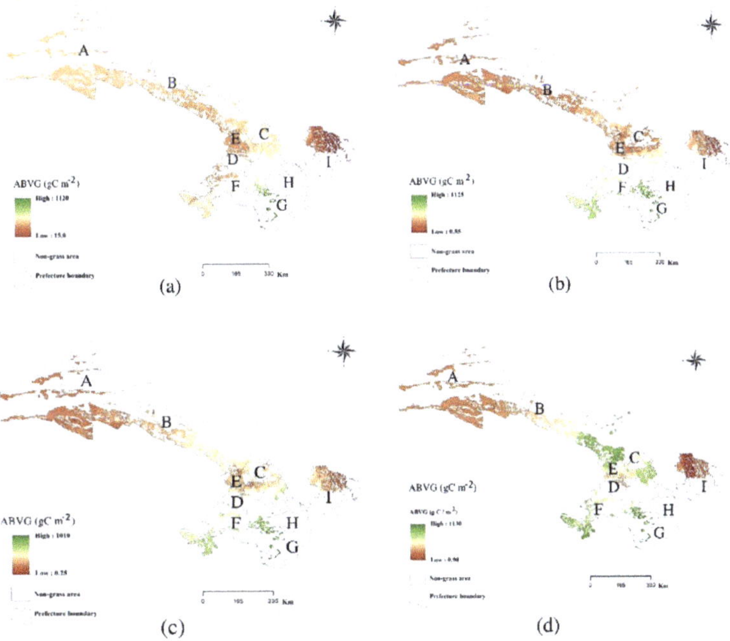

Figure 4. Spatial and temporal distribution of the aboveground biomass in the grasslands of Gansu from 1968 to 2018: (**a**)1968 to 1978, (**b**) 1979 to 1989, (**c**) 1990 to 2000, (**d**) 2001 to 2018.

During the first decade, the aboveground biomass was relatively low, especially in zone I (Figure 4). Vegetation health in the northwest deteriorated in the second and third decades, but improved between 2000 and 2018. During this period, ABVG increased significantly, especially in the alpine regions. However, it deteriorated drastically again in zone I. In this zone, we simulated intensive erosion and consistent land tilling practices.

We calculated the percentage change in SOC between 1968 and 2018 of SOCS in all 81 sites to determine the SOC dynamics. We then interpolated the values using the Kriging interpolation method. Figure 5 shows the results.

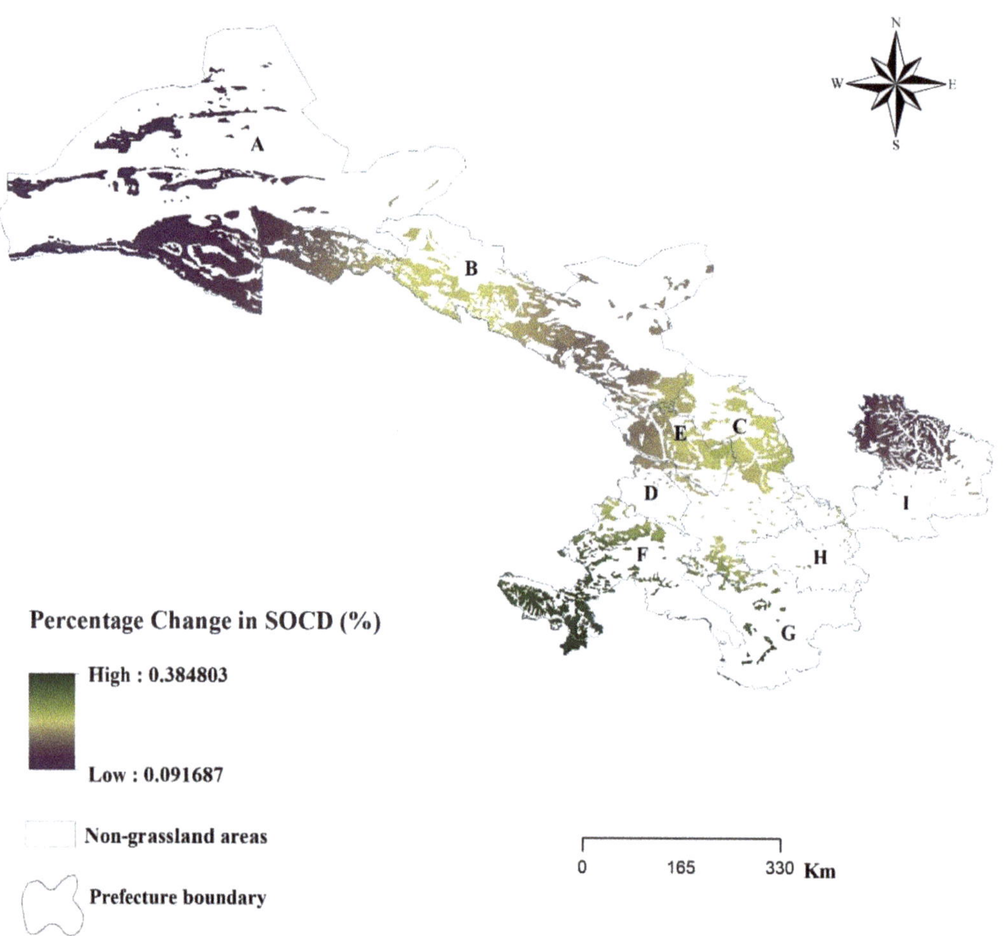

Figure 5. Percentage change in SOCD in Gansu grasslands between 1968 and 2018.

As illustrated by Figure 5, the change in SOCD is not evenly distributed. The southeast and central parts accumulated more SOC than the northwest and zone I to the south. These areas of least increase underwent constant tilling and other anthropogenic interferences. As a result, SOC could not accumulate as much as it did in the southeast.

3.2.3. SOCD and ABCG Distribution by Soil Type

For each soil group, we computed the average SOCS and ABVG between 1968 and 2018 across all of the study sites. Figure 6 illustrates the distribution of SOCS and ABVG among the various soil types between 1968 and 2018.

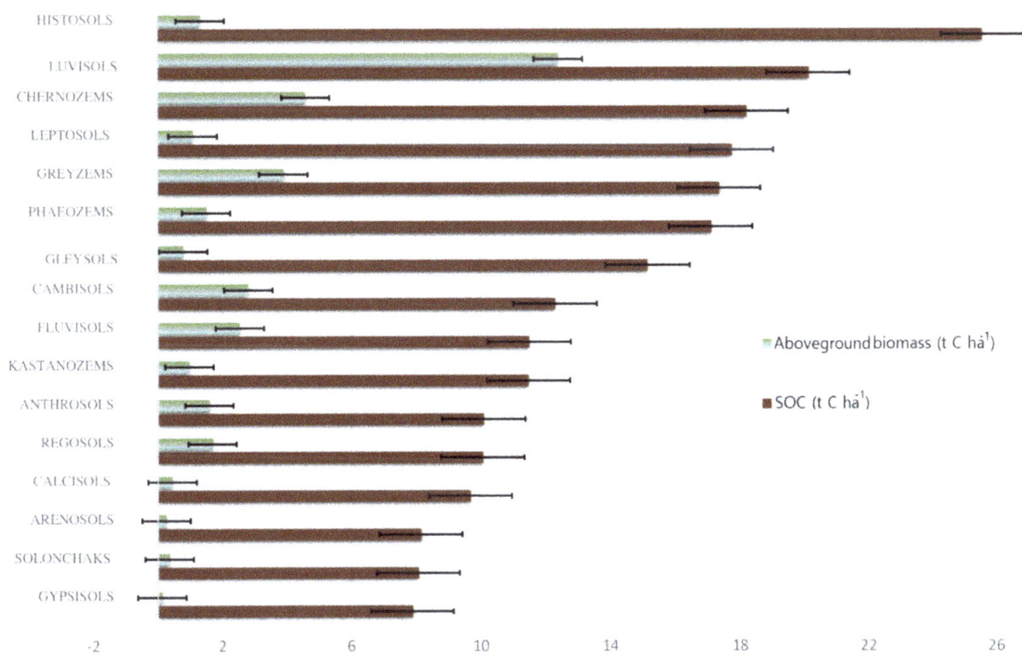

Figure 6. SOC distribution of SOC and aboveground biomass by grassland type.

Luvisols have the highest aboveground biomass, followed by Chernozems. In particular, Luvisols are found only in the southeastern part. In addition, Arenosols, Calcisols, and Anthrosols found in the northwest and the central part of the province have very low values of ABVG. There is also a huge variation in the distribution of ABVG among the various soil groups. Among the soil groups with the highest concentration of SOC are Histosols and Luvisols, localized in the Gannan alpine region. Following closely are Chernonozems, which are mainly found in the southeast, with very little coverage in the central part. Gypsisols and Solonchaks, found only in the northwest, have the least ABVG, followed by Calcisols, Arenosols, and Anthrosols, which are mostly localized in the northwest and central parts. It can be seen that extreme values of ABVG are found in soils that are localized in either the southeast or northwest. Soils found either throughout the province or only in the central part have values in between. Typical examples are Regosols, Phaeozems, and Kastanozems.

Among the soil groups, Hostosols sequestered the most SOC (52.33×10^6 t C), representing 11.95% (Figure 7). Gypsisols, Solonchaks, and Arenosols altogether accounted for about 11% of C storage. This is mainly due to their low coverage and low SOCD.

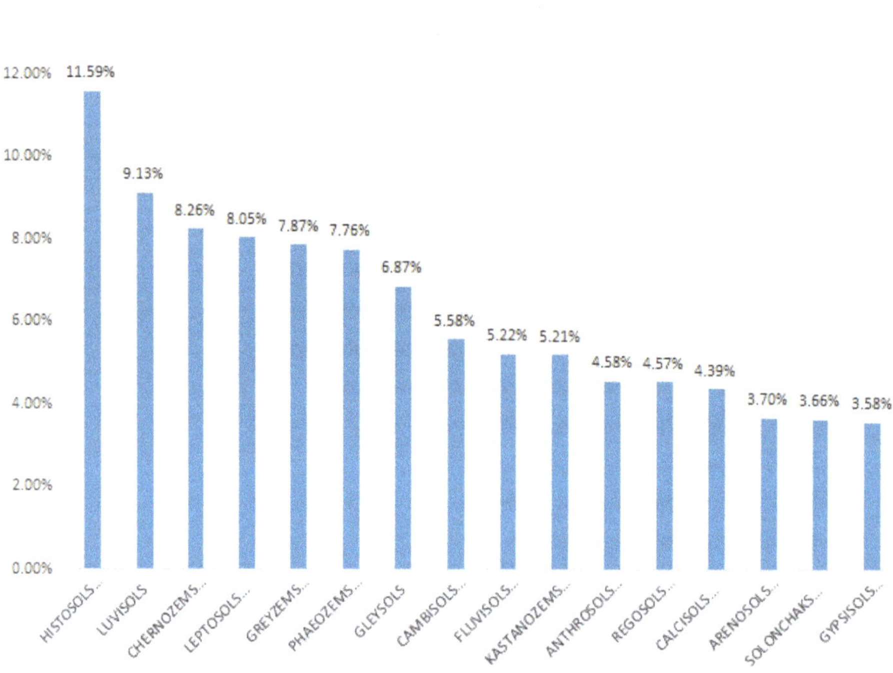

Figure 7. Percentage distribution of SOCS by soil type in Gansu Province, China, between 1968 and 2018.

Gypsisols, Solonchaks, Arenosols, and Calcisols each sequestered 17.44×10^6 t C, representing 4% of the total SOC.

4. Discussion

We adapted the CENTURY model to the grasslands of Gansu and successfully estimated the amount of carbon sequestered in the soils (0–20 cm), as well as the aboveground biomass. Our measured and observed SOCD and ABVG had correlation coefficients of 0.76 and 0.50, respectively, at $p < 0.01$ (Tables 6 and 7). Histosols had the highest SOC sequestration rate (25 t C/ha), with Gypsisols having the least (7.8 t C/ha). Aboveground biomass ranged from 0.002 to 1.130 kg C m^{-2} among the 16 soil types across the province (Figure 5). The average aboveground biomass density estimated was $2.95 \times 102\ 0.415 \pm 1.15 \times 102$ g C/m^2, which was consistent with Xu et al. (0.69 ± 0.20 kg C m^{-2}) [41]. Precipitation was correlated with SOC and aboveground biomass with a correlation efficiency of 0.57 and 0.89, respectively, at $p < 0.01$.

Compared with other regional studies, the SOCD of Gansu is lower than the average value of 48.6 t C ha^{-1} in the arid region in Western China [17]. This suggests that the grasslands of Gansu Province have a huge SOC sequestration potential. Our findings are consistent with the majority of SOC studies in the province and its environs. For example, Guo et al. [42] estimated the grassland SOCD in Inner Mongolia to be 19.9 t C/ha, a value within our estimated values in Gansu. According to their study, SOC values were higher in the southeast (21.56 t C/ha) but lower in the northwest (13.77 t C/ha) part of Gansu Province, a pattern we also discovered in our study.

The spatial distribution of the aboveground biomass was similar to that of the SOC density (Figures 4 and 5). The aboveground biomass and SOC densities declined from the southeast to the northwest. The majority of the soils found in the south are relatively fertile (Catling D., 1992) [43], giving rise to huge variations in SOC and ABVG distributions.

Another reason could be from the differences in management practices and soil properties, such as the pH. The function of pH in biomass production cannot be overemphasized. Neina, D. (2019) [44] refers to soil pH as the master soil variable, observing that it controls the solubility, mobility, and bioavailability of trace elements, which determines their translocation in plants. It also affects the soil enzymes and microbial activities (Marinos, R. and Bernhardt, E.) [45], as well as biodegradation. Therefore, in our model, we placed special emphasis on the variation of pH among the various soils. According to Neina, McCauley, and 2009 [46], D. (2019) [47] the activities of soil microorganisms and crop growth are the greatest near neutral conditions, but the pH ranges vary for each type of microorganism and plant. They claim that very acidic soils (less than 5) cause microbial activities and numbers to be considerably lower than in the soils that are more neutral [46]. According to esf.edu [47], most minerals and nutrients are more soluble or available in acid soils than in neutral or slightly alkaline soils, but adds that optimum pH conditions for individual crops will vary. In their research, Zhou, et al. (2019) [48] observed that soil carbon is negatively correlated with soil pH, demonstrating that relatively low pH benefits the accumulation of organic matter. According to McCauley, A., Jones, C., and Jacobsen, J. (2009) [46], soil pH strongly affects soil functions and plant nutrient availability by influencing the chemical solubility and availability of the plant essential nutrients, as well as the organic matter decomposition.

The majority of soils in the province have optimal levels of pH 5–8 for plant growth (Table 1). For instance, the northwestern part is dominated by Euteric soils, which are neutral to slightly acid, with a good availability of nutrients, fine texture, moderate to high cation exchange capacity, and low to moderate organic matter content, and the natural fertility is fairly high [43].

Soils in the northwest have the right level of pH for plant growth, much like soils in the southeast. However, the positive effect of pH is nullified by the absence of sufficient plants. For example, Marinos and Bernhardt (2018) [45] found that microbial respiration and SOC solubility were strongly stimulated by increased soil pH, but only in the presence of plants, stating that a soil pH increase of 0.76 units increased soil respiration by 19% in the organic soil horizon and 38% in the mineral soil horizon, whereas in unplanted pots, soil pH had no effect on microbial respiration. The presence of the right pH alone cannot trigger plant growth. They observed that while increased soil pH enhanced plant-mediated heterotrophic respiration, it had no effect on plant growth. In contrast, soil Ca enrichment increased the relative growth rate of plants by 22%, but had no impact on microbial respiration. Plant biomass is higher in the southeast, because the majority of the soils are Ca rich. We therefore suggest that policy-makers lay special emphasis on enriching the calcium content of soils in the northwest of the province to facilitate plant growth.

Another major factor inhibiting plant growth in the northwest is the climate. Xu et al., 2018 [41], found that precipitation positively correlated with vegetation biomass and SOC, whereas temperature was negatively correlated with SOC. The northwestern part of the province has very low precipitation, thereby affecting vegetation biomass production. These factors lead to low SOC and vegetation biomass [46] in the northwest. The low amount of precipitation in the northwest also leads to little leaching of base cations, resulting in a relatively high degree of base saturation [46]. According to the same author, cation and anion exchange capacity (the soil's ability to retain and supply nutrients to a crop) is directly affected by soil pH. When the soil pH is high (i.e., low concentration of H+), more base cations will be on the particle exchange sites and will thus be less susceptible to leaching. An increase in precipitation causes increased leaching of base cations and the soil pH is lowered [46]. However, when the soil pH is lower (i.e., higher concentration of H+), more H+ ions are available to "exchange" the base cations, thereby removing them from exchange sites and releasing them into the soil water. As a result, exchanged nutrients are either taken up by the plant or lost through leaching or erosion. The southeastern part is dominated by Calcaric soils and some amount of Euteric soils. Calcaric means the presence of calcium carbonate. They have high natural fertility but are potentially alkaline [43].

According to Céspedes-Payret et al. (2017) [49], bulk density values are not significantly correlated with SOC content. Our model could not predict aboveground biomass as much as it did SOC because of the inherent problem of the century model not being able to model the finer timescale of ABVG accumulation. Our finding also indicates that precipitation affects aboveground biomass more than it does SOC.

Another factor contributing to the spatial variation is differences in climate. The southeastern part of the province has more conducive conditions for plant growth than the northwest. For example, rainfall increases from the northwest to the southeast. However, temperature decreases in the same direction. The northwestern part is icy cold throughout most of the year. According to Xu et al., 2018 [41], the main factors that affect the spatial distribution of aboveground biomass are climate and soil nutrients. They observed that these factors account for 68.16% of the total variation in aboveground biomass in a GLM analysis. Among these factors, climate was the most important influencing factor, explaining 50.49% of the total variance. In addition, they observed that soil texture plays a significant role in the total variance in the spatial patterns of SOC density for the 0–20 cm soil layers. The climate of Gansu Province is very diverse and, therefore, precipitation and temperatures vary greatly [29]. The century model was able to account for the variation in precipitation and temperature, and showed that carbon density was influenced greatly by climate. The aboveground biomass density was generally higher in high-precipitation areas. The temperate arid regions had the lowest aboveground biomass density. High values of aboveground biomass were estimated in the southeastern part compared with the northwest part. Nevertheless, the amount of ABVG simulated by our model in the northwest was greater than the measured values. We attribute this to the overgeneralization of parameters and conditions throughout the simulation. Most SOC was concentrated in the alpine regions. The arid regions, located in the northwest, had the lowest SOC density at 0–20 cm. This was consistent with the findings of Xu et al., 2018 [41].

Between 1968 and 2018, the entire region was a carbon sink and since then, the SOC has been on a gradual incline. Beginning in 1968, the SOC showed a sharp decrease, but began to increase in 1971 and has been increasing annually. The southeastern part saw a rather gradual increase in SOCD, while the majority of the decrease took place in the arid regions and the central part of the province. Even though the province has generally been a carbon sink for the past 50 years, a few areas have been shown to lose SOC. This was mainly due to overgrazing and other management practices, which our model accounted for. For instance, zones A, B, and I (Figures 1 and 4) lost aboveground biomass rapidly due to extensive land tilling (National Research Council, 1992) [50,51]. The FAO observed that soil loss due to wind erosion was 10 to 15 cm and sand loss was 20 to 150 cm in the northwest [51]. Therefore, our model was calibrated to simulate more erosion in the arid regions. According to FAO and historical records, overgrazing accounts for over 20% of decertified areas. The incorporation of intensive grazing into our simulation contributed to its accuracy. However, low-intensity grazing was simulated between 1988 and 1998. During this period, the government established plantations for desertification control and measures were put in place to regulate grazing [51].

According to Xu et al., 2011 [52], there has been a large-scale reclamation of arid land in Northwest China over the past 50 years, converting the natural desert landscape into an anthropogenic oasis. SOC is greatly influenced by anthropogenic activities (Breuer et al., 2006 [53], and Kasel and Bennett, 2007 [54]). Therefore, Xu et al., 2011, speculate that drastic human activities may have caused the radical change in SOC. As a result, we observed a relatively stable increase in SOC in the alpine regions than in areas influenced by human activities. In arid regions, we found that SOC storage decreased most of the time between 1968 and 2000, and gradually increased afterward. The rapid loss of SOC in semi-arid regions under cultivation has been reported by several researchers. For example, Elberling et al. (2003) [55] reported that up to 24% of SOC in the upper 1 m layer has been lost over the 40 years since the Savannah began to be cultivated in semiarid Senegal in West Africa. Similarly, Ogle et al. (2004) [56] observed that the level of SOC in tropical regions was reduced to 32% due to a 20-year tillage. A similar phenomenon was observed by Su et al. (2004) [57] in the semiarid Horqin sandy steppe of northern China. They also observed that in just 3 years of cultivation, sandy grassland lost up to 38% of SOC. In most of these instances, the loss was rapid within the first few years, as it was in the case of the Nigerian semiarid Savannah. (Jaiyeoba, 2003) [58].

Our model was not without flaws. These were partly due to inevitable assumptions and the oversimplification of reality. We did not account for land cover change over the research period, for example. The area of grassland used could only match that in 2010 and therefore did not account for the change. Secondly, the monthly timeframe of the century model fails to adequately depict the precise timescale of biomass growth (Yu et al., 2014) [59]. This undoubtedly affected the biomass and SOC stock estimation. Furthermore, as shown in the literature, the chemical properties of the same soil in various places might differ dramatically. Wherever necessary, we used the average values across all of the areas where those soils were found. A major drawback faced was the fact that at most sites, historical data were hard to find, especially data on management practices such as fertilization and irrigation. In such circumstances, we used our discretion based on extant management practices in those areas. Another instance of oversimplification of reality is the model's failure to account for altitude. Altitude affects vegetative carbon and SOC, according to Xu et al., 2018 [41]. They observed that vegetation carbon declined with altitude, whereas SOC density increased with increasing latitude. Wang et al. [60] also observed a similar pattern in China. However, the CENTURY model does not account for this important factor. Other researchers such as Oelbermann have also found some operational problems with the model. They found that the model estimated lower SOC than the measured SOC values because it failed to account for changes in soil bulk density [61].

We recommend CENTURY developers include the effects of altitude in order to increase the model's estimation accuracy. We also recommend that authorities in charge give attention to Ca modification practices, especially in the northwest, alongside intensive plant growth. In addition, Slaton et al., 2001 [62], and McCauley, A., Jones, C., and Jacobsen, J. (2009) [46] suggest the addition of amendments to soils to modify the soil pH. Suggested amendments are elemental sulfur, ferrous sulfate (FeSO4), aluminum sulfate ($Al_2(SO_4)_3$), ammonium (NH_4+)-based fertilizers, soil organic matter (SOM), and NH4+-based fertilizers. The acidifying effects of fertilization may be more than compensated for by other land use practices. For instance, a study by Jones, C.A., J. Jacobsen, and S. Lorbeer, 2002 [63], found that over a 20-year period, fertilized and cultivated soils in Montana experienced, on average, higher pH values than non-fertilized/non-cultivated soils. They explained that in fertilized/cultivated soils, practices such as crop removal during harvest and tilling decrease SOM levels and subsequent acid production. Additionally, tillage increases surface and sub-surface soil mixing, moving $CaCO_3$ from the sub-surface closer to the soil surface.

5. Conclusions

The application of the CENTURY model to simulate the SOC dynamics in Gansu Province was successful. Over the past 50 years, the region has been a carbon sink for the most part. However, intensive grazing and anthropogenic activities in the arid regions made the area a slight carbon source in the early years of the simulation. The region became a carbon sink due to government interventions after the second part of the simulation. We estimated that the grasslands of Gansu stored $436.098 \times 106 \pm 87.25 \times 104$ t C of soil SOC at 0–20cm, and accounted for about 0.468% of the total national SOC carbon inventory and 1.54% of the total national grassland SOC stock. We also estimated an average SOC density of 15.75 t C/ha across the province. The most carbon is sequestered in the alpine and southeastern regions of the province.

The SOC sequestration potential differs among the soil types. For instance, Histosols sequestered the most amount of SOC, followed by Luvislols and Chernozems, all of which are found mostly in the south. Gypsisols, Solonchaks, and Arenosols, all found in the northwest, sequestered the least SOC. For the past 50 years, the province has been a carbon sink. However, a few areas were shown to lose SOC. The SOC loss was mainly due to overgrazing and other management practices. In general, our estimated SOCD was less than the national average, indicating the grassland of the province has a huge carbon sequestration potential.

Author Contributions: Conceptualization, M.Z. and S.N.; methodology, S.N.; software, S.N.; validation, M.Z., S.N. and T.N.; formal analysis, S.N.; investigation, S.N.; resources, X.W.; data curation, S.N.; writing—original draft preparation, S.N.; writing—review and editing, M.Z.; visualization, T.N.; supervision, M.Z.; project administration, M.Z.; funding acquisition, M.Z. All authors have read and agreed to the published version of the manuscript.

Funding: This research was funded by the Young Tutor Support Fund Project of the Gansu Agriculture University, no. GAU-QDFC-2019-03; the Research Fund of Higher Education of Gansu, China, no. 2019A-051; and the Natural Science Foundation of Gansu Province, China, no. 1606RJZA077 and 1308RJZA262.

Institutional Review Board Statement: Not applicable.

Informed Consent Statement: Not applicable.

Data Availability Statement: Publicly available datasets were analyzed in this study. This data can be found here: http://www.ecosystem.csdb.cn/ (accessed on 12 March 2020); www.worldclim.org (accessed on 12 March 2020); http://koeppen-geiger.vu-wien.ac.at/present.htm (accessed on 12 March 2020); http://www.fao.org/soils-portal/data-hub/soil-maps-and-databases/harmonized-world-soil-database-v12/en/ (accessed on 12 March 2020).

Conflicts of Interest: The authors declare no conflict of interest. The funders had no role in the design of the study; in the collection, analyses, or interpretation of data; in the writing of the manuscript; or in the decision to publish the results.

References

1. Masson-Delmotte, V.; Zhai, P.; Pörtner, H.O.; Roberts, D.; Skea, J.; Shukla, P.R.; Pirani, A.; Moufouma-Okia, W.; Péan, C.; Pidcock, R.; et al. *Global Warming of 1.5 °C. An IPCC Special Report on the Impacts of Global Warming of 1.5 °C above pre Industrial Levels and Related Global Greenhouse Gas Emission Pathways, in the Context of Strengthening the Global Response to the Threat of Climate Change, Sustainable Development, and Efforts to Eradicate Poverty;* IPCC: Geneva, Switzerland, 2018.
2. Han, R.Q.; Zhen, D.; Dai, E.F.; Wu, S.H.; Zhao, M.H. Response of production potential to climate fluctuation in major grain regions of China. *Resour. Sci.* **2014**, *36*, 2611–2623.
3. Cline, W.R. The impact of global warming of agriculture: Comment. *Am. Econ. Rev.* **1996**, *86*, 1309–1311.
4. Tanaka, S.K.; Zhu, T.; Lund, J.R.; Howitt, R.E.; Jenkins, M.W.; Pulido, M.A.; Tauber, M.; Ritzema, R.S.; Ferreira, I.C. Climate warming and water management adaptation for California. *Clim. Chang.* **2006**, *76*, 361–387. [CrossRef]
5. Yang, X.M. Using agricultural soils to fix organic carbon—Alleviating global warming and improving soil productivity. *Soil Environ. Sci.* **2000**, *9*, 311–315.
6. Yang, X.M.; Zhang, X.P.; Fang, H.J. The significance of carbon sequestration in agricultural soils to alleviate global warming. *Sci. Geogr. Sin.* **2003**, *23*, 101–106.
7. Intergovernmental Panel on Climate Change. *Climate Change 1995: Impact, Adaptation and Mitigation: Scientific Technical Analyses;* Working Group 1; Cambridge University Press: Cambridge, UK, 1996.
8. Lal, R. Soil Erosion and Gaseous Emissions. *Appl. Sci.* **2020**, *10*, 2784. [CrossRef]
9. Sornpoon, M.; Bonnet, S.; Garivait, S. Effect of open burning on soil carbon stock in sugarcane plantation in Thailand. *World Acad. Sci. Eng. Technol.* **2014**, *8*, 822–827.
10. Ray, R.L.; Griffin, R.W.; Fares, A.; Elhassan, A.; Awal, R.; Woldesenbet, S.; Risch, E. Soil CO_2 emission in response to organic amendments, temperature, and rainfall. *Sci. Rep.* **2020**, *10*, 5849. [CrossRef]
11. Wang, K.F. *Coupling Development of Triple-Microbe Model and Its Simulation of Global Soil Organic Carbon and Microbial Carbon;* Northwest A & Forestry University: Xianyang, China, 2017.
12. Bohn, H.L. Estimate of organic carbon in world soils. *Soil Sci. Soc. Am. J.* **1982**, *46*, 1118–1119. [CrossRef]
13. Jobbágy, E.G.; Jackson, R.B. The vertical distribution of soil organic carbon and its relation to climate and vegetation. *Ecol. Appl.* **2000**, *10*, 423–436. [CrossRef]
14. Wang, S.Q.; Zhou, C.H. Estimating soil organic carbon reservoir of terrestrial ecosystem in China. *Geogr. Res.* **1999**, *18*, 349–356.
15. Luo, W.; Zhang, H.H.; Chen, J.J.; Liu, Y.; Li, D.Q. Storage and spatial distribution of soil organic carbon in Guangdong Province. *Ecol. Environ. Sci.* **2018**, *27*, 1593–1601.

16. Han, B.; Wang, X.K.; Ou, Y.Z.Y. Distribution and change of agro-ecosystem carbon pool in Northeast of China. *Chin. J. Soil Sci.* **2004**, *35*, 1140–1147.
17. Zhang, J.; Lin, M.; Ao, Z.Q. Estimation of soil organic carbon storage in arid areas of Western China. *J. Arid Area Resour. Environ.* **2018**, *32*, 132–137.
18. Yang, A.G.; Miao, Z.H.; Qiu, F.F.; Yang, Q.C.; Wang, Z.M.; Mao, D.H. Estimation and spatial distribution of organic carbon storage in topsoil of Sanjiang Plain based on GIS. *Bull. Soil Water Conserv.* **2015**, *35*, 155–158.
19. Xiao, X.M.; Wang, Y.F.; Chen, Z.Z. Dynamics of primary productivity and soil organic matter of typical steppe in the Xilin River Basin of Inner Mongolia and their responses to climate change. *J. Integr. Plant Biol.* **1996**, *38*, 45–52.
20. Zhang, Y.Q.; Tang, Y.H.; Jiang, J. Dynamic characteristics of soil organic carbon in grassland ecosystems of the Qinghai-Tibet Plateau. *Sci. Sin.* **2006**, *36*, 1140–1147.
21. Althoff, T.D.; Menezes, R.S.C.; Pinto, A.D.S.; Pareyn, F.G.C.; Carvalho, A.L.D.; Martins, J.C.R.; Carvalho, E.X.D.; Silva, A.S.A.D.; Dutra, E.D.; Sampaio, E.V.D.S.B. Adaptation of the CENTURY model to simulate C and N dynamics of Caatinga dry forest before and after deforestation. *Agric. Ecosyst. Environ.* **2018**, *254*, 26–34. [CrossRef]
22. Tang, M.; Wang, S.; Zhao, M.; Falyu, Q.; Liu, X. Simulated Soil Organic Carbon Density Changes from 1980 to 2016 in Shandong Province Dry Farmlands Using the CENTURY Model. *Sustainability* **2020**, *12*, 5384. [CrossRef]
23. Available online: https://www.chinadaily.com.cn/m/gansu/2013-10/24/content_17055494.htm (accessed on 15 February 2020).
24. Kottek, M.; Grieser, J.; Beck, C.; Rudolf, B.; Rubel, F. World map of the Köppen-Geiger climate classification updated. *Meteorol. Z.* **2006**, *15*, 259–263. [CrossRef]
25. Nachtergaele, F.; van Velthuizen, H.; Verelst, L.; Batjes, N.H.; Dijkshoorn, K.; van Engelen, V.W.P.; Fischer, G.; Jones, A.; Montanarella, L.; Petri, M. The harmonized world soil database. In Proceedings of the 19th World Congress of Soil Science, Soil Solutions for a Changing World, Brisbane, Australia, 1–6 August 2010; pp. 34–37.
26. Driessen, P.; Deckers, J.; Spaargaren, O.; Nachtergaele, F. *Lecture Notes on the Major soils of the World*; Food and Agriculture Organization (FAO): Rome, Italy, 2000.
27. Harmonized World Soil Database Version 1.2 (HWSD V1.2). Available online: http://www.fao.org/soils-portal/data-hub/soil-maps-and-databases/harmonized-world-soil-database-v12/en/ (accessed on 6 January 2020).
28. Available online: http://www.ecosystem.csdb.cn/ (accessed on 1 August 2020).
29. Available online: www.worldclim.org (accessed on 20 March 2020).
30. Pattern, N.C.S. Climatic Research Unit, University of East Anglia. Available online: https://crudata.uea.ac.uk/cru/data/ncp (accessed on 10 August 2020).
31. Available online: http://koeppen-geiger.vu-wien.ac.at/present.htm (accessed on 15 February 2020).
32. Rusong, W. The frontiers of urban ecological research in industrial transformation. *Acta Ecol. Sin.* **2000**, *20*, 830–840.
33. Shi, X.Z.; Yu, D.S.; Warner, E.D.; Pan, X.Z.; Petersen, G.W.; Gong, Z.G.; Weindorf, D.C. Soil database of 1:1,000,000 digital soil survey and reference system of the Chinese genetic soil classification system. *Soil Surv. Horiz.* **2004**, *45*, 129–136. [CrossRef]
34. Kodešová, R.; Alias, L.J. Available online: http://www.koedoe.co.za (accessed on 10 August 2020).
35. Spawn, S.A.; Gibbs, H.K. Global Aboveground and Belowground Biomass Carbon Density Maps for the Year 2010. *ORNL DAAC* **2020**. [CrossRef]
36. Parton, W.J. The CENTURY Model. In *Evaluation of Soil Organic Matter Models*; Springer: Berlin/Heidelberg, Germany, 1996; pp. 283–291.
37. Available online: https://www.nrel.colostate.edu/projects/century/ (accessed on 10 June 2020).
38. Bortolon, E.S.O.; Mielniczuk, J.; Tornquist, C.G.; Lopes, F.; Bergamaschi, H. Validation of the Century model to estimate the impact of agriculture on soil organic carbon in Southern Brazil. *Geoderma* **2011**, *167*, 156–166. [CrossRef]
39. Bockstaller, C.; Girardin, P. How to validate environmental indicators. *Agric. Syst.* **2003**, *76*, 639–653. [CrossRef]
40. Gomes, A.G.; Varriale, M.C. *Modelagem de Ecossistemas, Uma Introducao*; UFSM: Santa Maria, Brazil, 2004.
41. Xu, L.; Yu, G.; He, N.; Wang, Q.; Gao, Y.; Wen, D.; Li, S.; Niu, S.; Ge, J. Carbon storage in China's terrestrial ecosystems: A synthesis. *Sci. Rep.* **2018**, *8*, 1–13. [CrossRef]
42. Guo, L.H.; Gao, J.B.; Wu, S.H.; Hao, C.Y.; Zhao, D.S. Spatial-temporal change of soil organic carbon and its susceptibility to climate change in Inner Mongolia Grassland 1981–2010. *Res. Environ. Sci.* **2016**, *29*, 1050–1058.
43. Catling, D. The Soils. In *Rice in Deep Water*; Palgrave Macmillan: London, UK, 1992. [CrossRef]
44. Neina, D. The Role of Soil pH in Plant Nutrition and Soil Remediation. *Appl. Environ. Soil Sci.* **2019**, *2019*, 1–9. [CrossRef]
45. Marinos, R.E.; Bernhardt, E.S. Soil carbon losses due to higher pH offset vegetation gains due to calcium enrichment in an acid mitigation experiment. *Ecology* **2018**, *99*, 2363–2373. [CrossRef]
46. McCauley, A.; Jones, C.; Jacobsen, J. Soil pH and organic matter. *Nutr. Manag. Module* **2009**, *8*, 1–12.
47. Available online: https://www.esf.edu/pubprog/brochure/soilph/soilph.htm (accessed on 10 March 2021).
48. Zhou, W.; Han, G.; Liu, M.; Li, X. Effects of soil pH and texture on soil carbon and nitrogen in soil profiles under different land uses in Mun River Basin, Northeast Thailand. *Peer J.* **2019**, *7*, e7880. [CrossRef]
49. Céspedes-Payret, C.; Bazzoni, B.; Gutiérrez, O.; Panario, D. Soil organic carbon vs. bulk density following temperate grassland afforestation. *Environ. Process.* **2017**, *4*, 75–92. [CrossRef]
50. National Research Council. *Grasslands and Grassland Sciences in Northern China*; National Academies Press: Washington, DC, USA, 1992.

51. Overview of Land Desertification Issues and Activities in the People's Republic of China. Available online: http://www.fao.org/3/w7539e/w7539e03.htm (accessed on 12 March 2020).
52. Xu, W.; Chen, X.; Luo, G.; Lin, Q. Using the CENTURY model to assess the impact of land reclamation and management practices in oasis agriculture on the dynamics of soil organic carbon in the arid region of North-western China. *Ecol. Complex.* **2011**, *8*, 30–37. [CrossRef]
53. Breuer, L.; Huisman, J.; Keller, T.; Frede, H.-G. Impact of a conversion from cropland to grassland on C and N storage and related soil properties: Analysis of a 60-year chronosequence. *Geoderma* **2006**, *133*, 6–18. [CrossRef]
54. Kasel, S.; Bennett, L. Land-use history, forest conversion, and soil organic carbon in pine plantations and native forests of south eastern Australia. *Geoderma* **2007**, *137*, 401–413. [CrossRef]
55. Elberling, B.; Touré, A.; Rasmussen, K. Changes in soil organic matter following groundnut–millet cropping at three locations in semi-arid Senegal, West Africa. *Agric. Ecosyst. Environ.* **2003**, *96*, 37–47. [CrossRef]
56. Ogle, S.M.; Conant, R.T.; Paustian, K. Deriving grassland management factors for a carbon accounting method developed by the intergovernmental panel on climate change. *Environ. Manag.* **2004**, *33*, 474–484. [CrossRef] [PubMed]
57. Su, Y.Z.; Zhao, H.l. Soil properties and plant species in an age sequence of Caragana microphylla plantations in the Horqin Sandy Land, north China. *Ecol. Eng.* **2003**, *20*, 223–235. [CrossRef]
58. Jaiyeoba, I.A. Changes in soil properties due to continuous cultivation in Nigerian semiarid Savannah. *Soil Tillage Res.* **2003**, *70*, 91–98. [CrossRef]
59. Yu, S.S.; Dou, S.; Yang, J.M. Application of central model in soil organic carbon research. *Soils Crop.* **2014**, *3*, 10–14.
60. Wang, S.Q. Analysis on spatial distribution characteristics of soil organic carbon reservoir in China. *Acta Geogr. Sin.* **2000**, *55*, 533–544.
61. Oelbermann, M.; Voroney, R.P. An evaluation of the century model to predict soil organic carbon: Examples from Costa Rica and Canada. *Agrofor. Syst.* **2011**, *82*, 37–50. [CrossRef]
62. Slaton, N.A.; Norman, R.J.; Gilmour, J.T. Oxidation rates of commercial elemental sulfur products applied to an alkaline silt loam from Arkansas. *Soil Sci. Soc. Am. J.* **2001**, *65*, 239–243. [CrossRef]
63. Jones, C.A.; Jacobsen, J.; Lorbeer, S. Metal concentrations in three Montana soils following 20 years of fertilization and cropping. *Commun. Soil Sci. Plant Anal.* **2002**, *33*, 1401–1414. [CrossRef]

Article

Temperature Variations and Possible Forcing Mechanisms over the Past 300 Years Recorded at Lake Chaonaqiu in the Western Loess Plateau

Keke Yu [1,2], Le Wang [1], Lipeng Liu [1], Enguo Sheng [3], Xingxing Liu [2,4] and Jianghu Lan [2,4,5,*]

1   College of Geography and Environment, Baoji University of Arts and Sciences, Baoji 721013, China; yukezan2005@163.com (K.Y.); lengfeng1995@126.com (L.W.); bwlllp@163.com (L.L.)
2   State Key Laboratory of Loess and Quaternary Geology, Institute of Earth Environment, Chinese Academy of Sciences, Xi'an 710061, China; liuxx@ieecas.cn
3   Department of Resource and Environment, Zunyi Normal College, Zunyi 563002, China; shengeg1985@126.com
4   Center for Excellence in Quaternary Science and Global Change, Chinese Academy of Sciences, Xi'an 710061, China
5   China-Pakistan Joint Research Center on Earth Sciences, CAS-HEC, Islamabad 45320, Pakistan
*   Correspondence: lanjh@ieecas.cn or lanjh115@163.com; Tel./Fax: +86-29-6233-6295

Abstract: Understanding the synchronicity of and discrepancy among temperature variations on the western Loess Plateau (WLP), China, is critical for establishing the drivers of regional temperature variability. Here we present an authigenic carbonate-content timeseries spanning the last 300 years from sediments collected from Lake Chaonaqiu in the Liupan Mountains, WLP, as a decadal-scale record of temperature. Our results reveal six periods of relatively low temperature, during the intervals AD 1743–1750, 1770–1780, 1792–1803, 1834–1898, 1930–1946, and 1970–1995, and three periods of relatively high temperature during 1813–1822, 1910–1928, and since 2000. These findings are consistent with tree-ring datasets from the WLP and correlate well with extreme cold and warm events documented in historical literature. Our temperature reconstruction is also potentially representative of large-scale climate patterns over northern China and more broadly over the Northern Hemisphere. The Pacific Decadal Oscillation (PDO) might be the dominant factor affecting temperature variations over the WLP on decadal timescales.

Keywords: authigenic carbonate; temperature variations; Lake Chaonaqiu; western loess plateau

Citation: Yu, K.; Wang, L.; Liu, L.; Sheng, E.; Liu, X.; Lan, J. Temperature Variations and Possible Forcing Mechanisms over the Past 300 Years Recorded at Lake Chaonaqiu in the Western Loess Plateau. *Sustainability* 2021, 13, 11376. https://doi.org/10.3390/su132011376

Academic Editor: Baojie He

Received: 15 September 2021
Accepted: 11 October 2021
Published: 14 October 2021

Publisher's Note: MDPI stays neutral with regard to jurisdictional claims in published maps and institutional affiliations.

1. Introduction

Anthropogenic global warming is a critical issue of broad scientific and socio-economic concern [1]. Since the mid-20th century, severe heatwaves of increasing duration and intensity have impacted many regions of the globe [2–5]. For example, the 2010 heatwave in western Russia caused a widespread decline in ecosystem productivity, while concurrently increasing respiration [6]. It is therefore important to investigate temperature variations in different regions in order to gain further understanding of 20th century warming in the context of the previous several hundred years or the previous millennium. Numerous studies focusing on the last millennium has greatly improved our understanding of climate change and the relative roles of natural and anthropogenic forcings [7–17]. It is well known that temperature varies on different timescales and the corresponding forcings are also variable [18]. This would reasonably result in variable temperature patterns in different regions on different timescales. Because most large-scale temperature curves are generated from data averaged over broad geographic areas, differences in regional temperature variations could possibly be masked. This may limit our understanding of regional temperature variations and weaken the reliability of regional climatic predictions. Therefore, it is crucial to master the details in temperature variations for different regions and shed light on the underlying dynamics.

65

Located at the juncture of the East Asia monsoon (EAM) and northwestern arid zone, the ecological frangibility and environmental sensitivity of the western Loess Plateau (WLP) make this an ideal region for studying global climatic changes. Mastering the similarities and differences in temperature variations along the WLP transect is thus vital to understand the mechanisms of temperature and precipitation variations. Despite the dearth of long-term meteorological data, scientists have made great efforts to reconstruct the paleoclimates by tree-rings throughout the WLP [19–22]. Yet, although common features are evident among these previous reconstructions, there are notable discrepancies in both the timing and magnitude of reconstructed events that remain unresolved. For example, Liu et al. [19] has reconstructed temperatures over the past 100 years at Huangling based on δ^{13}C in tree rings. Using the tree ring width data at Kongtong Mt., Song et al. [22] has further reconstructed temperature variations over the past 283 years. However, the possible forcing mechanisms have not been comprehensively discussed [19–21]. To help address these inconsistencies, we extracted temperature proxy indices from sediments collected from Lake Chaonaqiu in the Liupan Mountains, WLP, and compared these data with existing records of decadal temperature variability along a transect across the WLP. We focused specifically on the phase relationship of decadal climatic variations over the past three centuries and explored possible forcing mechanisms for these changes.

2. Materials and Methods

2.1. Background and Sampling

Lake Chaonaqiu (2430 m elevation; also known as Lake Tianchi) is a small alpine barrier freshwater lake in the Liupan Mountains, WLP, located ~30 km northeast of Zhuanglang County (Figure 1A). With an area of 0.02 km^2 and maximum water depth of 9 m, the lake is fed primarily by rainfall and drains seasonally via a topographic low on its western shore. The underlying bedrock throughout the lake basin is red sandstone. The salinity and pH of lake water are 0.17 g/L and 7.83, respectively [23], and total phosphorus (TP) and nitrogen (TN) concentrations are 40.2 µg/L and 1096.8 µg/L [24], respectively. Mean annual precipitation at Lake Chaonaqiu is 615 mm, with a mean annual air temperature of 3.4 °C [25].

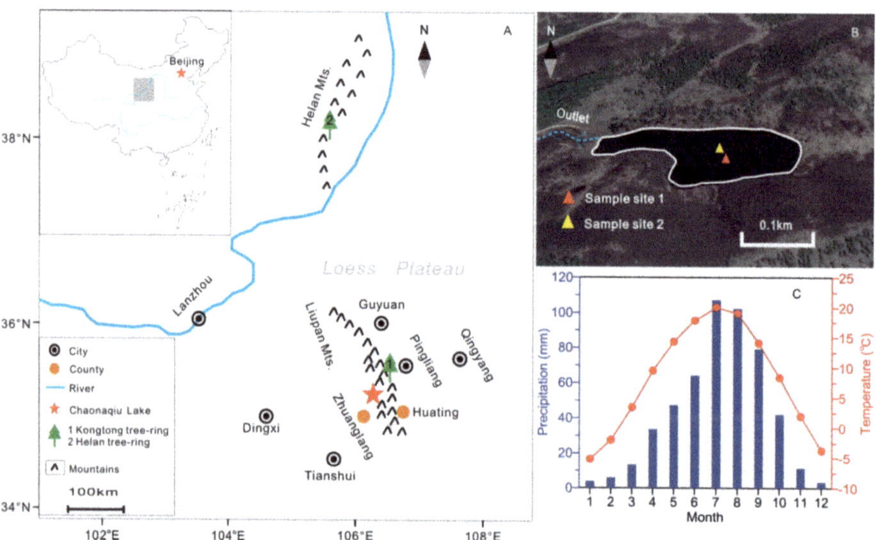

Figure 1. Overview of the study site. (**A**) Location of the study area on the western Chinese Loess Plateau, and other sites mentioned in the text. (**B**) Aerial view of Lake Chaonaqiu and sample site locations. (**C**) Monthly mean precipitation (blue bars) and monthly mean temperature (red dotted line) recorded at Zhuanglang meteorological station since 1960.

We used a UWITEC gravity corer to collect four surface sediment cores at two sample sites located in the center of the lake (35°15′53.08″ N, 106°18′35.99″ E) during September 2012. The sediment profiles were undisturbed, and the sediment–water interface remained clear. Cores CNQ12-1 and CNQ12-4 were extracted from Sites 1 and 2, respectively (Figure 1B). Core CNQ12-1 (~73 cm long) was subsampled in the field at one-centimeter intervals to quantify the core's mass depth. Core CNQ12-4 was also subsampled at centimeter intervals, but in the laboratory rather than in the field. Owing to minor compaction during transportation and storage, and material loss during subsampling, we were unable to establish an accurate mass depth, and therefore age model, for core CNQ12-4.

2.2. Methods

For both cores CNQ12-1 and CNQ12-4, we measured ^{137}Cs and ^{210}Pb$_{ex}$ radioactivity via high-resolution, multi-channel gamma-ray spectrometry using an Ortec Hyperpure Germanium (HPGe) well detector (GWL-250-15), with an experimental error of <10% and detection limit of 0.1 Bg kg^{-1} (at 99% confidence; [26]). The bulk carbonate content (carb%) of core CNQ12-1 was determined by titration with diluted perchloric acid (HClO$_4$; 0.1 mol L^{-1}), with an analytical precision better than 0.5% [27,28]; we also measured the bulk carbonate content of core CNQ12-4 to crosscheck our results from the first core. We selected a total of twenty-six representative sediment samples and one surface soil sample for X-ray diffraction (XRD) analysis. Samples were ground to a grain size of <74 µm (<200 mesh) before measurements, after which XRD patterns were obtained using a PANalytical X'Pert Pro MPD diffractometer with CuKα radiation, and a Ni filter set at 40 kV and 40 mA intensity. Diffraction patterns were scanned from 3° to 70° 2θ, using a step size of 0.02° [29]. Finally, elemental compositions of odd-numbered samples from core CNQ12-1 were determined using an X-ray fluorescence spectrometer (XRF, Axios advanced (PW4400); [30]). All measurements were performed at the Institute of Earth Environment, Chinese Academy of Sciences (IEECAS), Xi'an, China.

3. Results

3.1. Chronology

Given that anthropogenic radionuclide ^{137}Cs is deposited from the atmosphere within a year, the point of maximum fallout as recorded in our cores provides a time marker for the year 1964 (all dates reported hereon are given in years AD; [17,26,27]). The ^{137}Cs curves for cores CNQ12-1 and CNQ12-4 both exhibit unimodal distributions with pronounced ^{137}Cs peaks (Figure 2a,b; [31]), a pattern that is similar to the classic pattern of global atmospheric ^{137}Cs fallout [26,32]. This close alignment confirms the reliability of the 1964-time marker in core CNQ12-1, where it occurs at a mass depth of 4.09 g·cm^{-2}, or average geometric depth of 17.5 cm (Figure 2a; [31]).

The ^{210}Pb$_{ex}$ curves for cores CNQ12-1 and CNQ12-4 exhibit clear subordinate fluctuations superimposed upon long-term logarithmic trends (Figure 2a,b; [31]). Such subordinate fluctuations might imply that biological activity has exerted a washing effect on ^{210}Pb$_{ex}$ concentrations, similar to results obtained from Lake Chenghai in Yunnan Province [33]. Considering the influence of biological washing on the accuracy of the ^{210}Pb$_{ex}$ age model, we chose not to employ ^{210}Pb$_{ex}$ radioactivity as a basis for generating core chronologies [31].

The chronology for core CNQ12-1 is well established, based on a constant mass accumulation rate of 0.0852 g·cm^{-2}·yr^{-1}, and spans the period 1743–2012 (268 years; Figure 2c; [31]). Figure 2 also depicts the ^{137}Cs–^{210}Pb (CRS model) ages of Chen et al. [24]; the age-control point at 39.75 cm was removed owing to the anomalously large dating error. As shown in Figure 2, the age-control points of Chen et al. [24] correlate well with our dating model. In addition, previous work at Lake Chaonaqiu has shown that the calibrated ^{14}C age of 620 cal yr BP in core GSA07 occurs at 162 cm depth [34–36], in agreement with our ^{137}Cs age model. Together, the close alignment of these multiple age constraints confirms the reliability of our chronology.

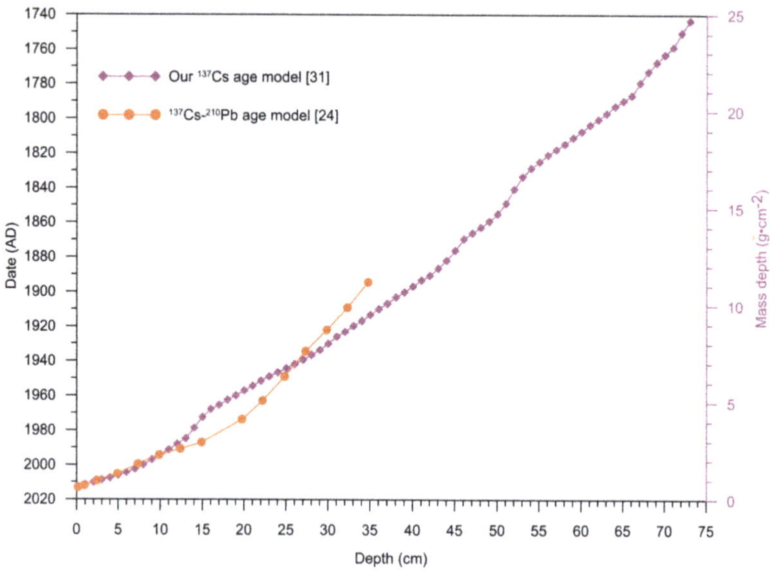

Figure 2. Chronologies for cores CNQ12-1 [31] and GS14A [24] from Lake Chaonaqiu.

3.2. Proxy Indices

As shown in Figure 3, carbonate contents of cores CNQ12-1 and CNQ12-4 range from 0.42% to 6.90%, with average values of 2.88% and 3.38%, respectively. We note the synchronicity of carbonate content between the two cores (Figure 3), which highlights the reliability of the CNQ12-1 carbonate curve. The relatively low carbonate content of the Lake Chaonaqiu sediments likely reflects the influence of two geochemical factors. First, given that precipitation of chemically deposited carbonate is directly affected by salinity [27,37], the relatively low salinity of Lake Chaonaqiu (0.17 g/L; [23]) is not conducive to carbonate precipitation, and thus carbonate sedimentation is minimal. In contrast, Lake Qinghai in central China has a high salinity (14.53 g/L; [38]) due to extensive evaporation, resulting in effective deposition of Ca^{2+} with $HCO_3{}^-$ and $CO_3{}^{2-}$ and a correspondingly high (>22%) carbonate content of lake sediments [37]. Second, the bedrock underlying the Lake Chaonaqiu catchment is dominated by red sandstone, which potentially restricts the input of Ca^{2+} and thus limits carbonate precipitation. Where catchments are underlain by limestone, such as Lake Sayram in northwest China [27] and Lake Lugu in southwest China [12,39], runoff supplies Ca^{2+} that is readily deposited with $HCO_3{}^-$ and $CO_3{}^{2-}$, resulting in high (40.4% and 24.62%, respectively) overall sediment-carbonate contents.

Our XRD results indicate that the mineralogic composition of core CNQ12-1 is dominated by quartz, albite, biotite, and calcite (Figure 4). Among the 26 sediment samples tested, the calcite signal exhibits an average strength of 1765 counts, with maximum (2451 counts) and minimum (1522 counts) values occurring at line depths of 1 and 73 cm, respectively (Figure 5). With the notable exception of calcite, the high degree of similarity in primary peaks and mineralogic compositions between lake sediments and surface soils suggests that surface runoff of terrestrial clastic material is the principal source of sediment in Lake Chaonaqiu. By this scenario, the calcite content of the lake sediments reflects subsequent chemical deposition.

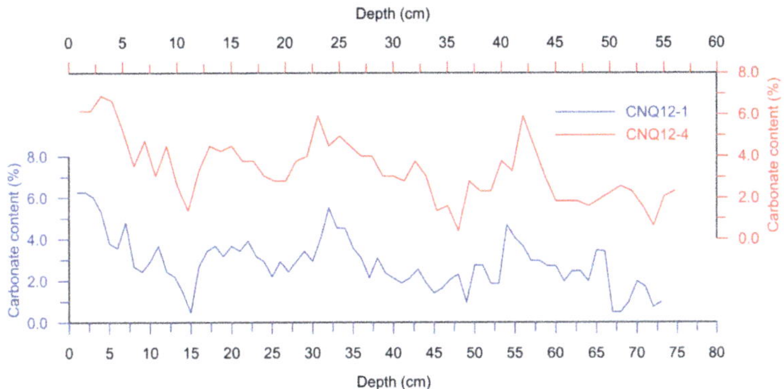

Figure 3. Carbonate contents of the two Lake Chaonaqiu sediment cores.

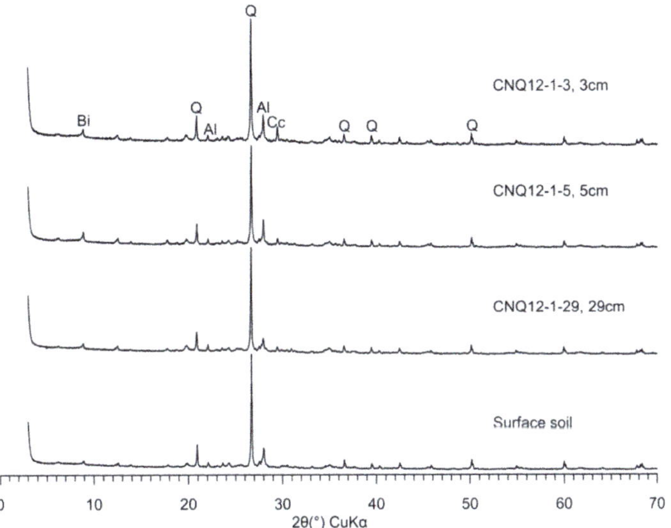

Figure 4. Results of XRD analysis for core CNQ12-1 and surface soils. Q—Quartz; Bi—Biotite; AL—Albite; Cc—Calcite.

The XRF chemical element results reveal an average Ca concentration of 2.44%, with maximum (5.29%) and minimum (1.04%) values occurring at line depths of 1 and 67 cm, respectively (Figure 5). Considering the similarities between Ca concentrations and carbonate content, their shared synchrony with the calcite signal strength (Figure 5), and the absence of calcite from surface soils (Figure 4), we propose that the bulk of carbonate in the Lake Chaonaqiu sediments is authigenic and that the Ca element is derived primarily from calcite.

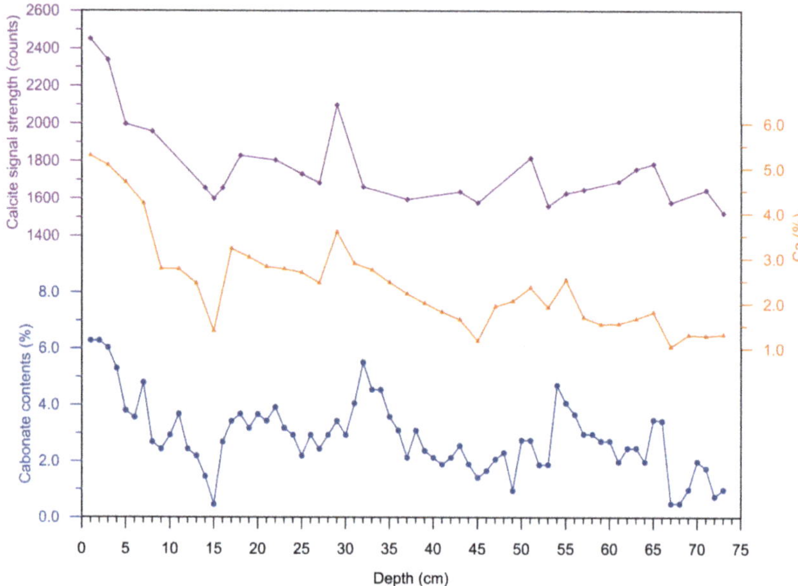

Figure 5. Change curves for carbonate content, elemental Ca concentration, and calcite signal strength for core CNQ12-1.

4. Discussion

4.1. Climatic Significance of the Authigenic Carbonate

The rate of carbonate precipitation in a lake is generally controlled by the ratio of evaporation to precipitation (E/P), with higher E/P resulting in carbonate supersaturation in the water column, elevated carbonate precipitation rates, and a greater overall carbonate content in lake sediments [27,37,40–44]. Moreover, the content of authigenic carbonate in lake sediments can reveal the dominant role of evaporation or precipitation in controlling E/P values. For example, while Lan et al. [27] reported that authigenic carbonate precipitation in Lake Sayram results from extensive summer evaporation, the authors also observed that evaporation should weaken as precipitation increases, resulting in unsaturation of lake water and a decline in carbonate sedimentation. Therefore, carbonate contents in Lake Sayram sediments can be used as an indicator of regional precipitation [27]. Although similar carbonate-derived paleoenvironmental interpretations have also been presented for Lake Bosten [40,44], Lake Dali [42], and Lake Sasikul [45], other studies have explored the role of temperature in controlling E/P values. At Lake Qinghai, for instance, since temperature is the dominant factor that controls the evaporation, and the influence of temperature on the salinity of lake water (primarily influenced by regional E/P) may be stronger than that of precipitation. Therefore, carbonate content in Lake Qinghai sediments can be used as an indicator of regional temperatures [37,43]. Similar inferences have been made for Lake Daihai [41].

Located on the margin of the Asian summer monsoon (ASM) region, our site in the Liupan Mountains experiences diurnal and annual temperature variability; mean annual evaporation (~1102.80 mm; [46]) is 1.8 times the mean annual precipitation (~615 mm; [25]). Because the surface of Lake Chaonaqiu is typically frozen between November and March [46], chemical carbonate precipitation results from extensive summer evaporation. Therefore, authigenic carbonates formed in Lake Chaonaqiu by chemical and biochemical processes are closely associated with the evaporation of the lake water, which primarily controls the chemical composition and salinity of the lake water. Given that the evaporation of lake water is mainly controlled by temperature and atmospheric relative humidity, and humidity is determined by temperature and atmospheric precipitation, we infer that the formation of authigenic carbonates in Lake Chaonaqiu can be linked predominantly with temperature. In addition, upon comparing the carbonate content with mean annual temperature data from Pingliang and Zhuanglang meteorological stations (Figure 6), we observed a positive correlation between them (Figure 6A,B), reinforcing our view that variations in authigenic carbonate content in Lake Chaonaqiu sediments can be employed as an indicator of regional temperature. When temperature rises, evaporation is enhanced, leading to an increase in Ca^{2+} and HCO_3^- concentrations of lake water and promoting carbonate supersaturation, and thus resulting in higher carbonate contents in Lake Chaonaqiu sediments. Conversely, unsaturation causes the carbonate content of lake sediments to decline when temperature drops. Therefore, we interpret the carbonate content in Lake Chaonaqiu sediments as a proxy index for temperature changes in this region, with increased carbonate content is related to higher temperature, and vice versa.

4.2. Temperature Variations at Lake Chaonaqiu over the Past 300 Years

As shown in Figure 6, the carbonate content of lake sediments is relatively low for the periods 1743–1750, 1770–1780, 1792–1803, 1834–1898, 1930–1946, and 1970–1995, which we interpret as reflecting cooler temperatures at Lake Chaonaqiu. It is noteworthy that low values in the period of 1834–1898 coincided with the final cold stage (1830–1890) of the Little Ice Age (LIA) in China [9,47,48]. According to Wen [49], several extremely cold events in Zhuanglang County have been described in historical documents (No. 1 and 18 in Table 1). For example, "On 1 October, the 7th year of the region of Emperor Tongzhi, Qing Dynasty (the traditional Chinese calendar, equivalent to 14 November 1868), Zhuanglang County was seriously impacted by a snowstorm, which buried roads and crushed vegetation."

The carbonate record also exhibits elevated values in 1813–1822, 1910–1928, and since 2000, which reflect periods of relatively higher temperatures in the Lake Chaonaqiu region. According to Wen [49], extremely warm events in Huating County (~40 km southeast of Lake Chaonaqiu) were also reported in historical accounts (No. 19 in Table 1): "In Autumn, the 3rd year of the reign of Emperor Xuantong, Qing Dynasty (1911), vegetation bloomed again in Huating".

4.3. Temperature Variations on the Western Loess Plateau over the Past 300 Years

The pattern of temperature variability at Lake Chaonaqiu over the last few centuries is similar to those reconstructed for other regions in China [7,50]. For example, the comparison of our carbonate dataset to tree-ring records from Kongtong Mountain [22], Helan Mountain [20], and the mid-eastern Tibetan Plateau [51] reveals a considerable degree of convergence among the various datasets over the past 300 years (Figure 6). Specifically, variations in carbonate content at Lake Chaonaqiu are broadly synchronous with fluctuations in tree-ring width, confirming that our record is a robust indicator of regional temperature. Due to dating uncertainties, sampling resolution, and site characteristics, the cold periods 1743–1750, 1770–1780, and 1970–1995 are not represented in the tree-ring records. Nonetheless, three cold periods (1792–1803, 1834–1898, and 1930–1946) and two warm intervals (1813–1822 and 1910–1928) are clearly documented in tree-ring and lake carbonate records alike (Figure 6). Although tree-ring-inferred temperatures exhibit subordinate fluctuations during the 1834–1898 cold episode, most likely owing to site-specific factors, the majority of regional cold extremes occurred during the cold intervals (Figure 6). This pattern suggests that the thermal signature of the LIA in China [9,47,48] was prevalent throughout the WLP, and indicates that this regional variability is captured in the Lake Chaonaqiu sedimentary record on a decadal scale.

Table 1. Extreme cold and warm events on the WLP identified in historical literature [49,52].

No.	Solar Calendar Dates	Description
1	31 July 1744	The 9th year of the reign of Emperor Qianlong, Qing Dynasty: On 25 July, uncovered oats in Haiyuan, Guyuan, and Huating Counties were damaged by frost.
2	24 April1748	The 13th year of the reign of Emperor Qianlong, Qing Dynasty: On the evening of 1 March, seedlings were killed by frost in Gangu, while on the same night, heavy snow fell on the suburbs in Guyuan.
3	1749	The 14th year of the reign of Emperor Qianlong, Qing Dynasty: Longde, Guyuan, and other counties were affected by summertime and autumn frosts.
4	December 1773	The 38th year of the reign of Emperor Qianlong, Qing Dynasty: In November, relief aid was provided to refugees of frost and famine in Jingchuan.
5	1776	The 41st year of the reign of Emperor Qianlong, Qing Dynasty: Chongxin, Jingchuan, and Lingtai Counties were affected by frost.
6	4 June 1777	The 42nd year of the reign of Emperor Qianlong, Qing Dynasty: On 2 May, Dingxi experienced frost.
7	December 1783	The 48th year of the reign of Emperor Qianlong, Qing Dynasty: In November, wildflowers were in full bloom in Zhenyuan. ★
8	November 1817	The 22nd year of the reign of Emperor Jiaqing, Qing Dynasty: In November, peach trees were blooming in Zhenyuan. ★
9	December 1818	The 23rd year of the reign of Emperor Jiaqing, Qing Dynasty: In November, peaches and winter jasmine were blooming in Zhenyuan. ★
10	1837	The 18th year of the reign of Emperor Daoguang, Qing Dynasty: On 12 January (equivalent to 5 March 1838), grain rations and seeds were provided to frost refugees in Guyuan and Longde.
11	19 December 1840	The 20th year of the reign of Emperor Daoguang, Qing Dynasty: On 26 November, the old and new taxes were postponed for frost refugees in Longde.
12	December 1841	The 21st year of the reign of Emperor Daoguang, Qing Dynasty: In November, the old taxes were postponed due to frost in Guyuan and Lingtai.
13	14 November 1868	The 7th year of the reign of Emperor Tongzhi, Qing Dynasty: On 1 October, Zhuanglang County was seriously impacted by a snowstorm, which buried roads and crushed vegetation.
14	April 1871	The 10th year of the reign of Emperor Tongzhi, Qing Dynasty: In April, Lingtai County suffered heavy frost, resulting in serious damage to seedlings.
15	1873	The 12th year of the reign of Emperor Tongzhi, Qing Dynasty: During summer, Lingtai County suffered heavy frost, resulting in the loss of seedlings and crops.
16	30 May 1884	The 10th year of the reign of Emperor Guangxu, Qing Dynasty: On 8 April heavy sleet fell in Jingchuan.
17	May,1890	The 16th year of the reign of Emperor Guangxu, Qing Dynasty: In April, crops in Jingyuan were killed by frost.
18	September 1902	The 28th year of the reign of Emperor Guangxu, Qing Dynasty: In August, seedlings in Guangleli (an ancient placename), Zhuanglang County, were damaged by heavy frost.
19	1911	The 3rd year of the reign of Emperor Xuantong, Qing Dynasty: In Autumn, vegetation bloomed again in Huating. ★
20	1915	The 4th year of the Republic of China: In Autumn, vegetation bloomed again in Huating. ★

Table 1. *Cont.*

No.	Solar Calendar Dates	Description
21	1930	The 19th year of the Republic of China: Longde was affected by black frost, resulting in famine refugees.
22	1932	The 21st year of the Republic of China: Piangliang County was affected by frost during the first half of the year. Black frost during the autumn damaged seedlings in Lingtai, impacting an area spanning >100 miles from north to south.
23	2 May 1933	The 22nd year of the Republic of China: On 8 April, Jingyuan County was impacted by a blizzard and intense cold, with >5 feet (equivalent to 155 cm) of snow falling in the Longshan Mountains.
24	1938	The 27th year of the Republic of China: It took quite a long time for the heavy snow in Huating to melt.
25	1940	The 29th year of the Republic of China: Black frost killed seedlings in Lingtai, Huanting, Chongxin, Pingliang, Zhenyuan, Zhuanglang, and Jingning Counties.
26	1941	The 30th year of the Republic of China: Black frost killed seedlings in Pingliang, Huanting, Chongxin, Jingning, and Longde Counties.
27	1942	The 31st year of the Republic of China: Black frost killed crops in Huanting, Zhuanglang, Pingliang, and Jingning Counties.
28	1943	The 32nd year of the Republic of China: Frost caused extensive damage in Baishui (and thirteen other towns in Pingliang County), Jingfchuan and Shuiluo (and eleven other towns in Zhuanglang County), and Jingning.
29	1945	The 34th year of the Republic of China: Pingliang suffered frost. On 4 October, early frost in Zhenyuan resulted in extensive ice cover (frost of 0.3 cm depth) and killed >80% of the buckwheat crop. In May, black frost occurred in Chengguan and five other towns in Huating County, damaging the majority of the sprouting wheat crop.
30	1947	The 36th year of the Republic of China: On 9 and 10 October heavy frost occurred in Longpan town in Pingliang County, killing the foxtail millet and buckwheat crops. Black frost occurred in Ankou and Longyan towns, Huating County, resulting in the destruction of sprouting wheat and other crops. On 15 May frost damaged crops in Shenshangu town, Chongxin County.
31	1971	On 25 September early frost occurred in Pingliang and other regions, with minimum ground temperatures of $-4\,°C$ to $-1\,°C$.
32	1972	Between 13 and 15 May late frosts occurred in Dingxi, Pingliang, and Qingyang, with minimum ground temperatures of $-5°C$ and widespread cotton, corn, and crop failures.
33	1974	Late frost ends on 9–10 May in Gansu Province, and is delayed for 15–22 days in Tianshui and Longdong. Such an occurrence is rare in recent decades.
34	1976	The average temperature for July in Pingliang dropped by $0.1–1.0\,°C$.
35	1979	
36	1987	Extreme cold events occurred in 1979, 1987, 1993, and 1995 in Ningxia Province (1949~2000).
37	1993	
38	1995	

Notes: black stars indicate extreme warm events.

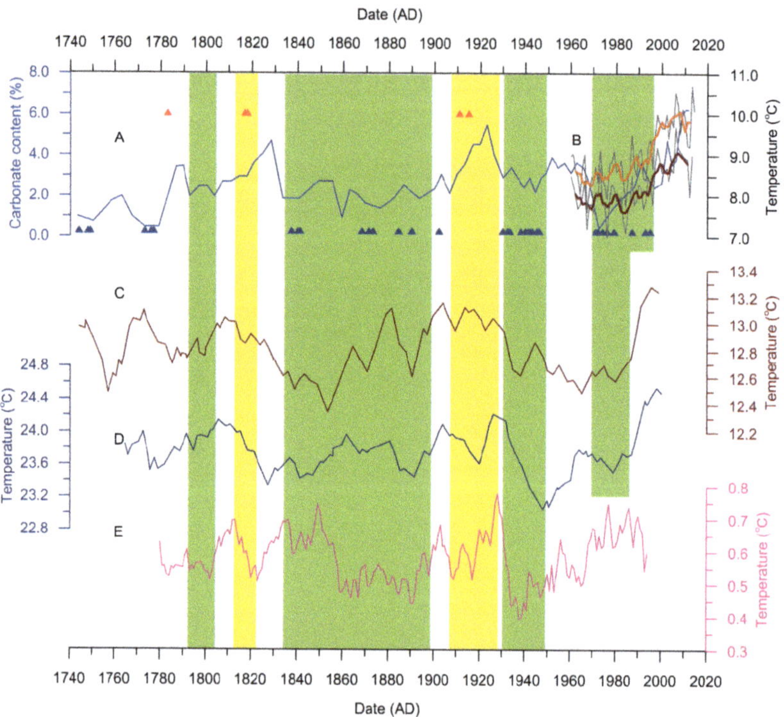

Figure 6. Comparison of the core CNQ12-1 carbonate record and tree-ring-inferred temperatures on the WLP. A. Carbonate content (*this paper*), extreme cold events (blue triangles) and warm events (red triangles) [49,52]. B. Annual mean temperature from Pingliang (orange line; 11-year running average) and Zhuanglang (dark red line; 11-year running average) meteorological stations since 1960. C. February–September air temperatures reconstructed from the Kongtong tree-ring record [22]. D. May–July air temperatures reconstructed from the Dulan–Wulan tree-ring record [51]. E. January–August air temperatures reconstructed from the Helan tree-ring record [20]. The green and yellow shadings indicate cold intervals and warm intervals, respectively.

A total of 33 extreme cold or frost events (details see Table 1) on the WLP recorded in historical literature [49,52] occurred during the six cold periods described above. For example, one of them stated (No. 23 in Table 1): "On 8 April the 22nd year of the Republic of China (the traditional Chinese calendar, equivalent to 2 May 1933), Jingyuan County was impacted by a blizzard and intense cold, with more than five feet (equivalent to 155 cm) of snow falling in the Longshan Mountains."

In addition to cold events, five episodes of extreme warmth (details see No. 7, 8, 9, 19, 20 in Table 1) on the WLP were also recorded in the historical literature [49], specifically during the intervals of 1813–1822 and 1910–1928. For example, one of them stated (No. 8 in Table 1): "In November, the 22nd year of the reign of Emperor Jiaqing, Qing Dynasty (November, 1817), peach trees were blooming in Zhenyuan". In general, peach trees bloom in March or April and lie dormant in November. However, when the temperature rises suddenly in November (a meteorological phenomenon known as *Daochunyang*), dormancy is interrupted and the peach trees can bloom again as in the spring. This out-of-season blooming of peach trees, along with the rejuvenation of vegetation (details see No. 20 in Table 1) that is typically dying off during the autumn, is indicative of a warm climate.

4.4. Possible Forcing Mechanisms of WLP Temperature Variations over the Past 300 Years

It is well known that the sun is the ultimate source of energy for Earth's climate [53–57], and myriad studies have demonstrated the close link between solar irradiance and terrestrial temperature variability [9,10,12,43]. For example, the low temperatures over the northern Tibetan Plateau (NTP) are broadly synchronized with the classical solar minima during the past several hundred years [9,10,12]. Moreover, atmospheric circulation, such as the "El Niño–Southern Oscillation" (ENSO) [8,58,59] and Pacific Decadal Oscillation (PDO) [22,60–62] might also influence regional temperature variations. For instance, ENSO-induced changes in regional hydrological cycles are expected to alter patterns of latent heating, thereby impacting temperatures on the southeastern margin of the Tibetan Plateau (S-ETP; [59]). Tollefson [61] argued that warm PDO phases coincide with periods of rapid global warming (e.g., 1920s to 1940s; 1980s and 1990s). As shown in Figure 7A, we note similar trends in low-frequency variability between the Chaonaqiu temperature record and total solar irradiance (TSI; [55]) curves, potentially indicating that TSI is a key driver of WLP temperatures on centennial timescales. This is not the case, however, on decadal/multi-decadal timescales. We speculate that the low altitude of the WLP relative to the TP makes the former less sensitive to changes in solar radiation, despite the weak influence of the ASM in this region. If so, atmospheric circulation might be the dominant factor impacting temperatures on decadal timescales on the WLP.

The PDO has been described by some as a long-lived El Niño-like pattern of North Pacific climate variability [63], which influences climate throughout the Pacific basin [60,63]. Indeed, 50 years of statistical data demonstrates a clear PDO signature both in atmospheric circulation over East Asia and decadal climate fluctuations in China, whereby the warm PDO phase coincides with elevated temperatures and reduced precipitation in northern China, and vice versa [60]. Based on monthly average precipitation and temperature data, Ma [62] also identified elevated precipitation and depressed temperatures in northern China during cold PDO phases, with the opposite being true during warm phases. To further evaluate the role of PDO on temperature changes in Lake Chaonaqiu, we compared our record with (i) temperature data from northern China [64], (ii) temperature anomalies in northwestern/northern China [65], (iii) Northern Hemisphere temperature [8], and (iv) the PDO index [66] for the past 300 years. As illustrated in Figure 7, the cold periods in the Lake Chaonaqiu record correspond to depressed temperatures in northern China and the Northern Hemisphere in general, with negative temperature anomalies in northwestern/northern China and cold PDO phases and vice versa (Figure 7). This suggests that the temperature variations inferred from Lake Chaonaqiu can represent the patterns of climate variations over the past 300 years in northern China, possibly even on a hemispheric scale. More importantly, it reveals that the PDO is closely related to decadal climate change in northern China. This result aligns with earlier studies [22,60] that suggest that a developing El Niño event during the cold PDO phase is related to more precipitation and lower temperatures in northern, northeastern, northwestern [60], and central China [22].

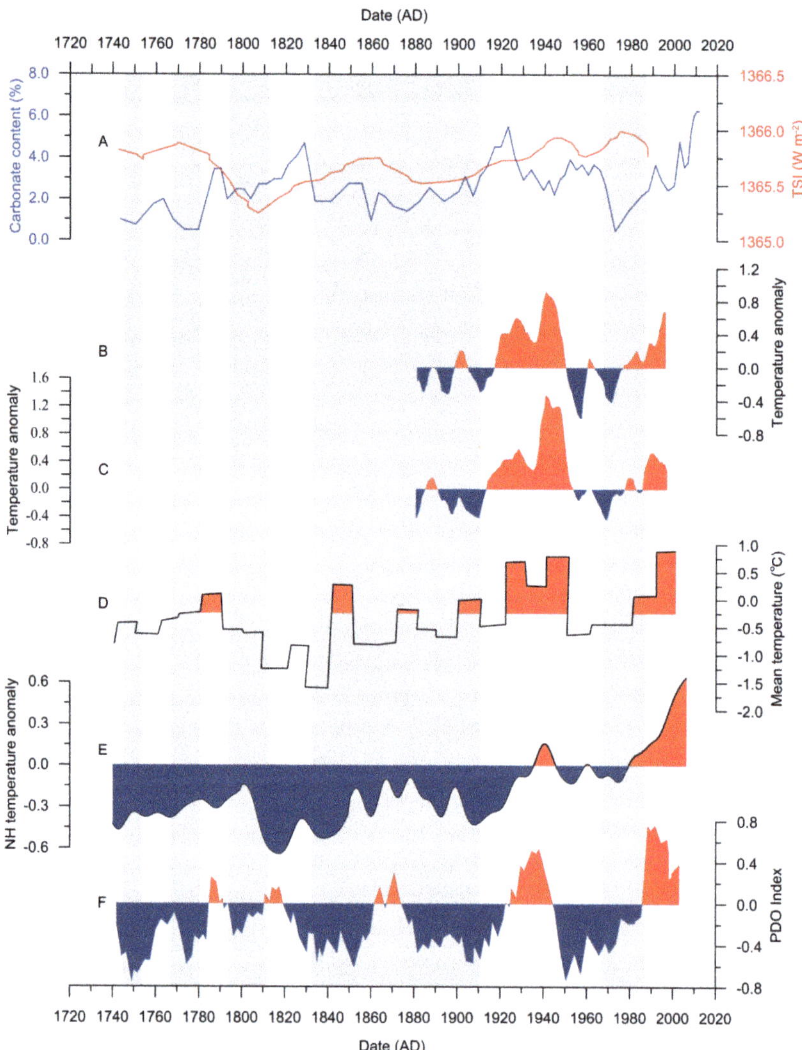

Figure 7. Temperature variations on the WLP and potential drivers. (**A**) Authigenic carbonate content in Lake Chaonaqiu sediments (blue line; this study) and total solar irradiance (TSI) reconstructed from polar ice ^{10}Be (red line; [55]). (**B**) Temperature anomalies in northern China [65]. (**C**) Temperature anomalies in northwestern China [65]. (**D**) Temperature in northern China [64]. (**E**) Northern Hemisphere (NH) temperature anomalies [8]. (**F**) PDO index [66]. Gray shadings highlight intervals of decreased temperature and cold PDO phase.

5. Conclusions

This study employed the content of authigenic carbonate in Lake Chaonaqiu sediments to reconstruct regional temperature variability on decadal timescales, and to investigate possible forcing mechanisms for such variability over the past 300 years. Our results revealed six cold and three warm periods that are consistent with other paleoclimate records from throughout the WLP and northern China, and more broadly across the Northern Hemisphere. This close agreement suggests that temperature variations recorded in the Lake Chaonaqiu region are representative not only of northern China, but also the Northern Hemisphere. We propose that the PDO is the dominant factor influencing temperature variability on the WLP on decadal timescales.

Author Contributions: Methodology, K.Y., L.W., L.L., E.S. and X.L.; Writing—original draft preparation, K.Y.; Writing—review and editing, K.Y. and J.L. All authors have read and agreed to the published version of the manuscript.

Funding: This work was funded by the Natural Science Foundation of China (41701032), the Project of State Key Laboratory of Loess and Quaternary Geology (SKLLQG1625), the Ph.D. Research Project of Baoji University of Arts and Sciences (ZK2017046), and the Youth Innovation Promotion Association of the Chinese Academy of Sciences (2021411).

Institutional Review Board Statement: Not applicable.

Informed Consent Statement: Not applicable.

Data Availability Statement: Data of this research are available in the East Asian Paleoenvironmental Science Database (http://paleodata.ieecas.cn/index_EN.aspx).

Conflicts of Interest: The authors declare no conflict of interest.

References

1. IPCC. *Special Report on Global Warming of 1.5*; Cambridge University Press: Cambridge, UK, 2018.
2. Della-Marta, P.M.; Haylock, M.R.; Luterbacher, J.; Wanner, H. Doubled length of western European summer heat waves since 1880. *J. Geophys. Res.* **2007**, *112*, D15103. [CrossRef]
3. Perkins, S.E.; Alexander, L.V.; Nairn, J.R. Increasing frequency, intensity and duration of observed global heatwaves and warm spells. *Geophys. Res. Lett.* **2012**, *39*, L20714. [CrossRef]
4. IPCC. Climate Change 2013: The Physical Science Basis. In *Contribution of Working Group I to the Fifth Assessment Report of the Intergovernmental Panel on Climate Change*; Cambridge University Press: Cambridge, UK, 2013.
5. Boeck, H.; Dreesen, F.E.; Janssens, I.A.; Nijs, I. Climatic characteristics of heat waves and their simulation in plant experiments. *Glob. Chang. Biol.* **2010**, *16*, 1992–2000. [CrossRef]
6. Bastos, A.; Gouveia, C.M.; Trigo, R.M.; Running, S.W. Analysing the spatio-temporal impacts of the 2003 and 2010 extreme heatwaves on plant productivity in Europe. *Biogeosciences* **2014**, *11*, 3421–3435. [CrossRef]
7. Wang, S.W.; Wen, X.Y.; Luo, Y.; Dong, W.J.; Yang, Z.B. Reconstruction of temperature series of China for the last 1000 years. *Chin. Sci. Bull.* **2007**, *8*, 958–964. [CrossRef]
8. Mann, M.E.; Zhang, Z.; Rutherford, S.; Bradley, R.S.; Hughes, M.K.; Shindell, D.; Ammann, C.; Faluvegi, G.; Ni, F. Global signatures and dynamical origins of the Little Ice Age and Medieval Climate Anomaly. *Science* **2009**, *326*, 1256–1260. [CrossRef] [PubMed]
9. He, Y.; Zhao, C.; Wang, Z.; Wang, H.; Song, M.; Liu, W.; Liu, Z. Late Holocene coupled moisture and temperature changes on the northern Tibetan Plateau. *Quat. Sci. Rev.* **2013**, *80*, 47–57. [CrossRef]
10. He, Y.X.; Liu, W.G.; Zhao, C.; Wang, Z.; Wang, H.Y.; Liu, Y.; Qin, X.Y.; Hu, Q.H.; An, Z.S.; Liu, Z.H. Solar influenced late Holocene temperature changes on the northern Tibetan Plateau. *Chin. Sci. Bull.* **2013**, *58*, 1053–1059. [CrossRef]
11. Li, Y.K.; Liu, G.N.; Chen, Y.X.; Li, Y.N.; Harbor, J. Timing and extent of Quaternary glaciations in the Tianger Range, eastern Tian Shan, China, investigated using 10Be surface exposure dating. *Quat. Sci. Rev.* **2014**, *98*, 7–23. [CrossRef]
12. Xu, H.; Sheng, E.; Lan, J.; Liu, B.; Yu, K.; Che, S. Decadal/multi-decadal temperature discrepancies along the eastern margin of the Tibetan Plateau. *Quat. Sci. Rev.* **2014**, *89*, 85–93. [CrossRef]
13. Aichner, B.; Feakins, S.J.; Lee, J.E.; Herzschuh, U.; Liu, X. High resolution leaf wax carbon and hydrogen isotopic record of late Holocene paleoclimate in arid Central Asia. *Clim. Past* **2015**, *11*, 619–633. [CrossRef]
14. Liu, X.; Herzschuh, U.; Wang, Y.; Kuhn, G.; Yu, Z. Glacier fluctuations of Muztagh Ata and temperature changes during the late Holocene in westernmost Tibetan Plateau, based on glaciolacustrine sediment records. *Geophys. Res. Lett.* **2015**, *41*, 6265–6273. [CrossRef]
15. Stoffel, M.; Khodri, M.; Corona, C.; Guillet, S.; Poulain, V.; Bekki, S.; Guiot, J.; Luckman, B.H.; Oppenheimer, C.; Lebas, N. Estimates of volcanic-induced cooling in the Northern Hemisphere over the past 1500 years. *Nat. Geosci.* **2015**, *8*, 784–788. [CrossRef]
16. Snyder, C.W. Evolution of global temperature over the past two million years. *Nature* **2016**, *538*, 226–228. [CrossRef] [PubMed]
17. Lan, J.; Xu, H.; Sheng, E.; Yu, K.; Wu, H.; Zhou, K.; Yan, D.; Ye, Y.; Wang, T. Climate changes reconstructed from a glacial lake in High Central Asia over the past two millennia. *Quat. Int.* **2018**, *487*, 43–53. [CrossRef]
18. Jones, P.D.; Mann, M.E. Climate over past millennia. *Rev. Geophys.* **2004**, *42*, RG2002. [CrossRef]
19. Liu, Y.; Wu, X.; Leavitt, S.W.; Hughes, M.K. Stable carbon isotope in tree rings from Huangling, China and climatic variation. *Sci. China (Ser. D)* **1996**, *39*, 152–161. [CrossRef]

20. Cai, Q.; Liu, Y. Temperature variability since 1776 inferred from tree-rings of Pinus tabulaeformis in Mt. Helan. *Acta Geogr. Sin.* **2006**, *61*, 929–936. (In Chinese with English abstract)
21. Cai, Q.F.; Liu, Y.; Song, H.M.; Sun, J.Y. Tree-ring-based reconstruction of the April to September mean temperature since 1826 AD for north-central Shaanxi Province,China. *Sci. China Ser. D Earth Sci.* **2008**, *51*, 1099–1106. [CrossRef]
22. Song, H.M.; Liu, Y.; Li, Q.; Hans, L. Tree-ring derived temperature records in the central Loess Plateau, China. *Quat. Int.* **2013**, *283*, 30–35. [CrossRef]
23. Xu, H.; Yu, K.K.; Lan, J.H.; Sheng, E.G.; Liu, B.; Ye, Y.D.; Hong, B.; Wu, H.X.; Zhou, K.E.; Yeager, K.M. Different responses of sedimentary δ15N to climatic changes and anthropogenic impacts in lakes across the Eastern margin of the Tibetan Plateau. *J. Asian Earth Sci.* **2016**, *123*, 111–118. [CrossRef]
24. Chen, J.; Liu, J.; Xie, C.; Chen, G.; Chen, J.; Zhang, Z.; Zhou, A.; Rühland, K.; Smol, J.P.; Chen, F. Biogeochemical responses to climate change and anthropogenic nitrogen deposition from a 200-year record from Tianchi Lake, Chinese Loess Plateau. *Quat. Int.* **2018**, *493*, 22–30. [CrossRef]
25. Zhou, A.F.; Sun, H.L.; Chen, F.H.; Zhao, Y.; An, C.B. High-resolution climate change in mid-late Holocene on Tianchi Lake,Liupan Mountain in the Loess Plateau in central China and its significance. *Chin. Sci. Bull.* **2010**, *55*, 2118–2121. [CrossRef]
26. Lan, J.; Wang, T.; Chawchai, S.; Cheng, P.; Xu, H. Time marker of ^{137}Cs fallout maximum in lake sediments of Northwest China. *Quat. Sci. Rev.* **2020**, *241*, 106413. [CrossRef]
27. Lan, J.H.; Xu, H.; Yu, K.K.; Sheng, E.G.; Zhou, K.E.; Wang, T.L.; Ye, Y.D.; Yan, D.N.; Wu, H.X.; Cheng, P.; et al. Late Holocene hydroclimatic variations and possible forcing mechanisms over the eastern Central Asia. *Sci. China Earth Sci.* **2019**, *62*, 112–125. [CrossRef]
28. Lan, J.; Xu, H.; Lang, Y.C.; Yu, K.K.; Zhou, P.; Kang, S.G.; Wang, X.L.; Wang, T.L.; Cheng, P.; Yan, D.N.; et al. Dramatic weakening of the East Asian summer monsoon in northern China during the transition from the Medieval Warm Period to the Little Ice Age. *Geology* **2020**, *48*, 307–312. [CrossRef]
29. Li, Y.; Song, Y.G.; Zeng, M.X.; Lin, W.W.; Orozbaev, R. Evaluating the paleoclimatic significance of clay mineral records from a late Pleistocene loess-paleosol section of the Ili Basin, Central Asia. *Quat. Res.* **2018**, *89*, 660–673. [CrossRef]
30. Xu, H.; Zhou, X.; Lan, J.; Liu, B.; Sheng, E.; Yu, K.; Cheng, P.; Wu, F.; Hong, B.; Yeager, K.M. Late Holocene Indian summer monsoon variations recorded at Lake Erhai, Southwestern China. *Quat. Res.* **2015**, *83*, 307–314. [CrossRef]
31. Yu, K.K.; Xu, H.; Lan, J.H.; Sheng, E.G.; Liu, B.; Wu, H.X.; Tan, L.C.; Yeager, K.M. Climate change and soil erosion in a small alpine lake basin on the Loess Plateau, China. *Earth Surf. Proc. Land.* **2017**, *42*, 1238–1247. [CrossRef]
32. Xu, H.; Liu, X.Y.; An, Z.S.; Hou, Z.H.; Liu, D.B. Spatial pattern of modern sedimentation rate of Qinghai Lake and a preliminary estimate of the sediment flux. *Chin. Sci. Bull.* **2010**, *55*, 621–627. [CrossRef]
33. Wan, G.J.; Chen, J.A.; Wu, F.C.; Xu, S.Q.; Bai, Z.G.; Wan, E.Y.; Wang, C.S.; Huang, R.G.; Yeager, K.M.; Santschi, P.H. Coupling between 210Pbex and organic matter in sediments of a nutrient-enriched lake: An example from Lake Chenhai, China. *Chem. Geol.* **2005**, *224*, 223–236. [CrossRef]
34. Zhao, Y.; Chen, F.; Zhou, A.; Yu, Z.; Ke, Z. Vegetation history, climate change and human activities over the last 6200 years on the Liupan Mountains in the southwestern Loess Plateau in central China. *Palaeogeogr. Palaeoclimatol. Palaeoecol.* **2010**, *293*, 197–205. [CrossRef]
35. Sun, H.; James, B.; Osamu, S.; Zhou, A. Mid- to- late Holocene hydroclimatic changes on the Chinese Loess Plateau: Evidence from n-alkanes from the sediments of Tianchi Lake. *J. Paleolimnol.* **2018**, *60*, 511–523. [CrossRef]
36. Zhang, C.; Zhou, A.F.; Zhang, H.X.; Zhang, Q.; Zhang, X.N.; Sun, H.L.; Zhao, C. Soil erosion in relation to climate change and vegetation cover over the past 2000 years as inferred from the Tianchi lake in the Chinese Loess Plateau. *J. Asian Earth Sci.* **2019**, *180*, 103850. [CrossRef]
37. Xu, H.; Ai, L.; Tan, L.; An, Z. Stable isotopes in bulk carbonates and organic matter in recent sediments of Lake Qinghai and their climatic implications. *Chem. Geol.* **2006**, *235*, 262–275. [CrossRef]
38. Lanzhou Branch of Chinese Academy of Sciences. *Evolution of Recent Environment in Qinghai Lake and Its Prediction*; Science Press: Beijing, China, 1994.
39. Sheng, E.G.; Yu, K.K.; Xu, H.; Lan, J.H.; Liu, B.; Che, S. Late Holocene Indian summer monsoon precipitation history at Lake Lugu, northwestern Yunnan Province, southwestern China. *Palaeogeogr. Palaeoclimatol. Palaeoecol.* **2015**, *438*, 24–33. [CrossRef]
40. Chen, F.H.; Huang, X.Z.; Zhang, J.W.; Holmes, J.A.; Chen, J.H. Humid little ice age in arid central Asia documented by Bosten Lake, Xinjiang. *Sci. China Ser. D Earth Sci.* **2006**, *49*, 1280–1290. [CrossRef]
41. Xiao, J.L.; Wu, J.; Si, B.; Liang, W.; Nakamura, T.; Liu, B.; Inouchi, Y. Holocene climate changes in the monsoon/arid transition reflected by carbon concentration in Daihai Lake of Inner Mongolia. *Holocene* **2006**, *16*, 551–560. [CrossRef]
42. Xiao, J.; Si, B.; Zhai, D.; Itoh, S.; Lomtatidze, Z. Hydrology of Dali Lake in central-eastern Inner Mongolia and Holocene East Asian monsoon variability. *J. Paleolimnol.* **2008**, *40*, 519–528. [CrossRef]
43. Xu, H.; Liu, X.; Hou, Z. Temperature variations at Lake Qinghai on decadal scales and the possible relation to solar activities. *J. Atmos. Sol. Terr. Phys.* **2008**, *70*, 138–144. [CrossRef]
44. Zhang, C.; Feng, Z.; Yang, Q.; Gou, X.; Sun, F. Holocene environmental variations recorded by organic-related and carbonate-related proxies of the lacustrine sediments from Bosten Lake, northwestern China. *Holocene* **2010**, *20*, 363–373. [CrossRef]
45. Lei, Y.; Tian, L.D.; Bird, B.W.; Hou, J.Z.; Ding, L.; Oimahmadov, I.; Gadoev, M. A 2540-year record of moisture variations derived from lacustrine sediment (Sasikul Lake) on the Pamir Plateau. *Holocene* **2014**, *24*, 761–770. [CrossRef]

46. Cheng, J.M.; Yu, Z.J.; Zhu, R.B.; Jin, J.W.; Jing, Z.B. *Comprehensive Scientific Investi-Gation Report of Liupan Mountains National Nature Reserve*; Science Press: Beijing, China, 2013; pp. 8–9.
47. Wang, S.W.; Gong, D.Y. Climate in China during the four special periods in Holocene. *Progr. Nat. Sci.* **2000**, *10*, 325–332. (In Chinese)
48. Yang, B.; Braeuning, A.; Johnson, K.R.; Shi, Y.F. General characteristics of temperature variation in China during the last two millennia. *Geophys. Res. Lett.* **2002**, *29*, 1324. [CrossRef]
49. Wen, K.G. Chinese meteorological disasters ceremony. In *Gansu Volume*; Dong, A.X., Ed.; Meteorology: Beijing, China, 2005; pp. 288–308.
50. Ge, Q.S.; Zhang, X.Z.; Hao, Z.X.; Zheng, J.Y. Rates of temperature change in China during the past 2000 years. *Sci. China* **2011**, *54*, 1627–1634. [CrossRef]
51. Liu, Y.; An, Z.S.; Linderholm, H.W.; Chen, D.L.; Tian, H. Annual temperatures during the last 2485 years in the Mid-Eastern Tibetan Plateau inferred from tree rings. *Sci. China Ser. D Earth Sci.* **2009**, *52*, 348–359. [CrossRef]
52. Wen, K.G. Chinese meteorological disasters ceremony. In *Ningxia Volume*; Xia, P.M., Ed.; Meteorology: Beijing, China, 2007; pp. 207–226.
53. Beer, J.; Mende, W.; Stellmacher, R. The role of the sun in climate forcing. *Quat. Sci. Rev.* **2000**, *19*, 403–415. [CrossRef]
54. Shindell, D.T. Solar forcing of regional climate change during the maunder minimum. *Science* **2001**, *294*, 2149–2152. [CrossRef] [PubMed]
55. Steinhilber, F.; Beer, J.; Frhlich, C. Total solar irradiance during the Holocene. *Geophys. Res. Lett.* **2009**, *36*, L19704. [CrossRef]
56. Gray, L.J. Solar influence on climate. *Rev. Geophys.* **2010**, *48*, 1–53. [CrossRef]
57. Yeo, K.L.; Solanki, S.K.; Krivova, N.A.; Rempel, M.; Witzke, V. The Dimmest State of the Sun. *Geophys. Res. Lett.* **2021**, *47*, e2020GL090243. [CrossRef]
58. Graham, N.E. Simulation of recent global temperature trends. *Science* **1995**, *267*, 666–671. [CrossRef]
59. Xu, H.; Hong, Y.; Bing, H.; Zhu, Y.; Yu, W. Influence of ENSO on multi-annual temperature variations at Hongyuan, NE Qinghai-Tibet plateau. *Int. J. Climatol.* **2010**, *30*, 120–126. [CrossRef]
60. Zhu, Y.M.; Yang, X.Q. Relationships between Pacific Decadal Oscillation (PDO) and climate variabilities in China. *Acta Meteorol. Sin.* **2003**, *61*, 641–653, (In Chinese with English abstract).
61. Tollefson, J. Climate change: The case of the missing heat. *Nature* **2014**, *505*, 276–278. [CrossRef] [PubMed]
62. Ma, Z.G. The interdecadal trend and shift of dry/wet over the central part of North China and their relationship to the Pacific Decadal Oscillation (PDO). *Chin. Sci. Bull.* **2007**, *52*, 2130–2139. [CrossRef]
63. Mantua, N.J.; Hare, S.R. The Pacific Decadal Oscillation. *J. Oceanogr.* **2002**, *58*, 35–44. [CrossRef]
64. Wang, S.W.; Gong, D.Y.; Zhu, J.H. Twentieth-century climatic warming in China in the context of the Holocene. *Holocene* **2001**, *11*, 313–321. [CrossRef]
65. Wang, S.W.; Ye, J.L.; Gong, D.Y.; Zhu, J.H.; Yao, T.D. Construction of mean annual temperature series for the last one hundred years in China. *J. Appl. Meteorol. Sci.* **1998**, *9*, 392–401, (In Chinese with English abstract).
66. Shen, C.; Wang, W.C.; Gong, W.; Hao, Z. A Pacific Decadal Oscillation record since 1470 AD reconstructed from proxy data of summer rainfall over eastern China. *Geophys. Res. Lett.* **2006**, *33*, L03702. [CrossRef]

 *sustainability*

Article

Absolute Contribution of the Non-Uniform Spatial Distribution of Atmospheric CO_2 to Net Primary Production through CO_2-Radiative Forcing

Jing Peng [1,*], Li Dan [1], Jinming Feng [1], Kairan Ying [1], Xiba Tang [2] and Fuqiang Yang [1]

[1] CAS Key Laboratory of Regional Climate-Environment for Temperate East Asia, Institute of Atmospheric Physics, Chinese Academy of Sciences, Beijing 100029, China; danli@tea.ac.cn (L.D.); fengjm@tea.ac.cn (J.F.); yingkr@tea.ac.cn (K.Y.); yangfq@tea.ac.cn (F.Y.)

[2] Laboratory of Cloud-Precipitation Physics and Severe Storms (LACS), Institute of Atmospheric Physics, Beijing 100029, China; tangxb_6@mail.iap.ac.cn

* Correspondence: pengjing@tea.ac.cn

Abstract: Atmospheric concentrations of CO_2 are the most important driver of the Earth's climate and ecosystems through CO_2-radiative forcing, fueling the surface temperature and latent heat flux on half-century timescales. We used FGOALS-s2 coupled with AVIM2 to estimate the response of net primary production (NPP) to spatial variations in CO_2 during the time period 1956–2005. We investigated how the induced variations in surface temperature and soil moisture influence NPP and the feedback of the oceans and sea ice on changes in NPP. The spatial variations in the concentrations of CO_2 resulted in a decrease in NPP from 1956 to 2005 when we included ocean and sea ice dynamics, but a slight increase in NPP without ocean and sea ice dynamics. One of the reasons is that the positive feedback of sea temperature to the surface temperature leads to a significant decrease in tropical NPP. Globally, the non-uniform spatial distribution of CO_2 absolutely contributed about 14.3% $\pm$ 2.2% to the terrestrial NPP when we included ocean and sea ice dynamics or about 11.5% $\pm$ 1.1% without ocean and sea ice dynamics. Our findings suggest that more attention should be paid to the response of NPP to spatial variations in atmospheric CO_2 through CO_2-radiative forcing, particularly at low latitudes, to better constrain the predicted carbon flux under current and future conditions. We also highlight the fundamental importance of changes in soil moisture in determining the pattern, response and magnitude of NPP to the non-uniform spatial distribution of CO_2 under a warming climate.

Keywords: non-uniform CO_2; CO_2-radiative forcing; net primary production; surface temperature; soil moisture; sea surface temperature

Citation: Peng, J.; Dan, L.; Feng, J.; Ying, K.; Tang, X.; Yang, F. Absolute Contribution of the Non-Uniform Spatial Distribution of Atmospheric CO_2 to Net Primary Production through CO_2-Radiative Forcing. *Sustainability* **2021**, *13*, 10897. https://doi.org/10.3390/su131910897

Academic Editor: Baojie He

Received: 12 August 2021
Accepted: 8 September 2021
Published: 30 September 2021

Publisher's Note: MDPI stays neutral with regard to jurisdictional claims in published maps and institutional affiliations.

1. Introduction

Atmospheric CO_2 concentrations are one of the most important drivers of climate in earth system models [1,2]. It can affect the global average surface temperature, the global water cycle, global sea-levels, the loss of Arctic sea ice and the atmospheric vapor pressure deficit through radiative forcing [3–5]. Changes in CO_2 concentrations indirectly affect the growth of land plants and the productivity of terrestrial vegetation (gross primary production) [6]—for example, the gross primary production of terrestrial vegetation decreased after the late 1990s as a result of the increased vapor pressure deficit [5].

There is large uncertainty in the magnitude and spatial distribution of the radiative effect of atmospheric CO_2 concentration ($\pm20\%$) based on multi-model simulations [7,8]. These models come from the fifth stage of the coupled model comparison project (CMIP5) [9] using the uniformity of atmospheric CO_2 concentration without spatial varying. Although the spatial heterogeneity of atmospheric CO_2 concentrations varies widely on a regional scale [10,11], the contribution of spatially non-uniform CO_2 to the Earth's climate and the terrestrial carbon cycle through CO_2-radiative forcing is largely unknown [12].

81

Several studies have reported the impacts of varying atmospheric CO_2 concentrations directly through CO_2-radiative forcing on climate state [13,14]. Furthermore, it can also disturb the carbon balance [15] and accumulation of carbon on land [16]. For instance, the variations in surface temperature, precipitation and soil moisture [17] influence the rate of photosynthesis and limit net primary production (NPP) and soil respiration [15]. Thus, changes in the Earth's climate through CO_2-radiative forcing therefore constrain the processes of the carbon cycle [2].

However, predictions of the rise in temperature under historical conditions remain highly uncertain, ranging from about 0.4 to 1.3 °C, mainly due to historical radiative forcing of greenhouse gases [18]. One of the uncertainties results from the lack of a clear understanding about the feedback of spatial variations in atmospheric CO_2 and the lack of reliable measurements of this process in the field. Modeling approaches are therefore vital to constrain climate variables and the carbon fluxes resulting from the spatial varying of CO_2 concentration.

On the other hand, most conventional methodologies do not consider the influence of spatial varying of CO_2 concentrations on the Earth's climate system and biogeochemical cycles. We would focus to quantify the response of terrestrial NPP to the spatial varying of CO_2 concentration through CO_2-radiative forcing. To further assess the significance of the non-uniform spatial distribution of CO_2 on NPP, we compared simulations with fully coupled atmosphere–land–ocean–sea ice components (simulation A) and coupled atmosphere–land components (simulation B) in the Flexible Global Ocean–Atmosphere–Land System Model Spectral Version 2 (FGOALS-s2) [19] coupled with the Atmosphere–Vegetation Interaction Model (AVIM) [20]. Specifically, we evaluated NPP in FGOALS-AVIM with the implementation of these two different simulations forced by a uniform or non-uniform spatial distribution of atmospheric CO_2 with or without active ocean and sea ice during the time period 1956–2005. We addressed three principal questions: (1) how much does the climate-driven feedback caused by the non-uniform spatial distribution of CO_2 through CO_2-raiative forcing affect climate variables and NPP; (2) how does the estimated NPP with and without ocean and sea ice dynamics differ as a result of the non-uniform spatial distribution of CO_2 through the radiative forcing associated with spatial variations in the climate variables; and (3) what are the implications of a non-uniform spatial distribution of CO_2 on the global NPP during the time period 1956–2005.

2. Methods and Simulations

The Student's *t*-test used in this study is used to estimate of the statistical significance between the variables from different simulations, which are calculated as follows:

$$t_i = \frac{\overline{x_i} - \overline{y_i}}{S_{Pi}\sqrt{\frac{1}{n_1} + \frac{1}{n_2}}} \tag{1}$$

where $\overline{x_i}$ or $\overline{y_i}$ is the averaged NPP from simulations with spatial varying of CO_2 concentration or without spatial varying of CO_2 concentration of the i-th grid cell during 1956−2005, S_{Pi} is the weighted average of the standard deviations of NPP from the two simulations during the same period in the i-th grid cell. n_1 or n_2 is the equivalent sample sizes of two simulations.

Based on the absolute contribution of each region (each grid cell or PFT) to global NPP, we further classify global NPP into different categories (Figure 1). To this end, we adopted an index to score each geographic location based on consistency. The magnitude of the absolute change in local NPP flux relative to the global NPP under CO_2 uniform distribution was calculated as the following, which was also estimated by [21]:

$$C_j = \frac{\sum_{t=1}^{n}\left|\Delta NPP_{jt}\right|}{\sum_{t=1}^{n} NPP_t} \tag{2}$$

where $|\Delta NPP_{jt}|$ is the absolute change in NPP caused by the non-uniform spatial distribution of CO_2 relative to a uniform spatial distribution of CO_2 through CO_2-radiative forcing for the jth region at the tth time (units: g C yr^{-1}) and NPP_t is the NPP for the tth time period using forcing by a uniform spatial distribution of CO_2 for the whole terrestrial ecosystem. C_j is the contribution of a non-uniform spatial distribution CO_2 through CO_2-radiative forcing for the jth region relative to the global NPP using forcing by a uniform spatial distribution of CO_2.

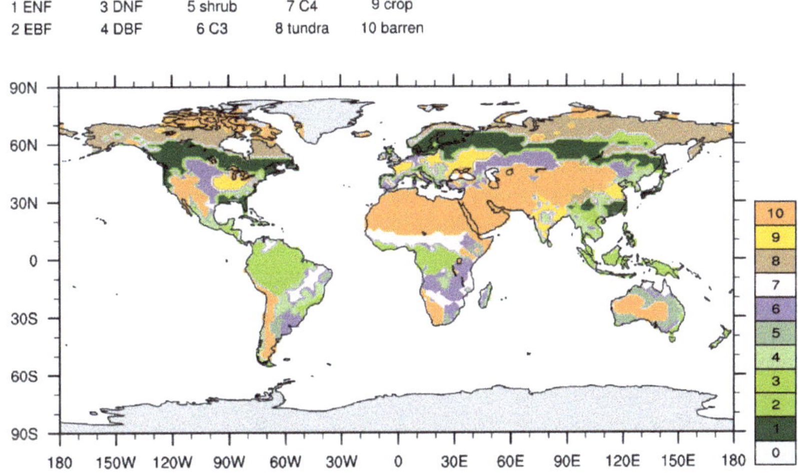

Figure 1. Geographical distribution of the 10 plant function types in FGOALS-AVIM. ENF, evergreen needle leaf forest; EBF, evergreen broadleaf forest; DBF, deciduous broadleaf forest.

FGOALS that we used here participated in the CMIP6 and was used as an assessment tool in the 6th Assessment Report of the Intergovernmental Panel on Climate Change. It is a fully coupled earth system model and is composed of four separate models simultaneously simulating the Earth's atmosphere (SAMIL2), oceans (LICOM2), land surface (AVIM) and sea ice (CSIM). FGOALS includes one central coupler component (CPL6). It has an interactive carbon cycle model in the land component and an ecosystem–biogeochemical module in the ocean component. The simulated atmospheric CO_2 concentrations are fully coupled to the land CO_2 fluxes and are used directly to compute radiative forcing. The release of methane from the melting of permafrost potentially has a huge impact on warming, but FGOALS-AVIM currently has only very simple carbon permafrost models and no release of marine methane is included.

We used FGLOAL-AVIM at a resolution of ($2.81° \times 1.66°$) [22]. This version of FGOALS has previously been used to assess the effects of CO_2-radiative forcing on the total cloud fraction, temperature and water vapor under different abruptly quadrupling CO_2 concentrations [23]. Evaluation of the mean summer evapotranspiration and mean annual runoff in the Lake Baikal basin [24], and major global biogeochemical fluxes and pool sizes of carbon [25–28].

Time series of the atmospheric CO_2 concentrations (i.e., fossil fuel burning, cement manufacturing and gas flaring in oilfields) from CMIP6 are available from 1850 to 2014 at a monthly resolution. The atmospheric CO_2 variable from CMIP6 only considers spatial variations in terms of latitude rather than longitude. We therefore used the Open-Data Inventory for Anthropogenic Carbon dioxide (ODIAC) [29] at a resolution of ($1° \times 1°$) to establish a quantitative relationship between the carbon emissions in each pixel and the latitudinally averaged carbon emissions to fully reflect the spatial heterogeneity of atmospheric CO_2 and to describe the impact of anthropogenic carbon emissions on its spatial distribution in the atmosphere. We assumed that this quantitative relationship is also applicable to the relationship between the atmospheric CO_2 variable of each pixel and the latitudinally averaged atmospheric CO_2 from CMIP6. The CO_2 variable was produced by fully considering the spatial heterogeneity, which reflected the effect of anthropogenic carbon emissions. The CO_2 variables for 1850–2005 were generated by re-gridding from a spatial resolution of ($1° \times 1°$) to ($2.81° \times 1.66°$). The CO_2 variable was then used as input data for our model.

To quantify the effects of a non-uniform spatial distribution of CO_2 on the carbon flux through CO_2-radiative forcing, we carried out two different simulations using FGOALS-AVIM with one of the two different components used to estimate NPP. We ran each experiment for a simulated 156 years from 1850 to 2005 (Table 1). Simulation A1 includes the non-uniform space–time-varying atmospheric concentrations of CO_2 based on the active atmosphere–land–ocean–sea ice components. Simulation A2 fixes only one atmospheric CO_2 level at the global scale with only the time-varying atmospheric concentrations of CO_2 without the spatial variations described by [30] and fully considers the interactions of the atmosphere–land–ocean–sea ice components. Simulation B1 uses the non-uniform time–space-varying atmospheric concentrations of CO_2 and the coupled atmosphere–land interactions without ocean and sea ice dynamics. Simulation B2 uses the uniform time-varying atmospheric concentrations of CO_2 and the coupled atmosphere–land interactions without ocean and sea ice dynamics.

Table 1. Forcing data for simulations.

Simulation	CO_2 Forcing	Component
A1	Spatial and temporal variations	Fully coupled atmosphere–land–ocean–sea ice
B1	Only temporal variations	Fully coupled atmosphere–land–ocean–sea ice
A2	Spatial and temporal variations	Only atmosphere–land coupling
B2	Only temporal variations	Only atmosphere–land coupling

The difference in the land NPP between simulations A1 and A2 or simulations B1 and B2 represents the effect of the non-uniform spatial distribution of atmospheric concentrations of CO_2. It is well known that physiological forcing is a robust driver influencing terrestrial plants and the accumulation of carbon on land [17,31], but we did not consider physiological forcing in the land component, in which atmospheric concentrations of CO_2 were fixed at the current level of ~355 ppmv.

Simulated NPP from TRENDY used in this study can be freely downloaded from TRENDY丨Dynamic Global Vegetation Model Projects (ceh.ac.uk) (see Table 2). The selected models mainly include CLM4C, CLM4CN, LPJ, LPJGUESS, OCN, SDGVM, TRIFFID and CABLE. The natural response of NPP across the terrestrial ecosystem is one of our focuses. Thus, S2 experiment from TRENDY was chosen, which did not include the impact of land use. For more detailed information about TRENDY simulation protocol could be seen in [32,33]. Previous studies have proved that results of TRENDY perform well and are also widely used to estimate the global carbon budget [6]. Compared with the observed data, TRENDY reproduced variations in carbon fluxes of the terrestrial ecosystem to a certain extent [34,35].

Table 2. Datasets used in this study.

Variable	Project	Period	Website	Ref.
NPP	TRENDY	1956–2005	TRENDY丨Dynamic Global Vegetation Model Projects (ceh.ac.uk)	[32]
CO_2 emission	ODIAC	1980–2014	https://db.cger.nies.go.jp/dataset/ODIAC/DL_odiac_v1.7.html (accessed on 19 August 2019).	[29]

3. Results

Under historical conditions from 1956 to 2005, NPP from simulations A1, A2, B1, and B2 was estimated. The spatial distribution of PFT is shown in Figure 1. In the following results, at the PFT level, the changes in NPP caused by the non-uniform distribution of CO_2 would be given.

3.1. Estimates of CO_2 Concentrations and NPP

There was a large spatial heterogeneity in atmospheric concentrations of CO_2 between the northern and southern hemispheres in the time period 1956–2005. The non-uniform spatial distribution of CO_2 was lower in Australia, most of Africa and southern South America, but higher in southern Europe, the eastern USA and southeast Asia. During the same time period, simulations A1 and B1 were driven by the same CO_2-radiative forcing and simulations A2 and B2 were enforced by the same CO_2 data. The spatial variations in CO_2 ranged between 339 and 351 ppmv (Figure 2a). Greater differences between simulations A1 and A2 or B1 and B2 were mainly located in southern Europe, the eastern USA and southeast Asia (Figure 2b).

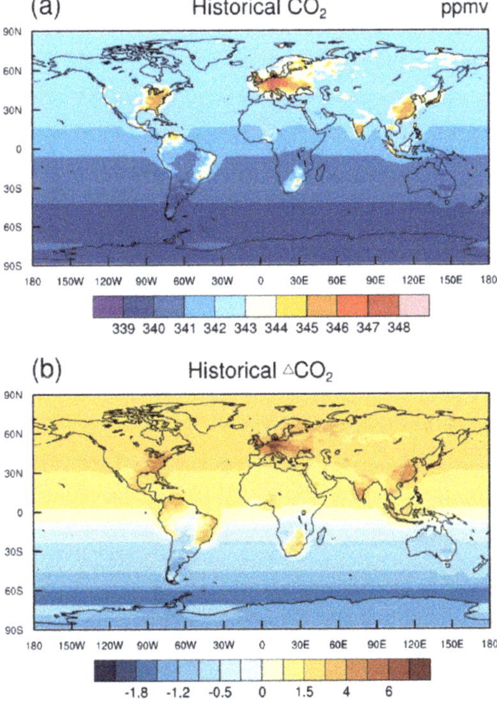

Figure 2. (**a**) Spatial variation in the mean annual atmospheric CO_2 concentrations during the time period 1956-2005 estimated by FGOALS-AVIM. (**b**) Mean annual changes in CO_2 concentrations during the time period 1956–2005 estimated by simulation A1 relative to A2 or simulation B1 relative to B2.

The NPP was estimated for simulations A1, A2, B1 and B2 during the time period 2000–2005 under current conditions. This period was chosen so that our results were comparable with NPP data from the TRENDY datasets. The TRENDY dataset consists of an ensemble of seven process-based terrestrial ecosystem model simulations [36]. We selected experiment S2 without land use changes and fires and including CLM4C, CLM4CN, LPJ, LPJ-GUESS, OCN, SDGVM and TRIFFID [37] from 1901 to 2010.

Our estimated land NPP for the four simulations of 51.39–56.47 Pg C yr^{-1} was lower than the estimates of about 70 Pg C yr^{-1} of [38], but higher than the 41 Pg C yr^{-1} estimated by [39]. The estimated NPPs of the four simulations were closer to the observed estimate of about 56.02 Pg C yr^{-1} of [40]. Globally, the mean annual NPP in the historical period from CMIP5 was about 62.6 Pg C yr^{-1} [41], whereas it was 55.53 ± 1.69 Pg C yr^{-1} from the TRENDY dataset. The mean annual NPP in the historical period simulated by FGOALS-AVIM was 53.80 ± 2.54, 53.75 ± 2.95, 58.75 ± 1.46 and 59.36 ± 2.73 Pg C yr^{-1} from simulations A1, A2, B1 and B2, respectively (Figure 3).

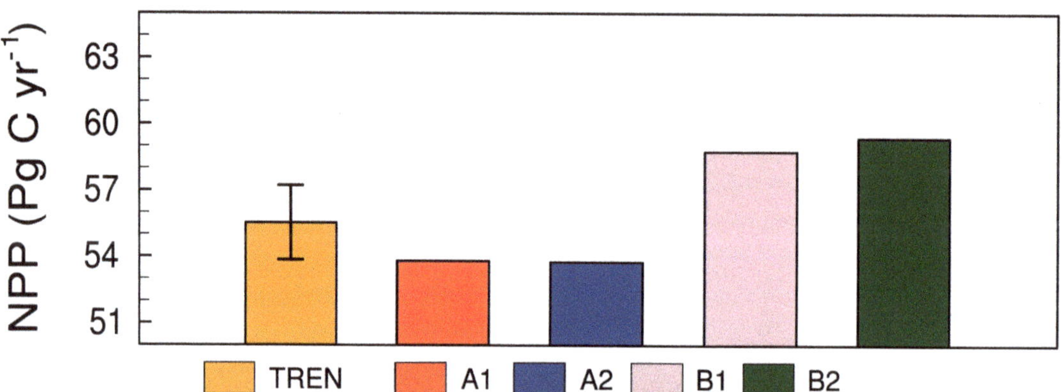

Figure 3. From 1956 to 2005, comparison of different estimates of net primary production based on the TRENDY ensemble (TREN, orange bar) taking into consideration the spatial heterogeneity of atmospheric CO_2 concentrations in the fully coupled atmosphere–land–ocean–sea ice components of FGOALS-AVIM (A1, red bar), the fully coupled atmosphere–land–ocean–sea ice components of FGOALS-AVIM without the spatial heterogeneity of atmospheric CO_2 concentrations (A2, blue bar), the spatial heterogeneity of atmospheric CO_2 concentrations with only atmosphere–land coupling in FGOALS-AVIM (B1, prink bar) and only atmosphere–land coupling without the spatial heterogeneity of atmospheric CO_2 concentrations (B2, green bar).

3.2. Estimates of Climatic Variables under Present Conditions

Over the simulation period 1956–2005, the non-uniform spatial distribution of CO_2 through CO_2-radiative forcing resulted in a global increase in low cloudiness of about 0.002% ± 0.005% (Figure 4a), a net solar flux of 0.52 ± 0.55 W m^{-2} including the atmosphere–land–ocean–sea ice dynamics (Figure 4e) and a terrestrial latent heat flux of 0.36 ± 0.97 W m^{-2} at the global scale (Figure 4c). At the global scale, the climate variables varied considerably among the different components in response to the non-uniform spatial distribution of CO_2 through CO_2-radiative forcing. During the time period 1956–2005, the non-uniform spatial distribution of CO_2 through CO_2-radiative forcing based on only atmosphere and land interactions caused an increase in the net solar flux of about 0.45 ± 0.31 W m^{-2} and a decrease in global low cloudiness of 0.001% ± 0.003%, especially over the ocean (Figure 4b; Table 3). By contrast, the simulated global latent heat flux based on atmosphere and land interactions decreased by about −0.64 ± 0.93 W m^{-2} (Figure 4d).

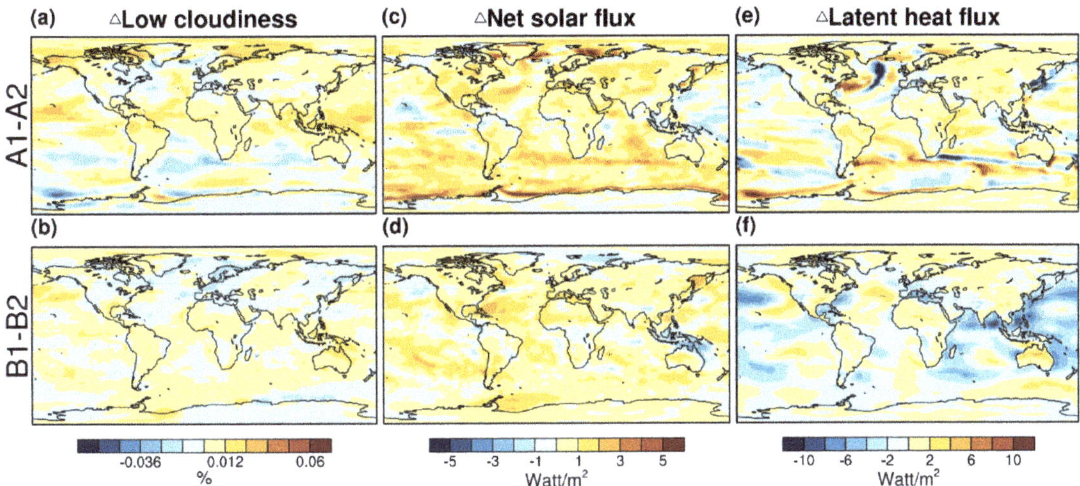

Figure 4. (**a**,**b**) Response of the mean annual low cloudiness to the spatial heterogeneity of CO_2 concentrations over the time period 1956–2005 estimated from simulation A1 or B1 relative to A2 or B2, respectively. (**c**,**d**) Spatial changes in the mean annual net solar flux at the Earth's surface estimated from simulation A1 or B1 relative to A2 or B2, respectively. (**e**,**f**) Spatial changes in the mean annual latent heat flux estimated from simulation A1 or B1 relative to A2 or B2, respectively.

Table 3. Response of variables to CO_2-radiative forcing as a result of the non-uniform spatial distribution of CO_2.

Variable	Range	A1–A2	B1–B2
CO_2	Global	-0.0007 ± 0.0037	
Precipitation (mm yr^{-1})	Land	6.77 ± 21.45	-9.26 ± 14.72
	Ocean	5.89 ± 14.15	-9.58 ± 9.92
Surface temperature (°C)	Land	0.57 ± 0.45	0.02 ± 0.05
	Ocean	0.45 ± 0.38	0.07 ± 0.08
Latent heat flux (W m^{-2})	Land	0.59 ± 1.23	-0.64 ± 0.93
	Ocean	0.38 ± 0.80	
Sensible heat flux (W m^{-2})	Land	-0.29 ± 0.46	
	Ocean	-0.20 ± 0.42	
Net radiation flux at Earth's surface (W m^{-2})	Land	0.79 ± 0.69	
	Ocean	0.59 ± 0.63	
Soil evaporation (kg m^{-2} yr^{-1})	Land	3.99 ± 7.52	
Vegetation evaporation (kg m^{-2} yr^{-1})	Land	0.43 ± 4.90	
Soil transpiration (kg m^{-2} yr^{-1})	Land	1.91 ± 5.70	
Surface runoff	Land	3.00 ± 6.18	
Soil moisture (kg m^{-2})	Land	0.03 ± 0.04	

To further analyze the causes of the differences in the estimated NPP resulting from the non-uniform spatial distribution of CO_2 through CO_2-radiatvie forcing, we estimated the changes in the vital variables (e.g., surface temperature, precipitation, evapotranspiration and soil moisture in the different components) for the historical time period 1956–2005. Non-uniform CO_2 from simulations fully coupled atmosphere–land–ocean–sea ice components caused an increase in surface temperature ranging from 0.57 ± 0.45 °C in the terrestrial ecosystem, where it from simulations only coupled atmosphere and land without active ocean and sea ice components led to less increased surface temperature of 0.02 ± 0.05 °C. The changes in the surface temperature at the regional level correspond well to the changes in the non-uniform spatial distribution of CO_2 considering only the atmosphere–land interactions (Figure 5b). The responses of the surface temperature in tropical regions to the non-uniform spatial distribution of CO_2 through CO_2-radiative forcing based on the fully coupled atmosphere–land–ocean–sea ice components were up to >0.2 °C, especially in the Amazon basin, central Africa and tropical Asia. By contrast, the simulations with only the atmosphere–land interactions simulated a small increase or even decrease in these regions.

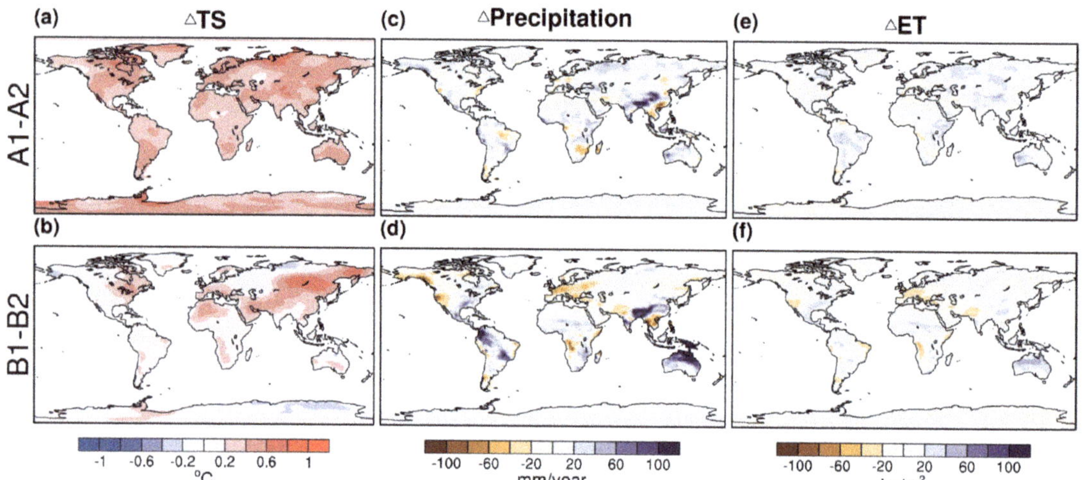

Figure 5. (**a**,**b**) Response of the mean annual surface temperature to the spatial heterogeneity of CO_2 concentrations over the time period 1956–2005 estimated from simulation A1 or B1 relative to A2 or B2, respectively. (**c**,**d**) Spatial changes in precipitation estimated from simulation A1 or B1 relative to A2 or B2, respectively. (**e**,**f**) Spatial changes in the mean evapotranspiration estimated from simulation A1 or B1 relative to A2 or B2, respectively.

In the time period 1956–2005, precipitation over the land surface increased in most of the northern hemisphere, but not in southern Europe, the eastern USA and southeast Asia. This was a response to the non-uniform spatial distribution of CO_2 through CO_2-radiative forcing (Figure 5c) compared with the uniform spatial distribution of CO_2 in the same period. The simulated increase in precipitation influenced by non-uniform spatial distribution of CO_2 considering the full atmosphere–land–ocean–sea ice interactions was much greater than that which considered only atmosphere–land interactions, apart from in the Amazon basin and tropical Asia. This was a result of the larger increase in the net radiation flux from surface warming and latent heat flux over land, especially in the northern hemisphere.

Correspondingly, the responses of the simulated change in evapotranspiration by the two differently coupled components differed considerably in both the direction and magnitude of change for most terrestrial ecosystems (Figure 5d,e). The two different components predicted an increase in evapotranspiration affected by the non-uniform spatial distribution of CO_2 through CO_2-radiative forcing relative to the uniform spatial distribution of CO_2 for most land surfaces. The simulated response of evapotranspiration to changes in the non-uniform spatial distribution of CO_2 differed at the regional scale. For example, evapotranspiration north of 50° N in the northern hemisphere was predicted to increase as a response to the uniform spatial distribution of CO_2 through CO_2-radiative forcing based on the fully coupled atmosphere–land–ocean–sea ice model, but to decrease in southern Europe, the eastern USA and southeast Asia. By contrast, evapotranspiration influenced by the non-uniform spatial distribution of CO_2 considering only the atmosphere–land interactions showed a much greater increase than the fully coupled atmosphere–land–ocean–sea ice interactions at low latitudes and in the eastern USA.

3.3. Responses of NPP to Spatial Varying of CO_2 Concentrations through Radiative Forcing

To quantify the impact of the non-uniform spatial distribution of CO_2 on NPP, we simulated the NPP for the time period 1956–2005 with or without the uniform spatial distribution of CO_2 after 1850. Specifically, we ran FGOALS-AVIM with CO_2 held at the same value at the global level for each land grid point in simulation A2 or B2 and compared the results with the non-uniform spatial distribution of CO_2 for each land grid point in simulation A1 or B1 for the time period 1956–2005.

We estimated that the decrease in NPP caused by the non-uniform spatial distribution of CO_2 through CO_2-radiative forcing during the time period 1956–2005 was -3.39 ± 12.33 g C m^{-2} yr^{-1} or about -0.5 ± 1.88 Pg C yr^{-1} for the fully coupled atmosphere–land–ocean–sea ice components (simulation A1 minus A2). The increase was about 0.88 ± 8.86 g C m^{-2} yr^{-1} or 0.1 ± 1.35 Pg C yr^{-1} when considering only atmosphere–land interactions (Figure 6a,b).

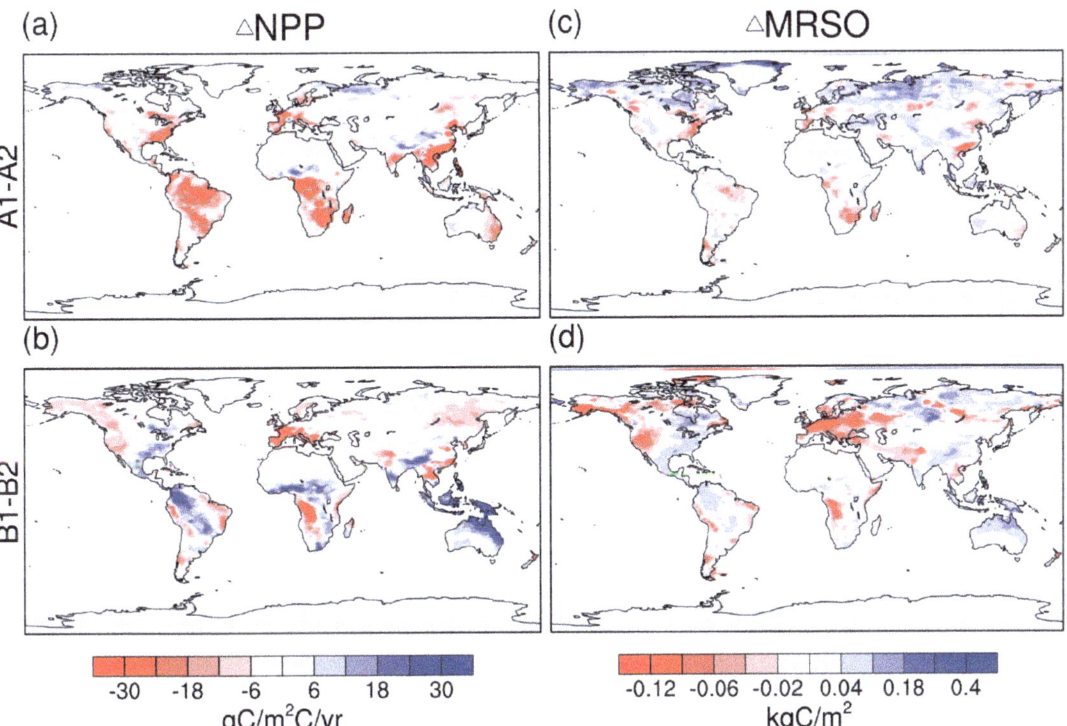

Figure 6. (**a**,**b**) Response of the mean net primary production to the spatial heterogeneity of CO_2 concentrations over the time period 1956–2005 estimated from simulation A1 or B1 relative to A2 or B2, respectively. (**c**,**d**) Spatial changes in soil moisture estimated from simulation A1 minus A2 or B1 minus B2, respectively.

The fully coupled atmosphere–land–ocean–sea ice components estimated a decrease in NPP caused by the non-uniform spatial distribution of CO_2 through CO_2-radiative forcing for terrestrial ecosystems (Figure 6a), whereas the coupled active atmosphere–land components predicted an increase (Figure 6b). The mechanism contributing to the estimated change in NPP was similar between the two simulations. With the simulated decrease in soil moisture (Figure 6c), the model based on the fully coupled atmosphere–land–ocean–sea ice components estimated a decrease in NPP for southern Europe, the eastern USA, southeast Asia, the Amazon basin and central Africa (Figure 6a). However, the simulation considering only the atmosphere–land interaction estimated an increase in soil moisture (Figure 4d) for the eastern USA, the Amazon basin and central Africa (Figure 6d). The estimated increase in NPP for these regions using only the active atmosphere–land components was attributed to the relatively higher increase in NPP (Figure 6b).

We used the present vegetation cover for the period of the 1990s based on the plant functional types (PFTs) of the International Geosphere Biosphere Program (IGBP) data [42]. At the PFT-level, the NPP simulated by simulations A due to spatial varying in CO_2 concentrations were reduced for EBF, DBF, C4 and crop, but increased for tundra. The variations in NPP simulated from simulations A and B were feeble for ENF. Globally, the changes of annual NPP from 1956 to 2005 resulting from varying in CO_2 concentrations was ~ -0.5 Pg C yr^{-1} from simulations A or estimates of ~0.1 Pg C yr^{-1} from simulations B (Supplementary Materials Figure S1).

The non-uniformity of CO_2 leads to substantial changes in surface temperature and water fluxes, and the changes caused by non-uniformity of CO_2 in NPP depend on changes in surface temperature and soil moisture, rather than the impact of CO_2, which may lead to changes in the spatial distribution of NPP (Figure 7). Figure 7b,e show how much change in NPP is locally related to changes in surface temperature, especially in the tropics and the Southern Hemisphere. Obviously, these two variables are significantly correlated, which indicates that changes in surface temperature can explain the changes in NPP in these regions. For example, considering that the fully coupled land-atmosphere-ocean-ice interaction, rising surface temperature leads to negative changes of NPP in tropics due to non-uniform CO_2. On the contrary, in most terrestrial ecosystems, there is a significantly positive correlation between soil moisture and local NPP ($p < 0.05$) (Figure 7c,f). Therefore, our results indicate that the competition between the impacts of surface temperature and soil moisture on NPP is one of the mechanisms leading to changes in NPP in the tropics.

Although the absolute change in CO_2 at the global scale was estimated to be only about 0.3%, the model based on coupled components A1 minus A2 or B1 minus B2 estimated an additional change in NPP caused by the non-uniform spatial distribution of CO_2 by 14.26% or 11.51% during the time period 1956–2005 (Figure 8). This reflects that the impact of CO_2 heterogeneity on carbon cycle was amplified by the climate effect. In addition, the difference between the two different coupled components probably reflects the contribution of the ocean and sea ice to the climate system. With global warming caused by the non-uniform spatial distribution of CO_2 through CO_2-radiative forcing, the model based on the fully coupled components A1 minus A2 estimated absolute variations in NPP caused by the non-uniform spatial distribution of CO_2 relative to the global NPP driven by a uniform spatial distribution of CO_2 ranging from 0.97 to 1.53% for C3 and from 2.92 to 4.03% for crops. The simulation with only atmosphere–land interactions ranged from 1.76 to 2.23% for C3 and from 2.56 to 3.32% for crops. For different PFTs (e.g., crops), the absolute changes in CO_2 from 0.05 to 0.10% were estimated to be higher than those for other PFTs. The impact of the non-uniform spatial distribution of CO_2 was largest for deciduous broadleaf forest, for which simulation A1 minus A2 estimated higher absolute changes in NPP from 2.53 to 4.65% for evergreen broadleaf forest due to the non-uniform spatial distribution of CO_2, compared with simulation B1 minus B2 from 1.93 to 3.05%. For crops, both simulations A1 minus A2 and B1 minus B2 estimated high absolute changes ranging from 2.56 to 3.32%.

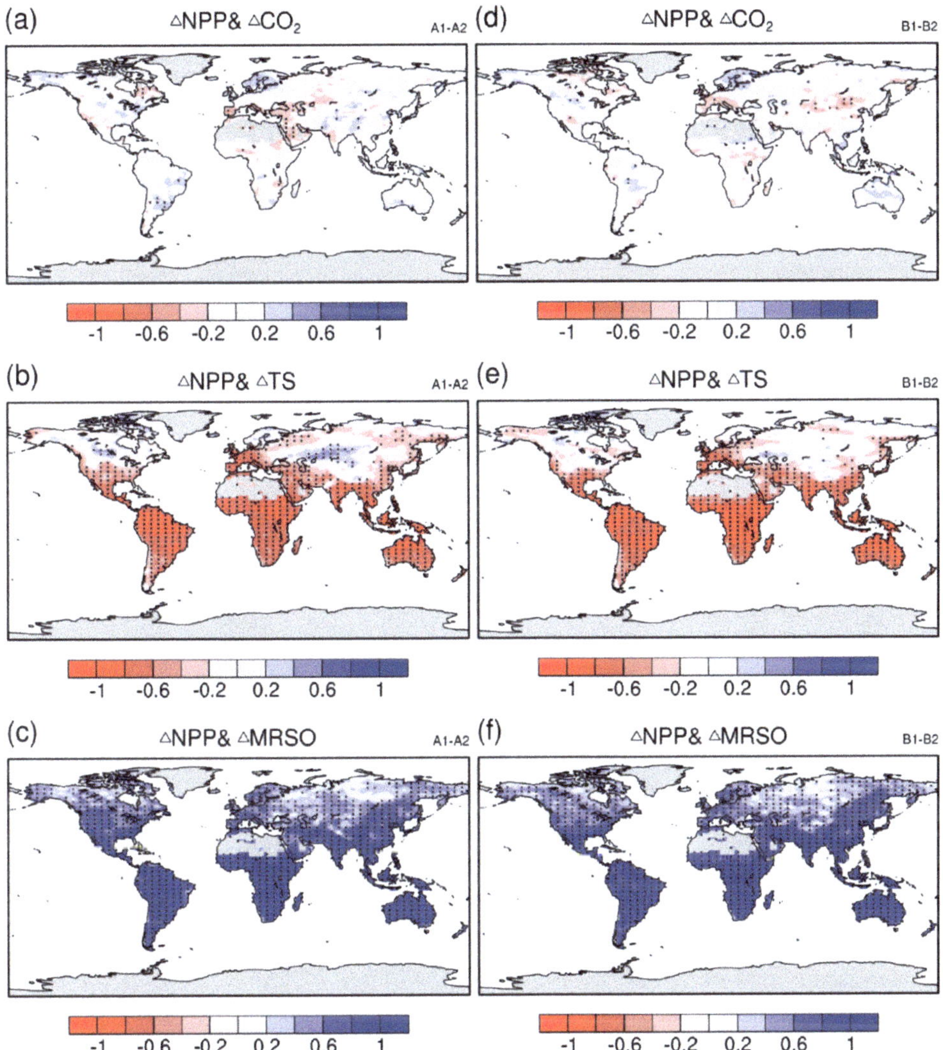

Figure 7. (**a–c**) Linear correlation coefficients of changes in CO_2 concentrations, the surface temperature (TS), soil moisture (MRSO) to NPP over the time period 1956–2005 estimated from simulation A1 relative to A2, respectively. (**d–f**) Linear correlation coefficients of changes in CO_2 concentrations, the surface temperature (TS), soil moisture (MRSO) to NPP over the time period 1956–2005 estimated from simulation B1 relative to B2, respectively. Dotted areas are areas with statistically significant changes at the 5% level using Student's *t*-test.

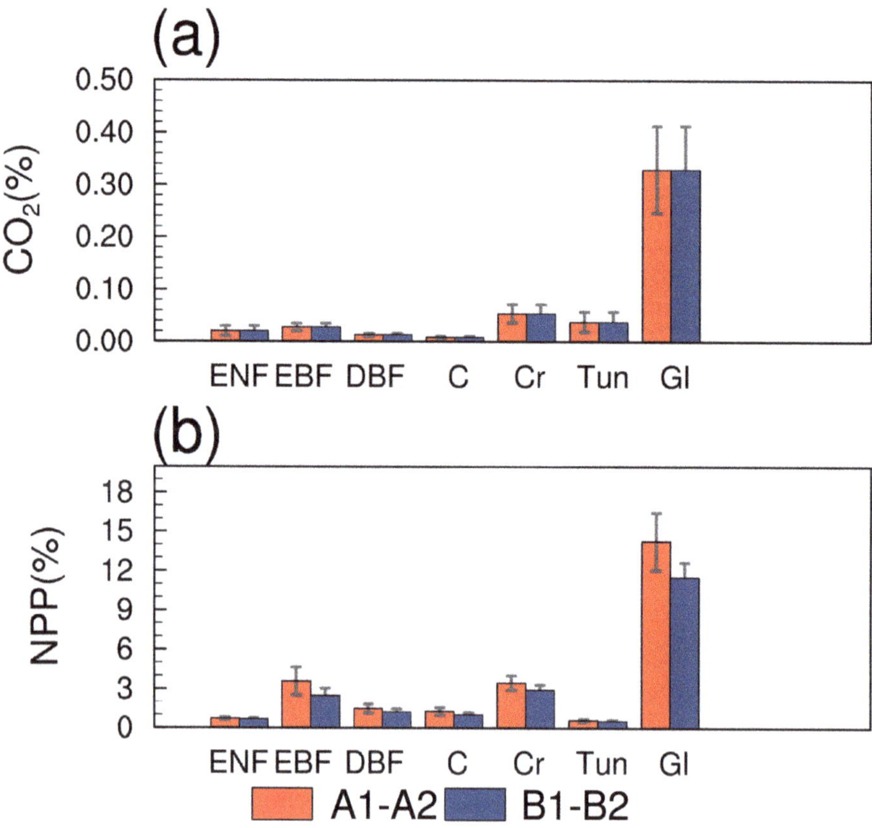

Figure 8. Changes in (**a**) CO$_2$ concentrations (units: %) and (**b**) net primary production (NPP) (units: %) over the time period 1956–2005 from simulation A1 or B1 relative to A2 or B2 as estimated using FGOALS-AVIM over different plant function types. ENF, evergreen needle leaf forest; EBF, evergreen broadleaf forest; DBF, deciduous broadleaf forest; C, C4; Cr, crop; Gl, global. Error bar are ±SD calculated from model simulations over the same time period.

4. Discussion

At present, few studies have considered effect of radiative forcing of spatial varying of CO$_2$ concentrations in the atmosphere on NPP. Our results showed that the non-uniform spatial distribution of CO$_2$ through CO$_2$-radiative forcing has important implications for carbon fluxes. We estimated that the non-uniform spatial distribution of CO$_2$ contributed to a decrease in NPP, although most other studies have not considered this. The CMIP5 provides an ensemble mean annual NPP of 62.6 Pg C yr^{-1} over the time period 1995–2004 [41]. However, this analysis did not consider the feedbacks from the non-uniform spatial distribution of CO$_2$ to the Earth's climate system. Our analysis showed an effect of the non-uniform spatial distribution of CO$_2$ on NPP without considering land use of up to 14.26 ± 2.22% of the NPP simulated by a fully coupled model. This highlights the tremendous leverage of a non-uniform spatial distribution of CO$_2$ on the global carbon cycle and we recommend a more substantial focus on understanding this process in the biosphere through the influence of the climate system based on coupled models that include dynamical sea ice and ocean components.

In the past 30 years, there has been an increase in NPP at the global scale, which is likely driven by CO_2 fertilization [34,35]. However, the increase in CO_2 radiative forcing limits this positive effect [1]. The exact impact of spatial varying in CO_2 concentration on the carbon cycle through radiative forcing is not yet fully understood. Our study illuminates that from 1956 to 2005, the spatial varying in the CO_2 concentration resulted in a reduction of NPP, which mainly occurred in the tropics and the middle-latitudes of the Northern Hemisphere. Nevertheless, as shown in Figure 7, a relationship between variations in CO_2 concentration and NPP could not pass a significant level. It might result due to not considering the impact of CO_2 fertilization. On the contrary, spatial varying in CO_2 concentration through the radiation effect influences temperature, precipitation and soil moisture, which ultimately affects NPP. The northern hemisphere accounts for nearly 68% of the global area, which has a higher CO_2 concentration than the globally averaged level. Higher temperature has a negative impact on NPP in tropics and mid-latitudes of the Northern Hemisphere [33,38,43,44]. In addition, considering ocean feedback, an increase in temperature could reduce NPP in these regions. Conversely, when the ocean feedback is shielded, increases in temperature is slight in tropics. At the same time, in this condition, the increase in the soil moisture would offset the negative impact of warming in tropics.

To estimate the contribution of the ocean and sea ice influenced by the spatial varying of CO_2 concentrations through CO_2-radiative forcing to the terrestrial NPP, we carried out simulations B1 and B2, in which the effect of CO_2-radiative forcing was assessed using fixed the sea surface temperature (SST) in 1980. The additional increase in SST was estimated to be $0.45 \pm 0.38\,°C$ using the fully coupled model, compared with $0.07 \pm 0.08\,°C$ based on only the coupled active atmosphere–land components. This can be understood by the positive feedback from a greater increase in the SST (Figure 9a) to surface temperature from simulations A1 minus A2 relative to from simulations B1 minus B2 without varying SST. The response of NPP in low-latitudes of terrestrial ecosystem to changes in surface temperature might be asymmetric. This may also lead to asymmetric NPP changes related to SST changes. For example, in some tropical terrestrial regions, increased SST has a negative impact on NPP under warmer changes [45]. Therefore, more detailed quantification is needed to understand the asymmetric response of NPP to SST changes.

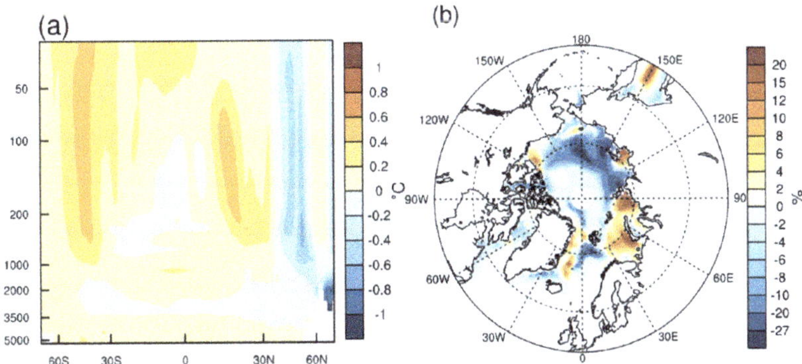

Figure 9. Modeled patterns for zonal changes in (**a**) ocean temperature with depth (units: °C) and (**b**) the fraction of ice area (units: %) in the time period 1956–2005 in FGOALS-AVIM simulation A1 relative to A2.

The spatial response of soil moisture differed widely, although the SST had a key influence soil moisture [46]. Under warmer conditions, a non-uniform spatial distribution of CO_2 decreased soil moisture across the eastern USA, southern Europe and southeast Asia in the fully coupled model. This is consistent with previous research [47], which showed that an increased SST in the tropics would increase the depletion of soil moisture in the eastern USA, southern Europe, southeast Asia and the tropics.

At high latitudes (e.g., the Arctic), the decrease in the area of ice as a result of the non-uniform spatial distribution of CO_2 through CO_2-radiative forcing is >2% (Figure 9b). An additional increase in precipitation has been shown in most of this region. This change can be explained by emergent positive feedbacks between the increasing area of melted ice and the increased water vapor at high latitudes in the fully coupled model. Recent studies have found that a decrease in Arctic sea ice has an important role in the Earth's climate and ecosystems [48]. Specifically, stronger warming results in an increased loss of sea ice, which, in turn, simulates more melting through a reduced albedo [48]. Ice-loss feedback could amplify changes in the Earth's climate at high latitudes [48,49]. We estimated larger increases in surface temperature and precipitation for tundra regions using the fully coupled model including sea ice dynamics than the model without sea ice dynamics. There is a positive feedback from an increase in melted sea ice in a warmer and wetter climate, which enhances the NPP by the non-uniform spatial distribution of CO_2 through CO_2-radiative forcing in the tundra regions.

This finding is consistent with a previous study of CO_2-radiative forcing and has important implications for our understanding of the changes in carbon flux associated with climate change [50]. Changes in the non-uniform spatial distribution of CO_2 concentrations influenced changes in the heat budget (Figure 4), then continued to regulate the spatial distribution of surface climate variables. Changes in the atmospheric circulation (Figure 10) introduced by the non-uniform spatial distribution of CO_2 also could influence the water balance and, in turn, affect soil moisture. Therefore, the patterns and responses of the water budget at the regional scales triggered by the non-uniform spatial distribution of CO_2 determine the patterns and responses in NPP. In particular, near-global increases in specific humidity do not mean that this increase necessarily results in an increase in precipitation. The simulation using the fully coupled model estimated that weaker westerlies reduced the input of humid air from the Atlantic Ocean to southern Europe, which reduced precipitation and then increased the depletion of soil moisture. The eastward radial circulation in the eastern USA was strengthened. Such a circulation pattern does not favor precipitation. The eastward radial circulation was also enhanced in East Asia and suppressed the monsoon from the Eastern Pacific Ocean, limiting precipitation and inducing droughts.

CO_2-physiological forcing could also have significant impact on the spatial variation of NPP [51]. However, it was not included in our results, because our main focus is on the impacts of CO_2 radiative forcing on NPP. For terrestrial processes, atmospheric CO_2 was uniformly applied to all PFTs at a fixed level of 355 ppmv. The impact of a spatial variation in CO_2 concentrations was masked in our analysis. Results may be different if we include a variation in the spatiotemporal distribution of atmospheric CO_2 through CO_2-physiological forcing in the land component. This needs further investigation.

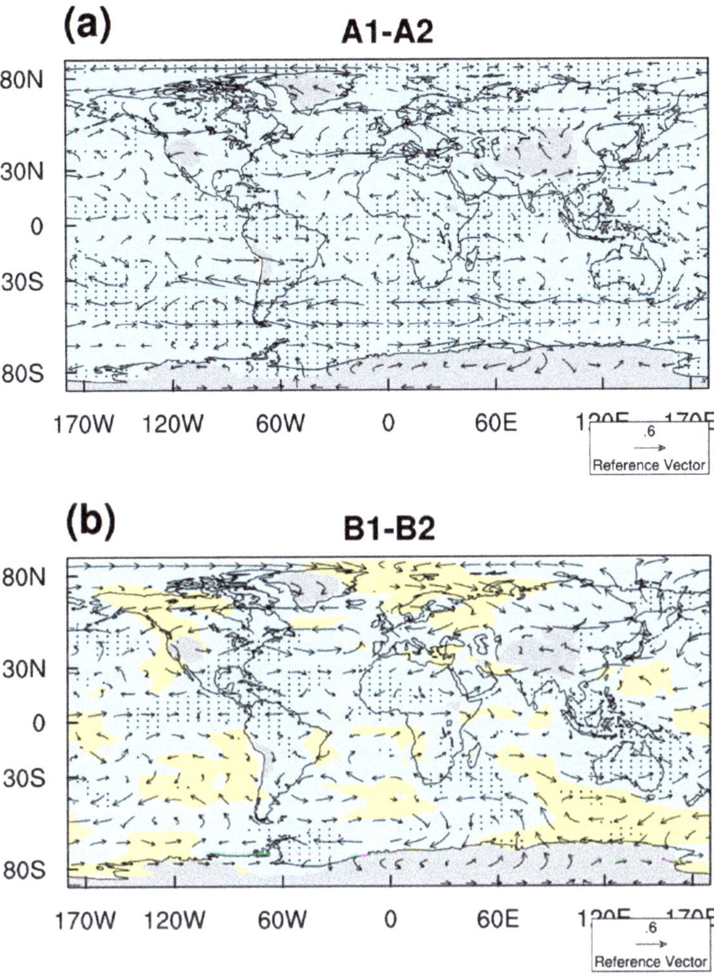

Figure 10. (**a**,**b**) Spatial changes in the mean specific humidity (units: g kg^{-1}) and 850 hPa winds (vectors; m s^{-1}) over the time period 1956–2005 estimated using FGOALS-AVIM from simulation A1 or B1 relative to A2 or B2, respectively. Dotted areas are areas with statistically significant changes at the 5% level using Student's *t*-test.

5. Conclusions

This study investigated the regulations of spatial varying in atmospheric CO_2 concentration through radiative forcing to annual NPP. The results implied that soil moisture resulting from spatial varying in CO_2 concentration dominates the changes of annual NPP, and the increased SST substantially reduced NPP in tropics. As a consequence, spatial varying of atmospheric CO_2 concentration did change the spatial pattern of NPP under the historical conditions from 1956 to 2005. In addition, our results showed the current offline terrestrial models only considering land and atmosphere interactions did not reproduce the feedback of ocean and sea ice to changes in NPP. In general, this study represented the key role of spatial varying in CO_2 concentration in regulating variations in NPP.

Supplementary Materials: The following are available online at https://www.mdpi.com/article/10.3390/su131910897/s1, Figure S1: (a) NPP and (b) soil moisture (MRSO) for the globe (GL), evergreen needle leaf forest (ENF); evergreen broadleaf forest (EBF); deciduous broadleaf forest (DBF); crop (C) and tundra (Tun) during the period 1956–2005 caused by the impact of spatial varying of CO_2 concentrations. Horizontal lines in the bars show the maximum and minimum NPP or soil moisture with considering feedbacks of ocean and sea ice (orange) and without considering feedbacks of ocean and sea ice (light blue) from 1956 to 2005.

Author Contributions: Conceptualization, methodology, writing—original draft preparation, J.P.; formal analysis, investigation, funding acquisition, L.D.; writing—review and editing, supervision, J.F.; writing—review and editing, supervision, K.Y.; formal analysis, project administration, X.T.; project administration, data curation, F.Y. All authors have read and agreed to the published version of the manuscript.

Funding: This work was funded by the National Natural Science Foundation of China (Grant Nos 41630532, 41975112 and 41575093) and the National Key Research and Development Program of China (Grant No. 2016YFA0602501).

Institutional Review Board Statement: Not applicable.

Informed Consent Statement: Not applicable.

Data Availability Statement: Data available on request due to restrictions.

Conflicts of Interest: The authors declare no conflict of interest.

References

1. Friedlingstein, P.; Meinshausen, M.; Arora, V.K.; Jones, C.D.; Anav, A.; Liddicoat, S.K.; Knutti, R. Uncertainties in CMIP5 Climate Projections due to Carbon Cycle Feedbacks. *J. Clim.* **2013**, *27*, 511–526. [CrossRef]
2. Friedlingstein, P. Carbon cycle feedbacks and future climate change. *Philos. Trans. R. Soc. A Math. Phys. Eng. Sci.* **2015**, *373*, 20140421. [CrossRef] [PubMed]
3. Govindasamy, B.; Caldeira, K. Geoengineering Earth's radiation balance to mitigate CO_2-induced climate change. *Geophys. Res. Lett.* **2000**, *27*, 2141–2144. [CrossRef]
4. Etminan, M.; Myhre, G.; Highwood, E.J.; Shine, K.P. Radiative forcing of carbon dioxide, methane, and nitrous oxide: A significant revision of the methane radiative forcing. *Geophys. Res. Lett.* **2016**, *43*, 12614–12623. [CrossRef]
5. Yuan, W.; Zheng, Y.; Piao, S.; Ciais, P.; Lombardozzi, D.; Wang, Y.; Ryu, Y.; Chen, G.; Dong, W.; Hu, Z.; et al. Increased atmospheric vapor pressure deficit reduces global vegetation growth. *Sci. Adv.* **2019**, *5*, eaax1396. [CrossRef] [PubMed]
6. Schimel, D.; Stephens, B.B.; Fisher, J.B. Effect of increasing CO_2 on the terrestrial carbon cycle. *Proc. Natl. Acad. Sci. USA* **2015**, *112*, 436–441. [CrossRef] [PubMed]
7. Huang, Y.; Xia, Y.; Tan, X. On the pattern of CO_2 radiative forcing and poleward energy transport. *J. Geophys. Res. Atmos.* **2017**, *122*, 10578–10593. [CrossRef]
8. Myhre, G.; Shindell, D.; Bréon, F.-M.; Collins, W.; Fuglestvedt, J.; Huang, J.; Koch, D.; Lamarque, J.-F.; Lee, D.; Mendoza, B.; et al. Anthropogenic and Natural Radiative Forcing. In *Climate Change 2013: The Physical Science Basis. Contribution of Working Group I to the Fifth Assessment Report of the Intergovernmental Panel on Climate Change*; Cambridge University Press: Cambridge, UK, 2013.
9. Taylor, K.E.; Stouffer, R.J.; Meehl, G.A. An overview of CMIP5 and the Experiment Design. *Bull. Am. Meteorol. Soc.* **2011**, *93*, 485–498. [CrossRef]
10. Nassar, R.; Napier-Linton, L.; Gurney, K.R.; Andres, R.; Oda, T.; Vogel, F.; Deng, F. Improving the temporal and spatial distribution of CO_2 emissions from global fossil fuel emission data sets. *J. Geophys. Res. Atmos.* **2013**, *118*, 917–933. [CrossRef]
11. Falahatkar, S.; Mousavi, S.M.; Farajzadeh, M. Spatial and temporal distribution of carbon dioxide gas using GOSAT data over IRAN. *Environ. Monit. Assess.* **2017**, *189*, 627. [CrossRef]
12. Wang, Y.; Feng, J.; Dan, L.; Lin, S.; Tian, J. The impact of uniform and nonuniform CO_2 concentrations on global climatic change. *Theor. Appl. Climatol.* **2019**, *139*, 45–55. [CrossRef]
13. Friedlingstein, P.; Cox, P.; Betts, R.; Bopp, L.; von Bloh, W.; Brovkin, V.; Cadule, P.; Doney, S.; Eby, M.; Fung, I.; et al. Climate–Carbon Cycle Feedback Analysis: Results from the C4MIP Model Intercomparison. *J. Clim.* **2006**, *19*, 3337–3353. [CrossRef]
14. Ramaswamy, V.; Collins, W.; Haywood, J.; Lean, J.; Mahowald, N.; Myhre, G.; Naik, V.; Shine, K.P.; Soden, B.; Stenchikov, G.; et al. Radiative Forcing of Climate: The Historical Evolution of the Radiative Forcing Concept, the Forcing Agents and their Quantification, and Applications. *AMS Meteorol. Monogr.* **2019**, *59*, 14.1–14.101. [CrossRef]
15. Ballantyne, A.; Smith, W.; Anderegg, W.; Kauppi, P.; Sarmiento, J.; Tans, P.; Shevliakova, E.; Pan, Y.; Poulter, B.; Anav, A.; et al. Accelerating net terrestrial carbon uptake during the warming hiatus due to reduced respiration. *Nat. Clim. Chang.* **2017**, *7*, 148–152. [CrossRef]
16. Cox, P.; Pearson, D.; Booth, B.; Friedlingstein, P.; Huntingford, C.; Jones, C.; Luke, C. Sensitivity of tropical carbon to climate change constrained by carbon dioxide variability. *Nature* **2013**, *494*, 341–344. [CrossRef] [PubMed]

17. Peng, J.; Dan, L.; Dong, W. Are there interactive effects of physiological and radiative forcing produced by increased CO_2 concentration on changes of land hydrological cycle? *Glob. Planet. Chang.* **2014**, *112*, 64–78. [CrossRef]
18. IPCC. Climate Change 2014: Synthesis Report. In *Contribution of Working Groups I, II and III to the Fifth Assessment Report of the Intergovernmental Panel on Climate Change*; IPCC: Geneva, Switzerland, 2014; p. 151.
19. Zhou, T.; Song, F.; Chen, X. Historical evolution of global and regional surface air temperature simulated by FGOALS-s2 and FGOALS-g2: How reliable are the model results? *Adv. Atmos. Sci.* **2013**, *30*, 638–657. [CrossRef]
20. Dan, L.; Cao, F.; Gao, R. The improvement of a regional climate model by coupling a land surface model with eco-physiological processes: A case study in 1998. *Clim. Chang.* **2015**, *129*, 457–470. [CrossRef]
21. Ahlström, A.; Raupach, M.R.; Schurgers, G.; Smith, B.; Arneth, A.; Jung, M.; Reichstein, M.; Canadell, J.G.; Friedlingstein, P.; Jain, A.K.; et al. The dominant role of semi-arid ecosystems in the trend and variability of the land CO_2 sink. *Science* **2015**, *348*, 895–899. [CrossRef]
22. Bao, Q.; Lin, P.; Zhou, T.; Liu, Y.; Yu, Y.; Wu, G.; He, B.; He, J.; Li, L.; Li, J.; et al. The Flexible Global Ocean-Atmosphere-Land system model, Spectral Version 2: FGOALS-s2. *Adv. Atmos. Sci.* **2013**, *30*, 561–576. [CrossRef]
23. Grise, K.M.; Polvani, L.M. Understanding the Time Scales of the Tropospheric Circulation Response to Abrupt CO_2 Forcing in the Southern Hemisphere: Seasonality and the Role of the Stratosphere. *J. Clim.* **2017**, *30*, 8497–8515. [CrossRef]
24. Törnqvist, R.; Jarsjö, J.; Pietroń, J.; Bring, A.; Rogberg, P.; Asokan, S.M.; Destouni, G. Evolution of the hydro-climate system in the Lake Baikal basin. *J. Hydrol.* **2014**, *519*, 1953–1962. [CrossRef]
25. Ji, J.; Hu, Y. A simple land surface process model for use in climate study. *J. Meteorol. Res.* **1989**, *3*, 342–351.
26. Ji, J.; Huang, M.; Li, K. Prediction of carbon exchanges between China terrestrial ecosystem and atmosphere in 21st century. *Sci. China Ser. D-Earth Sci.* **2008**, *51*, 885–898. [CrossRef]
27. Wang, J.; Bao, Q.; Zeng, N.; Liu, Y.; Wu, G.; Ji, D. Earth System Model FGOALS-s2: Coupling a dynamic global vegetation and terrestrial carbon model with the physical climate system model. *Adv. Atmos. Sci.* **2013**, *30*, 1549–1559. [CrossRef]
28. Peng, J.; Dan, L. The Response of the Terrestrial Carbon Cycle Simulated by FGOALS-AVIM to Rising CO_2. In *Flexible Global Ocean-Atmosphere-Land System Model*; Springer: Berlin/Heidelberg, Germany, 2014; pp. 393–403.
29. Oda, T.; Maksyutov, S.; Andres, R.J. The Open-source Data Inventory for Anthropogenic CO_2, version 2016 (ODIAC2016): A global monthly fossil fuel CO_2 gridded emissions data product for tracer transport simulations and surface flux inversions. *Earth Syst. Sci. Data* **2018**, *10*, 87–107. [CrossRef]
30. Peng, J.; Wang, Y.; Houlton, B.Z.; Dan, L.; Pak, B.; Tang, X. Global Carbon Sequestration Is Highly Sensitive to Model-Based Formulations of Nitrogen Fixation. *Glob. Biogeochem. Cycles* **2020**, *34*, e2019GB006296. [CrossRef]
31. Peng, J.; Dan, L. Impacts of CO_2 concentration and climate change on the terrestrial carbon flux using six global climate-carbon coupled models. *Ecol. Model.* **2015**, *304*, 69–83. [CrossRef]
32. Peng, J.; Dan, L.; Yang, F.; Tang, X.; Wang, D. Global and regional estimation of carbon uptake using CMIP6 ESM compared with TRENDY ensembles at the centennial scale. *J. Geophys. Res. Atmos.* **2012**, *126*, e2021JD035135. [CrossRef]
33. Peng, J.; Peng, J.; Dan, L.; Ying, K.; Yang, S.; Tang, X.; Yang, F. China's Interannual Variability of Net Primary Production Is Dominated by the Central China Region. *J. Geophys. Res. Atmos.* **2021**, *126*, e2020JD033362. [CrossRef]
34. Sitch, S.; Friedlingstein, P.; Gruber, N.; Jones, S.D.; Murray-Tortarolo, G.; Ahlström, A.; Doney, S.C.; Graven, H.; Heinze, C.; Huntingford, C.; et al. Recent trends and drivers of regional sources and sinks of carbon dioxide. *Biogeosciences* **2015**, *12*, 653–679. [CrossRef]
35. Friedlingstein, P.; Friedlingstein, P.; O'sullivan, M.; Jones, M.W.; Andrew, R.M.; Hauck, J.; Olsen, A.; Peters, G.P.; Peters, W.; Pongratz, J.; et al. Global Carbon Budget 2020. *Earth Syst. Sci. Data* **2020**, *12*, 3269–3340. [CrossRef]
36. Zhang, X.; Rayner, P.; Wang, Y.; Silver, J.D.; Lu, X.; Pak, B.; Zheng, X. Linear and nonlinear effects of dominant drivers on the trends in global and regional land carbon uptake: 1959 to 2013. *Geophys. Res. Lett.* **2016**, *43*, 1607–1614. [CrossRef]
37. Piao, S.; Sitch, S.; Ciais, P.; Friedlingstein, P.; Peylin, P.; Wang, X.; Ahlström, A.; Anav, A.; Canadell, J.; Cong, N.; et al. Evaluation of terrestrial carbon cycle models for their response to climate variability and to CO_2 trends. *Glob. Chang. Biol.* **2013**, *19*, 2117–2132. [CrossRef]
38. Piao, S.; Ciais, P.; Friedlingstein, P.; de Noblet-Ducoudré, N.; Cadule, P.; Viovy, N.; Wang, T. Spatiotemporal patterns of terrestrial carbon cycle during the 20th century. *Glob. Biogeochem. Cycles* **2009**, *23*. [CrossRef]
39. Thornton, P.; Zimmermann, N. An Improved Canopy Integration Scheme for a Land Surface Model with Prognostic Canopy Structure. *J. Clim.* **2007**, *20*, 3902–3923. [CrossRef]
40. Zhao, M.; Heinsch, F.A.; Nemani, R.R.; Running, S.W. Improvements of the MODIS terrestrial gross and net pri-mary pro-duction global data set. *Remote Sens. Environ.* **2005**, *95*, 164–176. [CrossRef]
41. Wieder, W.; Cleveland, C.C.; Smith, W.; Todd-Brown, K. Future productivity and carbon storage limited by terrestrial nutrient availability. *Nat. Geosci.* **2015**, *8*, 441–444. [CrossRef]
42. Loveland, T.R.; Reed, B.C.; Brown, J.F.; Ohlen, D.O.; Zhu, Z.; Yang, L.W.M.J.; Merchant, J.W. Development of a global land characteristics database and IGBP DISCover from 1 km AVHRR data. *Int. J. Remote Sens.* **2000**, *21*, 1303–1330. [CrossRef]
43. Fernández-Martínez, M.; Sardans, J.; Chevallier, F.; Ciais, P.; Obersteiner, M.; Vicca, S.; Canadell, J.G.; Bastos, A.; Friedlingstein, P.; Sitch, S.; et al. Global trends in carbon sinks and their relationships with CO_2 and temperature. *Nat. Clim. Chang.* **2019**, *9*, 73–79. [CrossRef]

44. Piao, S.; Wang, X.; Park, T.; Chen, C.; Lian, X.; He, Y.; Bjerke, J.; Chen, A.; Ciais, P.; Tømmervik, H.; et al. Characteristics, drivers and feedbacks of global greening. *Nat. Rev. Earth Environ.* **2020**, *1*, 14–27. [CrossRef]
45. Kim, I.-W.; Stuecker, M.F.; Timmermann, A.; Zeller, E.; Kug, J.-S.; Park, S.-W.; Kim, J.-S. Tropical Indo-Pacific SST influences on vegetation variability in eastern Africa. *Sci. Rep.* **2021**, *11*, 10462. [CrossRef] [PubMed]
46. Mei, R.; Wang, G. Impact of Sea Surface Temperature and Soil Moisture on Summer Precipitation in the United States Based on Observational Data. *J. Hydrometeorol.* **2011**, *12*, 1086–1099. [CrossRef]
47. Dai, A. Increasing drought under global warming in observations and models. *Nat. Clim. Chang.* **2013**, *3*, 52–58. [CrossRef]
48. Meier, W.N.; Hovelsrud, G.K.; Van Oort, B.E.; Key, J.R.; Kovacs, K.M.; Michel, C.; Haas, C.; Granskog, M.; Gerland, S.; Perovich, D.K.; et al. Arctic sea ice in transformation: A review of recent observed changes and impacts on biology and human activity. *Rev. Geophys.* **2014**, *52*, 185–217. [CrossRef]
49. Dai, A.; Luo, D.; Song, M.; Liu, J. Arctic amplification is caused by sea-ice loss under increasing CO_2. *Nat. Commun.* **2019**, *10*, 121. [CrossRef]
50. Cao, L.; Bala, G.; Caldeira, K.; Nemani, R.; Ban-Weiss, G. Importance of carbon dioxide physiological forcing to future cli-mate change. *Proc. Natl. Acad. Sci. USA* **2010**, *107*, 9513–9518. [CrossRef]
51. Swann, A.L.S.; Hoffman, F.M.; Koven, C.D.; Randerson, J.T. Plant responses to increasing CO_2 reduce estimates of climate impacts on drought severity. *Proc. Natl. Acad. Sci. USA* **2016**, *113*, 10019–10024. [CrossRef]

sustainability

Review

Building Climate Change Adaptation and Resilience through Soil Organic Carbon Restoration in Sub-Saharan Rural Communities: Challenges and Opportunities

Alex Taylor [1,*], Maarten Wynants [1], Linus Munishi [2], Claire Kelly [1], Kelvin Mtei [3], Francis Mkilema [2], Patrick Ndakidemi [2], Mona Nasseri [4], Alice Kalnins [1], Aloyce Patrick [2], David Gilvear [1] and William Blake [1]

[1] School of Geography, Earth and Environmental Sciences, University of Plymouth, Plymouth PL4 8AA, UK; maarten.wynants@plymouth.ac.uk (M.W.); claire.kelly@plymouth.ac.uk (C.K.); alice.kalnins@plymouth.ac.uk (A.K.); david.gilvear@plymouth.ac.uk (D.G.); william.blake@plymouth.ac.uk (W.B.)

[2] School of Life Sciences and Bio-Engineering, The Nelson Mandela African Institution of Science and Technology, P.O. Box 447, Arusha 23311, Tanzania; linus.munishi@nm-aist.ac.tz (L.M.); francis.mkilema@nm-aist.ac.tz (F.M.); patrick.ndakidemi@nm-aist.ac.tz (P.N.); patrickaloyce@gmail.com (A.P.)

[3] School of Materials, Energy, Water and Environmental Science, The Nelson Mandela African Institution of Science and Technology, P.O. Box 447, Arusha 23311, Tanzania; kelvin.mtei@nm-aist.ac.tz

[4] Ecological Design Thinking, Schumacher College, Totnes TQ9 6EA, UK; Mona.Nasseri@schumachercollege.org.uk

* Correspondence: alex.taylor-34@plymouth.ac.uk; Tel.: +44-(0)-1752-584709

Citation: Taylor, A.; Wynants, M.; Munishi, L.; Kelly, C.; Mtei, K.; Mkilema, F.; Ndakidemi, P.; Nasseri, M.; Kalnins, A.; Patrick, A.; et al. Building Climate Change Adaptation and Resilience through Soil Organic Carbon Restoration in Sub-Saharan Rural Communities: Challenges and Opportunities. *Sustainability* **2021**, *13*, 10966. https://doi.org/10.3390/su131910966

Academic Editors: Jun Yang, Ayyoob Sharifi, Baojie He and Chi Feng

Received: 13 August 2021
Accepted: 30 September 2021
Published: 2 October 2021

Abstract: Soil organic carbon (SOC) is widely recognised as pivotal in soil function, exerting important controls on soil structure, moisture retention, nutrient cycling and biodiversity, which in turn underpins a range of provisioning, supporting and regulatory ecosystem services. SOC stocks in sub-Saharan Africa (SSA) are threatened by changes in land practice and climatic factors, which destabilises the soil system and resilience to continued climate change. Here, we provide a review of the role of SOC in overall soil health and the challenges and opportunities associated with maintaining and building SOC stocks in SSA. As an exemplar national case, we focus on Tanzania where we provide context under research for the "Jali Ardhi" (Care for the Land) Project. The review details (i) the role of SOC in soil systems; (ii) sustainable land management (SLM) techniques for maintaining and building SOC; (iii) barriers (environmental, economic and social) to SLM implementation; and (iv) opportunities for overcoming barriers to SLM adoption. We provide evidence for the importance of site-specific characterisation of the biophysicochemical and socio-economic context for effective climate adaptation. In particular, we highlight the importance of SOC pools for soil function and the need for practitioners to consider the type of biomass returns to the soil to achieve healthy, balanced systems. In line with the need for local-scale site characterisation we discuss the use of established survey protocols alongside opportunities to complement these with recent technologies, such as rapid in situ scanning tools and aerial surveys. We discuss how these tools can be used to improve soil health assessments and develop critical understanding of landscape connectivity and the management of shared resources under co-design strategies.

Keywords: Soil organic carbon; systems thinking; whole system; interdisciplinary; sustainable land management

1. Introduction

Sustainable management of soils is critical to underpin food, energy and water security for future generations. In sub-Saharan Africa (SSA), these facets are threatened by increasing rates of soil degradation underpinned by a complex interaction between natural vulnerability and anthropogenic activities [1]. In SSA, detrimental land management practices can be traced across a series of governance shifts through history, often stemming

Sustainability **2021**, *13*, 10966. https://doi.org/10.3390/su131910966   https://www.mdpi.com/journal/sustainability

from disruption to indigenous systems at the start of the colonial period, and a change to rigid structures, which lacked adaptive capacity at the community scale [2]. Together with changing land use, a rise in global surface temperature is linked to increased occurrence of extreme weather events, which is predicted to intensify with continued warming [3,4]. This exacerbates soil degradation resulting in a destabilised soil system leading to fracturing of the provisioning, regulatory and supporting ecosystem services provided by healthy systems [5]. Soil degradation can be broadly classified as physical (e.g., loss of organic matter, impacts to soil structure and aggregate stability with subsequent erosion), chemical (e.g., contamination, nutrient depletion or salinization) and biological (loss of biological diversity), and the rate and processes by which these occur is highly site-specific in relation to both socio-economic drivers and environmental conditions [6]. In East Africa, soil loss has been largely attributed to erosion by water with an estimated mean soil loss of around 6.3 t/ha/yr, half of which is likely to be derived from croplands [7]. Efforts to reduce erosion by water must consider landscape connectivity in a holistic manner, which considers social and economic frameworks alongside an understanding of landscape hydrology [8–10]. Crucially, assessment and development of sustainable land management (SLM) must be embedded within a participatory approach, drawing on local knowledge and building practical systems of change from within the local community, ultimately fostering community-led management, which is not reliant upon but supported by external services [10–15].

Soil organic carbon (SOC) is a fundamental component for healthy soil function, exerting important controls on soil structure, water retention, biodiversity and nutrient cycling [16]. Maintaining and building effective SOC pools can, therefore, play a key role in climate change adaptation strategies, offering an opportunity to build resilience with triple-win (productivity-ecosystem-livelihood) benefits [17]. However, with conversion of natural vegetation to agriculture and subsequent intensive practice, biomass outputs often outweigh inputs, leading to a deficit in soil organic matter (SOM) and associated SOC [18,19]. In turn this impacts upon soil structure and drainage properties, facilitating overland flow and further soil and SOC loss, which lowers fertility and increases susceptibility to drought. Owing to SOC importance, building and maintaining healthy SOC pools is acknowledged in the United Nations Convention to Combat Desertification (UNCCD) as essential for unlocking the full potential of soil ecosystem services and achieving many targets under the global Sustainable Development Goals (SDG) framework [16]. Focus on SOC also serves to address United Nations Climate Change Conference of the Parties 2021 (COP26) *action on adaptation* agendas. The pivotal role of SOC in soil function has placed it as a key component in soil degradation assessments (e.g., Bunning et al. [20]), particularly given its sensitivity to land use change [19,21]. In 2017, the sub-Saharan Africa (SSA) Soil Fertility Prioritization Summit was designed to identify key barriers to improving soil fertility and, working with stakeholders across SSA, develop interdisciplinary, evidence-based strategies to overcome these barriers and implement change [22]. Summarising the findings of the summit, Stewart et al. [23] argued that previous attempts to overcome soil fertility issues in Africa have not adequately invested in overall soil health since the approaches have not considered broader social, regulatory and economic factors influencing the soil fertility supply chain, emphasising the importance of holistic, evidence-driven approaches to alleviating poverty associated with low soil fertility. A survey undertaken at the summit revealed that stakeholders regarded SOC deficiency as a key limiting factor in improving soil fertility in SSA, with identified barriers to improvement including the need for site-specific research, access to training and extension services, gender equality issues and access to local soil assessment tools.

Against that background we explore the role of SOC in underpinning climate change adaptation and resilience in SSA. We focus on Tanzania as an exemplar national case in the context of a continuing research programme, the "Jali Ardhi" (Care for the Land) Project [13,15,24], which adopts a participatory and interdisciplinary approach to agropastoral land management assessment. Herein, this contribution aims to provide an overview of the importance of SOC components for soil functioning followed by methods to improve and maintain SOC. In addition, an evaluation of barriers to change and opportunities for improved integrated assessment and SLM implementation is provided.

2. The Role of SOC in Climate Smart Agriculture

There is increasing recognition of the role of SOC functions in ecosystem service provision [16], with regard to regulatory services, such as carbon sequestration, flood regulation and contaminant immobilisation; cultural services, such as recreation; supporting services via healthy nutrient and microbial pools, and resulting provisioning services in the form of food, fuel and water supplies [5,25]. Healthy soil function supported by SOC can, therefore, be used as tool for adapting to climate change and building resilience in rural communities through appropriate SLM practice [26,27]. SOC is often conceptualised as comprising different pools, each with distinct roles and residence times in the soil system. The 'active' or labile pool is derived from organic matter (OM), which is highly susceptible to microbial decomposition and rapidly cycled (mineralised) (~1–2 years) to underpin soil nutrient supplies. This pool is largely influenced by the addition of recent material with relatively low C:N ratios and low molecular weight compounds, which can be readily decomposed and assimilated by soil microbes [28]. The 'slow' SOC pool relates to carbon which has been assimilated into microbial biomass and partially stabilised via mineral matrix interactions or within soil aggregates [29,30], with decadal residence timescales. A further highly stable or 'passive' pool comprises recalcitrant material with residence times > 100 years, exerting an important influence on soil cation exchange capacity (CEC) owing to a large number of charged functional groups [16]. The active SOC component, whilst critical for nutrient cycling, is also important for ensuring SOC stabilisation through microbial uptake and subsequent cycling to organo-mineral complexes in the slow pool [29]. Both the slow and passive pools play a key role in developing healthy soil structure with associated benefits such as aggregate stability and improved moisture retention. A functioning soil system maintains an effective balance between nutrient release and subsequent carbon loss via mineralisation, and SOC stabilisation.

Numerous studies highlight the decline of SOC under intensive agricultural systems in SSA [26,30–32] owing to a destructive spiral of degradation in which a lack of biomass return to the soils under cropping regimes, or loss of vegetation via overgrazing or land use change, impacts upon key soil functions, leaving soils vulnerable to erosion by wind and water, further exacerbating SOC reduction [17]. The soil functions underpinned by SOC are well documented (Figure 1) and include maintaining healthy soil structure and moisture retention, enhancing CEC, pH buffering, provision of nutrient pools and microbial diversity [33]. SOC depletion can, therefore, lead to physical, chemical and biological degradation. For example, Pabst et al. [31] assessed the impact of changing land use upon organic carbon and microbial activity in a range of systems on the southern slopes of Kilimanjaro, Tanzania. The authors studied representative systems such as savannah, coffee plantations, maize plantations and traditional home gardens, with findings highlighting depleted organic carbon and less stable microbial activity under intensive cropping systems, owing to a lack of organic matter supply and exacerbated by lower rainfall conditions in drier areas. Healthy microbial activity and carbon pools in home gardens reflected the use of agroforestry techniques, which involved adding mixed residues to soils, thus increasing available substrate for microbes.

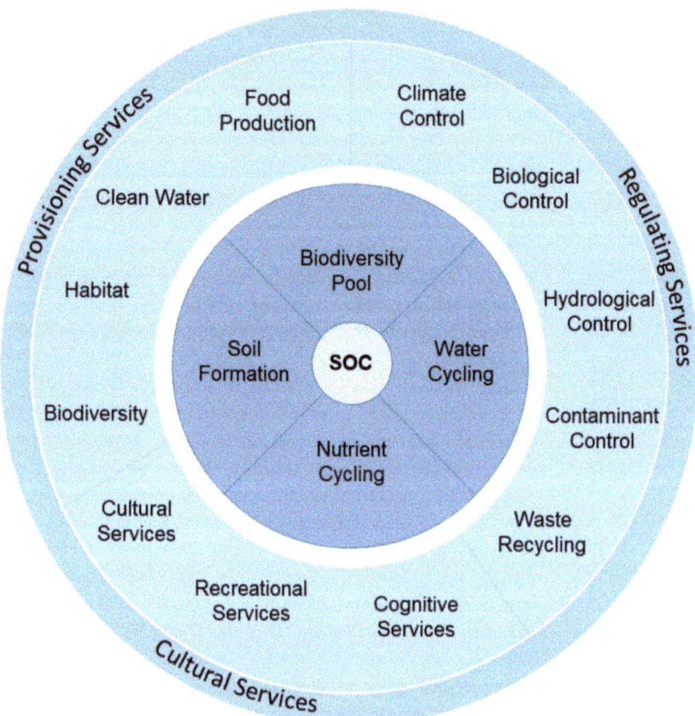

Figure 1. Soil functions (inner circle) and ecosystem services (outer circle) underpinned by soil organic carbon. Adapted from Laban et al. [34].

In many smallholder agricultural communities there is often a trade-off between available biomass to return to soils and the need for fodder, with crop residues often used for the latter [18]. It is important, therefore, that systems are developed whereby OM can be effectively allocated to soils to build the SOC resource during cropping regimes, whilst maintaining needs for fodder and fuel. In a study of SOC in a range of Tanzanian soils, Winoiecki et al. [21] demonstrated a linear relationship between SOC and total nitrogen, and showed that cultivation practices had led to depletion of soil C by around 50% in comparison to semi natural sites. Similarly Solomon et al. [19] showed clearing woodland for cultivation of maize and beans without fertiliser inputs led to > 50% reduction in N and C in Tanzanian Luvisols, with improvements shown for homestead systems where manures and fallow rotations were applied. This is in line with the findings of others whereby, in time, routine allocation of OM to the soil can increase yields, potentially offsetting shortfalls in biomass allocation [35]. At the global scale, meta-analyses suggest a positive linear relationship between SOC and cereal yields to around 2% SOC, beyond which yield improvements tend to plateau [36]. Analyses show a general tendency for higher yields with effective combination of quality SOC inputs and mineral fertiliser [36,37]. In dryland environments, achieving SOC concentrations to around the 2% threshold is likely to be challenging although, where SOC is < 1%, there is evidence to suggest that relatively minor increases can have a positive effect on yield [36].

Despite the well-documented benefits of OM allocation to soils in SSA, interactions between OM and the surrounding environment are complex, with functional benefits depending upon a number of physicochemical factors and appropriate management decisions. The active SOC pool is readily influenced by the addition of OM with relatively low C:N ratios, such as legume crops, and it is this rapidly cycled component which has been shown to have a positive effect on crop yield, and which is more sensitive to land use change [19,31,33]. In contrast, studies focusing upon the relationship between different SOC pools and crop yield have shown negative relationships between mineral associated (slow) SOC pools and yield in experimental plots, possibly owing to organo-mineral complexes reducing plant available nutrient supplies [30]. This highlights the need to consider the role of specific SOC pools in land management planning rather than total SOC per se, through additions of effective types of OM. However, long-term soil structural benefits obtained from stabilised SOC pools are particularly important in the context of soil aggregate formation and stability. Aggregate stability is often a key focus in land degradation assessment and SOC can promote soil aggregation through a variety of mechanisms, such as clay-humic complexes with polyvalent cations, and via the binding effects of microbial products [38]. Hydrophobicity of organic substances can also reduce the wetting of aggregates and promote stability [33]. Additionally, the type of OM amendments and soil textural class are both important factors in determining the size of aggregates [38]. Other studies highlight the importance of SOC interactions with sesquioxides, particularly Al-sesquioxides, which most likely exert an important influence on SOC stability and soil aggregation in low activity clay soils in SSA [39].

It follows, then, that management programmes require a thorough assessment of (i) soil properties and (ii) the types and quantity of OM available for application. With regard to the latter, and in relation to benefits to soil fertility, OM can be classified according to its ability to provide available nutrients and, thus, its benefits to crop yield. Palm et al. [40] suggested a plant residue quality index for tropical agroecosystems based upon the nitrogen, lignin and polyphenol content of a range of residues, deriving predictors of available nutrient release to underpin farm applications of OM. Those classed as higher quality residues generally displayed lower lignin and phenol and higher nitrogen content, and were likely to be effective as direct land applications. It was suggested that those with higher lignin and phenol content should be applied in conjunction with higher quality residue or mineral fertilisers. Integrated soil fertility management (ISFM) approaches are a common component of SLM programmes in SSA, and the quantity and types of OM applied, and their interaction with mineral fertilisers is of key importance in determining crop yields [37,41,42]. For example, Gram et al. [37] assessed the effect of OM type and mineral N applications upon maize yields in SSA. Here, the authors applied the OM quality index of Palm et al. [40] with the addition of a subgroup for manures, to include materials such as animal manure and composts. Highest yields were achieved where mineral N was combined with high quality OM, and where 50% of the available N was derived from OM. Interestingly, positive linear relationships were found between sole applications of OM and yield for the high quality OM and manure. A linear response was not shown for the low quality residues (e.g., maize and groundnut residues), which when applied alone, showed a limited effect upon yield and relied upon a higher application of mineral N for any marked yield increase. At higher application rates the use of lower quality residues led to a decline in yield potentially owing to N immobilising effects of phenols [41]. Additionally, sole application of mineral fertiliser did not show a linear response in yield likely owing to thresholds in plant uptake beyond applications of ~100kgN/ha. Overall agronomic efficiency (application rate versus yield) and yield response were found to be more stable with sole applications of quality OM, likely reflecting wider benefits associated with other factors such as moisture retention. Only applications of OM had a positive effect on SOC and it was assumed that any biomass increase from sole applications of mineral N was not returned to the soil, and that excess available N may have facilitated SOC decline.

Clearly the type and quantity of OM amendments needs consideration in relation to the local environment and the challenge in many smallholder systems is stimulating and maintaining biomass inputs to, ideally, achieve a balanced system, less reliant upon external inputs. Sole applications of OM can lead to significant yield improvements but the quality of inputs is of key importance. Attention must, therefore, be upon fostering SLM practice which can provide a broad range of quality organic matter of sufficient quantity to underpin food security. The application of lower quality residues in the context of soil erosion control should not be ignored, however, given the potential for residue barriers to reduce soil loss and, in turn, maintain SOC stocks [43,44]. In time, soils can achieve healthy microbial communities to assist with nutrient supply, root growth and disease resistance as well as structural improvements to aid moisture retention, infiltration and aeration. These factors combine to support resilience to uncertain climatic regimes. Building soil health in degraded systems is, however, likely to take time and it can often take a number of years to accrue benefits associated with SOC targeted approaches [27]. It is, therefore, important to incorporate carefully calculated 'quick win' incentives within management schemes to maintain perceptions of effectiveness and cost–benefit [45].

Understanding the role of SOC highlights the importance of undertaking detailed local–level land evaluations to assess land suitability and potential within an integrated management context. That is, one in which the physicochemical requirements are considered alongside social and economic structures, to enable the selection of optimal SLM practices [20]. It is crucial that resource planning is undertaken with stakeholders at all stages, supported by an enabling and policy-responsive framework. Ziadat et al. [46] describe the importance of a land resource planning (LRP) approach to foster the adoption of appropriate land use measures, aligned with the local ecological, economic and social context to support resilience to climate change and promote sustainable development. This integrated approach involves targeting key areas for change, assessing land potential, identifying appropriate SLM measures and implementing change, supported by appropriate financial mechanisms and performance monitoring. Cataloguing SLM measures in relation to a range of environments is an important decision support component, offering stakeholders a means to access data relating to potential SLM options, their effectiveness and, importantly, challenges and opportunities relating to their implementation. An example of such a database is the World Overview of Conservation Approaches and Technologies (WOCAT) Global SLM Database; a search engine enabling case study information to be accessed for a range of SLM technologies and approaches. Below, we provide a synthesis of typical SLM practices applied within agro-pastoral communities in SSA, documenting their application and effectiveness in terms of enhancing SOC and climate change resilience.

3. Common SLM Methods to Maintain and Enhance SOC

There is a general lack of empirical data comparing the effects of land management techniques upon SOC in SSA, hampered further by differences in the approaches to measuring and reporting SOC across existing studies [47]. The effects of land management are also likely to be highly varied according to a range of environmental factors, presenting challenges for comparison between studies and highlighting the importance for site-specific assessments [48,49]. Nevertheless, the impacts of agriculture and poor management practice upon SOC are well documented and, site-specific variability aside, there is a broad body of evidence to support both the importance of SOC in dryland environments and the conservation measures that can be implemented to provide positive changes to soil health [34,50]. Sustainable management practices require an integrated approach to provide synergy between soil moisture, nutrients and structure to maintain and build SOC stocks and prevent net loss. Here, we summarise a range of approaches applied in rural communities, which specifically document the effects of land management upon SOC.

Diverse climatic zones in SSA naturally influence background SOC stocks such that upland humid areas with higher rainfall and biomass will often see greater SOC relative to semi-arid lowlands, where rainfall shortage and fluctuation is a limiting factor [36]. In semi-arid rangeland environments, livestock grazing pressure can lead to loss of total vegetation cover, leaving soils exposed and vulnerable to water and wind erosion. In turn this impacts upon the root systems and seed banks of native plants and stimulates encroachment by invasive species [10]. Management practice is focused upon restoration of native vegetation cover by improving mobility across varied grazing lands or through enclosure schemes, which may be combined with reseeding of native species. Exclusion zones can be implemented by physical barriers, an agreed set of communal rules or, in some cases, established bylaws in more extensive grazing lands (often hundreds of hectares) [15]. Mixed systems may also effectively combine stall feeding with soil-stabilising plants used on hillslope cropland [51]. Where management strategies are functioning, the benefits associated with enclosure systems are well reported in the literature, with evidence of successful regeneration of palatable, perennial vegetation [17,52]. In turn this can effectively increase the SOC stock in comparison to overgrazed open areas [51] (Table 1), and improve drought resilience through provision of fodder and diversification of income [53,54]. Traditional mobile corral systems (Boma) have also been shown to provide benefits for vegetation cover and SOC even after short periods of use [51,55].

In croplands, reduced tillage, planting of cover crops and effective crop rotation form the basis of conservation agriculture (CA), which has been shown to provide benefits associated with improved SOC [50]. High erosion risk crops, such as maize, can be interplanted with a variety of cover crops, most notably legumes (e.g., *Mucuna pruriens*; *Lablab purpureus*), to increase vegetation cover both during crop growth and post-harvest. Cover crops have been shown to have multiple benefits in cropping systems by reducing rainsplash erosion, increasing SOM, improving overall fertility and providing an additional source of palatable produce, particularly useful in mixed farming systems [56,57]. Application of OM mulches (e.g., crop residues) can also be beneficial for reducing runoff and providing improvements to soil health and crop yields [45,58,59]. As previously discussed, however, the quality of OM is an important consideration [47] and difficulties of returning crop residues to the soil often arise where fodder is required [18]. The use of multiple cropping systems and agroforestry schemes may provide additional fodder sources, enabling crop biomass to be reinvested in the soil, and a balanced livestock-cropping system also presents opportunities for manure application to soils, which may have benefits for plant nutrient uptake [18,59].

In the literature, the benefits associated with CA appear to be highly site-specific, linked to climate and soil type (e.g., Swanepoel et al. [49]), and small-scale management of available OM resources. For example, Castellanos-Navarrete et al. [60] showed that integrated applications of crop residue and manure did not provide adequate nutrients to cropland, with large nutrient losses from manure supplies owing to poor storage practices. Residues and manures are also often inadequate for supplying plant available P, and in SSA effective P application via OM mulch requires careful consideration of the type and source of available plant material [61]. In the short-term, some studies report that CA has overall limited influence upon crop yield and food security for the smallholder in SSA (e.g., Corbeels et al. [62]), often linked to failure to adapt systems to local settings in terms of environmental factors and socio-economic needs and constraints [63]. In the context of climate change resilience, however, the role of CA for SOC management requires a shift in profitability assessment to include longer-term environmental, social and economic gains likely to be seen with well-adapted CA systems, particularly in drylands where drought resilience is going to be of increasing importance [49,62,63].

SLM can be implemented to reduce soil and SOC loss by various runoff interception techniques; forms of physical barriers designed to directly break the structural connectivity components in the landscape and, in doing so, act as rainwater harvesting (RWH) methods. These methods can draw upon in situ, micro-scale structures, capturing water and sediment at the planting zone [64–66] or be larger, macro-scale structures, which are designed to channel water into the cropped area [64,67]. Both approaches have the capacity to increase and maintain SOC by reducing soil loss, capturing sediment and associated SOM, and by improving moisture and nutrient retention, which benefits microbial pools and biomass. Water storage structures, such as reservoirs, can also help to maintain biomass during the dry season. To offset labour costs associated with larger structures, novel techniques utilise existing structures acting as runoff pathways, such as roads and tracks, to harvest overland flow [68].

Table 1. Examples of quantified impacts of sustainable land management (SLM) practice on SOC in sub-Saharan Africa (SSA).

Intervention	Location	Aridity	Impact on SOC	Sample Depth (cm)	Land Management Timeframe (yr)	Associated Yield Impact	Source
Agroforestry	Amhara and Oromia regions, Ethiopia	Semi-arid - subhumid	Total SOC increase 28% relative to traditional maize cropping. $p < 0.05$	0–15	8.5 (median)	n/a	[69]
Agroforestry	Mt Kilimanjaro (Machame to Lake Chala), Tanzania	Semi-arid - humid	Total SOC ~23% increase relative to conventional agriculture	Horizon-based sampling	nd	nd	[31]
Enclosure	Dida-Hara, Ethiopia	Arid - semi-arid	Evidence of total SOC increase 8–30% relative to grazed areas. No statistical significance ($p > 0.05$)	0–30	Enclosure timescales: <20; 20–30; >30	n/a	[70]
Enclosure	Chepareria Ward, Kenya	Semi-arid	Mean total SOC increase by 27% relative to open grazing. POC increase up to 55%. $p < 0.001$	0–40	Enclosure timescales: 3–10; 10–20; >20	n/a	[71]
Enclosure	Amhara and Oromia regions, Ethiopia	Semi-arid - subhumid	Total SOC increase 68%. $p < 0.05$	0–15	13 (median)	n/a	[69]
Fanya-juu	Makueni County, Kenya	Semi-arid	Total SOC increase ~30% relative to conventional agriculture	0–85	35–40	Improved maize yield based on farmer estimates	[66]
Grow Biointensive Sustainable Agriculture (GBSA)	Kilmambogo; Thika; Muranga, Kenya	Semi-arid - subhumid	Total SOC (assumed Walkley and Black method) increase 30% over study period. $p < 0.05$	nd	4	~70% increase in maize; 60% increase in sweet potato	[35]
Homestead (manure application and fallow)	Naberera, Tanzania	Semi-arid	Total C increase > 100% relative to conventional cultivation. $p < 0.05$	0–10	10	nd	[19]

Table 1. *Cont.*

Intervention	Location	Aridity	Impact on SOC	Sample Depth (cm)	Land Management Timeframe (yr)	Associated Yield Impact	Source
Legume intercropping	Salima; Dowa; Balaka; Nkhotakota, Malawi	Sub humid - humid	Total SOC increase 33–73% relative to conventional cropping	0–100	10	nd	[72]
Mulching	Lushoto District, Tanzania	Semi-arid - subhumid	Total SOC increase ~15–30% relative to control (no mulch). $p < 0.05$	0–30	2–3	up to 56% greater maize yield	[73]
Reduced tillage	Buffelsvlei and Zeekoegat, South Africa	Subhumid	Site-specific. Up to 18% increase in total SOC in clay soil. $p < 0.001$	0–30	6–8	No clear links to increasing yield. Greater dependence upon climatic factors and soil type	[49]
Ridge tillage with organic matter (OM) incorporated	Mbozi District, Tanzania	Subhumid	Total SOC increase ~30% using ridges with OM	10–20	1–2	> 100% maize increase compared to traditional tillage; 45% increase in bean yield	[65]

4. Examples of Barriers to SLM Adoption in SSA

Land degradation in SSA has many drivers [74] and, in the past, a top-down approach to SLM has led to failures in effective implementation often owing to a lack of consideration for differing community structures and the environment at the local level. It is now generally recognised that SLM must be targeted to specific communities in participatory approaches, which consider the unique socio-economic and governance contexts as well as environmental factors [15,17,24]. These approaches focus on empowering local communities to identify, manage and ultimately reverse land degradation practices to restore soil health and boost resilience to climate change.

Adoption of SLM is heavily influenced by the perception of benefits associated with carrying out alternative practices, owing to a lack of capital which often evokes a risk averse attitude in rural communities [75]. The use of field-based training and trial plots has become a useful tool for influencing SLM uptake through direct, evidence-led schemes whereby benefits associated with sustainable land practice can be clearly demonstrated [76]. Education and awareness is a common barrier to SLM (Table 2) and, therefore, field-based projects also serve as an important outlet for training and experiential learning to provide the foundation for change at the community level. Once facilitated, there are numerous indicators that social capital, such as shared learning and trust, is important for underpinning community cohesion, with the development of community groups strongly facilitating education and diffusion of knowledge relating to SLM [76–78]. Such groups may also be effective in offsetting opportunity costs associated with labour shortages, another common barrier to SLM uptake [79]. In contrast, however, where systems are based on kinship networks, adoption of SLM can be hindered owing to the perceived insurance role of the network and the sharing of any individual gains during periods of hardship [78].

Although the perception of benefits is highly influential, of crucial importance is the nature in which these benefits are obtained throughout the year, with cash flow and fluidity identified as vital factors in SLM adoption. The ability of poorer households to generate benefits at key times of the year, rather than the magnitude of benefits per se, influences the choice (or combination of choices) of SLM practice [80]. Since many SLM practices accrue direct or indirect benefits with time [63] this often becomes a barrier where poor communities have to rely upon short term gains. This highlights the need for 'quick win' opportunities to work alongside longer-term goals in SLM programmes [53]. The situation can be exacerbated where land tenure is insecure, with rented land often less likely to be invested in for the long-term [78]. Insecure land tenure and resulting land conflicts can impact upon long-term soil health, hindered by a lack of land policy targeted at the local level, which fails to adequately address cultural norms and transboundary relationships between land users [81]. There are additional problems associated with willingness to contribute to the management of shared, common land resources, particularly in pastoral communities, where motivating individual actions for the benefit of a community can create challenges. Exploring the social dynamic in Massai communities in Tanzania, Rabinovich et al. [82] found willingness to protect shared resources was strongly linked to the development of group norms and community identity focused on sustainable practices, with participation in decision making and access to group discussions likely to play a key role in enhancing SLM adoption.

Numerous studies identify inadequate access to reliable, local markets as a barrier to income, in turn, impacting as an opportunity cost upon SLM adoption (Table 2). Problems associated with a lack of variety of produce can mean that markets become flooded with certain produce, leading to price reductions. Additionally, larger markets are often at distance from smallholders who then rely on intermediate traders to purchase their products. Rural communities often receive low prices in such situations and a lack of storage facilities for perishable produce enhances the need to sell upon harvest and further reduces the bargaining power of rural communities [83]. There are, however, contrasting findings with regard to the link between income and SLM adoption. In some examples, accessible local markets enable income generation and allocation of funds into soil conservation measures [84], or perceived income increases likelihood of SLM adoption [76]. In other areas, income is negatively correlated with SLM adoption, particularly where income is generated from off-farm activities, with surplus likely to be allocated to basic household needs rather than to SLM practice [85,86]. This shows the importance of considering opportunity costs associated with capital surplus rather than capital surplus per se. Relationships between income and SLM adoption can also be influenced by land unit distinction where more profitable fields in, for example, more favourable growing locations, are less likely to be targeted for SLM practice [86]. This could potentially create challenges with regard to land management that enhances landscape unit connectivity and subsequent loss of soil and nutrients to erosion [10].

Table 2. Common factors affecting adoption of sustainable land management (SLM) in examples from Tanzania and Kenya.

Intervention	Location	Adoption Factors	Limiting Factors	Source
Rainwater harvesting (RWH) techniques	Chome-Makanya Catchment, Tanzania	-Effective external financial support -Evidence-based benefits -Gradual introduction -Effective local-level governance	-Access to markets -Translation of policy into practice -Access to credit facilities -Willingness of extension officers to incorporate local/traditional knowledge	[75]
RWH, buffer strips, mulching, organic matter (OM) application	Lushoto, Tanzania	-Low risk practices -Evidence-based benefits-Cash flow and liquidity benefits	-Opportunity cost: labour and finance -Perceived risk -Timescale of return	[80]
Intercropping, mulching, manure application	Vihiga and Kakamega counties, Kenya	-Social capital: shared learning -Understanding of plot characteristics	-Education -Access to loans	[77]
Intercropping, reduced tillage, rotation, manure application	Karatu; Mbulu; Mvomero; Kilosa, Tanzania	-Land tenure security -Social capital: shared learning -Market integration -Timing of benefits -Low risk practices	-Tenure (land rental) -Market access -Distance to plot (from residence) -Kin networks	[78]
Mixed soil and water conservation measures	Usambara Mountains and Pare Mountains, Tanzania	-Awareness of soil degradation -Availability of family labour -Social capital: shared labour	-Off farm income	[85]
Terracing, RWH methods	Uporoto Mountains, Tanzania	n/a	-Labour -Reliance on erosion risk practices for short-term benefits	[79]
Conservation agriculture (CA)	Arusha Region, Tanzania	-Knowledge/training -Demonstrable benefits -Flexibility of CA practice -Good organisational structure and linkage of promoting groups -Compatibility with local customs/norms -Acceptability (village leaders)	-Cost and liquidity -Complexity of CA practices -Availability of social networks -Administrative set up at regional level -Market accessibility -Availability of quality control structures (i.e., CA practice monitoring)	[87]

Table 2. *Cont.*

Intervention	Location	Adoption Factors	Limiting Factors	Source
Terracing, Fanya juu, buffer strips, agroforestry, mulching	Usambara Highlands, Tanzania	-Access to extension services -Attitude/willingness	-Size of farm fields/plots -Opportunity costs: finance	[86]
Buffer strips, mulching, tree planting, contour ridges, stall feeding	Kondoa District, Tanzania	-Access to extension services -Household size (labour availability) -Crop income -Market access-Education and awareness	-Cessation of donor funding -Land shortages -Population pressure	[84]
Terracing, Fanya juu, buffer strips	Usambara Highlands, Tanzania	-Micro-credit schemes -Social capital: shared labour -Gradual introduction -Mixed cropping	-Opportunity costs: labour -Market accessibility	[88]
Micro-and macro-RWH	Same District, Tanzania	-Social capital: shared labour -Long-term external (NGO) support -Local governance structure -Field demonstrations	-Opportunity costs: finance; labour -Lack of technical support	[76]

5. Opportunities to Support SLM Adoption through Interdisciplinary and Cross-Sector Collaboration

Clearly successful regeneration and maintenance of SOC for healthy soils requires a detailed understanding of the complex interaction between people and landscape resources at the local-scale [17]. For context here, we describe the approach to assessing climate adaptation gaps and building resilience within the "Jali Ardhi" (Care for the Land) project in Tanzania [13]. Fundamental to this project, and elsewhere, is a participatory approach to developing sustainable land use programmes, aimed at empowering communities to build resilience to soil erosion challenges from the bottom-up, in a system that, whilst receiving external support, ultimately becomes self-sustaining. Kessler et al. [89] describe the importance of building resilience-based stewardship at the village level, underpinned by motivation and stewardship principles, which are both closely linked with awareness of the natural environment, and people-land-resource connectivity [10]. Kessler et al. [89] make a distinction between extrinsic motivation, relating to externally derived incentives, and intrinsic motivation, which is a self-driven appreciation for natural systems. Whilst extrinsic motivation can play an important role in the protection of soil resources (e.g., Kelly et al. [15]), the Jali Ardhi project also aims to build intrinsic motivation by reducing awareness barriers and highlighting landscape-people connections at the village-scale. The preceding sections have described the importance of SOC in underpinning soil health and, in turn, healthy functions and services, which support community resilience. It has also been highlighted that the effectiveness of SLM practices is highly site-specific and detailed environmental (social, economic and natural environment) surveys are, therefore, crucial to support planning. Such surveys should be used to characterise the site to ensure effective targeting of SLM practice and to provide a benchmark against which to assess the effectiveness of SLM. Although we recognise the importance of tailoring these surveys to the local setting, to support knowledge exchange it is important for practitioners to standardise assessment criteria as far as is practicable and the local level land resources assessment methodology (LADA-Local) detailed by Bunning et al. [20] provides one such format. This survey aligns closely with reporting templates developed by WOCAT and offers an opportunity for projects to contribute to the Global SLM Database, providing a valuable open-access resource for decision support.

Within the Jali Ardhi suite of projects we have identified opportunities to supplement the LADA-Local approach with other survey technologies to help to develop a broader understanding of the landscape features and, importantly, to help connect the land users' understanding of their activities to landscape processes and subsequent gain in terms of climate change resilience. For example, Blake et al. [10] describe the use of unmanned aerial vehicle (UAV) surveys and geographic information system (GIS) outputs to identify hydrological connectivity between landscape units. The resulting outputs highlighted the need to consider transboundary linkages across landscape units to reduce soil loss, which ultimately requires inter-community cooperation for effective mitigation. Identification of transboundary connections also offers an opportunity for practitioners to explore the development of payment for ecosystem services (PES) approaches to SLM [17,90,91]. Aerial outputs can provide an important platform for community-based resource mapping exercises and a catalyst for wider debate on community decision making, which are important components in participatory SLM programmes [20,46,89]. Other examples draw on GIS for effective targeting of mitigation practice by considering hydrological pathways and soil type in conjunction with planning for water storage zones [92].

More recently, rapid in situ assessment tools have been developed, which offer user friendly platforms to assist in the assessment of soil health and management. An example is the Land-Potential Knowledge System (LandPKS), a smart phone application technology which aids the collection of local-level biophysical data, including soil carbon and wider soil health characteristics, to support effective land use planning [93]. The LandPKS tool has been trialled in Tanzania with potential to be integrated into national and regional land planning policy frameworks [94]. The wider adoption of such approaches has obvious limitations in terms of access to smart phone services although there is evidence to support growing access to mobile phones amongst rural communities in Tanzania and their use in enhancing agricultural productivity [95]. An additional, complementary technique used in the Jali Ardhi project is rapid screening of SOC and nutrients using a near infrared soil scanner (AgroCares, Netherlands) [96], which enables high spatial resolution measurements of total SOC and nutrients to be obtained in situ. Both the LandPKS and soil scanner can be used, not only to complement the LADA-local process, but also to raise awareness of soil health and functioning as part of a participatory, 'citizen science' approach [96]. We, therefore, adopt a two-way process of undertaking the necessary site characterisations alongside joint learning with the local community using 'new' technologies, helping to negate barriers associated with awareness and building motivation and stewardship potential. Importantly, the combination of survey approaches applied helps to identify SOC pools by considering the active SOC component [20] as well as total SOC and soil nutrients (soil scanner). Potential solutions for land management improvements can then be supported by the use of LandPKS and OM indices (e.g., Palm et al. [40]), together with community collaboration to identify appropriate strategies for change. This approach does require collaboration and support from local institutions for supply and training in the use of the technologies, at least initially, to build knowledge and capacity at the local level. Quantitative analyses can also be integrated with more qualitative local-scale indicators such as soil colour, which can be adopted by communities and used more readily going forward.

The Jali Ardhi approach is summarised in Figure 2 and encapsulates the need for whole farm systems assessment to build resilience within rural communities [12]. The land-people-resource assessment step described above is followed by a selection of potential SLM options (the SLM 'toolbox') for the local environment, drawing upon survey outputs and valuable knowledge exchange databases such as the WOCAT Global SLM database, which provides a platform for cataloguing SLM options and, importantly, a range of mechanisms to offset the common barriers identified in Table 2. Hence, such databases can be effectively utilised for decision support. The process can also be complemented by input from local agricultural advisors. The biophysical survey results and community input derived in step one are then presented with the SLM options 'toolbox' to the community during a participatory co-design process to define achievable SLM targets. It is crucial that all stakeholders are actively involved in this process to identify barriers and solutions from the outset. If transboundary connectivity issues have been identified in step one, then this stage should also involve an inter-community discussion to identify common goals. It is also important for extension officers to take part in the co-design process to ensure future support is in line with community targets, and to help to overcome institutional barriers, particularly those associated with market access and effective credit strategies [12]. Implementation strategies can then be agreed within the community with field-led demonstrations playing an important role in facilitating community group formation and shared learning, and an opportunity to overcome risk aversion.

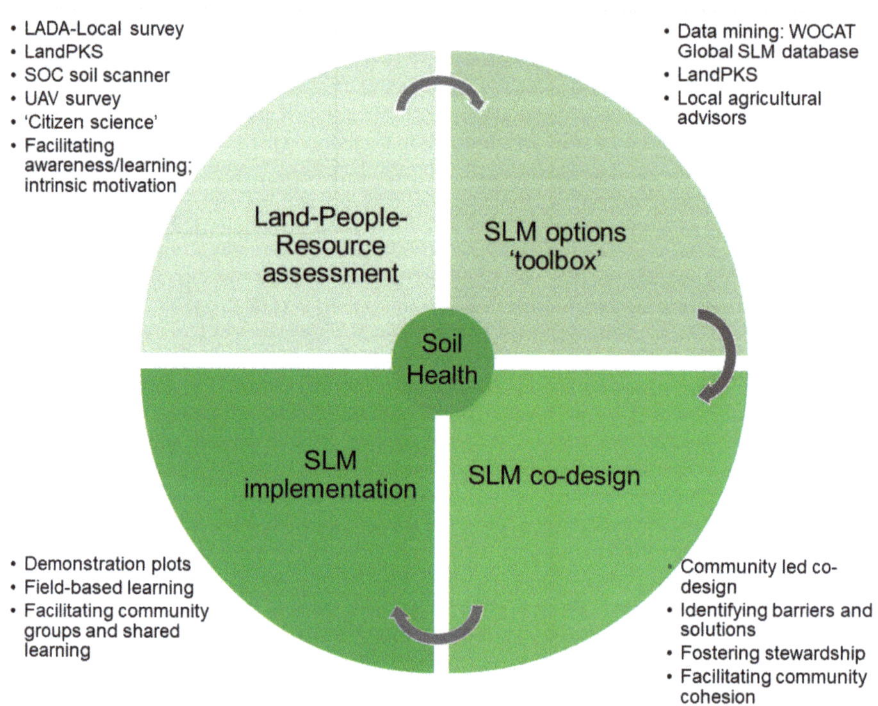

Figure 2. Pathway to change under the Jali Ardhi project.

6. Conclusions

Soil organic carbon plays a pivotal role in soil health and supports a range of ecosystem services crucial for building climate change responses in rural communities in SSA. SOC has been prioritised as a key component for improving soil fertility in SSA with failure to adequately assess local-level social, economic and biophysical factors identified as key barriers to change. Here, we focus upon the importance of SOC as a foundation for soil health and resilience with a key focus upon Tanzania as an exemplar national case. Significant SOC loss through unsustainable agricultural practice leads to soil degradation spirals, exacerbated by climate change, with rural communities often facing challenges associated with returning biomass back to soils to underpin healthy soil functions. Whilst returning adequate OM quantity to soil systems is crucial, practitioners should also focus upon OM quality as a foundation for nutrient pools and fertility, and stimulating soil stability within balanced, diverse farm systems. Here, we present a *whole-system approach* which draws upon standardised local level assessment procedures, supplemented by more recent technologies, which are utilised in participatory soil systems learning programmes alongside local biophysical assessments. Reliable community level assessments are used as the foundation for developing community co-design and implementation of SLM strategies to build resilience to a range of challenges through SOC pools. With predicted intensifying of climate challenges, the approach offers an opportunity to assess climate change adaptation gaps and promote the adoption of locally relevant SLM strategies. Restoring and maintaining soil ecosystems through effective SOC management can protect livelihoods and contribute to the *action on adaptation* targets under the United Nations Climate Change Conference of the Parties 2021 (COP26) agenda.

Author Contributions: Conceptualization, A.T. and W.B.; writing—original draft preparation, A.T., W.B., M.W., L.M.; writing—review and editing, M.W., L.M., C.K., K.M., F.M., P.N., A.P., M.N., A.K., D.G.; funding acquisition, W.B., C.K., P.N., L.M., K.M. All authors have read and agreed to the published version of the manuscript.

Funding: This research was funded by: Research England "QR" Global Challenges Research Funds allocated by the University of Plymouth [Ardhi na Kujifunza (Land and Learning)] building on UK Natural Environment Research Council grant NE/R009309/1 and Research Council UK Global Challenges Research Fund (GCRF) grant NE/P015603/1; Biotechnology and Biological Sciences Research Council grant BB/T012560/1. The study represents a contribution to the joint UN FAO/IAEA Coordinated Research Projects (CRP) "D15017: Nuclear Techniques for a Better Understanding of the Impact of Climate Change on Soil Erosion in Upland Agro-ecosystems" and "D1.50.18: 'Multiple Isotope Fingerprints to Identify Sources and Transport of Agro- Contaminants'".

Institutional Review Board Statement: Not applicable.

Informed Consent Statement: Not applicable.

Data Availability Statement: No new data were created or analyzed in this study. Data sharing is not applicable to this article.

Acknowledgments: The research team would like to acknowledge the commitment and support of the project communities in Arusha District, Tanzania.

Conflicts of Interest: The authors declare no conflict of interest. The funders had no role in the design of the study; in the collection, analyses, or interpretation of data; in the writing of the manuscript, or in the decision to publish the results.

References

1. Kiage, L.M. Perspectives on the assumed causes of land degradation in the rangelands of Sub-Saharan Africa. *Prog. Phys. Geogr. Earth Environ.* **2013**, *37*, 664–684. [CrossRef]
2. Wynants, M.; Kelly, C.; Mtei, K.; Munishi, L.; Patrick, A.; Rabinovich, A.; Nasseri, M.; Gilvear, D.; Roberts, N.; Boeckx, P.; et al. Drivers of increased soil erosion in East Africa's agro-pastoral systems: Changing interactions between the social, economic and natural domains. *Reg. Environ. Chang.* **2019**, *19*, 1909–1921. [CrossRef]
3. IPCC. *Climate Change 2021: The Physical Science Basis. Contribution of Working Group I to the Sixth Assessment Report of the Intergovernmental Panel on Climate Change*; Masson-Delmotte, V., Zhai, P., Pirani, A., Connors, L., Péan, C., Berger, S., Caud, N., Chen, Y., Goldfarb, L., Gomis, M.I., et al., Eds.; Cambridge University Press: Cambridge, UK, 2021; in press.
4. Nearing, M.A.; Pruski, F.F.; O'Neal, M.R. Expected Climate Change Impacts on Soil Erosion Rates: A Review. *J. Soil Water Conserv.* **2004**, *59*, 43–50.
5. Kihara, J.; Bolo, P.; Kinyua, M.; Nyawira, S.S.; Sommer, R. Soil health and ecosystem services: Lessons from sub-Sahara Africa (SSA). *Geoderma* **2020**, *370*, 114342. [CrossRef]
6. Tully, K.; Sullivan, C.; Weil, R.; Sanchez, P. The State of Soil Degradation in Sub-Saharan Africa: Baselines, Trajectories, and Solutions. *Sustainability* **2015**, *7*, 6523–6552. [CrossRef]
7. Fenta, A.A.; Tsunekawa, A.; Haregeweyn, N.; Poesen, J.; Tsubo, M.; Borrelli, P.; Panagos, P.; Vanmaercke, M.; Broeckx, J.; Yasuda, H.; et al. Land susceptibility to water and wind erosion risks in the East Africa region. *Sci. Total. Environ.* **2020**, *703*, 135016. [CrossRef] [PubMed]
8. Wynants, M.; Munishi, L.; Mtei, K.; Bodé, S.; Patrick, A.; Taylor, A.; Gilvear, D.; Ndakidemi, P.; Blake, W.H.; Boeckx, P. Soil erosion and sediment transport in Tanzania: Part I - sediment source tracing in three neighbouring river catchments. *Earth Surf. Process. Landforms* **2021**. [CrossRef]
9. Wynants, M.; Patrick, A.; Munishi, L.; Mtei, K.; Bodé, S.; Taylor, A.; Millward, G.; Roberts, N.; Gilvear, D.; Ndakidemi, P.; et al. Soil erosion and sediment transport in Tanzania: Part II - sedimentological evidence of phased land degradation. *Earth Surf. Process. Landforms* **2021**. [CrossRef]
10. Blake, W.H.; Kelly, C.; Wynants, M.; Patrick, A.; Lewin, S.; Lawson, J.; Nasolwa, E.; Page, A.; Nasseri, M.; Marks, C.; et al. Integrating land-water-people connectivity concepts across disciplines for co-design of soil erosion solutions. *Land Degrad. Dev.* **2020**, *32*, 3415–3430. [CrossRef]
11. Nkiaka, E.; Okpara, U.T.; Okumah, M. Food-energy-water security in sub-Saharan Africa: Quantitative and spatial assessments using an indicator-based approach. *Environ. Dev.* **2021**, 100655. [CrossRef]
12. Descheemaeker, K.; Oosting, S.J.; Tui, S.H.-K.; Masikati, P.; Falconnier, G.N.; Giller, K.E. Climate change adaptation and mitigation in smallholder crop–livestock systems in sub-Saharan Africa: A call for integrated impact assessments. *Reg. Environ. Chang.* **2016**, *16*, 2331–2343. [CrossRef]

13. Blake, W.H.; Rabinovich, A.; Wynants, M.; Kelly, C.; Nasseri, M.; Ngondya, I.; Patrick, A.; Mtei, K.; Munishi, L.; Boeckx, P.; et al. Soil erosion in East Africa: An interdisciplinary approach to realising pastoral land management change. *Environ. Res. Lett.* **2018**, *13*, 124014. [CrossRef]

14. Marques, M.J.; Schwilch, G.; Lauterburg, N.; Crittenden, S.; Tesfai, M.; Stolte, J.; Zdruli, P.; Zucca, C.; Petursdottir, T.; Evelpidou, N.; et al. Multifaceted Impacts of Sustainable Land Management in Drylands: A Review. *Sustainability* **2016**, *8*, 177. [CrossRef]

15. Kelly, C.; Wynants, M.; Munishi, L.K.; Nasseri, M.; Patrick, A.; Mtei, K.M.; Mkilema, F.; Rabinovich, A.; Gilvear, D.; Wilson, G.; et al. 'Mind the Gap': Reconnecting Local Actions and Multi-Level Policies to Bridge the Governance Gap. An Example of Soil Erosion Action from East Africa. *Land* **2020**, *9*, 352. [CrossRef]

16. FAO. *Soil Organic Carbon: The Hidden Potential*; Food and Agriculture Organisation of the United Nations: Rome, Italy, 2017.

17. Liniger, H.P.R.; Studer, M.; Hauert, C.; Gurtner, M. Sustainable Land Management in Practice: Guidelines and Best Practices for Sub-Saharan Africa; TerrAfrica, World Overview of Conservation Approaches and Technologies (WOCAT) and Food and Ag-riculture Organization of the United Nations (FAO). 2011. Available online: http://www.fao.org/3/i1861e/i1861e00.pdf (accessed on 24 May 2021).

18. Duncan, A.J.; Bachewe, F.; Mekonnen, K.; Valbuena, D.; Rachier, G.; Lule, D.; Bahta, M.; Erenstein, O. Crop residue allocation to livestock feed, soil improvement and other uses along a productivity gradient in Eastern Africa. *Agric. Ecosyst. Environ.* **2016**, *228*, 101–110. [CrossRef]

19. Solomon, D.; Lehmann, J.; Zech, W. Land use effects on soil organic matter properties of chromic luvisols in semi-arid northern Tanzania: Carbon, nitrogen, lignin and carbohydrates. *Agric. Ecosyst. Environ.* **2000**, *78*, 203–213. [CrossRef]

20. Bunning, S.; Mcdonagh, J.; Rioux, J. *Land Degradation Assessment in Drylands. Manual for Local Level Assessment of Land Deg-radation and Sustainable Land Managment. Part 2: Field Methodology & Tools*; Food and Agriculture Organisation of the United Nations: Rome, Italy, 2016.

21. Winowiecki, L.; Vågen, T.-G.; Massawe, B.; Jelinski, N.A.; Lyamchai, C.; Sayula, G.; Msoka, E. Landscape-scale variability of soil health indicators: Effects of cultivation on soil organic carbon in the Usambara Mountains of Tanzania. *Nutr. Cycl. Agroecosystems* **2016**, *105*, 263–274. [CrossRef]

22. Middendorf, B.J.; Pierzynski, G.M.; Stewart, Z.P.; Vara Prasad, P.V.; Middendorf, B.J.; Pierzynski, G.M.; Stewart, Z.P.; Prasad, P.V.V. *Sub-Saharan Africa Soil Fertility Prioritisation Report II. Summit Results*; SIIL Kansas State University: Manhattan, KS, USA, 2017.

23. Stewart, Z.P.; Pierzynski, G.M.; Middendorf, B.J.; Prasad, P.V.V. Approaches to improve soil fertility in sub-Saharan Africa. *J. Exp. Bot.* **2020**, *71*, 632–641. [CrossRef]

24. Rabinovich, A.; Kelly, C.; Wilson, G.; Nasseri, M.; Ngondya, I.; Patrick, A.; Blake, W.H.; Mtei, K.; Munishi, L.; Ndakidemi, P. "We will change whether we want it or not": Soil erosion in Maasai land as a social dilemma and a challenge to community resilience. *J. Environ. Psychol.* **2019**, *66*, 101365. [CrossRef]

25. Baveye, P.C.; Schnee, L.S.; Boivin, P.; Laba, M.; Radulovich, R. Soil Organic Matter Research and Climate Change: Merely Re-storing Carbon Versus Restoring Soil Functions. *Front. Environ. Sci.* **2020**, *8*, 161. [CrossRef]

26. Woodfine, A. Using Sustainable Land Managment Practices to Adapt and Mitigate Climate Change in Sub-Saharan Africa. Resource Guide Version 1.0, TerrAfrica Regional Sustainable Land Management. 2009. Available online: http://www.ipcinfo.org/fileadmin/user_upload/terrafrica/docs/SLM_SUB-SAHARAN_AFRICA.pdf (accessed on 1 August 2009).

27. Nyambo, P.; Chiduza, C.; Araya, T. Effect of conservation agriculture on selected soil physical properties on a haplic cambisol in Alice, Eastern Cape, South Africa. *Arch. Agron. Soil Sci.* **2020**, 1–14. [CrossRef]

28. Burns, R.G.; DeForest, J.; Marxsen, J.; Sinsabaugh, R.L.; Stromberger, M.E.; Wallenstein, M.; Weintraub, M.; Zoppini, A. Soil enzymes in a changing environment: Current knowledge and future directions. *Soil Biol. Biochem.* **2013**, *58*, 216–234. [CrossRef]

29. Cotrufo, M.F.; Wallenstein, M.D.; Boot, C.M.; Denef, K.; Paul, E. The Microbial Efficiency-Matrix Stabilization (MEMS) framework integrates plant litter decomposition with soil organic matter stabilization: Do labile plant inputs form stable soil organic matter? *Glob. Chang. Biol.* **2013**, *19*, 988–995. [CrossRef] [PubMed]

30. Wood, S.A.; Sokol, N.; Bell, C.W.; Bradford, M.A.; Naeem, S.; Wallenstein, M.D.; Palm, C.A. Opposing effects of different soil organic matter fractions on crop yields. *Ecol. Appl.* **2016**, *26*, 2072–2085. [CrossRef]

31. Pabst, H.; Gerschlauer, F.; Kiese, R.; Kuzyakov, Y. Land Use and Precipitation Affect Organic and Microbial Carbon Stocks and the Specific Metabolic Quotient in Soils of Eleven Ecosystems of Mt. Kilimanjaro, Tanzania. *Land Degrad. Dev.* **2015**, *27*, 592–602. [CrossRef]

32. Mkonda, M.Y.; He, X. Stocks and Ecological Significance of Soil Carbon in Tanzania. *Nat. Resour. Conserv.* **2016**, *4*, 42–51. [CrossRef]

33. Obalum, S.; Chibuike, G.; Peth, S.; Ouyang, Y. Soil organic matter as sole indicator of soil degradation. *Environ. Monit. Assess.* **2017**, *189*, 176. [CrossRef]

34. Laban, P.; Metternicht, G.; Davies, J. *Soil Biodiversity and Soil Organic Carbon: Keeping Drylands Alive*; IUCN: Gland, Switzerland, 2018.

35. Beeby, J.; Moore, S.; Taylor, L.; Nderitu, S. Effects of a One-Time Organic Fertilizer Application on Long-Term Crop and Residue Yields, and Soil Quality Measurements Using Biointensive Agriculture. *Front. Sustain. Food Syst.* **2020**, *4*, 4. [CrossRef]

36. Oldfield, E.E.; Bradford, M.A.; Wood, S.A. Global meta-analysis of the relationship between soil organic matter and crop yields. *SOIL* **2019**, *5*, 15–32. [CrossRef]

37. Gram, G.; Roobroeck, D.; Pypers, P.; Six, J.; Merckx, R.; Vanlauwe, B. Combining organic and mineral fertilizers as a climate-smart integrated soil fertility management practice in sub-Saharan Africa: A meta-analysis. *PLoS ONE* **2020**, *15*, e0239552. [CrossRef]

38. Verchot, L.V.; Dutaur, L.; Shepherd, K.D.; Albrecht, A. Organic matter stabilization in soil aggregates: Understanding the biogeochemical mechanisms that determine the fate of carbon inputs in soils. *Geoderma* **2011**, *161*, 182–193. [CrossRef]

39. Barthès, B.G.; Kouakoua, E.; Larré-Larrouy, M.-C.; Razafimbelo, T.M.; de Luca, E.F.; Azontonde, A.; Neves, C.; DE Freitas, P.L.; Feller, C.L. Texture and sesquioxide effects on water-stable aggregates and organic matter in some tropical soils. *Geoderma* **2008**, *143*, 14–25. [CrossRef]

40. A Palm, C.; Gachengo, C.N.; Delve, R.J.; Cadisch, G.; Giller, K. Organic inputs for soil fertility management in tropical agroecosystems: Application of an organic resource database. *Agric. Ecosyst. Environ.* **2001**, *83*, 27–42. [CrossRef]

41. Gentile, R.; Vanlauwe, B.; Chivenge, P.; Six, J. Interactive effects from combining fertilizer and organic residue inputs on nitrogen transformations. *Soil Biol. Biochem.* **2008**, *40*, 2375–2384. [CrossRef]

42. Bayu, T. Review on contribution of integrated soil fertility management for climate change mitigation and agricultural sustainability. *Cogent Environ. Sci.* **2020**, *6*. [CrossRef]

43. Kimiti, D.W.; Riginos, C.; Belnap, J. Low-cost grass restoration using erosion barriers in a degraded African rangeland. *Restor. Ecol.* **2016**, *25*, 376–384. [CrossRef]

44. Wolka, K.; Mulder, J.; Biazin, B. Effects of soil and water conservation techniques on crop yield, runoff and soil loss in Sub-Saharan Africa: A review. *Agric. Water Manag.* **2018**, *207*, 67–79. [CrossRef]

45. FAO. *Sustainable Land Management (SLM) in Practice in the Kagera Basin. Lessons Learned for Scaling up at Landscape Level—Results of the Kagera Transboundary Agro-Ecosystem Management Project (Kagera TAMP)*; Food and Agricultural Organisation of the United Nations: Rome, Italy, 2017.

46. Ziadat, F.; Bunning, S.; De Pauw, E. *Land Resource Planning for Sustainable Land Management*; Food and Agricultural Organisation of the United Nations: Rome, Italy, 2017.

47. Palm, C.; Blanco-Canqui, H.; DeClerck, F.; Gatere, L.; Grace, P. Conservation agriculture and ecosystem services: An overview. *Agric. Ecosyst. Environ.* **2014**, *187*, 87–105. [CrossRef]

48. Yayneshet, T.; Treydte, A. A meta-analysis of the effects of communal livestock grazing on vegetation and soils in sub-Saharan Africa. *J. Arid. Environ.* **2015**, *116*, 18–24. [CrossRef]

49. Swanepoel, C.; Rötter, R.; van der Laan, M.; Annandale, J.; Beukes, D.; du Preez, C.; Swanepoel, L.; van der Merwe, A.; Hoffmann, M. The benefits of conservation agriculture on soil organic carbon and yield in southern Africa are site-specific. *Soil Tillage Res.* **2018**, *183*, 72–82. [CrossRef]

50. Verhulst, N.; Francois, I.; Govaerts, B. Conservation Agriculture, Improving Soil Quality for Sustainable Production Systems? In *Advances in Soil Science:Food Security and Soil Quality*; Lal, R., Stewart, B.A., Eds.; CRC Press: Boca Raton, FL, USA, 2010; pp. 137–208.

51. Liniger, H.P.R.; Studer, M. Sustainable Rangeland Management in Sub-Saharan Africa - Guidelines to Good Practice; TerrAfrica; World Bank, Washington D.C.; World Overview of Conservation Approaches and Technologies (WOCAT), World Bank Group (WBG), Washington D.C., USA and Centre for Development and Environment (CDE), University of Bern, Bern, Switzerland. 2019. Available online: https://www.wocat.net/library/media/174/ (accessed on 13 September 2019).

52. Verdoodt, A.; Mureithi, S.M.; Van Ranst, E. Impacts of management and enclosure age on recovery of the herbaceous rangeland vegetation in semi-arid Kenya. *J. Arid. Environ.* **2010**, *74*, 1066–1073. [CrossRef]

53. Mureithi, S.M.; Verdoodt, A.; Njoka, J.T.; Gachene, C.K.K.; Van Ranst, E. Benefits Derived from Rehabilitating a Degraded Semi-Arid Rangeland in Communal Enclosures, Kenya. *Land Degrad. Dev.* **2016**, *27*, 1853–1862. [CrossRef]

54. Wairore, J.N.; Mureithi, S.M.; Wasonga, O.V.; Nyberg, G. Enclosing the commons: Reasons for the adoption and adaptation of enclosures in the arid and semi-arid rangelands of Chepareria, Kenya. *SpringerPlus* **2015**, *4*, 595. [CrossRef]

55. Veblen, K.E.; Porensky, L.M. Thresholds are in the eye of the beholder: Plants and wildlife respond differently to short-term cattle corrals. *Ecol. Appl.* **2019**, *29*, e01982. [CrossRef]

56. Sogbedji, J.M.; Van Es, H.M.; Agbeko, K.L. Cover Cropping and Nutrient Management Strategies for Maize Production in Western Africa. *Agron. J.* **2006**, *98*, 883–889. [CrossRef]

57. Kaye, J.P.; Quemada, M. Using cover crops to mitigate and adapt to climate change. A review. *Agron. Sustain. Dev.* **2017**, *37*, 4. [CrossRef]

58. Okeyo, A.; Mucheru-Muna, M.; Mugwe, J.; Ngetich, K.; Mugendi, D.; Diels, J.; Shisanya, C. Effects of selected soil and water conservation technologies on nutrient losses and maize yields in the central highlands of Kenya. *Agric. Water Manag.* **2014**, *137*, 52–58. [CrossRef]

59. Branca, G.; Lipper, L.; McCarthy, N.; Jolejole, M.C. Food security, climate change, and sustainable land management. A review. *Agron. Sustain. Dev.* **2013**, *33*, 635–650. [CrossRef]

60. Castellanos-Navarrete, A.; Tittonell, P.; Rufino, M.; Giller, K. Feeding, crop residue and manure management for integrated soil fertility management – A case study from Kenya. *Agric. Syst.* **2015**, *134*, 24–35. [CrossRef]

61. Margenot, A.; Rao, I.M.; Sommer, R. Phosphorus Fertilization and Managment in Soils of Sub-Saharan Africa. In *Soil Phos-phorus (Advances in Soil Science)*; Lal, R., Stewart, B.A., Eds.; CRC Press: Boca Raton, FL, USA, 2016; pp. 151–208.

62. Corbeels, M.; Naudin, K.; Whitbread, A.M.; Kühne, R.; Letourmy, P. Limits of conservation agriculture to overcome low crop yields in sub-Saharan Africa. *Nat. Food* **2020**, *1*, 447–454. [CrossRef]

63. Page, K.L.; Dang, Y.P.; Dalal, R.C. The Ability of Conservation Agriculture to Conserve Soil Organic Carbon and the Subsequent Impact on Soil Physical, Chemical, and Biological Properties and Yield. *Front. Sustain. Food Syst.* **2020**, *4*, 31. [CrossRef]

64. Mbilinyi, B.; Tumbo, S.; Mahoo, H.; Senkondo, E.; Hatibu, N. Indigenous knowledge as decision support tool in rainwater harvesting. *Phys. Chem. Earth Parts A/B/C* **2005**, *30*, 792–798. [CrossRef]

65. Malley, Z.J.; Mzimbiri, M.K.; Mwakasendo, J.A. Integrating local knowledge with science and technology in management of soil, water and nutrients: Implications for management of natural capital for sustainable rural livelihoods. *Int. J. Sustain. Dev. World Ecol.* **2009**, *16*, 151–163. [CrossRef]

66. Saiz, G.; Wandera, F.M.; Pelster, D.; Ngetich, W.; Okalebo, J.R.; Rufino, M.; Butterbach-Bahl, K. Long-term assessment of soil and water conservation measures (Fanya-juu terraces) on soil organic matter in South Eastern Kenya. *Geoderma* **2016**, *274*, 1–9. [CrossRef]

67. Taye, G.; Poesen, J.; van Wesemael, B.; Vanmaercke, M.; Teka, D.; Deckers, J.; Goosse, T.; Maetens, W.; Nyssen, J.; Hallet, V.; et al. Effects of land use, slope gradient, and soil and water conservation structures on runoff and soil loss in semi-arid Northern Ethiopia. *Phys. Geogr.* **2013**, *34*, 236–259. [CrossRef]

68. Woldearegay, K.; Van Steenbergen, F.; Perez, M.A.; Grum, B.; Van Beusekom, M. Water Harvesting from Roads: Climate Resilience in Tigray, Ethiopia. In Proceedings of the IRF Europe & Central Asia Regional Congress, Istanbul, Turkey, 15–18 September 2015.

69. Rimhanen, K.; Ketoja, E.; Yli-Halla, M.; Kahiluoto, H. Ethiopian agriculture has greater potential for carbon sequestration than previously estimated. *Glob. Chang. Biol.* **2016**, *22*, 3739–3749. [CrossRef]

70. Feyisa, K.; Beyene, S.; Angassa, A.; Said, M.Y.; de Leeuw, J.; Abebe, A.; Megersa, B. Effects of enclosure management on carbon sequestration, soil properties and vegetation attributes in East African rangelands. *Catena* **2017**, *159*, 9–19. [CrossRef]

71. Oduor, C.O.; Karanja, N.K.; Onwonga, R.N.; Mureithi, S.M.; Pelster, D.; Nyberg, G. Enhancing soil organic carbon, particulate organic carbon and microbial biomass in semi-arid rangeland using pasture enclosures. *BMC Ecol.* **2018**, *18*, 1–9. [CrossRef]

72. Simwaka, P.L.; Tesfamariam, E.H.; Ngwira, A.R.; Chirwa, P.W. Carbon sequestration and selected hydraulic characteristics under conservation agriculture and traditional tillage practices in Malawi. *Soil Res.* **2020**, *58*, 759. [CrossRef]

73. Mwango, S.B.; Msanya, B.M.; Mtakwa, P.W.; Kimaro, D.N.; Deckers, J.; Poesen, J. Effectiveness OF Mulching Under Miraba in Controlling Soil Erosion, Fertility Restoration and Crop Yield in the Usambara Mountains, Tanzania. *Land Degrad. Dev.* **2014**, *27*, 1266–1275. [CrossRef]

74. Kariuki, R.W.; Munishi, L.K.; Courtney-Mustaphi, C.J.; Capitani, C.; Shoemaker, A.; Lane, P.J.; Marchant, R. Integrating stakeholders' perspectives and spatial modelling to develop scenarios of future land use and land cover change in northern Tanzania. *PLoS ONE* **2021**, *16*, e0245516. [CrossRef]

75. Chikozho, C. Policy and institutional dimensions of small-holder farmer innovations in the Thukela River Basin of South Africa and the Pangani River Basin of Tanzania: A comparative perspective. *Phys. Chem. Earth Parts A/B/C* **2005**, *30*, 913–924. [CrossRef]

76. Tumbo, S.; Mutabazi, K.; Byakugila, M.; Mahoo, H. An empirical framework for scaling-out of water system innovations: Lessons from diffusion of water system innovations in the Makanya catchment in Northern Tanzania. *Agric. Water Manag.* **2011**, *98*, 1761–1773. [CrossRef]

77. Kanyenji, G.M.; Oluoch-Kosura, W.; Onyango, C.M.; Ng'Ang'A, S.K. Prospects and constraints in smallholder farmers' adoption of multiple soil carbon enhancing practices in Western Kenya. *Heliyon* **2020**, *6*, e03226. [CrossRef] [PubMed]

78. Kassie, M.; Jaleta, M.; Shiferaw, B.; Mmbando, F.; Mekuria, M. Adoption of interrelated sustainable agricultural practices in smallholder systems: Evidence from rural Tanzania. *Technol. Forecast. Soc. Chang.* **2013**, *80*, 525–540. [CrossRef]

79. Mwanukuzi, P.K. Impact of non-livelihood-based land management on land resources: The case of upland watersheds in Uporoto Mountains, South West Tanzania. *Geogr. J.* **2011**, *177*, 27–34. [CrossRef] [PubMed]

80. Emerton, L.; Snyder, K.A. Rethinking sustainable land management planning: Understanding the social and economic drivers of farmer decision-making in Africa. *Land Use Policy* **2018**, *79*, 684–694. [CrossRef]

81. Rubakula, G.; Wang, Z.; Wei, C. Land Conflict Management through the Implementation of the National Land Policy in Tanzania: Evidence from Kigoma Region. *Sustainability* **2019**, *11*, 6315. [CrossRef]

82. Rabinovich, A.; Heath, S.C.; Zhischenko, V.; Mkilema, F.; Patrick, A.; Nasseri, M.; Wynants, M.; Blake, W.H.; Mtei, K.; Munishi, L.; et al. Protecting the commons: Predictors of willingness to mitigate communal land degradation among Maasai pastoralists. *J. Environ. Psychol.* **2020**, *72*, 101504. [CrossRef]

83. Nakawuka, P.; Langan, S.; Schmitter, P.; Barron, J. A review of trends, constraints and opportunities of smallholder irrigation in East Africa. *Glob. Food Secur.* **2018**, *17*, 196–212. [CrossRef]

84. Shrestha, R.P.; Ligonja, P.J. Social perception of soil conservation benefits in Kondoa eroded area of Tanzania. *Int. Soil Water Conserv. Res.* **2015**, *3*, 183–195. [CrossRef]

85. Mbaga-Semgalawe, Z.; Folmer, H. Household adoption behaviour of improved soil conservation: The case of the North Pare and West Usambara Mountains of Tanzania. *Land Use Policy* **2000**, *17*, 321–336. [CrossRef]

86. Nyanga, A.; Kessler, A.; Tenge, A. Key socio-economic factors influencing sustainable land management investments in the West Usambara Highlands, Tanzania. *Land Use Policy* **2016**, *51*, 260–266. [CrossRef]

87. Ndah, H.T.; Schuler, J.; Uthes, S.; Zander, P.; Triomphe, B.; Mkomwa, S.; Corbeels, M. Adoption Potential for Conservation Agriculture in Africa: A Newly Developed Assessment Approach (QAToCA) Applied in Kenya and Tanzania. *Land Degrad. Dev.* **2015**, *26*, 133–141. [CrossRef]

88. Tenge, A.; De Graaff, J.; Hella, J. Financial efficiency of major soil and water conservation measures in West Usambara highlands, Tanzania. *Appl. Geogr.* **2005**, *25*, 348–366. [CrossRef]

89. Kessler, A.; Van Reemst, L.; Beun, M.; Slingerland, E.; Pol, L.; De Winne, R. Mobilising farmers to stop land degradation: A different discourse from Burundi. *Land Degrad. Dev.* **2020**, *32*, 3403–3414. [CrossRef]

90. Branca, G.; Lipper, L.; Neves, B.; Lopa, D.; Mwanyoka, I. Payments for Watershed Services Supporting Sustainable Agricultural Development in Tanzania. *J. Environ. Dev.* **2011**, *20*, 278–302. [CrossRef]

91. Capitani, C.; Mukama, K.; Mbilinyi, B.; Malugu, I.; Munishi, P.; Burgess, N.; Platts, P.; Sallu, S.; Marchant, R. From local scenarios to national maps: A participatory framework for envisioning the future of Tanzania. *Ecol. Soc.* **2016**, *21*, 4. [CrossRef]

92. Mbilinyi, B.; Tumbo, S.; Mahoo, H.; Mkiramwinyi, F. GIS-based decision support system for identifying potential sites for rainwater harvesting. *Phys. Chem. Earth Parts A/B/C* **2007**, *32*, 1074–1081. [CrossRef]

93. Herrick, J.E.; Bouvier, I.; Coetzee, M.; Elias, E.; Karl, J.W.; Liniger, H.; Matuszak, J.; Ndungu, L.W.; Obersteiner, M.; Shepherd, K.D.; et al. The land?potential knowledge system (landpks): Mobile apps and collaboration for optimizing climate change investments. *Ecosyst. Health Sustain.* **2016**, *2*, 01209. [CrossRef]

94. Quandt, A.; Herrick, J.; Bouvier, I. LandPKS: A New Mobile Tool for Sustainable Land-Use Planning and Management. In Proceedings of the 2018 World Bank Conference on Land and Poverty, Washington, DC, USA, 22 March 2018.

95. Quandt, A.; Salerno, J.D.; Neff, J.C.; Baird, T.D.; Herrick, J.E.; McCabe, J.T.; Xu, E.; Hartter, J. Mobile phone use is associated with higher smallholder agricultural productivity in Tanzania, East Africa. *PLoS ONE* **2020**, *15*, e0237337. [CrossRef] [PubMed]

96. Kelly, C. *Soils, Science and Community ActioN (SoilSCAN) to Reduce Land Degradation in East Africa*; University of Plymouth: Plymouth, UK, 2021; in press.

sustainability

Article

Analyzing the Spatial Heterogeneity of the Built Environment and Its Impact on the Urban Thermal Environment—Case Study of Downtown Shanghai

Jiejie Han [1,2,*], Xi Zhao [1], Hao Zhang [1,*] and Yu Liu [1,3]

[1] Laboratory for Applied Earth Observation and Spatial Analysis, Department of Environmental Science and Engineering, Jiangwan Campus, Fudan University, Shanghai 200438, China; 19210740064@fudan.edu.cn (X.Z.); liuyu1982@henu.edu.cn (Y.L.)

[2] Guangzhou Urban Planning & Design Survey Research Institute, Guangzhou 510030, China

[3] School of Civil Engineering and Architecture, Henan University, Kaifeng 475003, China

* Correspondence: fudan_jiejiehan@163.com (J.H.); zhokzhok@163.com (H.Z.)

Abstract: Ongoing urban expansion has accelerated the explosive growth of urban populations and has led to a dramatic increase in the impervious surface area within urban areas. This, in turn, has exacerbated the surface heat island effect within cities. However, the importance of the surface heat island effect within urban areas, scilicet the intra-SUHI effect, has attracted less concern. The aim of this study was to quantitatively explore the relationship between the spatial heterogeneity of a built environment and the intra-urban surface heat island (intra-SUHI) effect using the thermally sharpened land surface temperature (LST) and high-resolution land-use classification products. The results show that at the land parcel scale, the parcel-based relative intensity of intra-SUHI should be attributed to the land parcels featured with differential land developmental intensity. Furthermore, the partial least squares regression (PLSR) modeling quantified the relative importance of the spatial heterogeneity indices of the built environment that exhibit a negative contribution to decreasing the parcel-based intra-SUHI effect or a positive contribution to increasing the intra-SUHI effect. Finally, based on the findings of this study, some practical countermeasures towards mitigating the adverse intra-SUHI effect and improving urban climatic adaption are discussed.

Keywords: built-up environment; spatial heterogeneity; urban thermal environment; blue–green space; land use pattern

Citation: Han, J.; Zhao, X.; Zhang, H.; Liu, Y. Analyzing the Spatial Heterogeneity of the Built Environment and Its Impact on the Urban Thermal Environment—Case Study of Downtown Shanghai. *Sustainability* **2021**, *13*, 11302. https://doi.org/10.3390/su132011302

Academic Editors: Baojie He, Ayyoob Sharifi, Chi Feng and Jun Yang

Received: 15 August 2021
Accepted: 11 October 2021
Published: 13 October 2021

Publisher's Note: MDPI stays neutral with regard to jurisdictional claims in published maps and institutional affiliations.

1. Introduction

Since the era of the industrial revolution, driven by capital flows and labor transfers, urban agglomerations and metropolitan areas have become the preferred destinations for urban–rural and cross-border migrants worldwide [1–3]. At present, global urban areas house more than 50% of the total population and 70–90% of all economic activities [4], resulting in unprecedented urban expansion and explosive growth of urban populations. The United Nations Department of Economic and Social Affairs has reported that about 55% of the global population lives in urban areas. By 2050, global urban residents will increase by 2.5 billion, of which 255 million will be in China, such that the urban population will account for 68% of the country's total population [5].

During urban expansion, impervious surfaces, such as buildings, roads, squares, bridges, and parking lots, dominate the land-use structure of the built environment. They consequently occupy and replace the area proportion of the pre-development landscapes (e.g., water bodies and vegetation) [6]. The intensive human activities in cities lead to severe ecological degradation [7], which has the consequence of artificial modification of the urban climate; in particular, the urban heat island (UHI) effect and its influences on human health have been of wide concern. To restore the urban natural environment and

121

improve urban resilience, artificial blue and green space (BGS), including water bodies, vegetation, and recreational landscapes, can be created and managed in order to deliver critical ecosystem services (e.g., air purification and climate modification) within cities [8–10]. Unfortunately, in most cases, impervious surfaces dominate the urban built environment's land-use structure and landscape pattern. Intensive urban land development usually alters the surface thermal energy radiation, biological characteristics, and hydrological cycle, resulting in changes in the absorption and emission of solar radiation and the surface heat flux [11–13]. Meanwhile, due to its relatively small area proportion within cities, the BGS that promotes local cooling effects is insufficient to offset the overwhelming heat emissions from the impervious surfaces, which heat the lowest layer of the air [14]. Such surface-to-air heating processes have been proven to profoundly impact the urban thermal environment [15–17]; for instance, artificial modification of the urban climate is closely related to the UHI effect, as measured in terms of air temperature (AT) and land surface temperature (LST).

To date, existing studies on the urban thermal environment have mainly focused on the UHI effect, including the boundary layer urban heat island (BLUHI) and canopy urban heat island (CUHI), characterized by AT, as well as the surface urban heat island (SUHI) effect, characterized by LST. The AT can generally be easily measured and used for assessing human thermal comfort, but the sparsity of weather stations within cities and in the urban fringe makes it difficult to characterize the spatial variation of AT, BLUHI, and CUHI in the study area. In contrast, satellite- or air-borne thermal remote sensing platforms, with sizable spatial cover, can provide alternative approaches for monitoring urban climates.

Since the 1970s, spaceborne thermal remote sensing technology has become a practical approach for monitoring regional and local SUHI effects. Previous studies on the multi-scale SUHI effect using spaceborne thermal infrared (TIR) data have produced fruitful results, ranging from low-resolution(~km) sensors, such as the Geostationary Operational Environmental Satellite (GOES), the Advanced Very High-Resolution Radiometer (AVHRR), and the Moderate-Resolution Imaging Spectroradiometer (MODIS), to high-resolution (60–120 m) sensors, such as Landsat 5 (Theme Mapper, TM), Landsat 7 (Enhanced Theme Mapper Plus, ETM+), Landsat 8 Operational Land Imager/Thermal Infrared Sensor (OLI/TIRS), and the Advanced Spaceborne Thermal Emission Reflection radiometer (ASTER) [18–20]. At present, given that most urban residents live in intensively developed land lacking green infrastructure, few studies have been carried out on the intra-SUHI effect, which indicates the SUHI effect within urban functioning zones (UFZs) and is closely related to human health and urban climate adaptation. In the context of urban settings characterized by complex land-use structure and landscape configuration, the formation of the intra-SUHI effect largely depends on the fine-scale land developmental features (e.g., land-use types, floor area ratio, building distances and heights, and landscape patterns of specific land parcels) [21,22]. It should be noted that, for these above-mentioned spaceborne TIR sensors, due to the problem of mixing pixels of land surfaces, they are still too coarse to generate detailed information regarding the LST and heat flux and, thus, it is impossible to depict the fine-scale pattern of the urban thermal environment when using such data. In contrast, the airborne high-resolution TIR imaging systems, such as NASA's advanced thermal infrared and land application sensor (ATLAS) and unmanned aerial vehicle (UAV)-borne TIR cameras, can effectively detect the fine-scale thermal effect of urban settings. However, the apparent shortcomings of ATLAS (e.g., fixed navigation routes and low cost-effectiveness) and UAV systems (e.g., flexible navigation routes but short flying durations) limit their applicability in the practice of detecting the variations in the urban thermal environment. Alternatively, recent studies have attempted to combine thermally sharpened satellite-retrieved data with commercial high-resolution optical images in order to provide a practical approach for evaluating the intra-SUHI effect [23,24]. Such an approach can provide a better understanding of the relationship between the spatial heterogeneity of the built environment and the urban thermal environment, considering that the recent literature emphasizing such relationships is relatively scarce. Moreover, the robustness and applicability of such approaches in practice need further testing in a variety of case studies.

In this study, Shanghai—one of the fast-growing megacities suffering from extreme summertime heat events—was taken as a case study. Our research goals were (1) to quantitatively analyze the spatial heterogeneity of the built environment and its impact on the summertime intra-SUHI effect within the city, and (2), based on the findings of this study, to provide the operational choices for decision making towards enhancing urban planning practices, mitigating the intra-SUHI effect, and improving urban climate adaption.

2. Study Area

Shanghai is located between latitudes 30°40′ N–31°53′ N and longitudes 120°52′ E–122°12′ E, in the front of the alluvial plain of the Yangtze River Delta (see Figure 1). The whole city is low-lying and flat, dominated by plains (with an area of 93.91%) and an average altitude of 2.19 m. Shanghai is located in the north sub-tropical monsoon climate zone, with abundant sunshine, abundant rain, and four obvious seasons. Local vegetation types are dominated by evergreen broad-leaved forest and evergreen deciduous broad-leaved mixed forest [25]. At present, the wetland stock is 464,600 hectares, and the forest coverage rate of the whole city reaches 17.6%. In this study, four typical urban functioning zones (UFZs), which represent the socio-economic features and urban land-use patterns within downtown Shanghai (Table 1), were used to investigate the relationship between the urban built environment and intra-SUHI effect.

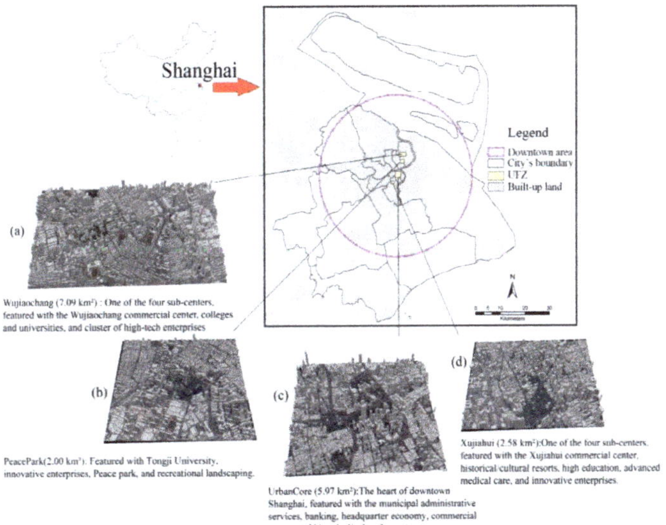

Figure 1. Location of the study area and four UFZs.

Table 1. Description of four UFZs within downtown Shanghai.

UFZ	Area (km^2)	Description
Wujiaochang	7.09	This UFZ includes a shopping center, university campus, research and development institutions, innovative enterprises, high-tech parks, and residential areas.
Peace Park	2.00	This UFZ includes parks and recreational landscapes, a university campus, research and development institutions, innovative enterprises, and residential areas.
Urban Core	5.97	This UFZ includes parks and recreational landscapes, the municipal administration, central business district, and residential areas.
Xujiahui	2.58	This UFZ includes parks, a commercial center, historical and cultural relics, higher education institutes, a health care center, high-tech enterprises, and residential areas.

3. Materials and Methods

3.1. Materials

Landsat 8 OLI/TIRS (Level 1T) and Quickbird imagery were used as the major data sets. Excluding the cloud contaminations, two cloud-free Landsat 8 OLI/TIRS images (path/row: 118/038, cloud cover < 10%) acquired during typical summer days (dated 13 August 2013 and 3 August 2015) were used for the retrieval of the LST and further exploration of the pattern of the urban thermal environment. Quickbird commercial high-resolution imagery covering the four UFZs was used to classify the fine-scale land-use structure. The auxiliary data sets included a commercial vector map data of Shanghai city (roads, buildings, land-uses, and so on) and an aerial remote sensing atlas of the Shanghai central area. A standard digital map of Shanghai downtown (Beijing Digital Space Technology Co., LTD, 2015, Beijing, China), an aerial atlas of Shanghai (Shanghai Academy of Surveying and Mapping 2015, Shanghai, China), Google Earth, and Baidu Map were used as further auxiliary data sets.

3.2. Methods

3.2.1. Land-Use Classification

To better depict the fine-scale land-use characteristics from the Quickbird high-resolution imagery, the object-oriented classification (OOC) method, which has higher classification accuracy than regular methods, such as spectral feature extraction and classification and regression trees (CART) [26], was used for land classification within the four UFZs.

For the Quickbird high-resolution imagery covering the four UFZs, classification with the OOC method was performed with the use of PIE-Basic® (version 6.0) software by PIESAT International Information Technology Limited. The overall accuracy of classification was 80.03%, and the accuracy of the post-classification was determined to be 93.01% with a manual check using Google Earth and Baidu Map layers, as well as a field survey. The validated land-use classification product and its classification scheme are shown in Figure A1 and Table A1, respectively.

In this study, considering the simplicity, representability, and availability of the data sets, the spatial heterogeneity of the built environment was measured using the two- and three-dimensional indices of urban morphology and land surfaces.

Table 2 lists several heterogeneity indices that may positively or negatively contribute to the urban thermal environment (for details, see Table A2). According to the National Standard for Urban Residential Planning and Design of China (GB50180-2018) [27], building heights, distances between buildings, and the SVF were calculated from data collected through in situ measurements. The land surfaces, including the impervious surfaces and BGS, were extracted from the land-use classification maps. Specifically, for BGS, the parcel-based land-use information was used to calculate the class-level pattern indices—namely, the mean patch size (MPS), the number of patches (NP), the largest patch index (LPI), and the SPLIT index—using Fragstats 4.2.1 software [28]. The three-dimensional green volume (3DGV) of the BGS, which refers to the total volume of vegetation with stratified layers, was estimated using several empirical models, ground measurements, and aerial photogrammetry [29].

3.2.2. Generation of Thermally Sharpened LST and Cross-Validation

The process for generating the thermally sharpened LST consisted of four steps. First, the 10 gray value thermal band Landsat 8 Level 1T data was converted to the top of atmosphere (TOA) radiance using the rescaling factors in the MTL file [30].

Secondly, co-Kriging interpolation was employed to generate the high-resolution TOA radiance by combining the Quickbird high-resolution land-use classification products and the raw TOA radiance [31]. Given the different resolutions of these two data sets, the high-resolution land-use classification products were resampled with multiple resolutions (1–9 m) and set as base maps to overlap and delimit the raw TOA radiance layer. We assumed that the same or similar surfaces would have the same or similar radiance values. Then, for each scene of the TOA radiance layer, tedious sampling of random points (ranging between 300 and 6000 points per km^2) was carried out in order to generate the co-Kriging interpolation results. By comparing the pairwise bias curve between the raw and interpolated TOA radiance values, we found that the threshold of 3000 points was reasonable, as the curve flattened and was nearly unchanged when the number of points was over 3000 [32]. Subsequently, the interpolated TOA radiance layer was resampled to multiple resolutions (1–9 m).

Table 2. Statistical description of indices used in this study.

Classification	Metrics	Abbr.	Unit	Range	Median	Mean	Sd
Urban morphology	Building height	Height	m	2.7–165.0	11.00	16.470	15.21
	Building spacing	Distance	m	4.3–107.1	19.82	23.030	14.72
	Sky view factor	SVF	%	2.48–26.38	13.19	12.053	6.0
	Area of impervious surface	ImperSurf	ha	0.47–24.96	5.08	6.016	4.16
	Area of blue–green space	BGS	ha	0.13–15.89	1.66	2.393	2.38
Land surface	Mean patch size	MPS	ha	0.00–0.44	0.01	0.030	0.05
	Largest patch index	LPI	-	8.61–99.25	34.40	40.210	22.800
	Number of patches	NP	-	8.00–840.00	96.500	124.340	110.870
	SPLIT	SPLIT	-	1.015–37.299	6.005	7.383	5.938
	3D green volume	3DGV	m^3	8001.702–834,316.313	92,085.870	133,118.807	135,579.551

Third, based on the multiple-resolution land-use maps, a surface emissivity correction for the land surfaces was performed according to empirical studies and laboratory testing [33,34]. The multiple-resolution interpolated TOA radiance and corrected surface emissivity layers were used to retrieve the thermally sharpened LST using the Range Transfer Equation (RTE) [35], which requires atmospheric correction for the thermal band [36].

Finally, cross-validation of the thermally sharpened LST products was performed by comparing the pixel-based root-mean-square error (RMSE) between the target LSTs (sharpened) and referencing LSTs [37]. To do so, all the thermally sharpened LST products were resampled to the same resolution as the unsharpened LST products (30 m). Then, by overlapping the 30 m unsharpened LST products and the LST products resampled from the sharpened products, the pixel-to-pixel 10-fold RMSEs were used for cross-validation of the thermally sharpened LST products. As shown in Figure A3, there were no significant differences in the RMSEs of the LST products (resampled from the 1–9 m sharpened LST products) and the original 30 m LST products; however, as they exhibited the best visual quality and the lowest RMSE, the 1 m resolution sharpened LST products were used for further analysis.

3.2.3. Calculation of Intra-SUHII

Rather than using the generalized concept of the UHI effect as measured by the LST difference between urban and rural areas, the intra-SUHI intensity in this study is defined as the LST difference between the impervious surfaces and the BGS (vegetated land and water bodies) in a given land parcel. The parcel-based intra-SUHII is calculated as follows [38]:

$$\text{Intra} - \text{SUHII} = \text{LST}_{\text{IS}} - \text{LST}_{\text{BGS}} \tag{1}$$

where the unit of intra-SUHII is Kelvin (K), LST_{IS} represents the average LST of the impervious surfaces, and LST_{BGS} represents the average LST of the BGSs.

At the land parcel level, according to the statistical description and analysis method, the intra-SUHIIs were divided into six levels, according to the percentile threshold on the cumulative probability curve of their normal distribution, as follows: Level 1—very low ($\leq$5%); Level 2—low (5–25%); Level 3—low to slightly high (25–50%); Level 4—medium-high (50–75%); Level 5—high (75–95%); Level 6—very high ($\geq$95%).

3.2.4. Statistical Analysis

This section mainly describes hierarchical cluster analysis (HCA) and partial least squares regression (PLSR) modeling. The former was used for the overall characterization of the land parcel clusters with differential land developmental intensity, while the latter was used for exploring the relationship between the intra-SUHII and multiple independent variables, particularly in the case of multi-collinearity. Essential data analysis was performed, including a normality test, outlier detection, Box–Cox transformation for skewed data, and Pearson's correlation. With the indices described in Table 2, HCA was performed using the Euclidean distance method. Seven typical land parcel clusters representing differential land developmental intensity were obtained (see Table A3).

The statistically significant correlation coefficients (see Table 4) revealed the existence of multi-collinearity between the independent variables. According to the result of Pearson correlation analysis, as there are many possible forms of PLSR models that involve the independent variables, it is time-consuming to establish the PLSR models. To avoid overfitting and determine a reasonable model, the leave-one-out (LOO) method was adopted, setting 90% of the randomly selected data as the training data and the rest as the test data. Finally, the optimal PLSR model, satisfying the highest determination coefficient (R^2) and minimum root mean square error (RMSE), was selected. The PLSR model, indicating the relationship between the intra-SUHII and the impervious surfaces, was written as follows:

$$\text{Intra-SUHII} = \alpha_1 + \beta_1 \cdot X_1 + \beta_2 \cdot X_2 + \beta_3 \cdot X_3 + \beta_4 \cdot X_4 + \beta_5 \cdot X_5 + \varepsilon_1 \tag{2}$$

where α_1 is the intercept/constant item, β_1–β_5 are the partial coefficients; X_1–X_5 are the height, distance, SVF, ImperSurf, and parcel area, respectively; ε_1 is the error term.

Similarly, the PLSR model, indicating the relationship between the intra-SUHII and the BGS, was written as follows:

$$\text{Intra-SUHII} = \alpha_1 + \beta_1 \cdot X_1 + \beta_2 \cdot X_2 + \beta_3 \cdot X_3 + \beta_4 \cdot X_4 + \beta_5 \cdot X_5 + \beta_6 \cdot X_6 + \beta_7 \cdot X_7 + \varepsilon_1 \tag{3}$$

where α_1 is the intercept/constant item, β_1–β_7 are the partial coefficients; X_1–X_7 are the 3DGV, BGS, MPS, NP, LPI, SPLIT, and parcel area, respectively; ε_1 is the error term.

In this study, all the essential statistical processes were performed using R statistical software 4.0.3 (developed at Bell Laboratories (formerly AT&T, now Lucent Technologies, Paris, France) by John Chambers and colleagues), and PLSR modeling was performed using the library 'pls' [39].

4. Results

4.1. Spatial Distribution Characteristics of the Urban Thermal Environment

Overall, Figure 2 reveals the spatial distribution of the urban thermal environment indicated with parcel-based LSTs and intra-SUHII levels on two summer days of 2013 and 2015, respectively. Figure 2a–h shows the remarkable pixel-based LST difference between the land parcels dominated with impervious surfaces and BGS. Figure 2 (a-1–h-1) shows the variations of the parcel-based intra-SUHII levels, which exhibit the similar spatial patterns shown in Figure 2a–h.

Table 3 shows the statistics of parcel-based LST associated with seven typical land parcel types featured with different land developmental intensities (see Table A3). As shown, Type I and VII parcels exhibited the lowest and second-lowest mean LSTs, followed by Type IV parcels. Apparently, these Type I, IV, and VII parcels, which featured dominant BGS and lower impervious surfaces, exhibited much lower mean LSTs than the other land parcel types. It is noteworthy that Type III, V, and VI parcels exhibited much higher mean LSTs, indicating that the adverse thermal effect is related to higher developmental intensity and lacking BGS. However, for Type V parcels, their temporally highest mean LSTs could be alleviated if newly BGS are created in their later management.

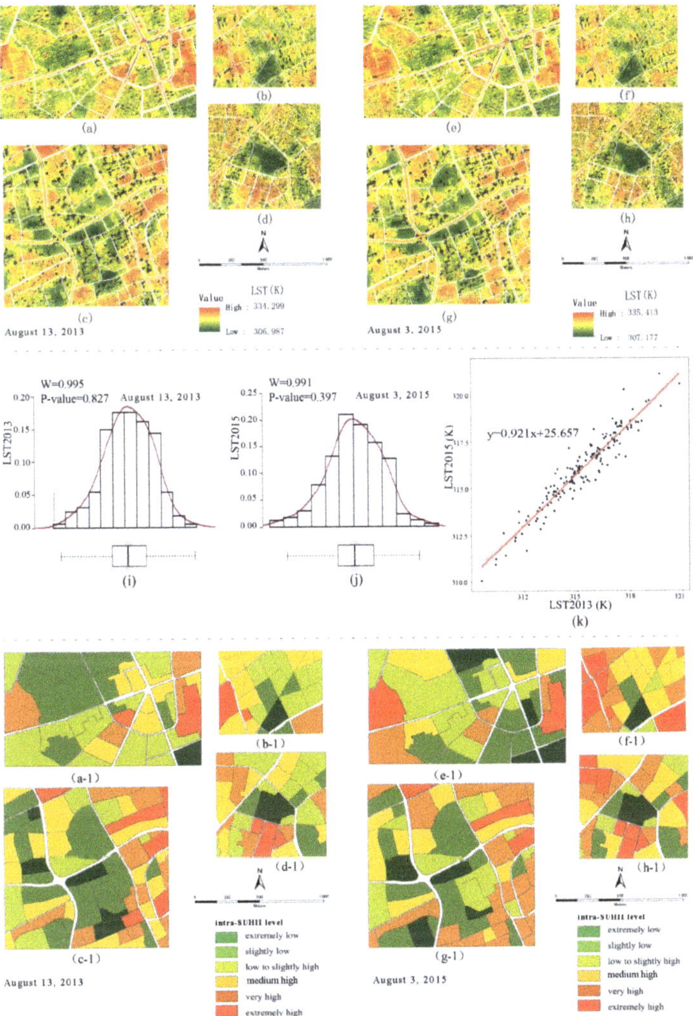

Figure 2. Spatial pattern of pixel-based LSTs and distribution pattern of parcel-based intra-SUHII levels at four UFZs (unit: K).

4.2. Relationship between the Spatial Heterogeneity of the Built Environment and the Urban Thermal Environment

Table 4 shows the significant positive and negative correlations between the spatial heterogeneity indices of the built environment, indicating the complicated relationships between the BGS and urban morphological characters. As shown, a competitive land-use structure exists between the ImperSurf and BGS across the land parcels. The significant negative coefficients between the ImperSurf and the BGS, LPI, MPS, and 3DGV indicate that the dominance of the ImperSurf is inclined to decrease the BGS, LPI, MPS, and 3DGV. The significant positive coefficient between the ImperSurf and the SPLIT indicates that the ImperSurf is inclined to increase the SPLIT since the dominance of the ImperSurf usually causes the absence of BGS and, consequently, an uneven pattern of BGS. On the other hand, the significant positive coefficients between the building distances and BGS, LPI, MPS, and 3DGV indicate that, to a certain extent, wider distances between buildings help shape a better BGS landscape configuration since the former provides available space for the creation of BGS. Moreover, the significant positive coefficients between the BGS, LPI, MPS, and 3DGV indicate the overall high standard of creation and management of the existing BGS, particularly under the pressure of ecological land scarcity in urban settings.

Table 3. LST statistics of seven characteristic types of land parcels.

Clustered Land Parcel Type	Min LST (K)	Max LST (K)	Mean LST (K)	Range (K)
Type I: Parks and recreational landscape	307.503	319.376	312.151	11.873
Type II: Mixture use of high-density residential and commercial areas	314.296	328.853	316.804	14.557
Type III: Poorly-planned old residential	315.793	320.829	319.193	5.036
Type IV: Well-planned modern residential	313.718	321.178	315.584	7.46
Type V: Mixture of land under construction and high-density low-rise (residential)	314.076	321.196	319.293	7.12
Type VI: Mixture of medium and high-density residential and commercial area	316.215	321.357	318.432	5.142
Type VII: University and college campus	309.212	316.346	313.324	7.134

Table 4. Pearson correlation coefficients between heterogeneity indices.

	3DGV	BGS	NP	LPI	SPLIT	MPS	Height	Distance	ImperSurf	SVF
BGS	0.733 **									
NP	0.105	0.063								
LPI	0.311 **	0.442 **	−0.417 **							
SPLIT	−0.301 **	−0.498 **	0.435 **	−0.756 **						
MPS	0.672 **	0.605 **	−0.493 **	0.591 **	−0.617 **					
Height	−0.090	−0.091	0.089	−0.092	0.135	−0.178 *				
Distance	0.290 **	0.293 *	−0.069	0.363 **	−0.236 **	0.392 **	0.123			
ImperSurf	−0.692 **	−0.953 **	0.065	−0.491 **	0.552 **	−0.662 **	0.100	−0.342 **		
SVF	−0.177 *	−0.123	0.156 *	−0.151 *	0.18 *	−0.242 **	0.969 **	−0.071	0.146	
Parcel-area	−0.733 **	−0.538 **	0.400 **	0.009	0.020	−0.402 **	−0.069	0.110	0.631 **	−0.101

Note: Except for height, the other indices were Box–Cox transformed. * and ** indicate the significance levels of 0.05 and 0.01, respectively.

Table 5 quantifies the relationship between the impervious surfaces' two- and three-dimensional indices and parcel-based intra-SUHII on two summer days. Herein, considering the independent variables were measured in different units, the standardized regression coefficients (S-Coefs) were used to interpret the results of the PLSR models. As shown, the PLSR models account for approximately 48.7–49.8% of the variance of parcel-based intra-SUHII in response to the independent variables. The positive and negative S-Coefs indicate their relative importance or strength in determining the variance of parcel-based intra-SUHII. The positive S-Coefs of ImperSurf indicate it exerted a much higher influence on increasing the parcel-based intra-SUHII. When controlling the other independent variables, each standard deviation increase in ImperSurf resulted in a 0.455~0.480 standard deviation increase in parcel-based intra-SUHII on two typical summer days. In contrast, the small positive S-Coefs of the parcel area indicate its very weak contribution to the increase in the parcel-based intra-SUHII. The negative S-Coefs of distance, height, and SVF indicate their descending ordinal of relative importance in negatively contributing to the variance of parcel-based intra-SUHII, as each standard deviation increase measured in these indices caused -0.368~-0.360, -0.135~-0.111, and -0.066~-0.082 standard deviation decreases in the parcel-based intra-SUHII, respectively.

Table 5. Coefficients of PLSR models focusing on impervious surfaces.

	Intra-SUHII2013		Intra-SUHII2015	
	Coef	**S-Coef**	**Coef**	**S-Coef**
Constant	7.673	0.000	6.583	0.000
Distance	−3.338	−0.368	−3.161	−0.360
Height	−0.180	−0.135	−0.144	−0.111
SVF	−0.028	−0.082	−0.021	−0.066
ImperSurf	0.000	0.455	0.000	0.480
Parcel area	0.128	0.042	0.216	0.072
Variance explained	48.7%		49.8%	

Note: Coef and S-Coef represent the unstandardized and standardized coefficients, respectively.

Table 6 shows the PLSR models account for approximately 41.7–43.1% of the variance of parcel-based intra-SUHII in response to the independent variables. The positive S-Coefs of SPLIT, parcel area, and MPS indicate their influence on increasing the parcel-based intra-SUHII. When controlling the other independent variables, each standard deviation increase in the SPLIT and parcel-area resulted in 0.186~0.198 standard deviation increases and 0.079~0.105 standard deviation increases in parcel-based intra-SUHII, respectively. In contrast, the smaller S-Coefs of MPS indicate its very weak importance in increasing the parcel-based intra-SUHII. Meanwhile, the descending sequence of negative S-Coefs of LPI, BGS, NP, and 3DGV indicate their differential relative importance in negatively contributing to the variance of parcel-based intra-SUHII, as each standard deviation increase measured in these indices caused -0.169~-0.159, -0.142~0.137, -0.064~-0.028, and -0.047~~-0.026 standard deviation decreases in the parcel-based intra-SUHII, respectively.

Table 6. Coefficients of PLSR models focusing on BGS.

	Intra-SUHII2013		Intra-SUHII2015	
	Coef	**S-Coef**	**Coef**	**S-Coef**
Constant	10.028	0.000	8.738	0.000
LPI	−1.288	−0.159	−1.323	−0.169
BGS	−0.374	−0.137	−0.375	−0.142
NP	−0.381	−0.064	−0.164	−0.028
3DGV	−0.098	−0.047	−0.053	−0.026
SPLIT	1.058	0.186	1.087	0.198
Parcel area	0.244	0.079	0.312	0.105
MPS	0.171	0.039	0.113	0.027
Variance explained	41.7%		43.1%	

Note: Coef and S-Coef represent unstandardized partial regression and standardized partial regression coefficients, respectively.

5. Discussion

The findings of this study show that, to an extent, the goal of alleviating the intra-SUHI effect can be achieved via optimizing land parcel design, for instance, by increasing the building distance/spacing and SVF, increasing the area proportion of BGS and 3DGV, and improving the spatial configuration of BGS. Taking the Cui-hu-tian-di (CHTD) modern residence (well planned) and the neighboring old residence (poorly planned) as examples, Figures 3 and 4 show the contrasting intra-SUHI effects between these two land parcels at the Urban Core UFZ. As can be seen, the CHTD modern residence, with higher building spacing (averaged 46 m) and higher BGS cover (39%), exhibits overall lower LSTs (ranging between 307 K and 318 K, and averaging 309 K). Moreover, together with the trees, the higher buildings with wide spacing help create shadow areas and ventilation corridors for cooling. In contrast, the neighboring old residence, with poorly planned building spacing (averaging 5 m) and lacking BGS, exhibits much higher LSTs (ranging between 309 and 330 K, and averaging 317 K). Obviously, for such an old residence, if future urban regeneration is performed with the land parcel design attributes of the CHTD modern residence, then the intra-SUHI effect will be substantially enhanced.

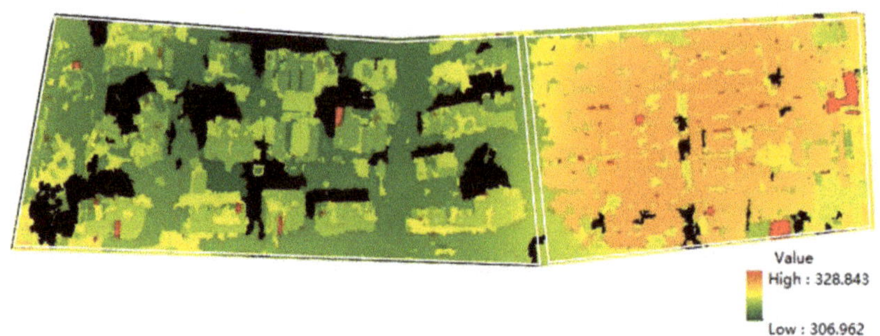

Figure 3. Comparison of fine-scale LSTs of two neighboring land parcels located at the Urban Core UFZ.

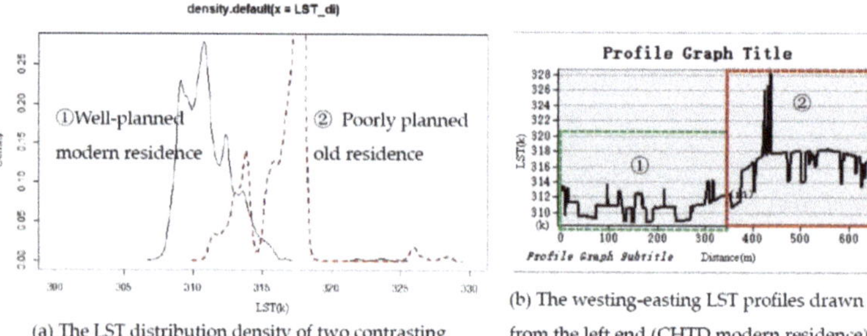

(a) The LST distribution density of two contrasting land parcels

(b) The westing-easting LST profiles drawn from the left end (CHTD modern residence) to the right end (old residence)

Figure 4. Comparison of fine-scale LSTs of two neighboring land parcels.

In the sense of urban resilience and urban planning, our findings exemplify the relative importance of land parcel design attributes in positive or negative contributions to the intra-SUHI effect. However, there are some shortcomings of this study. Firstly, due to the 16-day revisiting interval of the Landsat 8 satellite and cloud contamination [30], the TIRS data captured within the instantaneous field of view (IFOV) could not provide sufficient a time series of thermal images for studying the seasonal variation of the intra-SUHI effect. Secondly, until now, the satellite-borne thermal sensors could not generate the ~m resolution data. We used thermally sharpened LSTs, which lack in situ measurements for validation since there are no weather stations or long-term observation sites of the four UFZs in downtown Shanghai [6,37]. Thirdly, the focus of this study was on the possible linkage between solar radiation and the thermal effect of land surfaces in built environments, regardless of the influence of the micro-climatic conditions. Thus, the interpretation power of the PLSR models is somewhat low. Fourthly, the findings of this study ignore the influence of anthropogenic heat emission, considering the complicated relationship that exists between the LST-based intra-SUHI effect and the AT-based UHI effect at a small scale.

Given the above shortcomings, future research should focus on the multi-point distribution of data monitoring and improve the empirical research system. Data assimilation processes, including three-dimensional modeling, computational fluid dynamics (CFD), machine learning methods and other technologies, the long-term in situ observation data, thermal infrared remote sensing data, and numerical simulation results, can provide ensured outputs with cross-validation and improve the accuracy of the prediction modeling.

6. Conclusions

In this study, we quantitatively examined the spatial heterogeneity of the built environment and its impact on the summertime intra-SUHI effect using the high-resolution land-use classification products and thermally sharpened LSTs. The findings are summarized as follows:

1. There are remarkable variations of LSTs and intra-SUHII among seven typical land parcels with different land developmental intensities. Overall, land parcels featured with dominant BGS and lower impervious surfaces, particularly parks and recreational landscapes, a university/college campus with a higher green cover, and well-planned modern residences exhibited much lower mean LSTs than the other land parcel types with dense buildings and lacking BGS.

2. The PLSR models quantitatively revealed the relative importance of the main effect of the urban built environment in determining the variances of the urban thermal environment. The results show that the building distance/spacing, SVF, LPI, and BGS are major negative contributors to decreasing the variance of the parcel-based intra-SUHI effect. In contrast, the ImperSurf and SPLIT are major positive contributors to increasing the variance of the parcel-based intra-SUHI effect.

To sum up, based on the findings, this study provides some practical implications towards alleviating the adverse UHI effects via potentially optimizing the land parcel design attributes, particularly focusing on increasing the two- and three-dimensional indices of BGS and reducing the influence of impervious surfaces. Future urban decision-making processes of mitigating UHI effects and improving cities' adaption to climate change should sufficiently embody these key points and produce scientifically sound countermeasures.

Author Contributions: H.Z. conceived the central idea and designed the technical framework for this study; J.H., X.Z. and Y.L. conducted the data processing and analysis; J.H. wrote the initial manuscript; H.Z. reviewed and edited the manuscript. All authors have read and agreed to the published version of the manuscript.

Funding: This research received no funding and The APC was covered with MDPI 's discount vouchers (reviewer's token) assigned to H.Z.

Institutional Review Board Statement: Not applicable.

Informed Consent Statement: Not applicable.

Data Availability Statement: The authors thank the Geospatial Data Cloud site, Computer Network Information Center, and the Chinese Academy of Sciences (http://www.gscloud.cn) (accessed on 12 October 2017) for providing the free Landsat TM/OLI images. The authors are indebted to the R Foundation for Statistical Computing and Beijing Piesat Information Technology Co., Ltd., for free usage of the PIE 6.0 Remote Sensing image processing system (https://www.piesat.cn/en/index.html) (accessed on 12 October 2020).

Conflicts of Interest: The authors declare no conflict of interest.

Appendix A

Table A1. Land-use classification using the OOC method.

Categories	Land-Use	Introduction	Assigned Surface Emissivity [40]
Blue–green space (BGS)	Water	River, creeks, lakes, and ponds	0.9925
	Tree	Evergreen trees, deciduous trees, and a mixture of both	0.95
	Shrub	Forest nurseries, hedges, and ornamental plants	0.95
	Lawn	Green land, mainly turf	0.95

Table A1. Cont.

Categories	Land-Use	Introduction	Assigned Surface Emissivity [40]
Impervious surface	Plastic runway	Athletic tracks paved with plastic compounds, and so on	0.92
	Hard-top pavement	Asphalt and concrete pavement	0.85
	Cement pavement	Traffic road paved with cement mortar	0.90
	Demolition of open space	Closed construction site or temporarily vacant land for demolition	0.83
	Light-weighted steel roof	Light steel roofs, mostly mobile houses or simple houses	0.66
	Bituminous roof	Asphalt paper waterproof roofs, more common in low- and high-density old residential areas	0.85
	Glass curtain wall	Glass exterior wall of high-rises used as office premises	0.94
	Light-colored wall	Building walls furnished with light-colored coating materials	0.90
	Shadow	Shadow of buildings and tall trees	-

Table A2. The class-level landscape pattern indices of BGS.

Indices and Abbreviation	Unit	Formula	Introduction
LPI—Largest Patch Index	%	$LPI = \dfrac{\max\limits_{i=1}^{i} a_{ij}}{A_i}$ a_{ij}: Patch ij area; A_i: Total landscape area	Maximum patch percentageLandscape area ratio
NP—Number of Patches	-	$NP = n_i$ n_i: Total area of category i landscape elements	The number of patches in the study area
MPS—Mean Patch Size	ha	$MPS = A_i / NP$ A_i: Total landscape area; NP: number of patches	Mean patch size
SPLIT—Splitting Index	%	$SPLIT = \dfrac{D_i}{A_i}$ D_i: Distance index of landscape type i; A_i: Total landscape area	Degree of patch dispersion

Table A3. Illustrations of seven typical land parcel types with differential land developmental intensity.

Type	Introduction
Type I: Park and recreational landscape	

Park and recreational landscape featured with high vegetation cover (≥60%) and low impervious cover (ranging between 5 and 20%).

Table A3. *Cont.*

Type	Introduction
Type II: Mixture of high-density residential and commercial area	

Mixture of high-density low-rise, multi-story residential and commercial areas. Building cover varies between 49 and 60%.

Type III: Poorly planned old residential areas

Old residential areas with a construction age of more than 50 years because the overall area has not been properly planned (e.g., in terms of roads, greening, housing). Mainly includes two types, 2–3 high-density old residential areas with a building density of 50–88% (averaged 61%), or single-story bungalows or 3–6 floors of old residential apartments with a building density of 46–66% (approximately averaged 56%).

Type IV: Modern residential area

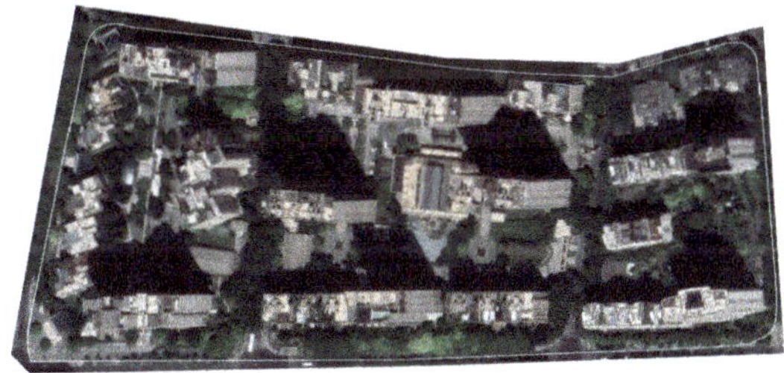

Modern residences with scientific and complete overall planning concepts, large distances between high-rise buildings, high vegetation coverage, and complete public service facilities. The building density varies between 25 and 35% (approximately averaged 26%).

Type V: Mixture of land under construction and high-density low-rise

Mixture of urban development land, including land to be demolished for reconstruction, land used for demolition and reconstruction, and land under construction. After the original building is demolished, temporary construction site housing is often built to facilitate construction.

Table A3. *Cont.*

Type	Introduction
Type VI: Mixture of medium-density residential and commercial area 	In the mixed area of medium-density residential communities and commercial districts, the building density varies between 27 and 51% (approximately averaged 40%).
Type VII: University campus with high green cover and low- to medium-density buildings 	Except for parks, areas with high green coverage, mainly low-rise, multi-story buildings, with building density varying between 15 and 30%.

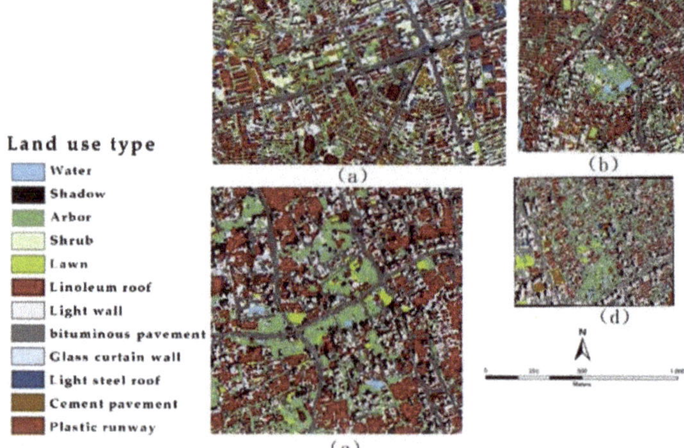

Figure A1. Land-use classification of the four UFZs. (**a–d**) represent Wujiaochang, Peace Park, Urban Core, and Xujiahui, respectively.

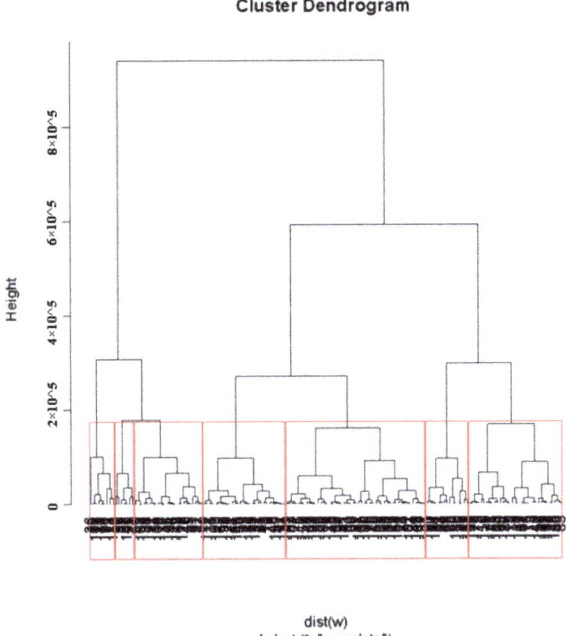

Figure A2. Hierarchical clustering dendrogram showing seven LUF types.

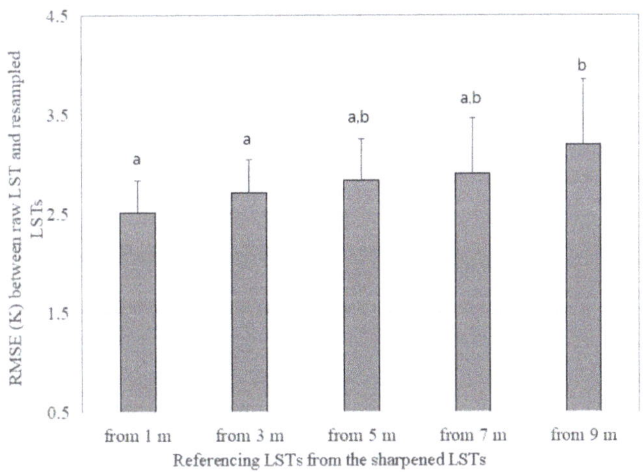

Figure A3. Pairwise RMSEs between the original 30 m LST and the resampled 30 m LSTs from the thermally sharpened products (1–9 m). Different symbols (a/b) indicate significant differences at the 0.05 level.

References

1. Li, J.-J.; Wang, X.-R.; Wang, X.-J.; Ma, W.-C.; Hao, Z. Remote sensing evaluation of urban heat island and its spatial pattern of the Shanghai metropolitan area, China. *Ecol. Complex.* **2009**, *6*, 413–420. [CrossRef]
2. Perz, S.G.; Xia, Y.; Shenkin, A. Global Integration and Local Connectivity: Trans-boundary Highway Paving and Rural-Urban Ties in the Southwestern Amazon. *J. Lat. Am. Geogr.* **2014**, *13*, 205–239. [CrossRef]

3. Wong, P.; Boon-Thong, L.; Leung, M. Hot Spots of Population Growth and Urbanisation in the Asia-Pacific Coastal Region. In *Global Change and Integrated Coastal Management*; Springer: Dordrecht, The Netherlands, 2006; pp. 163–195.
4. Schneider, A.; Friedl, M.A.; Potere, D. Mapping global urban areas using MODIS 500-m data: New methods and datasets based on 'urban ecoregions'. *Remote Sens. Environ.* **2010**, *114*, 1733–1746. [CrossRef]
5. United Nations; DESA. *Revision of World Urbanization Prospects*; DESE: New York, NY, USA, 2018; Volume 2021.
6. Zhang, H.; Li, T.-T.; Han, J.-J. Quantifying the relationship between land use features and intra-surface urban heat island effect: Study on downtown Shanghai. *Appl. Geogr.* **2020**, *125*, 102305. [CrossRef]
7. Shen, L. The Study of the Response Relationship of Land Cover and Its Changes in the City of Low Urban Background. Ph.D. Thesis, Zhejiang University, Hangzhou, China, 2013.
8. Wang, Y.; Bakker, F.; Groot, R.; Wörtche, H. Effect of ecosystem services provided by urban green infrastructure on indoor environment: A literature review. *Build. Environ.* **2014**, *77*, 88–100. [CrossRef]
9. Gobster, P.H. Alternative Approaches to Urban Natural Areas Restoration: Integrating Social and Ecological Goals. In *Forest Landscape Restoration*; USDA: Washington, DC, USA, 2012; pp. 155–176.
10. Wang, Y.; Ni, Z.; Peng, Y.; Xia, B. Local variation of outdoor thermal comfort in different urban green spaces in Guangzhou, a subtropical city in South China. *Urban For. Urban Green.* **2018**, *32*, 99–112. [CrossRef]
11. Mallick, J.; Rahman, A.; Singh, C.K. Modeling urban heat islands in heterogeneous land surface and its correlation with impervious surface area by using night-time ASTER satellite data in highly urbanizing city, Delhi-India. *Adv. Space Res.* **2013**, *52*, 639–655. [CrossRef]
12. Xiao, J.Y.; Zhang, Q.; Wang, Y.; Na, J.I.; Xing, L.I. Urban Surface Heat Flux Analysis Based on Remote Sensing: A Case Study of Shijiazhuang City. *Sci. Geogr. Sin.* **2014**, *34*, 338–343.
13. Stone, B.; Norman, J. Land use planning and surface heat island formation: A parcel-based radiation flux approach. *Atmos. Environ.* **2006**, *40*, 3561–3573. [CrossRef]
14. Hove, L.V.; Jacobs, C.; Heusinkveld, B.G.; Elbers, J.A.; Driel, B.V.; Holtslag, A. Temporal and spatial variability of urban heat island and thermal comfort within the Rotterdam agglomeration. *Build. Environ.* **2015**, *83*, 91–103. [CrossRef]
15. Chen, H.S.; Li, X.; Wenjian, H. Numerical simulation of the effect of land use change on regional climate in China in recent 20 years. *Atmos. Sci.* **2015**, *2*, 357–369.
16. Hua, W.; Haishan, C.; Xing, L. Land use/cover change and its climate effects in China: A review. *Adv. Earth Sci.* **2014**, *9*, 1025–1036.
17. Lin, Y.; Jim, C.Y.; Deng, J.; Wang, Z. Urbanization effect on spatiotemporal thermal patterns and changes in Hangzhou (China). *Build. Environ.* **2018**, *145*, 166–176. [CrossRef]
18. Weng, Q. Thermal infrared remote sensing for urban climate and environmental studies: Methods, applications, and trends. *ISPRS J. Photogramm. Remote Sens.* **2009**, *64*, 335–344. [CrossRef]
19. Bonafoni, S.; Anniballe, R.; Gioli, B.; Toscano, P. Downscaling Landsat Land Surface Temperature over the urban area of Florence. *Eur. J. Remote Sens.* **2016**, *49*, 553–569. [CrossRef]
20. Holderness, T.; Barr, S.; Dawson, R.; Hall, J. An evaluation of thermal Earth observation for characterizing urban heatwave event dynamics using the urban heat island intensity metric. *Int. J. Remote Sens.* **2013**, *34*, 864–884. [CrossRef]
21. Li, H. Thermal Environment of Medium Cities Supported by Multi-Source Remote Sensing Data. Ph.D. Thesis, Chengdu Univerisity of Technology (CDUT), Chengdu, China, 2012.
22. Bo, D.; Xijin, Z.; Shangming, D.; Qiang, Y. Spatial characteristics and countermeasures of urban thermal environment in Chengdu. *Sichuan Environ.* **2011**, *30*, 124–127.
23. Ma, H.; Li, T.; Jiang, Z.H.; Peng, G. Unexpected large-scale atmospheric response to urbanization in East China. *Clim. Dyn.* **2018**, *52*, 4293–4303. [CrossRef]
24. Ichinose, T.; Liu, K. Modelling the Relationship between the Urban Development and Subsurface Warming in Seven Asian Megacities. *Sustain. Cities Soc.* **2018**, *38*, 560–570. [CrossRef]
25. Doan, V.Q.; Kusaka, H.; Nguyen, T.M. Roles of past, present, and future land use and anthropogenic heat release changes on urban heat island effects in Hanoi, Vietnam: Numerical experiments with a regional climate model. *Sustain. Cities Soc.* **2019**, *47*, 101479. [CrossRef]
26. Huang, X. *Study on Hyperspectral Image Classification Based on Statistics*; UESTC: Chengdu, China, 2008.
27. Li, Y. Street level urban design qualities for walkability: Combining 2D and 3D GIS measures. *Comput. Environ. Urban Syst.* **2017**, *64*, 288–296.
28. McGarigal, K.; Cushman, S.; Ene, E. *FRAGSTATS: Classification Map Spatial Pattern Analysis Procedures*; USDA: Washington, DC, USA, 2012; Volume 2021.
29. Zhou, Y.; Zhou, J. The Urban Eco-environ-mental Estimating System Based on 3-dimension Vegetation Quantity. *Chin. Landsc. Archit.* **2001**, *17*, 77–79.
30. USDOI; USGS. *Landsat 8 (L8) Data Users Handbook*; Version 1.0; EROS: Sioux Falls, SD, USA, 2015.
31. Guo, Y.-J.; Han, J.-J.; Zhao, X.; Dai, X.-Y.; Zhang, H. Understanding the Role of Optimized Land Use/Land Cover Components in Mitigating Summertime Intra-Surface Urban Heat Island Effect: A Study on Downtown Shanghai, China. *Energies* **2020**, *13*, 1678. [CrossRef]

32. Zhang, H.; Jing, X.-M.; Chen, J.-Y.; Li, J.-J.; Schwegler, B. Characterizing Urban Fabric Properties and Their Thermal Effect Using QuickBird Image and Landsat 8 Thermal Infrared (TIR) Data: The Case of Downtown Shanghai, China. *Remote Sens.* **2016**, *8*, 541. [CrossRef]
33. Nichol, J.E. High-Resolution Surface Temperature Patterns Related to Urban Morphology in a Tropical City: A Satellite-Based Study. *J. Appl. Meteorol.* **1996**, *35*, 135–152. [CrossRef]
34. Weng, Q.; Lu, D.; Schubring, J. Estimation of Land Surface Temperature-Vegetation Abundance Relationship for Urban Heat Island Studies. *Remote Sens. Environ.* **2004**, *89*, 467–483. [CrossRef]
35. Jiménez-Muñoz, J.C.; Sobrino, J.A. A generalized single-channel method for retrieving land surface temperature from remote sensing data. *J. Geophys. Res. Atmos.* **2003**, *108*, 4688. [CrossRef]
36. He, L.; Yan, G.; Wang, Q.; Li, X. Models and Its Analysis about the Atmospheric Correction of Optical Remote Sensing Imagery. *Geo-Inf. Sci.* **2005**, *7*, 33–38.
37. Nurit, A.; William, P.K.; Martha, C.A.; Fuqin, L.; Christopher, M.U.N. A vegetation index based technique for spatial sharpen-ing of thermal imagery. *Remote Sens. Environ.* **2007**, *107*, 545–558.
38. Zhang, H.; Li, T.-T.; Liu, Y.; Han, J.-J.; Guo, Y.-J. Understanding the contributions of land parcel features to intra-surface urban heat island intensity and magnitude: A study of downtown Shanghai, China. *Land Degrad. Dev.* **2020**, *32*, 1353–1367. [CrossRef]
39. Mevik, B. The pls Package: Principal Component and Partial Least Squares Regression in R. *J. Stat. Softw.* **2007**, *18*, 2. [CrossRef]
40. *Uni-Trend Inc.*, version 1.08.14s; Uni-Trend Technology Co., Ltd.: Guangdong, China, 2014.

sustainability

MDPI

Article

Perceived Effects of Climate Change and Extreme Weather Events on Forests and Forest-Based Livelihoods in Malawi

Harold L. W. Chisale [1,2,*], Paxie W. Chirwa [1], Folaranmi D. Babalola [1,3] and Samuel O. M. Manda [4,5]

[1] Forest Science Postgraduate Programme, University of Pretoria, Pretoria 0028, South Africa; paxie.chirwa@up.ac.za (P.W.C.); babalola.fd@unilorin.edu.ng (F.D.B.)
[2] Department of Forestry, Bunda Campus, Lilongwe University of Agriculture and Natural Resources, Lilongwe P.O. Box 219, Malawi
[3] Department of Forest Resources Management, University of Ilorin, Ilorin 240003, Nigeria
[4] Biostatistics Research Unit, South African Medical Research Council, Pretoria 0001, South Africa; samuel.manda@mrc.ac.za
[5] Department of Statistics, University of Pretoria, Pretoria 0028, South Africa
* Correspondence: u18264248@tuks.co.za or chisale.harold2@gmail.com; Tel.: +27-845178998

Citation: Chisale, H.L.W.; Chirwa, P.W.; Babalola, F.D.; Manda, S.O.M. Perceived Effects of Climate Change and Extreme Weather Events on Forests and Forest-Based Livelihoods in Malawi. *Sustainability* **2021**, *13*, 11748. https://doi.org/10.3390/su132111748

Academic Editors: Baojie He, Ayyoob Sharifi, Chi Feng and Jun Yang

Received: 2 September 2021
Accepted: 13 October 2021
Published: 25 October 2021

Publisher's Note: MDPI stays neutral with regard to jurisdictional claims in published maps and institutional affiliations.

Abstract: The emerging risks and impacts of climate change and extreme weather events on forest ecosystems present significant threats to forest-based livelihoods. Understanding climate change and its consequences on forests and the livelihoods of forest-dependent communities could support forest-based strategies for responding to climate change. Using perception-based assessment principles, we assessed the effects of climate change and extreme weather events on forests and forest-based livelihood among the forest-dependent communities around the Mchinji and Phirilongwe Forest Reserves in the Mchinji and Mangochi districts in Malawi. Content analysis was used to analyze qualitative data. The impact of erratic rainfall, high temperatures, strong winds, flooding, and droughts was investigated using logistic regression models. The respondents perceived increasing erratic rainfall, high temperatures, strong winds, flooding, and droughts as key extreme climate events in their locality. These results varied significantly between the study sites ($p < 0.05$). Erratic rainfall was perceived to pose extended effects on access to the forest in both Phirilongwe in Mangochi (43%) and Mchinji (61%). Climate change was found to be associated with reduced availability of firewood, thatch grasses, fruits and food, vegetables, mushrooms, and medicinal plants ($p < 0.05$). Erratic rainfall and high temperatures were more likely perceived to cause reduced availability of essential forest products, and increased flooding and strong winds were less likely attributed to any effect on forest product availability. The study concludes that climate change and extreme weather events can affect the access and availability of forest products for livelihoods. Locally based approaches such as forest products domestication are recommended to address threats to climate-sensitive forest-based livelihoods.

Keywords: forest dependent communities; essential forest products; sensitivity; binary regression model; forest-based livelihoods; climate change

1. Introduction

Climate change and variability are some of the most overwhelming challenges facing humanity in the 21th century, thereby threatening the attainment of sustainable development goals [1]. Although the major impacts of climate change are evenly distributed around the globe, some parts of the world are projected to experience the worst impacts due to several factors. Sub-Saharan Africa, for instance, as elsewhere in developing countries, is more vulnerable to climate change due to poverty and poor infrastructure development [1–3]. In addition, sub-Saharan Africa is more dependent on rain-fed agriculture and land-based resource use such as forests, freshwater, and riverine systems as sources of potable water, fish, and transport [4]. Specifically, the Miombo woodlands of southern

139

Africa support the livelihoods of over 100 million rural and 50 million urban residents apart from sustaining the national economies of these countries [5].

The FAO [6] reports that global food production systems have greatly been affected by climate change. This problem is exacerbated by the low adaptive capacities of poor communities, which reduces their resilience [7]. As such, forest products have been used to bolster the low food production of the rural communities since time immemorial [5]. Thus, rural household livelihoods in most developing countries are highly dependent on forest resources. Oeba et al. [8] affirm that forests and tree-based systems complement agricultural production in providing better and more nutritionally balanced diets. In addition, forests provide wood fuel for cooking and a greater variety of food consumption choices such as wild food and vegetables, fruits, and fodder for livestock, particularly during lean seasons and periods of vulnerability [8–11]. To the marginalized groups such as forest communities, forests deliver a broad set of ecosystem services, which enhance and support crop production [12].

It has been estimated that approximately 20 per cent of the world population is forest-dependent [13]. In Malawi, as elsewhere in developing countries, the majority of the rural household livelihoods and the large proportion of the urban households are highly dependent on forest resources to meet their nutritional, energy, cultural, and medicinal needs [5,6,10,14]. Forest resources are crucial for rural development in Malawi. The dominant rural livelihood activities in Malawi, as elsewhere in Africa, are farming, animal husbandry, and harvesting and trade in forest resources [15]. For example, Makungwa [16] reported that 63% of the rural population in Malawi continues to rely on traditional medicine to cure ailments. This translates into 9.5 million people being dependent on traditional medicine in Malawi. However, in addition to forest degradation and deforestation, climate change and extreme weather events present a huge challenge to forest-based livelihoods. Concerning deforestation, Ngwira and Watanabe [17] estimated that about 30,000–40,000 hectares of forest land in Malawi is lost annually due to increased agricultural activities and excessive wood and charcoal biomass consumption.

Sein et al. [18] indicated that some of the extreme weather events that affect agriculture production include increased global temperatures and erratic rainfall (both unpredictable increase and decrease in rainfall amounts). Other studies have revealed that erratic rainfall as climate variability is the main trigger of some extreme weather events such as droughts and floods [1,18,19]. To support this assertion in the forestry sector, Ofoegbu et al. [11] revealed that forest-based households were very likely to perceive that drought reduces the availability of firewood in Vhembe, South Africa. In another study in Zambia, Robledo et al. [20] found that flooding positively influences mushroom reproduction and harvesting and negatively affects honey production. The study further revealed that drought, if not severe, can boost honey production due to its positive impacts on inducing flowering. However, extended droughts were revealed to kill bees, thereby negatively affecting honey production [20].

The risks and impacts of climate change and extreme weather on forest ecosystems are increasingly becoming serious threats to forest-dependent communities [1,21,22]. The observed and predicted impacts of climate change is projected to have an extensive range of consequences, which include droughts, floods hailstorms, and erratic rainfall, ultimately reducing crop productivity, among others [1,11,15,23,24]. These impacts also present significant threats to forests, livelihoods, and rural development, which may lead to increased poverty levels. In a recent assessment of the future impacts of climate change on Malawi forests, Edward et al. [15] reported that Malawi's current dry forests will be replaced by thorn woodland forests with a significant reduction in the living biomass of the forests. This presents many challenges and opportunities for individuals, households, and the wider society.

Klein [25] argued that even though climate change and variability are considered as a common occurrence, their manifestation is local. Ofoegbu et al. [11] call for the comprehensive understanding of the forest community's demographic features and their level of reliance on vulnerable forest resources as paramount. This understanding is envisaged to assist in construing how climate change would manifest in the forest community being considered. Malone [26] observes that climatic events and extremes produce different levels of socio-economic impact in the same community. Davison et al. [27] also argued that the variations in the climatic impacts emanating from similar climatic events do not solely depend on the location and time of the manifestation of the event but also people's level of interaction with the forest resources in their locality.

The concern of the impact of climate change, whether physical or socioeconomic in the forestry sector, has led to the urgent need to develop and implement national and regional forest-based strategies for responding to climate change [3]. Reducing vulnerability, increasing resilience, and improving adaptation to climate change is vital in various sectors, including health and forestry [22]. However, what shapes the vulnerability, resilience, and adaptability to climate change in the forestry sector is poorly understood. This is evident in how the policy documents are framed, leaving out the forest dependents' inputs at the local level. Therefore, it is important to understand how forest-dependent communities perceive and understand the impacts of climate change on the forest for their livelihood and sustenance. This will help to address their immediate needs, which will incentivize their full participation in the implementation of forest programs that address climate change and variability [28].

Studies have shown that forest-based livelihoods are insecure due to the long history of marginalization, exclusion, unclear property rights, and remoteness [5,6,9,29]. Taini et al. [30] assert that vulnerability assessments globally have been concentrated on dry regions, leaving forest people out. This phenomenon has not spared Malawi, where forest-dependent communities have not been adequately represented in the climate policy development process. Mostly, forest-dependent people are marginalized and considered unimportant, leading to their exclusion. For example, Velded et al. [31] reported that forest-dependent communities are ranked the lowest economically within the communities as compared to their fellow villagers who relied on agricultural and non-farming activities in the Chiradzulo district of Malawi. However, findings from the research conducted in Malawi and Zambia revealed that increasing agriculture production and productivity reduces the reliance of forest-dependent communities on forest resources for livelihoods, thereby contributing to forest conservation [32,33].

In Malawi, studies on the impact of climate change on forest ecosystems and the contribution of the forest ecosystems services to people's livelihoods are limited. However, amongst the available literature, Jumbe [34] noted that much of it dwells on the biological aspect, rendering the social aspect not much explored. In response to this gap, a proliferation of research studies emerged with much focus on the contribution of the forest ecosystems services to the livelihood of the people [9,31,35–38]. Recently, an attempt to link climate change and variability to forestry and forest-dependent communities has been made [15,32,39,40]. However, most of these studies fail to provide critical insights in terms of effectively analyzing the perceived vulnerability of forest-dependent communities and adaptation strategies at the household level. As a result, policy-makers are not fully aware of the vulnerability of the forest people to climate change and variability.

The impacts of climate change on the livelihoods of forest-dependent communities have been documented by various authors in Malawi [35], South Africa [11], India [21,41], Ethiopia [42], China [43], Mozambique [44], and Bhutan [45]. However, no study explicitly addresses the question of which forest products, amongst those used for livelihoods by forest-dependent communities, have been affected by which climate change and extreme weather events in space and time. We anticipate those policy makers may specifically devise deliberate climate change measures and policies targeting issues at the local level. This study was therefore designed to explore the local perceptions of the impacts of climate change and variability on forests and forest communities around the Mchinji and Phirilongwe Forest Reserves in the Mchinji and Mangochi districts, respectively. To address this objective, the paper is organized into the following main sections and themes: observed climate change and extreme events by the forest-dependent communities over the past 20 years, effects of the observed climate change and extreme weather events on forest access for forest-based livelihoods, and assessing the sensitivity of the priority forest products to identify the key climatic impact factors.

2. Materials and Methods

2.1. Study Location

The study was conducted at two sites in Malawi (Figure 1) involving communities around Mchinji and Phirilongwe Forest Reserves in Mchinji and Mangochi districts, respectively. Mchinji Forest Reserve is found between latitudes 13°51′26″ East and longitude 32°51′26″ South, whereas Phirilongwe Forest Reserve is found between the latitude of 14°34′45″ South and the longitude of 34°57′52″ East. In these two reserves, no government intervention or project is being implemented. According to GoM [46], Mchinji district has a total land area of 3131 km^2 with a total population of 602,305 people and a population density of 192 persons per square kilometre. Mangochi district has a total land area of 6729 km^2 with a total population of 1,148,611 people and a population density of 171 persons per square kilometre [46]. Mchinji forest reserve was gazetted in 1924 with a total forest area of 20,885 ha, whereas Phirilongwe forest reserve, situated on the western side of Mangochi district was gazetted in the year 1924 with a total forest area of 16,129 ha. Vegetatively, both Mchinji and Phirilongwe forest reserves and the surrounding customary forest are covered with Miombo woodland with Brachystegia as a dominant tree species. The common tree species in these reserves are Brachystegia julbernadia species such as *Julbernadia paniculata* (Benth) Troupin, *Julbernadia globiflora* (Benth), *Uapaca kirkiana* (Müll.Arg), *Pericopsis angolensis* (Baker) Meeuwen, and *Pterocarpus angolensis* DC. On the other hand, a major available non-timber forest product being harvested in the Phirilongwe forest reserve is the *Oxytenanthera abys-sinica* (A. Rich) Munro (local bamboo), which commonly grows naturally on the escarpment of the Phirilongwe Mountain.

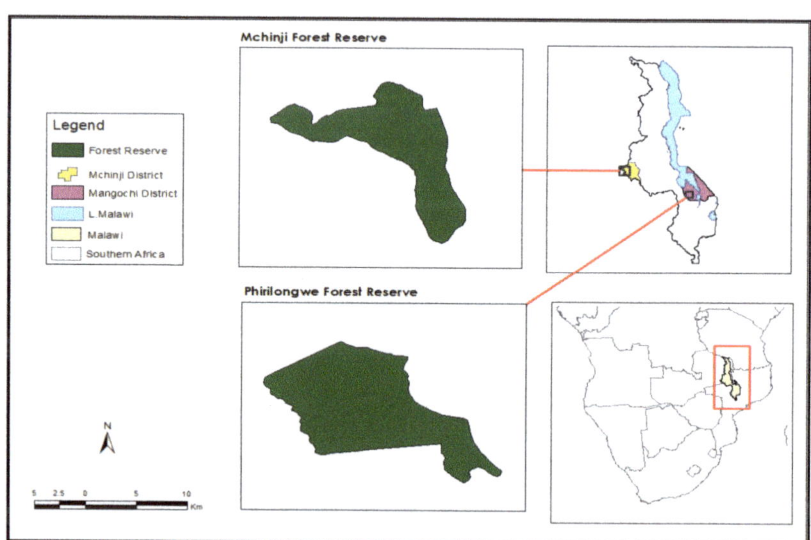

Figure 1. Map of the location of Mchinji and Phirilongwe Forest Reserves in Malawi.

2.2. Study Design and Statistical Analysis

We used a cross-sectional observational study design using a sample survey to collect data from select households. From each of the two districts considered in this study, one forest reserve was selected, namely, Mchinji Forest Reserve in Mchinji district and Phirilongwe Forest Reserve in Mongochi district. There were 134,799 households in Mchinji and

152,879 households in Mangochi. The lists of households surrounding Mchinji and Phirilongwe forest reserves were accessed from their respective district councils. For sample size calculations, we used the equation in Krejcie and Morgan [47] and considered a stratified sample design where the required number of households for each forest reserve was determined independently. There is a lack of data locally on levels of reduced availability of the essential forest products. Thus, we assumed for each of the six forest-based livelihood products, they were equally likely to be reduced or not reduced, so we set the prevalence of being reduced to be 50%, with a level of precision at 5% and confidence level at 95%. These assumptions were for both forest reserve communities. Hence, a total of 227 and 195 households were to be sampled from Mchinji and Mangochi, respectively. The number of households to be sampled in each district were further allocated into the respective traditional authorities and subsequent villages proportionally to the size of those forest communities. For Mchiniji, 71 households were allocated to T/A Mlonyeni, 75 households to T/A Nyoka, and 81 households to T/A Mkanda. For the Mangochi district, 64 households were from T/A Mponda, 64 households from T/A Chilipa, and 67 households from T/A Mtonda. Thus, we interviewed 422 household heads and/or their representatives in total, and the interviews were conducted between April and November 2019.

2.3. Data Collections

A household questionnaire was used to collect data on forest-dependent communities' perceived effects of climate change and extreme weather events on forests and forest-based livelihoods. The questions were adopted from the Climate Risk Assessment Guide Framework developed by the UNDP [48]. In the UNDP Risk Assessment Framework, the first part focuses on identifying the climate extreme events occurring in the study area. The questionnaire uses a rating technique in the assessment of the climate impact on forests. This assessment framework was also previously employed in various studies such as Lazo et al. [49], Williamson et al. [50], and Asherleaf et al. [51] in analyzing the impacts of climate change on Canadian forests. Recently, Ofoegbu et al. [11] and Basu [21] adopted the same rating techniques in assessing the impacts of climate change on the forest-based livelihood of Vhembe district and West Bengal in South Africa and India, respectively. In this paper, we only analyzed and used the data sets of the responses of participants whose ages were 35 years and above because the study had set 20 years as a recall period. Studies have shown that the probability of recalling major climate events in an area is increased by the age and experience of individuals [4,7,28,52]. Limuwa et al. [7] observed that a 20-year recalling period might be sufficient to validate the climate events of an area.

2.4. Data Analysis

Descriptive statistics for continuous data were expressed as means (SD) or as median and interquartile ranges for skewed distributions. Discrete or categorical data were summarized using frequencies and percentages. The independent t-test was used for the comparison of normally distributed data; otherwise, non-parametric alternatives were used.

The analysis of the perceived increase and decrease of each climate and the extreme event was performed to identify key priority climate hazards of the study sites. Erratic rainfall, serious floods, high temperatures, prolonged dry spells, hailstorm incidences, strong winds, and landslides were the climate variables and extreme weather events tested. In our study, we adopted the definition for climate variability by Thornton et al. [19] as the fluctuation to the natural climate system and the extreme weather events as the weather events significantly different from the usually considered normal pattern. These evaluated climate variables were compiled using the previous literature on climate extreme events in Malawi [53–56].

On the other hand, the main essential products tested were firewood, wild fruits and food, wild vegetables, bee honey, mushrooms, medicinal plants, and thatch grasses. These were the essential forest products that were revealed to contribute to forest-dependent communities' livelihoods. Each of these was taken as an outcome variable and was coded as 1 (reported reduced availability) and 0 (no change in the availability). Associations between discrete or categorical data were assessed using Chi-squared tests.

Associations between reduced availability (for each of the essential products) and potential predictor factors adverse climate and the extreme event (erratic rainfall, serious floods, high temperatures, prolonged dry spells, and strong winds) and sociodemographic factors (age, gender, employment, and education) were quantified by odds ratios (OR) with 95% confidence intervals (CI) from fitting multivariate logistic regression analyses. Thus, suppose Y_{ij} denotes the perceived reduction in essential forest product by the respondent, say in Mchinji, where $i = 1, 2, \ldots, 227; j = 1, 2, \ldots, F)$ and where F is the number of the essential products. Furthermore, let $P_{ij} = Prob(Y_{ij} = 1)$ be the probability that household i perceived product j to be in reduced availability, then the effects of climatic and adverse events and socioeconomic factors are modelled by a logit link function as follows:

$$\log(\frac{P_{ij}}{1 - P_{ij}}) = \beta_0 + \beta_C^T \times Climate\ Factors + \beta_{SES}^T \times SES\ Factors \tag{1}$$

where β_C^T and β_{SES}^T are vectors of regression coefficients for the climatic (weather) events and socio-economic factors. We used SPSS version 25 for all the statistical analyses. Qualitative data collected through focus group discussion and key informant interviews were analyzed using content analysis.

3. Results

3.1. Demographic Characteristics of the Respondents

The results of the demographic characteristics of the respondents revealed that Mchinji was dominated by male (53%) respondents, while Mangochi was dominated by female (56%) respondents (Table 1). This might be attributed to the fact that most men in Mangochi are fishermen and therefore spend most of their time on the lake while their male counterparts in Mchinji are mostly farmers. Concerning age >35 years, Mangochi had 76.92% compared to 68.3% in Mchinji. However, we only analysed the responses of participants whose ages were 35 and above to understand their local climate trends because the study had set 20 years as a recall period. In terms of household size, 45.7% (n = 195) of the households in Mangochi had a household size greater than 6 compared to 32.6% in Mchinji. The results also indicate that 84% of respondents in Mangochi were married compared to 75% in Mchinji. In terms of education, 33% of the study population in Mchinji had accessed secondary education compared to only 10% in Mangochi. Furthermore, 24% of the respondents in Mangochi had no formal education compared to only 8% in Mchinji.

Table 1. Demographic characteristics of the respondents.

	Proportion of Respondents in %		Chi-Square Results	
Variable	**Mchinji (n = 227)**	**Mangochi (n = 195)**	**χ^2**	**p-Value**
Age of respondents				
20–34	31.7	23.08	3.909	0.048
$\geq$35	68.3	76.92		
Gender				
Male	53.3	44.1	3.554	0.059
Female	46.7	55.9		
Marital status				
Single	4.8	4.1		
Married	75.3	83.6		
Separated	4	2.6	6.224	0.183
Divorced	7.9	3.1		
Widowed	7.9	6.7		

Table 1. Cont.

Variable	Proportion of Respondents in %		Chi-Square Results	
	Mchinji ($n = 227$)	Mangochi ($n = 195$)	X^2	p-Value
Level of Education				
No formal education	8.4	24.1	40.846	0.000
Primary	59	65.6		
Secondary	32.6	10.3		
Household size				
<3	7.9	6.2		
3 to 5	59.5	48.2	13.843	0.003
6 to 8	26.4	43.1		
>9	6.2	2.6		
Employment status				
Self-Employed	63	55.38	2.521	0.112
Unemployed	37	44.62		

3.2. Observed Climate Change and Extreme Weather Events

The results on the observed climate variability and change show that participants from both study sites perceived a general increase in all the climate extreme events apart from hailstorms and landslides in their locality (Table 2). Erratic rainfall, which refers to the unpredictable and out of season rainfall, was perceived to have increased over the past 20 years by 83.3% and 95.4% in Mchinji and Mangochi, respectively. The chi-square test reveals that these results are statistically significant ($p = 0.000$) across the study sites.

Table 2. Perceived climate change and extreme weather events across the sites.

Variable	Response	Proportion of Respondents (%)		Chi-Square Results	
		Mchinji ($n = 155$)	Mangochi ($n = 150$)	X^2	p-Value
Erratic Rainfall	Increase	83.3	95.4	17.699	0.000
	Decrease	11.9	4.6		
	Constant	4.8	0		
Flooding events	Increase	81.5	84.1	5.612	0.060
	Decrease	8.8	11.8		
	Constant	9.7	4.1		
High temperatures	Increase	71.4	79.5	8.020	0.018
	Decrease	9.7	11.3		
	Constant	18.9	9.2		
Prolonged dry spells	Increase	74.4	84.6	11.120	0.004
	Decrease	14.1	4.6		
	Constant	11.5	10.8		
Hailstorms	Increase	29.6	46.2	21.918	0.000
	Decrease	60.4	53.8		
	Constant	10.0	0.0		
Strong Winds	Increase	75.8	89.7	20.934	0.000
	Decrease	8.8	7.7		
	Constant	15.4	2.6		
Landslides	Increase	28.2	36.2	9.483	0.009
	Decrease	51.6	53.8		
	Constant	20.2	10.0		

Though not statistically different, flooding events have increased in frequency by 81.5% in Mchinji compared to 84% in Mangochi. On the other hand, incidences of high temperatures have increased by 79.5% in Mangochi compared to 71.4% in Mchinji. The other notable perceptions on climatic events in the study are the reduction in the incidences of hailstorms in Mchinji (60.4%) and Mangochi (53.8%) and landslide incidences in Mchinji (51.6%) and Mangochi (53.8%). The results further revealed a significant increase in the frequency of strong winds ($p = 0.000$) and prolonged dry spells ($p = 0.004$).

3.3. Effects of Observed Climate Change and Extreme Weather Events on Access to Forests

The results of the analysis of the observed extreme weather events to understand how they have affected access to essential forest products for the livelihood of the forest communities in the study sites are presented in Table 3. Generally, all the observed extreme weather events were perceived to have affected and reduced access to the forest for more than three months for essential forest products for livelihoods of 65–94% ($n = 150$) of forest-based households in Mangochi and 59–92% in Mchinji ($n = 155$). However, it was only erratic rainfall that was perceived to pose extended reduced access to the forest for essential forest products to 61.2% and 42.5% of forest-based households in Mchinji and Mangochi, respectively. Likewise, a small proportion of forest-based households in Mchinji (32.6%) and Phirilongwe in Mangochi (42.5%) perceived extended reduced access to the forest due to prolonged droughts. The results further record that high temperatures did not affect access to forests for the livelihoods of 41% of forest-based households in Mchinji and 35% in Mangochi. All these results were statistically significant ($p = 0.05$) apart from the results on prolonged drought. However, the results from both the Focus Group Discussions (FDGs) and key informant interviews recorded that increased high temperatures are not a concern for the forest-dependent communities in both sites.

Table 3. Perceived effects of climate variability and change on access to forests.

Climate Events	Responses	Proportion of Respondents in %		Chi-Square Results	
		Mchinji ($n = 155$)	Mangochi ($n = 155$)	X^2	*p*-Value
Erratic rainfall	not effected	8.4	6	15.137	0.001
	Temporary reduced access (3–4 months)	30.4	51.3		
	Extended reduced access (>5 months)	61.2	42.5		
Flooding	not effected	26.9	15.9	6.014	0.048
	Temporary reduced access (3–4 months)	45.8	60		
	Extended reduced access (>5 months)	27.3	24.1		
High temperatures	not effected	41	35.4	9.492	0.009
	Temporary reduced access (3–4 months)	36.6	47.7		
	Extended reduced access (>5 months)	22.4	16.9		
Prolonged Drought	not effected	18.9	12.3	1.802	0.406
	Temporary reduced access (3–4 months)	48.5	51.8		
	Extended reduced access (>5 months)	32.6	35.9		
Strong winds	not effected	27.8	11.8	19.745	0.000
	Temporary reduced access (3–4 months)	48	69.7		
	Extended reduced access (>5 months)	24.2	18.5		

3.4. Sensitivity of the Priority Forest Products to Key Climatic Impact Factors

The perceived threat of climate change and extreme weather events on essential forest products used for their livelihoods were investigated. Table 4a,b present the results of fitting a logistics regression on whether a particular essential product was threatened or not by the effects of key observed extreme climatic events. The results show that the likelihood of perceiving a reduction in the availability of firewood was more likely due to increasing erratic rainfall (OR = 4.965, CI = 2.5–9.86). On the other hand, increased flooding incidences were less likely to be perceived to result in reduced firewood availability (OR = 0.562, $p = 0.033$). The likelihood of perceiving reduced availability of wild fruits and food was more likely attributed to increased dry spells (OR = 1.979, CI = 1.136–3.449) and was less likely perceived as a result of increased flooding events (OR = 0.62, CI = 0.407–0.946). Similarly, the reduced availability of thatch grasses was more likely perceived as the adverse effects of increased erratic rains (OR = 7.584, p = 0.000) and increased high-temperature events (OR = 1.985, CI = 1.129–3.490). However, the likelihood of reduced availability of thatch grasses due to severe flooding was less likely perceived by the respondents (OR = 0.33, CI = 0.211–0.516). Forest-based communities further perceived the reduced availability of mushrooms due to the adverse effects of severe erratic rainfall (OR = 6.480, CI = 2.722–15.429). Nevertheless, the likelihood of reduced mushroom availability due to increased strong winds and flooding events were significantly less likely perceived by the communities OR = 0.544, p = 0.044 and OR = 0.395, p = 0.000, respectively. The likelihood of reduced availability of wild vegetables was more likely attributed to the increasingly erratic rainfall events (OR = 3.154, p = 0.010). However, communities perceived that wild vegetables were significantly less threatened by increasing flooding events (OR = 0.552, CI = 0.351–0.870). Reduction in availability of medicinal plants was more likely perceived to be a result of adverse effects of increasing erratic rainfall (OR = 5.992, p = 0.000) and high temperatures (OR = 2.436, CI = 1.136–4.376). On the other hand, increased flooding events were less likely to be perceived to cause a reduced availability of medicinal plants. The results of drought, education, and gender were not statistically significant at a 95% Confident interval. However, older respondents were less likely to report the reduced availability of fruits and food, thatch grasses, mushrooms, and vegetables. Self-employed forest residents were more likely to perceive the reduced availability of firewood, wild fruits and food, wild vegetables, and medicinal plants. Missing on the list of essential forest products is honey, where results for all predictors were statistically non-significant at a 95% Confident Interval, apart from districts in Mchinji where the reduced availability of honey was more likely perceived with OR = 3.692, CI = 2.211–6.168 and a p = 0.000. In addition, the likelihood of reporting the reduced availability of wild vegetables was significantly more perceived in the Mchinji district (OR = 1.684, p = 0.025).

Table 4. (a) Odd ratios for predictor variables of reduced firewood, fruits and food, and thatch grass. (b) Odds ratios for predictor variables of reduced mushrooms, wild vegetables, and medicinal plants.

	(a)		
	Firewood	**Wild Fruits and Food**	**Thatch Grass**
Independent Predictor	**Odds Ratio (95% CI)**	**Odds Ratio (95% CI)**	**Odds Ratio (95% CI)**
Age (≥35 years vs. <35 year)	0.623 (0.352–1.104) *	0.606 (0.381–0.963)	0.46 (0.286–0.755)
Gender (Male vs. Female)	0.986 (0.604–1.611) *	1.442 (0.950–2.186) *	1.095 (0.703–1.704) *
Uneducated (Yes vs. No)	1.572 (0.745–3.313) *	0.907 (0.508–1.620) *	1.053 (0.572–1.94) *
Employment (Yes vs. No)	1.659 (1.056–2.601)	1.796 (1.178–2.739)	1.054 (0.77–1.641) *
District (Mchinji vs. Mangochi)	0.63 (0.376–1.053) *	0.758 (0.496–1.160) *	1.108 (0.711–1.727) *
Erratic rainfall (Yes vs. No)	4.965 (2.215–16.205)	2.268 (1.141–4.51)	7.89 (2.892–21.328)
Flooding (Yes vs. No)	0.434 (0.277–0.678)	0.62 (0.407–0.946)	0.33 (0.211– 0.516)
High Temperatures (Yes vs. No)	2.436 (1.356–4.376)	0.695 (0.415–1.166) *	1.985 (1.129–3.49)
Strong winds (Yes vs. No)	1.752 (0.929–3.302) *	0.687 (0.390–1.208) *	1.599 (0.863–2.963) *
Drought (Yes vs. No)	0.748 (0.379–1.476) *	1.736 (0.982–3.070) *	0.602 (0.329–1.101) *

Table 4. Cont.

| | (b) | | |
| | Mushroom | Wild Vegetable | Medicinal Plant |
Independent Predictor	**Odds Ratio** **(95% CI)**	**Odds Ratio** **(95% CI)**	**Odds Ratio** **(95% CI)**
Age (≥35 years vs. <35 year)	0.51 (0.319–0,826)	0.547 (0.335–0.891)	0.746 (0.459–1.213) *
Gender (Mala vs. Female)	0.966 (0.628–1.487) *	0.739 (0.469–1.165) *	0.93 (0.596–1.452) *
Uneducated (Yes vs. No)	1.147 (0.631–2.087) *	0.616 (0.315–1.205) *	0.677 (0.36–1.274) *
Employ (Yes vs. No)	1.132 (0.732–1.751) *	2.44 (1.521–3.915)	1.659 (1.059–2.601)
District (Mchinji vs. Mangochi)	0.962 (0.622–1,487) *	1.684 (1.067–2.657)	1.093 (0.703–1.701) *
Erratic rainfall(Yes vs. No)	6.48 (2.72–15.43)	3.15 (1.31–7.594)	5.99 (2.215–16.206)
Flooding(Yes vs. No)	0.395 (0.256–0.61)	0.552 (0.351–0.87)	0.434 (0.277–0.678)
High temperatures(Yes vs. No)	1.642 (0.955–2.823) *	1.641 (0.917–2.936) *	2.436 (1.356–4.376)
Strong winds (Yes vs. No)	0.544 (0.301–0.984)	1.62 (0.836–3.136) *	0.916 (0.494–1.698) *
Drought (Yes vs. No)	0.777 (0.433–1.394) *	1.616 (0.837–3.120) *	1.744 (0.922–3.299) *

* not significant at 95% CI.

4. Discussion

This study set out to use perception-based assessment principles to assess the impact of climate change and extreme weather events on forests and forest-based livelihoods, adjusting for the influence of socioeconomic factors in Malawi. Two forest-dependent communities in two purposively chosen districts in Malawi were used. The section discusses the observed climate change and extreme events over the past 20 years, the effects of these observed climate change and extreme weather events on forest access for forest-based livelihoods, and the sensitivity of the priority forest products to identify the key climatic impact factors. For each of the six main essential products (firewood, wild fruits and food, wild vegetables, mushrooms, medicinal plants, and thatch grasses), a logistic regression model was used to identify its independent predictors.

4.1. Observed Climate Variability and Extreme Events

The study has found that the majority of the forest-dependent communities across the two study sites have perceived an increase in the assessed frequencies of various climatic factors and extreme weather events such as erratic rainfall, flooding events, strong winds, droughts, and high temperatures. These findings are in line with the results of the study by Fujisawa et al. [56], Edward et al. [15], Limuwa et al. [7], Munthali et al. [28], and Chisale [23]. Forest-based households have proven to know their local climate system in our study. This is a positive revelation as far as climate intervention adaptation is concerned. Studies have shown that perceiving local climatic changes is the first stage of the adaptation process to reduce the impacts of the perceived changing climate [21,57–59]. On the other hand, this study shows that forest-dependent communities failed to perceive the increase in the frequencies of hailstorms and landslides events of the past years. Although these might be construed as contradictory results to the findings of Msilimba and Holmes [60] and Omran et al. [61], this might be attributed to their interaction and their long term exposure to the extreme events and local climate and environment. This supports the proposition that, although climate change can be considered at regional and national level, its manifestation is always locally felt, thereby calling for in-depth empirical studies at a local level [21]. However, findings on reduced hailstorms and landslides events in Mchinji and Mangochi best explain and support the findings of Msilimba [62], which attributed hailstorms as the cause of landslides. Thus, reduced hailstorms result in reduced landslides. Furthermore, Msilimba [62] argued that landslides are frequently occurring in mountainous terrains and result in minimal socioeconomic impacts on the society and are thus not well noticed by the locals. This might also apply to the hailstorm that their impacts have not been well noticed by the forest-dependent communities in the studied sites.

4.2. Effects of Observed Climate Change and Extreme Events on Access to Forests

Our study has further revealed disparities in the perceptions of the effects of climate change on access to various forest products used for livelihood. Although these findings may expose the failure on the abilities of forest-based communities to correctly identify the impacts of climate change on their livelihood [11], it gives a true insight of what these communities consider as attributes of concern from climate change impacts for their livelihood in the study sites. Arndt et al. [63] argue that local people have experience and knowledge of local climatic patterns accumulated over the years, which might not be noticed by scientific research. It may be imperative to start harnessing the use of this accumulated knowledge and experience in real-time before they become obsolete. Local communities in this study generally perceived that all the observed extreme climate events affected their access to forest products for their livelihood for over three months. It was revealed during the focus group discussion that mushrooms and medicinal herbs have been heavily affected. In addition, honey production has dwindled due to the drying of rivers. On the other hand, erratic rainfall was perceived to have an extended impact, whereas temperature has not affected their access to forest products. The results of high temperature posing no risk on forest-based livelihood in our study corroborate the findings of Ofoegbu et al. [11] and contradict the empirical findings elsewhere [64,65]. These results suggest that increased temperatures are not of concern to the local communities in Malawi and parts of Southern Africa. Furthermore, local people will always be concerned with those climate attributes that directly affect their livelihoods [21]. However, it could also be attributed to the underestimation of the climate change impacts by the local communities, as proposed by [11], which increases their vulnerability levels to the non-perceived climate trends.

4.3. Essential Forest Products Sensitivity and Vulnerability to Climate Variability

The study has further shown that the forest-dependent communities perceived the sensitivity of some of their forest-based livelihoods to some specific climatic events. For example, respondents perceived that the reduced availability of most essential forest products such as firewood, forest fruits and vegetables, thatch grasses, and mushrooms were more likely due to the adverse effects of increased erratic rainfall and high temperatures. Nevertheless, increased flooding and strong winds were less likely perceived to cause reduced availability of essential forest products. This suggests that not all climatic events pose the same threats to forest-based livelihood. These results support the findings of the study by Ofoegbu et al. [11] and Basu [21] in Vhembe, South Africa and Bengal in India, respectively. In this context, the study also suggests that there are different ways through which climate change affects essential forest products for livelihoods, which are perceived differently by forest-based households. Particularly, high temperatures and erratic rainfall were the only climatic events that were perceived to pose significant threats to firewood. Generally, the rest of the essential forest products significantly perceived to be threatened are all non-wood forest products, such as wild fruits and vegetables as threatened by erratic rainfall and high temperatures. Unlike the findings of Ofoegbu et al. [11], where flooding and erratic rainfall were perceived to pose no significant threat to any forest products, our study unveiled that bee honey is perceived to be threatened by flooding, and thatch grass is threatened significantly by erratic rainfall. We may speculate that their findings in Vhembe were largely influenced by the prevailing climatic conditions of the area, which is conspicuously drier as compared to the Mchinji and Mangochi districts in our study. These results may support the findings of Chilongo [9] which indicated that most high-valued wood products of the forests, such as timber, with high potential to bail them out of poverty, are beyond the reach of the local forest-dependent households in Malawi. This might be the reason for the non-perception of the timber and the construction wood products' sensitivity to climate change and variability in our study.

Generally, the findings of the sensitivity of the various essential forest products to specific climatic events provide insights on the opportunity to develop strategies and interventions to manage the forests by taking into consideration the prevailing climatic events. As suggested by other scholars, these results support the proposition that the perception of the impacts of climate variability and extreme events on forest-based livelihoods and natural resources are more influenced by other socioeconomic factors [11,21,32,56,60]. Thus, forest-dependent communities are more likely to perceive the sensitivity of those forest products that contribute more to their social welfare. Specifically, it is the heightened interaction of the climate and the social-economic pressure that affects forest use and management. This suggests that the resilience of the forest-based livelihood cannot be considered in isolation from the socio-economic needs of the forest-dependent communities. There is a need to look at it holistically, employing the systems thinking model to completely address the sustainability of the forest-based livelihood.

5. Conclusions

We assessed the perceived effects of climate change and extreme weather events on forests and forest-based livelihoods of the forest-dependent communities around the Mchinji and Phirilongwe Forest Reserves in Malawi. The forest-dependent communities identified increasing incidences of erratic rainfall, flooding, high temperatures, prolonged dry spells, and strong winds as key climate variability and extreme events of the study sites. Generally, all five observed extreme climate events reduced the access of the forest to forest-dependent communities for varying periods. However, only erratic rainfall was perceived to pose an extended reduction in access to the forest for livelihood. Mixed results were revealed regarding the sensitivity of essential forest products to increased extreme climate events. Respondents perceived that the reduced availability of most essential forest products was more likely due to adverse effects of increasingly erratic rainfall and high temperatures. Nevertheless, increased flooding and strong winds were less likely perceived to cause the reduced availability of essential forest products. The study has shown that climate change and extreme weather events can affect the access and availability of forest products for livelihoods. We, therefore, recommend concerted efforts and systems approaches to addressing the sensitivity of identified forest-based livelihoods to climate change and socioeconomic pressures. We further call for site-based adaptation and mitigation measures targeting the identified vulnerable forest products such as forest product domestication and respective climate threats in these study sites. We recommend further studies to understand forest use as a climate change coping strategy and assessing the adaptive capacity of these forest-based households.

Author Contributions: Conceptualization, H.L.W.C. and P.W.C.; methodology, H.L.W.C. and S.O.M.M.; software, H.L.W.C.; validation, H.L.W.C., P.W.C., F.D.B. and S.O.M.M.; formal analysis, H.L.W.C.; investigation, H.L.W.C.; resources, H.L.W.C. and P.W.C.; data curation, H.L.W.C., P.W.C.; writing—original draft preparation, H.L.W.C.; writing—review and editing, H.L.W.C., P.W.C., F.D.B. and S.O.M.M.; visualization, H.L.W.C., P.W.C., F.D.B. and S.O.M.M.; supervision, P.W.C. and F.D.B.; project administration, H.L.W.C.; funding acquisition, H.L.W.C. and P.W.C. All authors have read and agreed to the published version of the manuscript.

Funding: This research was funded collaboratively by the Lilongwe University of Agriculture and Natural Resources (LUANAR) and the University of Pretoria under the Gradate Teaching Assistantship Programme of the RUFORUM and The APC was funded by LUANAR. Samuel Manda's time on this paper was supported by the South Africa Medical Research Council.

Institutional Review Board Statement: The study was conducted according to the guidelines of the Declaration of the University of Pretoria and approved by the Faculty of Natural and Agricultural Sciences Ethics Committee (protocol code: 180000128 and date of approval; 21 January 2019).

Informed Consent Statement: Informed consent was obtained from all subjects involved in the study.

Data Availability Statement: Data not publicly available.

Acknowledgments: This project was funded by the Regional Universities Forum for Capacity Building in Agriculture (RUFORUM) under the Graduate Teaching Assistantship (GTA). Precisely, the project was funded by the University of Pretoria and the Lilongwe University of Agriculture and

Natural Resources (LUANAR)—RUFORAM Bursary scheme. The RUFORUM Coordination office staff and the University of Pretoria Ruforum Coordinator, Frans Swanepoel, are hereby acknowledged. The forest-based households and the technical district staff both in Mchinji and Mangochi are hereby acknowledged for responding to the questionnaires. Thumbs up to all the research assistants who took part in the data collection of this project. Special mention to Solomon J. Sibale for study site map production and the study site shapefiles.

Conflicts of Interest: The authors declare no conflict of interest.

References

1. IPCC. Climate change 2014: Climate impacts, adaptation and vulnerability. In *Working Group II to Intergovernmental Panel on Climate Change Fifth Assessment Report*; Intergovernmental Panel on Climate Change: Geneva, Switzerland, 2014.
2. Kurukulasuriya, P.; Mendelson, R. Crop selection: Adaptation to Climate change in Africa. In *Environment and Technology Production Division*; International Food Policy Research Institute (IFPRI): Washington, DC, USA, 2006.
3. Chidumayo, E.; Okali, D.; Kowero, G.; Larwanou, M. (Eds.) *Climate Change and African Forest and Wildlife Resources*; African Forest Forum: Nairobi, Kenya, 2011.
4. IPCC. Climate change 2007: Climate impacts, adaptation and vulnerability. In *Working Group II to Intergovernmental Panel on Climate Change Fourth Assessment Report*; Intergovernmental Panel on Climate Change: Geneva, Switzerland, 2007; Available online: https://www.ipcc.ch/pdf/assessmentreport/ar4/wg2/ar4_wg2_full_report.pdf (accessed on 26 July 2018).
5. Ribeiro, N.S.; Grundy, I.M.; Goncalves, F.M.P.; Moura, I.; Santos, M.J.; Kamoto, J.; Barros, A.I.R.; Gandiwa, E. People in Miombo Woodlands: Socio-ecological dynamics. In *Miombo Woodlands in a Changing Environment: Securing the Resilience and Sustainability of People and Woodlands*; Ribeiro, N.S., Katerere, Y., Chirwa, P.W., Grundy, I.M., Eds.; Springer: Cham, Switzerland, 2020. [CrossRef]
6. FAO. *State of the World's Forests 2015*; FAO: Rome, Italy, 2015.
7. Limuwa, M.M.; Sitaula, B.K.; Njaya, F.; Storebakken, T. Evaluation of small-scale fishers' perceptions on climate change and their coping strategies: Insights from Lake Malawi. *Climate* **2018**, *6*, 34. [CrossRef]
8. Oeba, V.O.; Otor, S.C.; Kung'u, J.B.; Muchiri, M.N. Modelling determinants of tree planting and retention on farm for improvement of forest cover in central Kenya. *Int. Sch. Res. Not.* **2012**, *2012*, 867249. [CrossRef]
9. Chilongo, T.M.S. Forests and Livelihoods in Malawi: Looking Beyond Aggregate Income Shares. Unpublished Doctoral Disertion, School of Economics and Business, Norwegian University of Life Sciences, Ås, Norway, 2005.
10. Chirwa, P.W.; Syampungani, S.; Geldenhuys, C.J. The ecology and management of the miombo woodlands for sustainable livelihoods in southern Africa: The case for non-timber forest products. *South. For.* **2008**, *70*, 237–245. [CrossRef]
11. Ofoegbu, C.; Chirwa, P.W.; Babalola, F.D.; Francis, J. Perception-based analysis of climate change effect on forest-based livelihood: The case of Vhembe District in South Africa. *Jàmbá J. Disaster Risk Stud.* **2016**, *8*, 1–11. [CrossRef]
12. FAO. *State of the World's Forests 2011*; FAO: Rome, Italy, 2011.
13. FAO. *State of the World's Forests 2014*; FAO: Rome, Italy, 2014.
14. Senganimalunje, T.C.; Chirwa, P.W.; Babalola, F.D. Exploring the Role of Forests as Natural Assets in Rural Livelihoods and Coping Strategies Against Risks and Shocks in Dedza east, Malawi. *J. Sustain. For.* **2020**. [CrossRef]
15. Edward, M.; Henry, U.; Maggie, M.; William, M. Modelling of climate conditions in forest vegetation zones in Malawi. *World J. Adv. Res. Rev.* **2019**, *1*, 36–44.
16. Makungwa, S. *Enhancing Productivity and Estimation of Carbon in CDM Forestry Projects: A Malawi Case Study*; The University of Edinburg: Edinburgh, UK, 2015.
17. Ngwira, S.; Watanabe, T. An analysis of the causes of deforestation in Malawi: A case of Mwazisi. *Land* **2019**, *8*, 48. [CrossRef]
18. Sein, M.; Nomura, H.; Takahashi, Y.; Ogata, K.; Mitsuyasu, Y. Impact of erratic rainfall from climate change on pulse production efficiency in lower Myanmar. *Sustainability* **2018**, *10*, 1–16.
19. Thornton, P.K.; Ericksen, P.J.; Herrero, M.; Challinor, A.J. Climate variability and vulnerability to climate change: A review. *Glob. Chang. Biol.* **2014**, *20*, 3313–3328. [CrossRef]
20. Robledo, C.; Clot, N.; Hammill, A.; Riché, B. The role of forest ecosystems in communitybased coping strategies to climate hazards: Three examples from rural areas in Africa. *J. For. Policy Econ.* **2012**, *24*, 20–28. [CrossRef]
21. Basu, J.P. *Climate Change Adaptation and Forest Dependent Communities: An Analytical Perspective of Different Agro-Climatic Regions of West Bengal, India*; Springer: New York, NY, USA, 2017.
22. Nunes, A.R. Assets for health: Linking vulnerability, resilience and adaptation to climate change. *Tyndall Cent. Clim. Chang. Res. Work. Pap.* **2016**, *163*, 1–41.
23. Chisale, H. *Climate Change in Malawi and Its Implication on Natural Resource Base*; LAP Lambert Academic Publishing: Sunnyvale, CA, USA, 2013; ISBN-13: 978-3659471360.
24. Pangapanga, P.I.; Jumbe, C.B.; Kanyanda, S.; Thangalimodzi, L. Unravelling strategic choices towards droughts and floods' adaptation in southern Malawi. *Disaster Risk Reduct.* **2012**, *2*, 57–66. [CrossRef]
25. Klein, R.J. Approaches, methods and tools for climate change impact, vulnerability and adaptation assessment. In *Keynote Lecture to the In-Session Workshop on Impacts of, and Vulnerability and Adaptation to, Climate Change, Proceedings of the Twenty-First Session of the UNFCCC Subsidiary Body for Scientific and Technical Advice, Buenos Aires, Argentina, 6–14 December 2004*; UNFCCC: Bonn, Germany, 2004; Volume 8.

26. Malone, E.L. Vulnerability and resilience in the face of climate change: Current research and needs for population information. *Popul. Action Int.* **2009**, *31*, 16063171.
27. Davidson, D.J.; Williamson, T.; Parkins, J.R. Understanding climate change risk and vulnerability in northern forest-based communities. *Can. J. For. Res.* **2004**, *33*, 2252–2261. [CrossRef]
28. Munthali, C.K.; Kasulo, V.; Matamula, S. Smallholder farmer's perception on climate change in Rumphi District, Malawi. *J. Agric. Ext. Rural Dev.* **2016**, *8*, 202–210.
29. Locatelli, B.; Kanninen, M.; Brockhaus, M.; Colfer, C.P.; Murdiyarso, D.; Santoso, H. *Facing an Uncertain Future: How Forests and People Can Adapt to Climate Change*; Forest Perspectives no. 5; CIFOR: Bogor, Indonesia, 2008.
30. Tiani, A.M.; Besa, M.C.; Devisscher, T.; Pavageau, C.; Butterfield, R.; Bharwani, S.; Bele, M.Y. *Assessing Current Social Vulnerability to Climate Change: A Participatory Methodology*; Working Paper 169; CIFOR: Bogor, Indonesia, 2015.
31. Vedeld, P.; Angelsen, A.; Bojö, J.; Sjaastad, E.; Berg, G.K. Forest environmental incomes and the rural poor. *For. Policy Econ.* **2007**, *9*, 869–879. [CrossRef]
32. Fisher, M.; Chaudhury, M.; McCusker, B. Do forests help rural households adapt to climate variability? Evidence from Southern Malawi. *World Dev.* **2010**, *38*, 1241–1250. [CrossRef]
33. Bwalya, S.M. Household dependence on forest income in rural Zambia. *Zamb. Soc. Sci. J.* **2011**, *2*, 6.
34. Jumbe, C.B.L. Community Forest Management, Poverty and Energy Use in Malawi. Ph.D. Thesis, Department of Economics and Resource Management, Norwegian University of Life Sciences, Ås, Norway, 2005.
35. Fisher, M. Household welfare and forest dependence in Southern Malawi. *Environ. Dev. Econ.* **2004**, *9*, 135–154. [CrossRef]
36. Fisher, M.; Shively, G.E. Agricultural subsidies and forest pressure in Malawi's Miombo Woodlands. *J. Agric. Resour. Econ.* **2007**, 349–362.
37. Bandyopadhyay, S.; Shyamsundar, P.; Baccini, A. Forests, biomass use and poverty in Malawi. *Ecol. Econ.* **2011**, *70*, 2461–2471. [CrossRef]
38. Senganimalunje, T.C.; Chirwa, P.W.; Babalola, F.D.; Graham, M.A. Does participatory forest management program lead to efficient forest resource use and improved rural livelihoods? Experiences from Mua-Livulezi Forest Reserve, Malawi. *Agrofor. Syst.* **2016**, *90*, 691–710. [CrossRef]
39. USAID. *Malawi Climate Change Vulnerability Assessment 2013: Africa and Latin America Resilience to Climate Change Project (ARCC) Report*; USAID–Malawi: Washington, DC, USA, 2013.
40. GoM. *The Malawi Vulnerability Assessment Committee (MVAC)*; Bulletin No. 8; GoM: Lilongwe, Malawi, 2012; Volume 2.
41. Sharma, R.C. *Forest for Poverty Alleviation: Chhattisgarh Experience*; Food and Agriculture Organisation: Rome, Italy, 2005; Volume 109.
42. Mamo, G.; Sjaastad, E.; Vedeld, P. Economic dependence on forest resources: A case from Dendi District, Ethiopia. *For. Policy Econ.* **2007**, *9*, 916–927. [CrossRef]
43. Shougong, Z.; Weichang, L.; Wenming, L.; Huafeng, D. Community forestry in mountain development: A case study in Guizhou Province, China. In *Forests for Poverty Reduction: Changing Role for Research, Development and Training Institutions*; Sim, H.C., Appanah, S., Hooda, N., Eds.; FAO: Bangkok, Thailand, 2003.
44. Lynam, T.; Cunliffe, R.; Mapaure, I. Assessing the importance of woodland landscape locations for both local communities and conservation in Gorongosa and Muanza Districts, Sofala Province, Mozambique. *Ecol. Soc.* **2004**, *9*. [CrossRef]
45. Tshering, D. 23 *Forests for Poverty Alleviation—Case of Bhutan*; Food and Agriculture Organisation: Rome, Italy, 2005; Volume 157.
46. GoM. *Population and Housing Census, 2018*; National Statistical Office: Zomba, Malawi, 2018.
47. Krejcie, R.V.; Morgan, D.W. Determining sample size for research activities. *Educ. Psychol. Meas.* **1970**, *30*, 607–610. [CrossRef]
48. United Nations Development Programme (UNDP). *Climate Risk Assessment Guide—Central Asia, Project: Enabling Integrated Climate Risk Assessment for CCD Planning in Central Asia*; CAMP Alatoo Public Foundation: Almaty, Kazakhstan, 2013.
49. Lazo, J.K.; Kennell, J.C.; Fisher, A. Expert and Layperson perceptions of ecosystem risk. *Risk Anal.* **2000**, *2*, 179–193. [CrossRef]
50. Williamson, T.B.; Parkins, J.R.; McFarlane, B.L. Perceptions of climate change risk to forest ecosystems and forest-based communities. *For. Chron.* **2005**, *81*, 710–716. [CrossRef]
51. Asherleaf. *Climate Risk & Opportunity Management*; Asher Innovation: Mont-Liban, Lebanon, 2012; pp. 1–12.
52. Mittal, N.; Pope, E.; Whitfield, S.; Bacon, J.; Bruno Soares, M.; Dougill, A.J.; Homberg, M.V.D.; Walker, D.P.; Vanya, C.L.; Tibu, A.; et al. Co-designing indices for tailored seasonal climate forecasts in Malawi. *Front. Clim.* **2021**, *2*, 30. [CrossRef]
53. Haghtalab, N.; Moore, N.; Ngongondo, C. Spatio-temporal analysis of rainfall variability and seasonality in Malawi. *Reg. Environ. Chang.* **2019**, *19*, 2041–2054. [CrossRef]
54. Vincent, K.; Dougill, A.J.; Dixon, J.L.; Stringer, L.C.; Cull, T. Identifying climate services needs for national planning: Insights from Malawi. *Clim. Policy* **2017**, *17*, 189–202. [CrossRef]
55. Pauw, K.; Thurlow, J.; Van Seventer, D.E. The economic costs of extreme weather events: A hydrometeorological CGE analysis for Malawi. *Environ. Dev. Econ.* **2011**, *16*, 177–198. [CrossRef]
56. Fujisawa, M.; Gordes, A.; Heureux, A. Assessing the impacts of climate change on the agriculture sectors in Malawi. In *The MOSAICC Methodology for National Adaptation Planning*; FAO: Rome, Italy, 2020. [CrossRef]
57. Füssel, H.M.; Klein, R.J.T. Climate change vulnerability assessments: An evolution of conceptual thinking. *Clim. Chang.* **2006**, *75*, 301–329. [CrossRef]

58. Ishaya, S.; Abaje, I.B. Indigenous people's perception on climate change and adaptation strategies in Jema'a local government area of Kaduna State, Nigeria. *J. Geogr. Reg. Plan.* **2008**, *1*, 138–143.
59. Mertz, O.; Mbow, C.; Reenberg, A.; Diouf, A. Farmers' perceptions of climate change and agricultural adaptation strategies in rural Sahel. *Environ. Manag.* **2009**, *43*, 804–816. [CrossRef]
60. Msilimba, G.; Holmes, P. A landslide hazard assessment and vulnerability appraisal procedure; Vunguvungu/Banga catchment, Northern Malawi. *Nat. Hazards* **2005**, *24*, 99–216. [CrossRef]
61. Omran, A.; Schwarz-Herion, O. *The Impact of Climate Change on Our Life: The Questions of Sustainability*; Springer: New York, NY, USA, 2018.
62. Msilimba, G.G. The socioeconomic and environmental effects of the 2003 landslides in the Rumphi and Ntcheu Districts (Malawi). *Nat. Hazards* **2010**, *53*, 347–360. [CrossRef]
63. Arndt, C.; Schlosser, A.; Strzepek, K.; Thurlow, J. Climate change and economic growth prospects for Malawi: An uncertainty approach. *J. Afr. Econ.* **2014**, *23* (Suppl. 2), ii83–ii107. [CrossRef]
64. Naidoo, S.; Davis, C.; Van Garderen, E.A. *Forests, Rangelands and Climate Change in Southern Africa*; Food and Agriculture Organization: Rome, Italy, 2013; Volume 12.
65. Gauthier, S.; Bernier, P.; Burton, P.J.; Edwards, J.; Isaac, K.; Isabel, N.; Jayen, K.; Le Goff, H.; Nelson, E.A. Climate change vulnerability and adaptation in the managed Canadian boreal Forest. *Environ. Rev.* **2014**, *22*, 256–285. [CrossRef]

 *sustainability*

Review

A Systematic Literature Review of Inclusive Climate Change Adaption

Ha Pham * and Marc Saner

Department of Geography, Environment and Geomatics, University of Ottawa, Ottawa, ON K1N 6N5, Canada; msaner@uottawa.ca
* Correspondence: hpham048@uottawa.ca

Abstract: Inclusive approaches have been applied in many areas, including human resources, international development, urban planning, and innovation. This paper is a systematic literature review to describe the usage trends, scope, and nature of the inclusive approach in the climate change adaptation (CCA) context. We developed search algorithms, explicit selection criteria, and a coding questionnaire, which we used to review a total of 106 peer-reviewed articles, 145 grey literature documents, and 67 national communications to the United Nations Framework Convention on Climate Change (UNFCCC); 318 documents were reviewed in total. Quantitatively, the methodology reveals a slight increase in usage, with a focus on non-Annex 1 countries, gender issues, and capacity building. Qualitatively, we arranged the key insights into the following three categories: (1) inclusion in who or what adapts; (2) motivating inclusive processes; and (3) anticipated outcomes of inclusive CCA. We conclude, with the observation, that many issues also apply to Annex 1 countries. We also argue that the common language nature of the word 'inclusive' makes it applicable to other CCA-relevant contexts, including government subsidies, science policy, knowledge integration and mobilization, performance measurement, and the breadth of the moral circle that a society should adopt.

Keywords: climate change adaptation; climate justice; inclusiveness; national communications; systematic review

Citation: Pham, H.; Saner, M. A Systematic Literature Review of Inclusive Climate Change Adaption. *Sustainability* **2021**, *13*, 10617. https://doi.org/10.3390/su131910617

Academic Editor: Baojie He

Received: 23 August 2021
Accepted: 21 September 2021
Published: 24 September 2021

Publisher's Note: MDPI stays neutral with regard to jurisdictional claims in published maps and institutional affiliations.

1. Introduction

Climate change adaptation (CCA) designates the process of adjustment to actual or expected climate effects, intended to avoid harm or exploit beneficial opportunities [1]. Successful adaptation requires an approach in which all stakeholders are involved, to ensure that all needs will be considered and all outcomes will be just [2–4]. The terms "participation", "stakeholder engagement", "public involvement", "bottom up", and "community-based" have been referred to widely in the discourse of the adaptation [5,6]. A participatory approach to CCA has been advocated by many international organizations [1,7–9], and has been embedded in CCA policies worldwide [10–12]. The idea of inclusion is firmly embedded in this established debate.

Simply creating spaces for public participation does not ensure the broader involvement of stakeholders [1,13]. Pre-existing power asymmetries reinforce the existing privileges of some stakeholders and suppress minority perspectives [14–16]. In the case of CCA, this situation is particularly problematic, because less powerful populations are often harmed more severely by global warming. Adaptation solutions resulting from inequitable processes cannot be responsive to the needs of the weak, and adaptation outcomes often fall short in the criteria of equity, fairness, and justice [17,18]. In addition, the interests of future generations and of sentient non-human animals are often underrepresented. Scholars have described this situation as "non-inclusive participation", "adverse inclusion", or "limited inclusion" [19–21]. Applying the goal of inclusiveness as applied to opportunities, needs, risks, benefits, costs, and profits is even more ambitious than addressing the challenge of participation.

155

The concept of inclusion was popularized in classical social science research, with the ideas of social inclusion and an inclusive society [22,23]. In recent decades, it has been reconceived by researchers and policy makers to apply to contemporary issues, including inclusive growth or development [24,25], inclusive/exclusive governance or decision making [26,27], inclusive cities or urbanization [28–31], and inclusive innovation [32–34]. In these contexts, it is of the utmost importance to realize that inclusion is a universal value, referring to the right of people to access regular things and participate in mainstream society. Besides, inclusion opens a new way of thinking that influences our beliefs and actions. Inclusion is about people gaining social acceptance, having positive interactions with their peers, and being valued for who they are [22,23]. As such, it must be internally motivated and caused by embracing the belief that all people have both the right to belong and the responsibility to respect the right of others to belong [22]. Inclusion values diversity and provides real opportunities for people (both with and without disabilities) to improve their lives [24,27,31].

The interest in inclusive approaches to climate change began some time ago [2,15,35,36]. However, there has been no coherent understanding of what form this inclusion should take [6,37]. There is a need for both conceptual and empirical work on the issue of inclusive CCA. This literature review contributes to this work. We refer to 'inclusiveness' throughout this text, because it is more commonly used than the synonymous 'inclusivity', but the literature search covers all related concepts.

A number of previous systematic literature reviews have specifically focused on CCA, including characterizing adaptation actions [38,39], governance of adaptation [40], and adaptation in different locations or sectors [41,42]. However, there are no reviews focusing on the inclusiveness of CCA. The purpose of this literature review is to systematically chart the usage of 'inclusiveness' in the CCA context. We believe that it is a useful concept that can complement related ideas, such as climate justice, equity, participation, respect for diversity, bias, and discrimination.

Multiple perspectives on inclusiveness flow into the issue of climate justice that gained momentum in the late 1990s [43]. There are several definitions of climate justice, and they often express the idea of (a lack of) inclusiveness. Hughes [44] (p. 51, our emphasis), for example, proposes the following three criteria for climate justice: 1. representation of vulnerable groups in adaptation planning processes; 2. priority setting and framing that recognize the adaptation needs of the vulnerable groups; and 3. impacts of adaptation that enhance the freedoms and assets of vulnerable groups. The Routledge Handbook on Climate Justice has put the need to embrace equity and inclusiveness at the core of the discussions on climate justice and how it can be achieved [43]. Many existing studies on issues of inclusiveness in CCA planning and policy-making processes posit that adequate representation and participation of the most marginalized and vulnerable—in both developed and developing countries—will yield more recognition, procedural justice, and distributive justice, which inter alia can define climate justice [45–49].

To address the climate emergency that has recently entered mainstream debates, scholars are reconceptualizing climate justice in a more inclusive way, advocating for the reemergence of intra-generation justice and multispecies justice. This is a conceptual expansion of the use of the term, decentering the human and recognizing the human relationship with other inter- and intra-generational people and more-than-human beings [50–52]. More inclusive approaches move the climate justice discourse into multi-temporal and multi-scalar realms. In the context of CCA, this scope delineates what systemic transformations may involve (and with whom), how to adapt to inevitable and possibly intolerable losses, and how to prefigure and enact alternative and just futures.

Our present interest in exploring the 'inclusiveness' concept more deeply is two-fold. First, it is an integral part of "EDI" (equity, diversity, inclusion), which is increasingly important in various contexts. The integration of EDI into the daily life of public servants, academics, and employees of non-government organizations makes it hard to ignore. Second, and ultimately more important, is the great versatility of the common language concept 'inclusiveness'. It is understood without great theoretical background or metaphysical justification. Additionally, it is applicable to a wide and growing range of relevant contexts, as we will show.

We use the methodology of systematic literature reviews because it provides a transparent, reproducible and rigorous approach to dealing with large information sets [53]. This approach provides quantitative results, such as trends and numerical comparisons, as well as a systematic input into summaries and analyses [38,54]. Because it is systematic, it is also constrained. The search algorithms define what may be included, and the questionnaire for data collection defines how the analysis is structured [41,42]. The upside is that the greater transparency and reproducibility of the search render future updates very feasible and improve the disclosure of the value judgments made by us. The downside is that it limits on snowballing and the search for the best references to support the emerging story. We cover the most current peer-reviewed literature, grey literature, and policy documents, with 318 documents in total.

In this paper, we deal with the following two main research questions: (i) How has the concept of inclusive CCA been used in the literature? (ii) What are the main components of CCA proposed in the literature? In order to answer these questions, we will describe the data selection, collection, and analysis (Section 2), demonstrate the trend of using this concept in the literature (Section 3), reveal a framework of inclusive CCA (Section 4), argue for the importance of this concept in the national adaptation climate change policy of both developed and developing countries (Section 5), and conclude our paper by discussing the concept's main contributions, limitations, and recommendations for future research.

2. Materials and Methods

A systematic review refers to a focused review of the literature that seeks to answer (a) specific research question(s) using predefined eligibility criteria for document selection and explicitly outlined and reproducible methods [38,53,54]. Systematic reviews have been increasingly used in the environmental change research context [38–42,55,56].

2.1. Data Selection

The following three data sources were used: the peer-reviewed literature, grey literature (reports by consultants, governments, and NGOs), and national communications (NCs) to the United Nations Framework Convention on Climate Change (UNFCCC).

For *peer-reviewed articles*, we used the following search query in Scopus (https://www.scopus.com/) (accessed on 22 May 2021):

TITLE-ABS-KEY (inclusi* AND ("climate chang*" OR "changing climate" OR "climate warm*" OR "warm* climate" OR "global warm*" OR "global chang*" OR "environment* chang*" OR "environment* warm*" OR "warm* environment") AND (adapt* ORinterven*)) within the following subject areas: environmental science, social science, earth and planetary science, art and humanities, and multidiscipline.

A total of 614 articles were retrieved for title and abstract scanning to select papers with clear relevance to CCA and inclusiveness. Based on the inclusion/exclusion criteria (see Table 1 and Figure 1, below), 106 documents met the final relevance screening criteria and underwent data extraction.

This review also gives extensive consideration to *grey literature* and policy documents because restricting a review to only peer-reviewed literature can miss key trends and insights with significant implications for biasing the results [38,53]. Relevant grey literature was identified using the following focused search in Google:

("climate change" OR "changing climate" OR "global warming" OR "environmental change") AND ("adapt" OR "adaptation" OR "intervene" OR "intervention") AND ("inclusive" OR "inclusion" OR "inclusivity" OR "inclusiveness") filetype:pdf. We restricted included documents to PDF files only to limit hits to a manageable number. The titles and descriptions provided within the standard Google search engine were reviewed to determine the relevance of each result.

Table 1. Inclusion and exclusion criteria for document selection.

Inclusion Criteria	Exclusion Criteria
Text in English	Text in other languages
Full text available	Only abstract or partial text available
Human response to climate change	Biological response to climate change
Adaptive response to climate change	Vulnerability, mitigation only
Refers directly to inclusiveness	Does not refer to inclusiveness
Sufficient detail for data extraction	Insufficient detail for data extraction

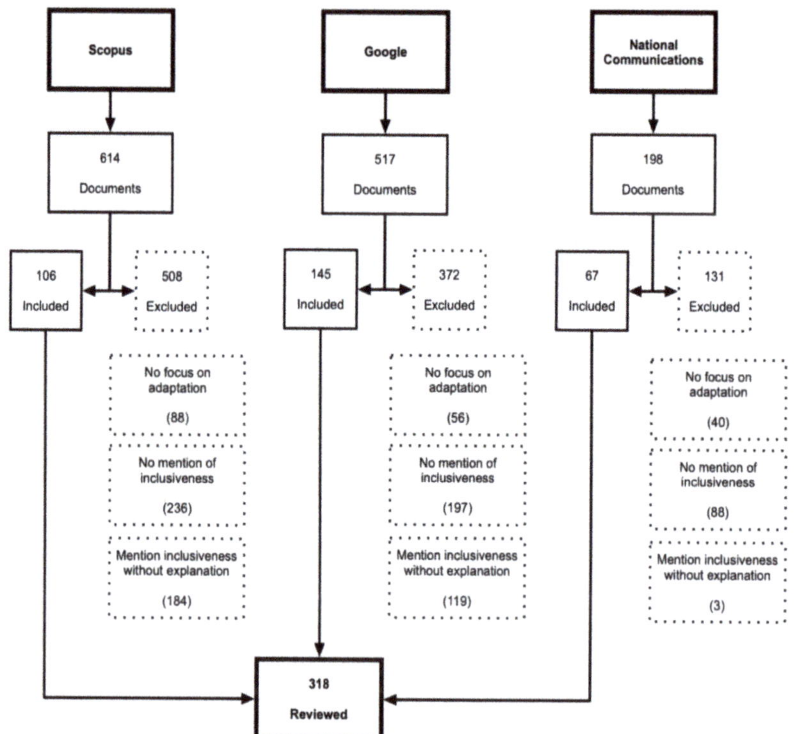

Figure 1. Document triage process.

A total of 517 search results met the initial screening (Mendeley has been used to remove duplicates compared to the Scopus search). A first page screen, followed by a full-text review, was applied to confirm eligibility using inclusion/exclusion criteria, and a total of 145 grey literature documents were retained and included for data extraction.

For *policy documents*, the most recent national communications were selected as the data source for this analysis. The NCs were considered the most appropriate data source for several reasons. First, national governments play key roles in adaptation planning and implementation by protecting vulnerable groups, supporting economic diversification, providing information, creating policy frameworks, making laws, and distributing financial support [1,57–60]. Second, NCs constitute a consistent source of English language information available for many developed and developing countries [42]. Third, national governments submit these documents to report their policy priorities and progress, which renders them official records [53].

A total of 198 NCs were extracted from the UNFCCC website, including 44 NCs of Annex 1 countries (NC6 reports of the US and Ukraine and NC7 reports of others), and the 154 most recent available NCs submitted by non-Annex 1 countries (the NC of Libya was not available). Annex 1 of the convention lists developed countries and economies in transition, whereas non-Annex 1 parties are mostly low-income, developing countries. All the reports were screened based on the inclusion and exclusion criteria, and a total of 67 NCs (22 from Annex 1 countries and 45 from non-Annex 1 countries) underwent data extraction.

2.2. Data Collection and Analysis

Following document screening, 318 articles from all three data sources (peer-reviewed, gray, and NC) were retained for full review (Figure 1, above). Peer-reviewed and grey literature were reviewed in full, while NCs were reviewed only where they concerned adaptation.

To achieve greater consistency, we developed an article review questionnaire (see Table 2, below) documenting how inclusive CCA is understood and occurs in the selected articles [39,53]. The general characteristics of the article in terms of authorship, year published, region of interest, and (conceptual/practical) approach provided a foundation for the quantitative portion of the systematic literature review. To support the qualitative portion, we structured the analysis of the literature by separating three themes, inspired by the adaptation assessment frameworks proposed by Smit and Pelling [4,61,62], as follows: (1) who or what adapts; (2) how to adapt (adaptation activities and adaptive capacity required to adapt); and (3) adaptation outcomes. Data were entered into a Microsoft Excel spreadsheet for descriptive statistics on quantitative trends.

Table 2. Questionnaire for data collection.

General Questions		
Lead author?		
Year published?		
Region of interest:	Annex 1? Non-Annex 1?	
Approach:	Conceptual? Practical (Example or Case)?	
Specific questions		
Who or what to include		
Scale:	Local and community? National? Regional? International scales? Necessity to cooperate across scales?	
Stakeholders:	A broader spectrum of actors?	Local people and local communities? Governments, including local and national levels? The private sector? Experts and research communities? NGOs and civil society? International actors, networks, and agencies?
	Vulnerable groups?	The poor? People with disabilities? The indigenous? Women and girls? Resource-dependent people?
	More than human–others?	

Table 2. *Cont.*

Specific questions	
Knowledge:	Traditional? Expert?
Techniques or tools:	Participatory (action) research approaches? Qualitative scenarios? A programming model? Adaptation design tool? Social ecological inventory (SEI)?

How to include (adaptation activities and capacity required to adapt)
Governance?   Institution?   Social capacity?

Outcomes
Consideration of the vulnerable groups?   Adequate participation?   Just results?   A status of resilience, inclusive development, and sustainability?

3. Usage of Inclusiveness in Climate Change Adaptation

3.1. Growing Usage

Of the 318 reviewed documents, only about 10% (29) date back to 2010 or earlier (see Figure 2, below). Ninety percent of the reviewed documents are dated 2011–2020, with an increasing trend that may have peaked in 2018. The data for 2021 are incomplete and only represent the first quarter—an extrapolation to the full year would lead to the same number as in the peak year 2018. We should note that an increase in the usage of "inclusive climate change adaptation" will likely be in line with any increase in the usage of the broader term "climate change adaptation".

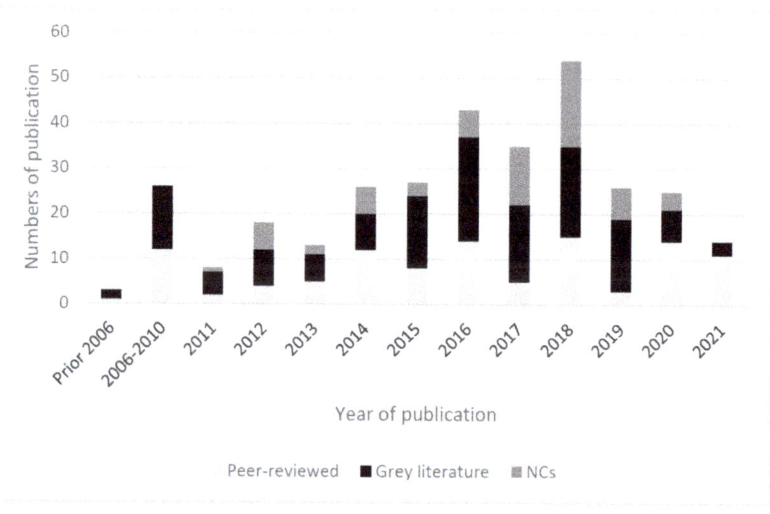

Figure 2. Number of publications by year.

3.2. Predominance of Practical Contexts

We used the following definitions to distinguish practical and conceptual contexts. Practical approaches include substantive reporting or discussion of an adaptation activity in place, excluding proposed strategies, empirical testing, and predictive modeling [39]. In contrast, conceptual approaches specify sequential relationships and feedback for adaptation processes in general, or for sectors or applications, providing the framework or structure for research, analyses, or modeling [62].

As shown in Figure 3, below, we categorized nearly two-thirds (199/318) of the reviewed literature in the practical category. This is agreement with Agrawal [63] and Oulu [59], who conclude that climate change adaptation is still a relatively new field, in which policy and practice tend to precede theory or advance simultaneously. They also argued that the lack of middle-range adaptation theories and comparative empirical studies is a glaring challenge. The results in Figure 3 may indicate that conceptual work in inclusive climate change adaption is lagging behind the progress made in practical research.

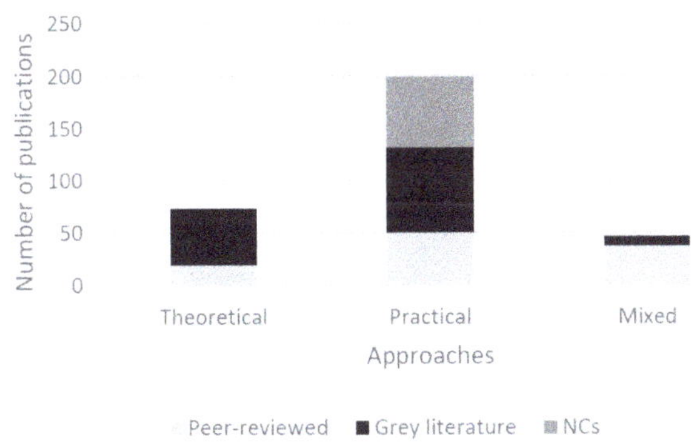

Figure 3. Theoretical vs. practical approaches.

Practical approaches include substantive reporting or discussion of an adaptation activity in place [39]. In other words, practical approaches describe cases or examples of how adaptation inclusiveness is happening in practice. This study is intended for a wide audience of development and CCA practitioners, to support their daily work. Moreover, the findings and good practice principles on inclusive CCA create good opportunities for researchers to test and complete the related theoretical issues that have not been well developed yet, and have applications extending well beyond the CCA field. Out of 199 documents using practical approaches, we identify 331 cases and examples. The following part presents some findings from our review of the cases and examples on inclusive CCA.

3.3. Predominance of Gender Issues

As shown in Figure 4, below, more than 43% (144/331) of the cases or examples in the reviewed literature are related to gender issues. Less than 20% referred to other vulnerable groups, such as the poor, people with disabilities, the indigenous, resource-dependent people, future generations, or non-human actors.

The significance of gender issues in adaptation inclusiveness can be explained in several ways.

First, vulnerability to climate change is not gender neutral. The inequitable distribution of rights, resources, and power increases the vulnerability of women, as do social rules and norms. Women often find themselves in a vicious cycle, in which limited access to resources amplifies their susceptibility to climate change, and vice versa [16,64].

Second, the terms "gender-inclusive adaptation" and "gender-responsive adaptation" have been used regularly by UN organizations and international donors, including UNDP, UNEP, UNFCCC, GCF, and ADB [65–70]. These organizations have developed very detailed toolboxes or checklists on how to integrate gender and climate change into policy and practice, and have encouraged all countries to integrate gender into national communication reporting [71–74]. This approach to CCA directly addresses the issues of gender inequality, provides strengthened supports to women, empowers them, and places them at the center of CCA processes. Better support to these groups would help to prevent further depletion of their resilience to climate change, and ensure that climate change does not accentuate or perpetuate existing gender inequities [75].

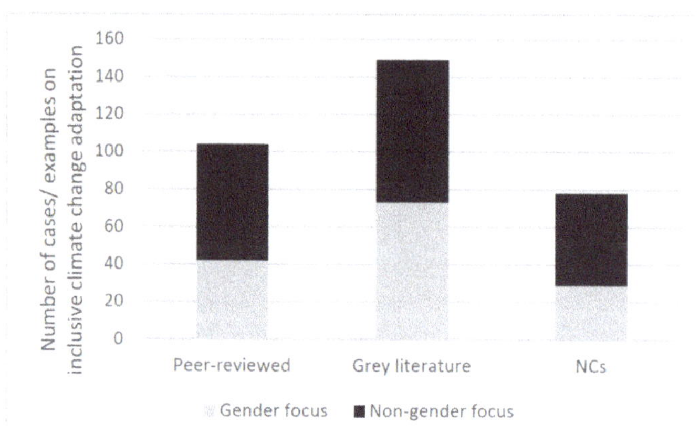

Figure 4. Cases with gender or non-gender foci.

3.4. Predominance of Non-Annex 1 Countries

In the CCA context, it is common to use the UNFCCC's classification of parties as "Annex 1" and "non-Annex 1" countries. As shown in Figure 5, below, nearly three-fourths (267/331) of the examples and cases originated in non-Annex 1 countries. As clarified by the UNFCCC, non-Annex 1 countries are mostly low-income, developing countries, while Annex 1 lists industrialized countries and economies in transition.

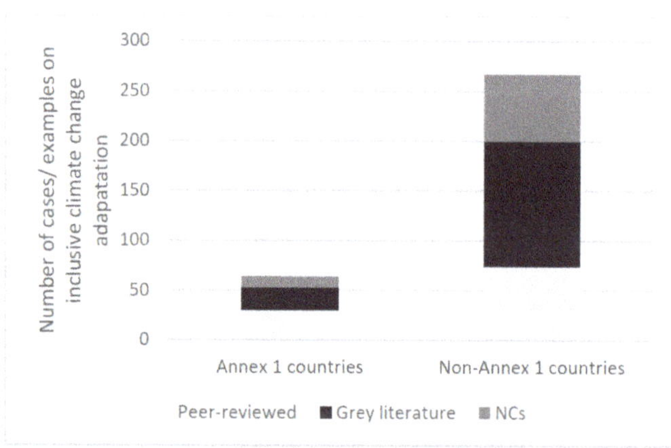

Figure 5. Cases in Annex 1 and non-Annex 1 countries.

Resurrección et al. [76] found similar results when conducting a review of the relevant peer-reviewed research and grey literature on gender and CCA, leaning more towards adaptation contexts in the global south, due, in large, to the availability of the literature. Populations that lack the resources for planned migration experience greater exposure to extreme weather events, particularly in developing countries with low income. Climate change can indirectly increase the risks of violent conflicts, by amplifying well-documented drivers of these conflicts, such as poverty and economic shocks [1].

3.5. Inclusiveness at the Intersection of Development, Gender and Capacity in National Communications

The non-Annex 1 countries considered inclusiveness to be a target/an aspiration. A group of countries (Bangladesh, Dominica, Belize, Indonesia, Kyrgyzstan, and Moldova) referred to inclusiveness in CCA as a broad-based strategy to empower every citizen or all stakeholders to participate fully and benefit from the adaptation process. In contrast, some countries emphasized the inclusion of related stakeholders in the adaptation process, such as local communities (Equatorial Guinea), the private sector (Fiji), multiple government entities (Jamaica), or the most vulnerable segments of the society (Namibia, Pakistan).

Another group of non-Annex 1 countries uses 'inclusiveness' in the context of CCA education. Efforts should be undertaken to develop materials and promote teacher training that is focused on climate change, linking education and awareness of climate change (Antigua and Barbuda, Bangladesh, Kuwait, Laos, Lesotho, South Africa).

Gender issues are at the center of inclusive CCA in many non-Annex 1 countries (Bosnia and Herzegovina, Jamaica, Nauru, Nigeria, Moldova, Tonga, Uganda). Gender inequalities intersect with climate risks and vulnerabilities, and climate change is likely to magnify the existing patterns of gender disadvantage. A meaningful adaptation or resilience-building effort would, therefore, involve the inclusion of a sex-disaggregated data management system, so gender can be mainstreamed into climate change planning and policies. In addition, there are capacity-building initiatives aimed at bridging the gender gap, by empowering women in climate change responses.

In NCs from Annex 1 countries, inclusiveness commonly focused on international development activities. According to the principle of forward-looking responsibility that is explicit in the UNFCCC's Article 4, the developed country parties shall support the development and enhancement of endogenous capacities and technologies of developing country parties, as well as assist them in meeting the costs of adaptation to climate change's adverse effects [77,78].

Annex 1 countries also recognize the intersectionality of development, gender, and inclusiveness. For example, the US and Australia provide adaptation leadership and training to women in Pacific countries and Peru, increasing their influence in driving solutions for CCA in international negotiations (UNFCCC) or municipal councils. The UK and Canada give funding to protect the poorest, most marginalized people across Bangladesh, Burma, Cuba, Nepal, Kenya, and Rwanda, from adverse climate effects, through poverty reduction and inclusive economic development. Japan and the UK provide financing to encourage multiple stakeholders' inclusion in CCA action in Vietnam and Nepal.

Italy, the Netherlands, Portugal, Switzerland, and the UK support developing partners in building national, institutional, community, and household capacity, to improve the inclusiveness in CCA. The UK aims to improve planning, budgeting, human resource management, performance management, and citizen engagement in Kenya. Switzerland and the Netherlands assist North Macedonia and Mozambique in the sustainable management of natural resources, through the practical application of conservation measures. The UK supports the capacity and systems development of financial service providers that serve the livelihoods and well-being of low-income people in Rwanda.

4. Three Components of Inclusive Climate Change Adaptation

In the adaptation assessment frameworks proposed by Smit and Pelling [4,61,62], the following four components are commonly distinguished: (1) Adaptation to what? (2) Who or what adapts? (3) How does adaptation occur? (4) How good is the adaptation? We converted these questions to the following three components of inclusive CCA: (1) inclusion in who or what adapts; (2) motivating inclusive processes; and (3) anticipated outcomes of inclusive CCA.

4.1. Inclusion in Who or What Adapts

The literature on inclusiveness in CCA focuses on local and community scales, but it also recognizes the importance of national, regional, international scales, as well as the necessity to cooperate across scales [6,18,36,79–87]. The implementation of projects conducted in a bottom-up process is normally facilitated by national policy, strategy, and, especially, financial resources directed by governments to support the local-level implementation of adaptation actions. International adaptation commitment and national adaptation strategies must be translated into local adaptation action programs and mostly implemented in the local context [88,89].

Who is included should remain a central question to be addressed in planning processes [36]. CCA calls for inclusive engagement across a broader spectrum of actors [90–95]. These stakeholders include, but are not limited to, local people and local communities; governments, including local and national levels; the leaders across administrations, ministries, and departments; experts and research communities; the private sector, NGOs, and civil society; and other international actors, networks, and agencies. Inclusive CCA favors vulnerable groups, and strengthens support of the poor, indigenous people, women, smallholder farmers, members of lower castes, and resource-dependent people [87,91,96–102]. They must be fully involved in decision-making processes for reasons of both justice and efficiency [6,75]. Some authors have considered non-humans and humans as an intimately coupled system within CCA [103–105].

Adaptation should be inclusive of both scientists and local communities, to form an integrated response to bridge traditional and expert knowledges, providing critical information that is key to the success of inclusive interventions [89]. Local forms of knowledge, including traditional knowledge, traditional ecological knowledge [83], indigenous knowledge [106], and experiential knowledge [99,107], have been highly recommended to bring a distinct and relevant point of view from vulnerable stakeholders to CCA. Conversely, scientific knowledge has many advantages, especially when attempting to understand biophysical processes at broad spatial and temporal scales.

Inclusiveness in CCA also refers to inclusive techniques or tools that range from participatory (action) research [82,108,109] to qualitative scenarios [110], inexact fuzzy multi-objective programming models, adaptation design tools, social ecological inventories (SEI), and ecological risk assessments (ERA) [111]. These approaches are means to explore the public's perception of, knowledge about, and participation in CCA [112,113].

4.2. Motivating Inclusive Processes

Although participation may be encouraged by law, established institutional frameworks might be reluctant to cede decision-making power. Inclusive processes for adaptation require inclusive institutions and can only be facilitated in organizational structures that foster stakeholder involvement in management [20,75,83,100,101,114]. One example of an institution for inclusive CCA is a multilevel and multisector institutional design. The multilevel institutional arrangements consider the local context and require a focus on effective cooperation across levels [115]. Chu et al. [37,116] indicated that more inclusive planning processes correspond to higher climate equity and justice outcomes in the short term, and an emphasis on building multisector governance institutions can enhance long-term program stability, while ensuring that diverse civil society actors have an ongoing voice in climate adaptation planning and implementation.

The literature on inclusive CCA also highlights governance as a significant contribution to inclusiveness, equity, and justice [84,88]. Several models of governance have been recommended to promote inclusiveness in CCA, including polycentric climate governance, collaborative climate governance, networked climate governance, and deliberative climate governance [92,117–119]. Ayers and Huq [120], Ayers [13], and Glavovic [121] argued for the potential contributions of inclusive governance approaches in CCA, including (i) creating safe arenas for public deliberation, to enable participants to explore and develop a shared understanding of adaptation concerns and to engage in different types of knowledge and knowledge claims; (ii) building a common purpose and stimulating participation in community activities; (iii) deepening the community problem-solving capacity, by improving participants' understanding and involving them constructively in community life, on a sustained basis; and (iv) facilitating intercommunity collaboration through cross-scalar and multilevel processes of authentic and inclusive dialogue, visioning, negotiation, and cooperation. Addressing risk and adapting to climate change cannot progress meaningfully without being framed in this broader governance milieu [121].

Adaptive capacity refers to the ability of systems, institutions, humans, and other organisms to adjust to potential damage, to take advantage of opportunities, or to respond to consequences [11]. While all types of capital (natural, social, and economic) are critical for building resilience and fostering adaptation to environmental stresses, inclusiveness in CCA refers mostly to social capital, defined as the value of relationships that facilitate cooperation and collective action [122]. At its core, social capital describes relations of trust, reciprocity, and exchange; the evolution of common rules; and the role of networks [14,123]. Social capital can be particularly important for promoting inclusiveness in adaptation processes to climate change threats, by enabling people to act collectively.

Lee [97] identified existing networks among community-based organizations, local groups, households, and individuals, emphasizing how these networks can bolster both farmer willingness and farmer ability to actively participate in CCA programs, provide the added benefit of increasing smallholder well-being and resiliency and adaptive capacity, and potentially contribute to adaptation inclusiveness, in terms of both broader inclusion and more just outcomes. Mittag [16] and Keessen et al. [124] related the concept of adaptation inclusiveness to the social norms of functioning communities, especially a sense of solidarity. Adaptation to climate change can be an inclusive and collective, rather than individual, effort [125]. Solidarity addresses how different actors can constructively work together to promote the resiliency and adaptation of a social system. The other factor of social capital that is strongly linked to inclusiveness in adaptations is place embeddedness or person–place bonds in local contexts [126]. Place identity contributes to shaping social values at a collective level, and opportunities for and/or barriers to collective action based on shared or divergent understandings of place. Person–place bonds can influence an individual's willingness to become engaged in collective climate change initiatives. The attempt to recognize and incorporate place identity into dialogue and decision making around adaptation will increase public participation, and foster trust between those engaging in planning and those impacted by decisions. When place-based meanings and values are incorporated into any planning process, the process and its outcomes become more place appropriate, as well as more relevant and useful to specific communities.

4.3. Anticipated Outcomes of Inclusive CCA

The literature covered in this review highlights four positive outcomes of inclusive CCA [37,47,115,116,127].

First, the social, economic, and political interests of the poor, underrepresented minorities, and other vulnerable groups are considered in the adaptation process. This is a positive outcome from the perspective of *climate justice*.

Second, the inclusion of all interests in the adaptation process improves *knowledge collection and management*, by involving the public in framing climate risks, vulnerabilities, and adaptation priorities. This facilitates access to climate information and knowledge, and addresses existing class, gender, caste, age, and wealth hierarchies in political decision making.

Third, formal or institutionalized adaptation projects and programs achieve just *results*, benefiting the greatest number of people, especially the most *vulnerable people* in these communities.

Fourth, building on the outcomes of inclusive adaptation processes, several authors have considered inclusiveness in CCA as a component of *resilience, development, and sustainability*. Inclusiveness allows local people and communities to improve their resources and capacity, making them more resilient [84]. Inclusiveness also fosters inclusive development, in which people's well-being is enhanced by advancing the equality of opportunity for all members of society, with attention to the poor, the vulnerable, and those disadvantaged groups who are normally excluded from the process of development [102]. Wamsler and Brink [128] referred to inclusive adaptation as a status of sustainability in adaptive and social systems. In this case, inclusiveness encourages the use of all potential adaptation measures, to ensure that all types of risk factors are addressed. A sustainable system can assist an individual, household, or community in reducing its level of risk, while maintaining or enhancing local adaptive capacities, both now and in the future, and thus not compromising the ability of future generations to meet their own needs.

5. Implications for National Climate Change Adaptation Policy

Most of the papers emphasized inclusive CCA in the context of developing or non-industrialized countries. There is a universal belief that developing countries are the most affected areas, and the poor in developing countries are the most vulnerable group to climate change adversity [8,11,129,130]. This belief derives from the dependence of that group on climate-sensitive sectors, such as agriculture, tourism, fisheries, and forestry; climate-sensitive infrastructure, such as houses, buildings, municipal services, and transportation networks; and limited adaptive capacity to cope with impacts [131] (p. 801). The global climate risk index 2020 found that all ten of the most affected countries during 1999–2018 were developing countries in the low-income or lower–middle-income country group. These results emphasize the vulnerability of poor countries to climatic risks [132]. Therefore, inclusive approaches should be utilized to engage and empower smallholders, women, and poor resource-dependent communities in developing countries [102].

However, the recent literature on adaptation has called attention to all vulnerable communities and the inequities arising from the uneven distribution of climate impacts, which are likely to be reflective of the conditions within developed countries. Adaptation to climate change consists of individual and collective choices that are undertaken at different levels of decision making, in the context of different social concerns and priorities, particularly the existing institutional frameworks for resources, wealth, and power distribution. All adaptation decisions thus compete for attention and resources with other pressing choices in society [133]. Adaptation is not a neutral process, but instead has equity dimensions that are part of the larger adaptive challenge of climate change, and are present in all types of countries and regions, including highly urbanized, developed countries, such as the United States, Canada, and many of the countries of Western Europe [134,135]. One highly cited example is Hurricane Katrina, which struck the United States in 2005, and the enduring legacy of racial segregation and poverty. Statistics showed that the storm's impacts weighed more heavily upon racial minorities and the poor, and the recovery of socially and economically vulnerable storm victims continues to lag behind that of mainstream society. The patterns of settlement exposed poor communities to increased damage and erected barriers to disaster precautions and reconstruction. In the other words, social and economic disparities heavily affected the impacts of Katrina on the most vulnerable groups, especially African Americans and the poor. Climate change impacts are expected to exacerbate poverty and create new poverty pockets in countries with increasing inequality, including developed countries [1].

Inclusive approaches to adaptation address gender inequality (for women), income inequality (for the poor), minority groups (for example, indigenous people), and underrepresented, disfavored, or marginalized groups (for example, communities of color, refugees, migrants, and the stateless). These groups clearly exist and are even growing in developed countries, due to the global economic crisis and recent migrant crisis. Involving less advantaged people properly in adaptation processes has been urgently required, not only to address human rights issues, but also to ensure the sustainable prosperity of industrialized nations. Therefore, the necessity of inclusive approaches emerges not only in developing countries, but also in industrialized countries [104]. However, inclusive approaches vary significantly in different contexts, due to the different characteristics of economic, social, and political systems.

6. Conclusions

The purpose of this literature review was to systematically chart the usage of 'inclusiveness' in the climate change adaptation context. 'Inclusion' has been gaining in interest and usage in several other contexts, in particular, human resources (EDI), as well as urban planning, international development, and innovation. A clear interest is also reflected in this literature review, although our quantitative analysis does not indicate a steep or consistent increase in usage. This review has shown common usage of the term in practical contexts, particularly development, knowledge mobilization, gender issues, marginalization, and poverty.

The common connotation of the word 'inclusion' is very broad and goes beyond public participation and even the idea of justice. Following the typology of CCA, provided by Smit et al. [62], the inclusion lens could be applied to all aspects of climate change and climate change adaption, as follows: (1) inclusive identification of the causes of CC; (2) inclusive identification of the effects of CC; (3) inclusive goals and processes of CCA; and (4) inclusive evaluation of CCA. The word can be used to diagnose both omission (neglect) and commission (discrimination) in a wide array of contexts, including subsidies for knowledge creation (science policy), processes during knowledge creation (HR), the translation of knowledge, the availability and accessibility of knowledge, the use of knowledge, the beneficiaries of providing solutions, the measurement of performance, the beneficiaries of providing performance measurement, corrective actions, and so forth.

Inclusion is also a central idea within the circle of moral attention that has historically widened. Over time, in western history, policies have been enacted, to include into the moral circle slaves, different races, women, sentient animals, and endangered species. Inclusive approaches to CCA could be used to emphasize the important role of the natural environment in adaptation and assessing the potential outcomes of human climate adaptation for the natural environment. Many societies have traditionally treated climate as a background for human activities, and climate change as an environmental problem or development issue, in which human beings attempt to stimulate, take advantage of, or harmonize with the non-human world. Inclusive adaptation, thus, provides a suitable intellectual framework and connotation to include non-anthropocentric viewpoints in debates and policy development.

Possibly the greatest value of 'inclusiveness' lies in the fact that it is a fairly clear, common language word. Compared to words such as 'justice', 'participation', 'equity', or even 'community' and 'democracy', it has relatively little metaphysical content. This makes it suitable for checklists and indicators for all policies and activities associated with CCA. Do we live up to the goals and promises of justice, impartiality, non-discrimination, equity, and diversity? If we consider inclusiveness, then we are off to good start.

We conclude this paper by discussing the limitations and suggestions for future research. One limitation lies in the data collection. In particular, we limited our search to documents in English that exclude the usage and understanding of inclusiveness expressed in non-English documentation. An artefact of the systematic method used here is that the uniformity and clarity of results is lower when compared to literature reviews that rely on cherry-picked sources. Future research will focus on broader conceptual analyses that are based on comparisons with other contexts, such as urban planning, innovation, and education. This broader conceptual understanding will then be applied to real-world case studies. The ultimate goal is to better understand the conditions and incentives that promote inclusiveness and climate justice.

Author Contributions: Conceptualizations, Methods, Data collection and Analysis, Writing First Draft, H.P.; Writing Review and Edit, Visualization, Implication and Conclusion—M.S. Both authors have read and agreed to the published version of the manuscript.

Funding: This research received no external funding.

Acknowledgments: We wish to thank participants at the first World Forum on Climate Justice (June 2019, Glasgow) for feedback on the talk by Ha Pham on 'inclusiveness' within the Pan-Canadian Framework on Clean Growth and Climate Change. We also thank Daniel Amyot for expert instruction on systematic literature reviews in a graduate seminar at the University of Ottawa (January 2019). All misconceptions and errors are entirely our own.

Conflicts of Interest: The authors declare no conflict of interest.

References

1. IPCC. *Climate Change 2014: Impacts, Adaptation and Vulnerability, Working Group II Contribution to the IPCC Fifth Assessment Report;* Cambridge University Press: Cambridge, UK, 2014. [CrossRef]
2. Dalby, S.; O'Lear, S. Towards ecological geopolitics. In *Reframing Climate Change: Constructing Ecological Geopolitics;* Dalby, S., O'Lear, S., Eds.; Routledge: London, UK; New York, NY, USA, 2016; pp. 203–216.
3. Lynch, P.; Duke, S. Climate Change. In *Environmental Politics: Scale and Power;* O'Lear, S., Ed.; Routledge: London, UK; New York, NY, USA, 2010; pp. 27–54. [CrossRef]
4. Pelling, M. *Adaptation to Climate Change: From Resilience to Transformation;* Routledge: London, UK; New York, NY, USA, 2011.
5. Farber, D.A.; Posner, B.E.A.; Weisbach, D. Climate justice: Climate change, resource conflicts, and social justice. In *Reframing Climate Change. Constructing Ecological Geopolitics;* O'Lear, S., Dalby, S., Eds.; Routledge: London, UK; New York, NY, USA, 2016; pp. 67–82.
6. Few, R.; Brown, K.; Tompkins, E.L. Public participation and climate change adaptation: Avoiding the illusion of inclusion. *Clim. Policy* **2007**, *7*, 46–59. [CrossRef]
7. UNFCCC. Report of the Ad Hoc Working Group on the Durban Platform for Enhanced Action-Synthesis report on the aggregate effect of the intended nationally determined contributions. In Proceedings of the Conference of the Parties Twenty-First Session Paris, Paris, France, 30 November–11 December 2015. Available online: https://unfccc.int/resource/docs/2015/cop21/eng/07.pdf (accessed on 2 July 2020).
8. IPCC. *Global Warming of 1.5 °C. An IPCC Special Report on the Impacts of Global Warming of 1.5 °C Above Pre-Industrial Levels and Related Global Greenhouse Gas Emission Pathways, in the Context of Strengthening the Global Response to the Threat of Climate Change, Sustainable Development, and Efforts to Eradicate Poverty;* Masson-Delmotte, V., Zhai, P., Pörtner, H.-O., Roberts, D., Skea, J., Shukla, P.R., Pirani, A., Moufouma-Okia, W., Péan, C., Pidcock, R., et al., Eds.; 2018; Available online: https://www.ipcc.ch/sr15/ (accessed on 20 July 2020).
9. UNFCCC. UN Climate Change Annual Report 2017. 2018. Available online: https://unfccc.int/resource/annualreport/ (accessed on 20 July 2020).
10. Government of Canada. Working Group on Adaptation and Climate Resilience. 2016. Available online: http://www.climatechange.gc.ca/Content/6/4/7/64778DD5-E2D9-4930-BE59-D6DB7DB5CBC0/WG_Report_ACR_e_v5.pdf (accessed on 20 July 2020).
11. Southern Voices on Adaptation. Joint Principles for Adaptation. 2015. Available online: http://southernvoices.net/images/Joint_Principles_for_Adaptation_version_3.pdf (accessed on 20 July 2020).
12. The United Kingdom Climate Impacts Program (UKCIP). Objective Setting for Climate Change. 2005. Available online: http://www.ukcip.org.uk/publications/ (accessed on 20 July 2020).
13. Patnaik, H.A. Gender and participation in community-based adaptation: Evidence from the decentralized climate funds project in Senegal. *World Dev.* **2021**, *142*, 105448. [CrossRef]
14. Bloomfield, D.; Collins, K.; Fry, C.; Munton, R. Deliberation and inclusion: Vehicles for increasing trust in UK public governance? *Environ. Plan. C Gov. Policy* **2001**, *19*, 501–513. [CrossRef]
15. Mittag, J. Perspectives on civic engagement in national strategies to combat climate change. *Democratization* **2012**, *19*, 994–1013. [CrossRef]
16. Reed, M.G. Stakeholder Participation for Environmental Management: A Literature Review. *Biol. Conserv.* **2008**, *141*, 2417–2431. [CrossRef]
17. Ayers, J. Resolving the Adaptation Paradox: Exploring the Potential for Deliberative Adaptation Policymaking in Bangladesh. *Glob. Environ. Politics* **2011**, *11*, 62–88. [CrossRef]
18. Nightingale, A.; Rankin, K. Politics of Social Marginalization and Inclusion: The Challenges of Adaptation to Climate Change. In *Inclusive Urbanization: Rethinking Policy, Practice and Research in the Age of Climate Change;* Shrestha, K., Ojha, H., McManus, P., Rubbo, A., Dhote, K., Eds.; Routledge: New York, NY, USA; London, UK, 2014; pp. 53–64.
19. Buechler, S. Gendered vulnerabilities and grassroots adaptation initiatives in home gardens and small orchards in Northwest Mexico. *AMBIO* **2016**, *45*, 322–334. [CrossRef]
20. Singh, C. Is participatory watershed development building local adaptive capacity? Findings from a case study in Rajasthan. *India Environ. Dev.* **2018**, *25*, 43–58. [CrossRef]
21. Wilk, J.; Jonsson, A.C.; Rydhagen, B.; Callejo, I.; Cerruto, N. Assessing vulnerability in Cochabamba, Bolivia and Kota, India: How do stakeholder processes affect suggested climate adaptation interventions? *Int. J. Urban. Sustain. Dev.* **2018**, *10*, 32–48. [CrossRef]

22. Sen, A. Social Exclusion: Concept, Application, and Scrutiny. Manila, Philippines: Asian Development Bank (ADB). 2000. Available online: http://hdl.handle.net/11540/2339 (accessed on 20 July 2020).
23. The United Nations Department of Economic and Social Affairs (UNDESA). 2009. Final Report of the Expert Group Meeting on Creating an Inclusive Society: Practical Strategies to Promote Social Integration. Available online: http://www.un.org/esa/socdev/egms/docs/2008/Paris-report.pdf (accessed on 20 July 2020).
24. Mello, L.; Dutz, M.A. (Eds.) *Promoting Inclusive Growth: Challenges and Policies*; OECD Publishing: Paris, France, 2012. [CrossRef]
25. OECD. *Policy Shaping and Policy Making: The Governance of Inclusive Growth*; OECD Publishing: Paris, France, 2015.
26. Gilman, H.R. More Inclusive Governance in the Digital Age (Data- Smart City Solutions). 2015. Available online: https://datasmart.ash.harvard.edu/assets/content/InclusiveGovernanceinthe (accessed on 2 July 2020).
27. Ison, R.L.; Wallis, P.J. Mechanisms for Inclusive Governance. In *Freshwater Governance for the 21st Century*; Karar, E., Ed.; Springer: Cham, Switzerland, 2017; pp. 159–185. [CrossRef]
28. Cruz, C.; Cadornigara, S.F.; Satterthwaite, D. Tools for Inclusive Cities: The Roles of Community-Based Engagement and Monitoring in Reducing Poverty (IIED Working Paper No. 10708). 2014. Available online: http://pubs.iied.org/pdfs/10708IIED (accessed on 2 July 2020).
29. Florian, S.; Michael, L. Inclusive Cities. Asian Development Bank-Urban. Development Series. 2011. Available online: https://www.adb.org/sites/default/files/publication/29053/inclusive-cities.pdf (accessed on 2 July 2020).
30. Shrestha, K.; Ojha, H.; McManus, P. *Inclusive Urbanization: Rethinking Policy, Practice and Research in the Age of Climate Change*; Routledge: New York, NY, USA; London, UK, 2015.
31. World Bank. World Bank Inclusive Cities Approach Paper. Report No: AUS8539. 2015. Available online: http://documents.worldbank.org/curated/en/402451468169453117/pdf/AUS8539-REVISED-WP-P148654-PUBLIC-Box393236B-Inclusive-Cities-Approach-Paper-w-Annexes-final.pdf (accessed on 20 July 2020).
32. Agola, N.; Hunter, A. *Inclusive Innovation for Sustainable Development: Theory and Practice*; Palgrave Macmillan: London, UK; Hunter: West Palm Beach, FL, USA, 2016.
33. Government of Canada. Canada-A Nation of Innovators. 2016. Available online: https://www.ic.gc.ca/eic/site/062.nsf/vwapj/InnovationNation_Report-EN.pdf/$file/InnovationNation_Report-EN.pdf (accessed on 20 July 2020).
34. Schillo, R.S.; Robinson, R.M.; Sainte-Marie, B. Inclusive Innovation in Developed Countries: The Who, What, Why, and How. *Technol. Innov. Manag. Rev.* **2017**, *7*, 34–46. Available online: https://timreview.ca/sites/default/files/article_PDF/SchilloRobinson_TIMReview_July2017.pdf (accessed on 20 July 2020). [CrossRef]
35. Tompkins, E.; Adger, W.N.; Brown, K. Institutional networks for inclusive coastal management in Trinidad and Tobago. *Environ. Plan. A* **2002**, *34*, 1095–1111. [CrossRef]
36. Archer, D.; Almansi, F.; DiGregorio, M.; Roberts, D.; Sharma, D.; Syam, D. Moving towards inclusive urban adaptation: Approaches to integrating community-based adaptation to climate change at city and national scale. *Clim. Dev.* **2014**, *6*, 345–356. [CrossRef]
37. Chu, E.; Anguelovski, I.; Carmin, J. Inclusive approaches to urban climate adaptation planning and implementation in the Global South. *Clim. Policy* **2016**, *16*, 372–392. [CrossRef]
38. Berrang-Ford, L.; Pearce, T.; Ford, J.D. Systematic review approaches for global environmental change research. *Reg. Environ. Chang.* **2015**, *15*, 755–769. [CrossRef]
39. Berrang-Ford, L.; Ford, J.D.; Patterson, J. Are we adapting to climate change? *Glob. Environ. Chang.* **2011**, *21*, 25–33. [CrossRef]
40. Biesbroek, G.R.; Termeer, C.; Klostermann, J.E.M.; Kabat, P. Analytical lenses on barriers in the governance of climate change adaptation. *Mitig. Adapt. Strateg. Glob. Chang.* **2015**, *19*, 1011–1032. [CrossRef]
41. Ford, J.D.; Pearce, T. What we know, do not know, and need to know about climate change vulnerability in the western Canadian Arctic: A systematic literature review. *Environ. Res. Lett* **2010**, *5*, 014008. [CrossRef]
42. Lesnikowski, A.C.; Ford, J.D.; Berrang-Ford, L.; Paterson, J.A.; Barrera, M.; Heymann, S.J. Adapting to health impacts of climate change: A study of UNFCCC Annex I parties. *Environ. Res. Lett.* **2011**, *6*, 044009. [CrossRef]
43. Jafry, T.; Mikulewicz, M.; Helwig, K. (Eds.) *Routledge Handbook of Climate Justice*; Routledge: London, UK, 2018. [CrossRef]
44. Hughes, S. Justice in urban climate change adaptation: Criteria and application to Delhi. *Ecol. Soc.* **2013**, *18*, 48. [CrossRef]
45. Bulkeley, H.; Gareth, A.; Edwards, S.; Fuller, S. Contesting climate justice in the city: Examining politics and practice in urban climate change experiments. *Glob. Environ. Chang.* **2014**, *25*, 31–40. [CrossRef]
46. Figueiredo, P.; Perkins, P.E. Women and water management in times of climate change: Participatory and inclusive processes. *J. Clean. Prod.* **2013**, *60*, 188–194. [CrossRef]
47. Jafry, T. Making the case for gender sensitive climate policy–lessons from South Asia/IGP. *Int. J. Clim. Chang. Strateg. Manag.* **2016**, *8*, 559–577. [CrossRef]
48. Nurhidayah, L.; McIlgorm, A. Coastal adaptation laws and the social justice of policies to address sea level rise: An Indonesian insight. *Ocean Coast. Manag.* **2019**, *171*, 11–18. [CrossRef]
49. Ingle, H.I.; Mikulewicz, M. Mental health and climate change: Tackling invisible injustice. *Lancet Planet Health* **2020**, *4*, e128–e130. [CrossRef]
50. Forsyth, T. Climate justice is not just ice. *Geoforum* **2014**, *54*, 230–232. [CrossRef]
51. Schlosberg, D. Theorizing environmental justice: The expanding sphere of a discourse. *Environ. Politics* **2013**, *22*, 37–55. [CrossRef]

52. Tschakert, P.; Schlosberg, D.; Celermajer, D.; Rickards, L.; Winter, C.; Thaler, M.; Stewart-Harawira, M.; Verlie, B. Multispecies justice: Climate-just futures with, for and beyond humans. *WIREs Clim. Chang.* **2021**, *12*, e699. [CrossRef]

53. Cooper, H.; Hedge, L.V. (Eds.) *The Handbook of Research Synthesis*; Russell Sage Foundation: New York, NY, USA, 1994.

54. Gough, D.; Thomas, J.; Oliver, S. Clarifying differences between review designs and methods. *Syst. Rev.* **2012**, *1*, 28–37. [CrossRef]

55. Lorenz, S.; Berman, R.; Dixon, J.; Lebel, S. Time for a systematic review: A response to Bassett and Fogelman's "Deja vu or something new? The adaptation concept in the climate change literature". *Geoforum* **2014**, *51*, 252–255. [CrossRef]

56. McLeman, R.A. Settlement abandonment in the context of global environmental change. *Glob. Environ. Chang. Hum. Policy Dimens* **2011**, *21*, S108–S120. [CrossRef]

57. Gill, G. *The Nature and Development of the Modern State*; Palgrave: London, UK, 2003.

58. Meadowcroft, J. *Climate Change Governance*; Policy Research Working Paper No. WPS 4941; World Bank Group: Washington, DC, USA, 2009. Available online: http://documents.worldbank.org/curated/en/210731468332049368/Climate-change-governance (accessed on 20 July 2020).

59. Oulu, M. Climate Change Governance: Emerging Legal and Institutional Frameworks for Developing Countries. In *Handbook of Climate Change Adaptation*; Leal Filho, W., Ed.; Springer: Berlin/Heidelberg, Germany, 2015. [CrossRef]

60. Pierre, J.; Peters, G. *Governance, Politics and the State*; Macmillan Press: London, UK, 2000.

61. Smit, B.; Burton, I.; Klein, R. The Science of Adaptation: A Framework for Assessment. *Mitig. Adapt. Strateg. Glob. Chang.* **1999**, *4*, 199–213. [CrossRef]

62. Smit, B.; Burton, I.; Klein, R.; Wandel, J. An anatomy of adaptation to climate change and variability. *Clim. Chang.* **2000**, *45*, 223–251. [CrossRef]

63. Agrawal, A. The Role of Local Institutions in Adaptation to Climate Change. 2008. Available online: http://www.icarus.info/wp-content/uploads/2009/11/agrawal-adaptation-institutions-livelihoods.pdf (accessed on 20 July 2020).

64. Butcher-Gollach, C. Planning, the urban poor and climate change in Small Island Developing States (SIDS): Unmitigated disaster or inclusive adaptation? *Int. Dev. Plan. Rev.* **2015**, *37*, 225–248. [CrossRef]

65. Aoyagi, M.; Suda, E.; Shinada, T. *Gender Inclusion in Climate Change Adaptation*; ADBI Working Paper, No. 309; Asian Development Bank Institute (ADBI): Tokyo, Japan, 2011.

66. Habtezion, S. Gender and Climate Change Asia and the Pacific. United Nations Development Programme. 2013. Available online: http://www.undp.org/content/dam/undp/library/gender/GenderandEnvironment/PB1-AP-Overview-Gender-and-climate-change.pdf (accessed on 20 July 2020).

67. Habtezion, S. Gender, Climate Change Adaptation and Disaster Risk Reduction. 2016. Available online: http://www.undp.org/content/dam/undp/library/gender/GenderandEnvironment/TrainingModules/Gender_Climate_Change_TrainingModule2AdaptationDRR.pdf (accessed on 20 July 2020).

68. Schalatek, L.; Nakhooda, S. Gender and Climate Finance. 2013. Available online: https://www.odi.org/sites/odi.org.uk/files/resource-documents/11046.pdf (accessed on 2 July 2020).

69. United Nations Development Program (UNDP). Resource Guide on Gender and Climate Change. 2009. Available online: https://www.undp.org/content/undp/en/home/librarypage/womens-empowerment/resource-guide-on-gender-and-climate-change.html (accessed on 2 July 2020).

70. UNDP. Gender and Adaptation: Africa. 2012. Available online: https://www.undp.org/content/undp/en/home/librarypage/womens-empowerment/gender_and_environmentenergy/gender_and_climatechange-africa.html (accessed on 20 July 2020).

71. Green Climate Fund. *GCF* Gender Equality and Social Inclusion Policy and Action Plan. 2018–2020. In Proceedings of the Meeting of the Board, Songdo, Incheon, Korea, 27 February–1 March 2018; Songdo Provisional agenda item 16 GCF/B.19/25. Available online: https://www.greenclimate.fund/sites/default/files/document/gcf-b19-25.pdf (accessed on 20 July 2020).

72. NDF & Asian Development Bank. Training Manual to Support Country-Driven Gender and Climate Change. 2015. Available online: https://pub.iges.or.jp/pub/training-manual-support-country-driven-gender (accessed on 20 July 2020).

73. UNDP. Gender Responsive National Communications National Communication. 2015. Available online: https://www.undp.org/content/undp/en/home/librarypage/womens-empowerment/gender-responsive-national-communications.html (accessed on 20 July 2020).

74. UNFCCC. Strengthening Gender Considerations in Adaptation Planning and Implementation in the Least Developed Countries. 2015. Available online: https://www4.unfccc.int/sites/NAPC/Documents%20NAP/UNFCCC_gender_in_NAPs.pdf (accessed on 20 July 2020).

75. Dulal, H.B.; Shah, K.U.; Ahmad, N. Social equity considerations in the implementation of Caribbean climate change adaptation policies. *Sustainability* **2009**, *1*, 363–383. [CrossRef]

76. Resurrección, B.P.; Bee, B.A.; Dankelman, I.; Park, C.M.Y.; Halder, M.; McMullen, C.P. *Gender-Transformative Climate Change Adaptation: Advancing Social Equity*; Background Paper to the 2019 Report of the Global Commission on Adaptation: Rotterdam, The Netherlands; Washington, DC, USA, 2019. Available online: https://www.sei.org/publications/gender-transformative-climate-change-adaptation-advancing-social-equity/#:~:text=Power%20and%20gender%20inequalities%20can%20constrain%20and%20undermine%20climate%20change%20adaptation.&text=Unfettering%20the%20agency%20of%20individuals,can%20help%20achieve%20this%20change (accessed on 20 July 2020).

77. United Nations Framework of Climate Change Convention (UNFCCC). United Nation Framework of Climate Change Convention. 1994. Available online: https://unfccc.int/resource/docs/convkp/conveng.pdf (accessed on 20 July 2020).

78. Adger, N.; Paavola, J.; Huq, S.; Mace, M.J. (Eds.) *Fairness in Adaptation to Climate Change*; MIT Press: Cambridge, MA, USA, 2006.
79. Yoseph-Paulus, R.; Hindmarsh, R. Addressing inadequacies of sectoral coordination and local capacity building in Indonesia for effective climate change adaptation. *Clim. Dev.* **2018**, *10*, 35–48. [CrossRef]
80. Köpsel, V.; Walsh, C. Coastal landscapes for whom? Adaptation challenges and landscape management in Cornwall. *Mar. Policy* **2018**, *97*, 278–286. [CrossRef]
81. Ayers, J. Understanding the Adaptation Paradox: Can. Global Climate Change Adaptation Policy Be Locally Inclusive? Ph.D. Thesis, London School of Economics, London, UK, 2010. Available online: http://etheses.lse.ac.uk/393/1/Ayers (accessed on 20 July 2020).
82. Fazey, I.; Kesby, M.; Evely, A.; Latham, I.; Wagatora, D.; Hagasua, J.E.; Reed, M.S.; Christie, M. A three-tiered approach to participatory vulnerability assessment in the Solomon Islands. *Glob. Environ. Chang.* **2010**, *20*, 713–728. [CrossRef]
83. Dowsley, M. Community clusters in wildlife and environmental management: Using TEK and community involvement to improve co-management in an era of rapid environmental change. *Polar Res.* **2009**, *28*, 43–59. [CrossRef]
84. Makina, A.; Moyo, T. Mind the gap: Institutional considerations for gender-inclusive climate change policy in Sub-Saharan Africa. *Local Environ.* **2016**, *21*, 1185–1197. [CrossRef]
85. Nagy, G.J.; Seijo, L.; Verocai, J.E.; Bidegain, M. Stakeholders' climate perception and adaptation in coastal Uruguay. *Int. J. Clim. Chang. Strateg. Manag.* **2014**, *6*, 63–84. [CrossRef]
86. Hill, M.; Wallner, A.; Furtado, J. Reducing vulnerability to climate change in the Swiss Alps: A study of adaptive planning. *Clim. Policy* **2010**, *10*, 70–86. [CrossRef]
87. Vij, S.; Biesbroek, R.; Groot, A.; Termeer, K.; Parajuli, B.P. Power interplay between actors: Using material and ideational resources to shape local adaptation plans of action (LAPAs) in Nepal. *Clim. Policy* **2018**, *19*, 1–14. [CrossRef]
88. Fenton, A.; Gallagher, D.; Wright, H.; Huq, S.; Nyandiga, C. Up-scaling finance for community-based adaptation. *Clim. Dev.* **2014**, *6*, 388–397. [CrossRef]
89. Ramirez-Villegas, J.; Khoury, C.K. Reconciling approaches to climate change adaptation for Colombian agriculture. *Clim. Chang.* **2013**, *119*, 575–583. [CrossRef]
90. McNamara, K.E.; Des Combes, H.J. Planning for Community Relocations Due to Climate Change in Fiji. *Int. J. Disaster Risk Sci.* **2015**, *6*, 315–319. [CrossRef]
91. Neuburger, M. Global discourses and the local impacts in Amazonia. Inclusion and exclusion processes in the Rio Negro region. *Erdkunde* **2008**, *62*, 339–356. [CrossRef]
92. Arriagada, R.A.; Aldunce, P.; Blanco, G.; Ibarra, C.; Moraga, P.; Nahuelhual, L.; O'Ryan, R.; Urquiza, A.; Gallardo, L. Climate change governance in the Anthropocene: Emergence of Polycentrism in Chile. *Elem. Sci. Anth.* **2018**, *6*, 1–13. [CrossRef]
93. Schmid, J.C.; Knierim, A.; Knuth, U. Policy-induced innovations networks on climate change adaptation-An ex-post analysis of collaboration success and its influencing factors. *Environ. Sci. Policy* **2016**, *56*, 67–79. [CrossRef]
94. Sprain, L. Paradoxes of Public Participation in Climate Change Governance. *Good Soc.* **2016**, *25*, 62–80. [CrossRef]
95. Ernst, K.M.; van Riemsdijk, M. Climate change scenario planning in Alaska's National Parks: Stakeholder involvement in the decision-making process. *Appl. Geogr.* **2013**, *45*, 22–28. [CrossRef]
96. Wani, K.A.; Ariana, L. Impact of climate change on indigenous people and adaptive capacity of bajo tribe, Indonesia. *Environ. Claims J.* **2018**, *30*, 302–313. [CrossRef]
97. Lee, J. Farmer participation in a climate-smart future: Evidence from the Kenya agricultural carbon market project. *Land Use Policy* **2017**, *68*, 72–79. [CrossRef]
98. Bhatasara S Rethinking climate change research in Zimbabwe. *J. Environ. Stud. Sci.* **2017**, *7*, 39–52. [CrossRef]
99. Abrha, M.G.; Simhadri, S. Local Climate Trends and Farmers' Perceptions in Southern Tigray, Northern Ethiopia. *Am. J. Environ. Sci.* **2015**, *11*, 262–277. [CrossRef]
100. Regmi, B.R.; Star, C. Identifying operational mechanisms for mainstreaming community-based adaptation in Nepal. *Clim. Dev.* **2014**, *6*, 306–317. [CrossRef]
101. Regmi, B.R.; Star, C.; Leal Filho, W. An overview of the opportunities and challenges of promoting climate change adaptation at the local level: A case study from a community adaptation planning in Nepal. *Clim. Chang.* **2016**, *138*, 537–550. [CrossRef]
102. Luo, X.; Muleta, D.; Hu, Z.; Tang, H.; Zhao, Z. Inclusive development and agricultural adaptation to climate change. *Curr. Opin. Environ. Sustain.* **2017**, *24*, 78–83. [CrossRef]
103. Polsky, C.; Neff, R.; Yarnal, B. Establishing vulnerability observatory networks to coordinate the collection and analysis of comparable data. In *Sustainable Communities on a Sustainable Planet: The Human-Environment Regional Observatory Project*; Yarnal, B., Polsky, C., O'Brien, J., Eds.; Cambridge University Press: Cambridge, UK, 2009; pp. 83–106.
104. Gupta, J.; Bavinck, M. Reprint of Inclusive development and coastal adaptiveness. *Ocean. Coast. Manag.* **2017**, *150*, 73–81. [CrossRef]
105. Robin, L. Environmental humanities and climate change: Understanding humans geologically and other life forms ethically. *Wiley Interdiscip. Rev. Clim. Chang.* **2018**, *9*, 7–9. [CrossRef]
106. Ford, J.D.; Cameron, L.; Rubis, J.; Maillet, M.; Nakashima, D. Including indigenous knowledge and experience in IPCC assessment reports. *Nat. Clim. Chang.* **2016**, *6*, 349–353. [CrossRef]
107. Hardy, R.D.; Milligan, R.A.; Heynen, N. Racial coastal formation: The environmental injustice of colorblind adaptation planning for sea-level rise. *Geoforum* **2017**, *87*, 62–72. [CrossRef]

108. Jonsson, A.C.; Rydhagen, B.; Wilk, J.; Feroz, A.; Rani, R. Climate change adaptation in urban india: The inclusive formulation of local adaptation strategies. *Global NEST. Int. J.* **2015**, *17*, 61–71.
109. Plante, S.; Vasseur, L.; Da Cunha, C. Engaging Local Communities for Climate Change Adaptation. In *Coastal Zones: Solutions for the 21st Century*; Baztan, J., Chouinard, O., Jorgensen, B., Tett, P., Vanderlinden, J., Vasseur, L., Eds.; Elsevier: Amsterdam, The Netherlands, 2015. [CrossRef]
110. Wesche, S.D.; Armitage, D.R. Using qualitative scenarios to understand regional environmental change in the Canadian North. *Reg. Environ. Chang.* **2014**, *14*, 1095–1108. [CrossRef]
111. Rodríguez-Romero, A.; Viguri, J.R.; Calosi, P. Acquiring an evolutionary perspective in marine ecotoxicology to tackle emerging concerns in a rapidly changing ocean. *Sci. Total Environ.* **2021**, *764*, 142816. [CrossRef]
112. Pielke, R.A.; Wilby, R.; Niyogi, D.; Hossain, F.; Dairuku, K. Dealing with Complexity and Extreme Events Using a Bottom-Up, Resource-Based Vulnerability Perspective. *Extrem. Events Nat. Hazards Complex. Perspect.* **2013**, *196*, 345–359. [CrossRef]
113. Reed, M.G.; Scott, A.; Natcher, D.; Johnston, M. Linking gender, climate change, adaptive capacity, and forest-based communities in Canada. *Can. J. For. Res.* **2014**, *44*, 995–1004. [CrossRef]
114. Fischer, H. Decentralization and the governance of climate adaptation: Situating community-based planning within broader trajectories of political transformation. *World Dev. Vol.* **2021**, *140*, 105335. [CrossRef]
115. Huntjens, P.; Zhang, T.; Nachbar, K. Climate Change and Implications for Security and Justice: The Need for Equitable, Inclusive, and Adaptive Governance of Climate Action. In *Just Security in an Undergoverned World*; Durch, W., Larik, J., Ponzio, R., Eds.; Oxford University Press: Oxford, UK, 2018. [CrossRef]
116. Chu, E.; Anguelovski, I.; Roberts, D. Climate adaptation as strategic urbanism: Assessing opportunities and uncertainties for equity and inclusive development in cities. *Cities* **2017**, *60*, 378–387. [CrossRef]
117. Bäckstrand, K. Accountability of Networked Climate Governance: The Rise of Transnational Climate Partnerships. *Glob. Environ. Politics* **2008**, *8*, 74–102. [CrossRef]
118. Hölscher, K.; Frantzeskaki, N.; McPhearson, T.; Loorbach, D. Tales of transforming cities: Transformative climate governance capacities in New York City, U.S. and Rotterdam, Netherlands. *J. Environ. Manag.* **2019**, *231*, 843–857. [CrossRef] [PubMed]
119. Pattberg, P. Public-private partnerships in global climate governance. *Wiley Interdiscip. Rev. Clim. Chang.* **2010**, *1*, 279–287. [CrossRef]
120. Ayers, J.; Huq, S. Adaptation, development and the community. In *Climate Adaptation Futures*; Palutikof, J., Boulter, S., Ash, A., Smith, M.S., Parry, M., Waschka, M., Guitart, D., Eds.; John Wiley & Sons, Ltd.: Hoboken, NJ, USA, 2013; pp. 203–214.
121. Glavovic, B.C. Adapting to climate change: Lessons from natural hazards planning. Waves of Adversity, Layers of Resilience: Floods, Hurricanes, Oil Spills and Climate Change in the Mississippi Delta. In *Adapting to Climate Change: Lessons from Natural Hazards Planning*; Glavovic, B.C., Smith, G.P., Eds.; Springer: Dordrecht, The Netherlands; Heidelberg, Germany; New York, NY, USA; London, UK, 2014; pp. 369–404. [CrossRef]
122. Ostrom, E.; Ahn, T.K. *Foundations of Social Capital*; Edward Elgar Pub.: Northampton, MA, USA, 2003.
123. Adger, W. Social Capital, Collective Action, and Adaptation to Climate Change. *Econ. Geogr.* **2003**, *79*, 387–404. [CrossRef]
124. Keessen, A.; Vink, M.J.; Wiering, M.; Boezeman, D.; Ernst, W. Solidarity in water management. *Ecol. Soc.* **2016**, *21*. [CrossRef]
125. Chatterton, P.; Featherstone, D.; Routledge, P. Articulating climate justice in Copenhagen: Antagonism, the commons, and solidarity. *Antipode* **2013**, *45*, 602–620. [CrossRef]
126. Fresque-Baxter, J.A.; Armitage, D. Place identity and climate change adaptation: A synthesis and framework for understanding. *Wiley Interdiscip. Rev. Clim. Chang.* **2012**, *3*, 251–266. [CrossRef]
127. Robinson, L.W.; Berkes, F. Multi-level participation for building adaptive capacity: Formal agency-community interactions in northern Kenya. *Glob. Environ. Chang.* **2011**, *21*, 1185–1194. [CrossRef]
128. Wamsler, C.; Brink, E. Moving beyond short-term coping and adaptation. *Environ. Urban* **2014**, *26*, 86–111. [CrossRef]
129. IPCC. Summary for Policymakers. In *IPCC Special Report on the Ocea and Cryosphere in a Changing Climate*; Pörtner, H.O., Roberts, D.C., Masson-Delmotte, V., Zhai, P., Tignor, M., Poloczanska, E., Mintenbeck, K., Alegriía, A., Nicolai, M., Okem, A., et al., Eds.; WMO: Geneva, Switzerland, 2019.
130. World Meteorological Organization (WMO). WMO Statement on the State of the Global Climate in 2019. 2020. Available online: https://library.wmo.int/doc_num.php?explnum_id=10211 (accessed on 20 July 2020).
131. Ford, J.D.; Berrang-Ford, L.; Bunce, A.; McKay, C.; Irwin, M.; Pearce, T. The status of climate change adaptation in Africa and Asia. *Reg. Environ. Chang.* **2015**, *15*, 801–814. [CrossRef]
132. Eckstein, D.; Künzel, V.; Schäfer, L.; Winges, M. *Global Climate Risk Index 2020: Who Suffers Most from Extreme Weather Events? Weather-Related Loss Events in 2018 and 1999 to 2018*; Brief Paper: Bonn, Germany, 2020.
133. Paavola, J.; Adger, W.N.; Huq, S. Multifaceted Justice in Adaptation to Climate Change. In *Fairness in Adaptation to Climate Change*; Adger, W.N., Paavola, J., Huq, S., Mace, M.J., Eds.; M.I.T. Press: Cambridge, MA, USA, 2006; pp. 263–277.
134. O'Brien, K. Global environmental change II: From adaptation to deliberate transformation. *Prog. Hum. Geogr.* **2012**, *36*, 667–676. [CrossRef]
135. Vancura, P.; Leichenko, R. Emerging Equity and Justice Concerns for Climate Change Adaptation. In *The Adaptive Challenge of Climate Change*; O'Brien, K., Selboe, E., Eds.; Cambridge University Press: Cambridge, UK, 2015; pp. 98–117. [CrossRef]

Article

Evaluation of Greenhouse Gas Emissions from Reservoirs: A Review

Ion V. Ion [1] and Antoaneta Ene [2,*]

[1] Thermal Engines and Environmental Engineering Research Center, Faculty of Engineering, "Dunarea de Jos" University of Galati, 800008 Galati, Romania; ion.ion@ugal.ro
[2] INPOLDE Research Center, Faculty of Sciences and Environment, ReForm-UDJG Multidisciplinary Platform, "Dunarea de Jos" University of Galati, 800008 Galati, Romania
* Correspondence: Antoaneta.Ene@ugal.ro

Abstract: In order to evaluate the greenhouse gas (GHG) emissions from a reservoir or from several reservoirs in a country or a climatic zone, simpler or more complex models based on measurements and analyses of emissions presented in the literature were developed, which take into account one or more reservoir-specific parameters. The application of the models in the assessment of GHG emissions from a multipurpose reservoir gave values that are more or less close to the average values reported in the literature for the temperate zone reservoirs. This is explained by the fact that some models only consider emissions caused by impoundment and not degassing, spillway emissions, and downstream emissions, or those that use different calculation periods. The only model that calculates GHG emissions over the life cycle that occur pre-impoundment, post-impoundment, from unrelated anthropogenic sources and due to the reservoir construction is the model used by the G-res tool. In addition, this tool is best suited for multipurpose reservoirs because it allocates GHG emissions for each use, thus facilitating the correct reporting of emissions. The G-res tool used to calculate GHG emissions from the Stânca-Costești Multipurpose Reservoir shows that this is a sink of GHG with a net emission of -5 g $CO_{2eq}/m^2/yr$ (without taking into account the emissions due to dam construction).

Keywords: emissions; greenhouse gas; multipurpose reservoirs; temperate climate

Citation: Ion, I.V.; Ene, A. Evaluation of Greenhouse Gas Emissions from Reservoirs: A Review. *Sustainability* **2021**, *13*, 11621. https://doi.org/10.3390/su132111621

Academic Editors: Baojie He, Ayyoob Sharifi, Chi Feng and Jun Yang

Received: 24 August 2021
Accepted: 17 October 2021
Published: 21 October 2021

Publisher's Note: MDPI stays neutral with regard to jurisdictional claims in published maps and institutional affiliations.

1. Introduction

Reservoirs are manmade lakes created by building dams on rivers for various purposes: flood control, electricity generation, irrigation, water supply, aquaculture, environmental services, recreational activities, navigation etc.

In freshwater ecosystems, several mechanisms are involved in the natural carbon cycle. They receive carbon from terrestrial ecosystems through drainage, capture the carbon through primary production, bury the carbon in sediments, emit GHG through biomass degradation and respiration, and transport the carbon downstream to the seas or oceans. GHG emissions can be increased by human activities around the ecosystem through sewage and agricultural pollution. Dams affect the natural carbon cycle in freshwater ecosystems through floods of terrestrial vegetation and soils. The flooded organic matter decomposes causing additional GHG emissions, especially in the first years after the reservoir creation. Flooding can also increase sedimentation and decomposition in the reservoir, due to longer water residence times, which can lead to higher GHG emissions [1]. In addition, reservoirs can have large fluctuations in the water level, especially hydroelectric reservoirs that store large volumes of water to be used during drought. It can, therefore, be said that artificial reservoirs differ from natural lakes by riverine nutrient inputs, the flooding of terrestrial organic carbon, and water-level fluctuations; they also may have different GHG emissions. Reservoirs present, from a social, economic and environmental point of view, not only advantages, but also disadvantages.

173

Reservoir use can serve single or multiple purposes. According to the International Commission on Large Dams (ICOLD), 70% of large reservoirs are designed for single-purpose usage. Around 11% of large reservoirs have been built only for hydropower generation and 14% for hydropower generation plus other uses. These high figures show why GHG emissions from reservoirs should be accounted for. In addition, the study of these emissions indicates ways to reduce them.

The main greenhouse gases emitted by a reservoir are CO_2, CH_4 and N_2O. They have a different global warming potential (GWP). For the time period of 100 years, GWP for CO_2 is 1; for CH_4, it is 34 times higher than that of CO_2, and for N_2O, it is 298 times that of CO_2 [2].

The CO_2 is generated by the decomposition of organic material and nutrients transported in the reservoir by affluent or by rainfall and overland flow, by the decomposition of dead organic matter stored in the soil of the reservoir, by the respiration of vegetation present in the reservoir, from CO_2 dissolved in water and from the oxidation of CH_4. The sediments in drawdown areas are also a source of CO_2 emission, due to their exposure to air during water level fluctuations.

The emission of CH_4 comes from the decomposition of organic matter and vegetation under anaerobic conditions in the soil or sediment layer of the reservoir.

Nitrous oxide (N_2O) arises as a by-product of the aerobic nitrification reaction or of the anaerobic denitrification that occurs in lake riparian areas. The few measurements of N_2O emission from reservoirs showed a variation similar to that of CH_4 in terms of generation. The contribution of N_2O to the total GHG emission expressed as an CO_2 equivalent is low, compared to CH_4 și CO_2 (N_2O—17 mg $CO_{2eq}/m^2/d$; CH_4—275 mg $CO_{2eq}/m^2/d$ and CO_2—1585 mg $CO_2/m^2/d$) [3].

GHG (CO_2 and CH_4) reaches the atmosphere through the following channels: diffusive flow from the reservoir surface, through degassing when passing through the hydraulic turbine and spillway (due to pressure drop), through diffusive flow at the downstream river surface. Methane can also reach the surface of the reservoir through bubbling in shallow areas of reservoir.

The main factors influencing GHG emissions are the carbon stock in soil and flooded biomass or that transported by the upstream rivers in reservoirs; the concentration of dissolved oxygen in the reservoir; water quality and nutrient content; the inflow and shape of the reservoir; the water depth and extension of the littoral zone; the wind speed at the reservoir surface; and the water temperature and configuration of dam intake and outlets [3]. These factors influence the biochemical processes of organic matter formation, respiration, methanogenesis, CH_4 oxidation, gas exchange between the reservoir and the atmosphere. The GHG measurements showed a variation in time and space within a reservoir and also a seasonal variation and a decrease in general with the age of the reservoir.

There are many studies on the evaluation of GHG emissions from reservoirs, which differ by the methodologies used, the lifespan considered, and the size and type of reservoir [1,4–39].

Most studies have analyzed GHG emissions from hydropower reservoirs. The few studies performed on natural lakes have shown that there are no significant differences between reservoir surface emissions from hydropower reservoirs, compared to non-hydropower reservoirs [10]. At hydropower reservoirs, there are also degassing emissions, downstream emissions and emissions from drawdown zones.

From the analysis of GHG emissions from 85 different hydroelectric reservoirs with a global distribution, it was observed that all the reservoirs are sources of CH_4 to the atmosphere, the majority (88%) are also a source of CO_2 (only 12% of reservoirs are net sinks of CO_2) and that there is a large variation in emissions [5].

Knowing the GHG emissions generated by reservoirs is an important factor in making decisions to finance future projects and discerning how environmentally friendly they are.

The purpose of this study is to review the main methodologies for assessing GHG emissions from a reservoir, find the most appropriate model to estimate GHG emissions from a multipurpose reservoir, and present the complete environmental performance of a reservoir placed in the temperate zone of Eastern Romania, a country for which data are limited. On the Romanian territory, there are about 400 large reservoirs totaling 6300 million m^3. The chosen Stânca-Costești reservoir is one of the representative reservoirs, being the second largest in the country.

The models used both for reservoirs for any purposes other than power generation and for hydropower reservoirs are presented. The results obtained are presented comparatively to help stakeholders in identifying and choosing the calculation method.

2. Assessment of GHG Emissions from the Reservoirs

Because measuring GHG emissions from reservoirs is not an easy task due to the large variation in time and space, especially of CH_4 emissions, several methodologies for their evaluation were developed. All methodologies are based on measurements of emissions from reservoirs located in different climatic zones. The different emission values reported in the literature show the influence of specific reservoir factors, such as type, climate zone, age and depth of the reservoir, neighboring land cover types, land uses before flooding and others.

Table 1 shows the estimated GHG emissions from different types of reservoirs and freshwater ecosystems [10].

Table 1. GHG emissions from reservoirs and freshwater ecosystems [10].

System Type	GHG Areal Rate, (mg/m^2/d)		CO$_2$ Equivalent Emissions, CO$_{2eq}$, (g/m^2/yr)
	CH$_4$	CO$_2$	
All reservoirs	160.8	1207.8	2436.38
	110–128	1822.68	2695.66–2253.76
Hydroelectric reservoirs	32.16–150	1412–2415	914.49–2742.98
Lakes	53.6	790.56	953.73
Ponds	36.18	1544.52	1012.74
Rivers	8–131	29,111.64	10,725.03–12,251.46
Wetlands	20–84	-	248.20–1042.44

In paper [16], an average global GHG emission is reported (reservoir surface plus drawdown area emissions and reservoir downstream emissions) from the hydropower reservoirs of 92 g CO_2/kWh and 5.7 g CH_4/kWh.

As stated in [32], the lifecycle GHG emissions from hydropower plants range from 1 g CO_{2eq}/kWh to 2200 g CO_{2eq}/kWh with an average value of 24 g CO_{2eq}/kWh.

A distribution by the source of GHG emission from hydropower reservoirs from different climatic zones is presented in Table 2 [16,28]. The large variation from one climate zone to another can be seen as well as the higher contribution of emissions from drawdown areas and the contribution of CO_2 emissions to the total emissions.

Table 2. GHG emissions from hydroelectric reservoirs [16,28].

Climate Zone	Carbon Emission, (mg/m^2/d)	
	CO$_2$	CH$_4$
Reservoir surface		
Boreal	753	9.1
Temperate	1500 [16]	20 [16]
	386 [28]	2.8 [28]
Tropical	3097	91.3
Drawdown area		
Temperate	2110	110
Tropical	3500 [16]	300 [16]
	13,000 [28]	235 [28]

The contribution to GHG emissions related to reservoir construction, meaning those from the activities related to dam construction (raw material extraction, equipment manufacturing, transportation, and building process of dam), is estimated to be (2.3–37.9) gCO$_{2eq}$/kWh [27].

In paper [26], an average global emission of 173 kg CO_2/MWh and 2.95 kg CH_4/MWh was estimated after the emissions from over 1400 hydroelectric power plants were analyzed. The study emphasizes the importance of analyzing each hydropower plant and the need for standardized measurement procedures, taking into consideration carbon burial, drawdown areas and methane bubbles. Additionally, in this paper [26], the following models were developed using generalized linear models:

- CO_2 emission expressed in kg CO_2/MWh as a function of the area-to-electricity ratio (ATER, km^2/GWh) and reservoir area (S, km^2):

$$CO_2 = -169.73 + 241.86 \times ATER + 120.34 \times \ln(S), \text{ kg } CO_2/\text{MWh} \qquad (1)$$

- CH_4 emission expressed in kg CH_4/MWh as a function of the reservoir age (A, years), area-to-electricity ratio (ATER, km^2/GWh) and maximum temperature (T_{max}, °C):

$$\ln(CH_4) = -9.81 - 0.75 \times \ln(A) + 1.18 \times \ln(ATER) + 4.5 \times \ln(T_{max}), \text{ kg } CH_4/\text{MWh} \qquad (2)$$

GHG emission expressed in mg C/m^2/d, which can also be used in the case of reservoirs with a use other than energy production, depending on the reservoir age (A, years), erosion rate (Er, t/ha/yr), reservoir area (S, km^2) and maximum temperature (T_{max}, °C):

$$CO_2 = 494.46 - 4.07 \times A + 8.09 \times Er, \text{ mg } CO_2/\text{m}^2/\text{d} \qquad (3)$$

$$\ln(CH_4) = -12.84 - 0.03 \times A + 0.21 \times \ln(S) - 0.01 \times Er + 4.88 \times \ln(T_{max}), \text{ mg } CH_4/\text{m}^2/\text{d} \qquad (4)$$

In [5], CO_2 and CH_4 emissions from 85 reservoirs between 68° N and 25° S were analyzed. The reservoir surface emissions were estimated, correlated to reservoir age (A, years) and latitude (L, °), dissolved organic carbon (DOC, mg/L) and mean reservoir depth (h, m). Based on the measured data, the CO_2 and CH_4 emissions were predicted, using the following multiple-regression equations:

$$\log(CO_2 + 400) = 3.06 - 0.16 \times \log(A) - 0.01 \times L + 0.41 \times \log(DOC), \text{ mg } CO_2/\text{m}^2/\text{d} \qquad (5)$$

$$\log(CH_4) = 1.33 - 0.36 \times \log(A) - 0.32 \times \log(h) + 0.39 \times \log(DOC) - 0.01L, \text{ mg } CH_4/\text{m}^2/\text{d} \qquad (6)$$

Ref. [38] presented a calculation tool called HydroCalculator, which, in addition to calculating financial indicators, also calculates greenhouse gas emissions. The biome carbon loss (BCL) model based on the initial carbon stock is used to calculate GHG emissions:

$$C_t = C_0\left(\frac{e^{-0.3t}}{5} + \frac{e^{-0.03t}}{3} + \frac{1}{2}\right), \text{ tonne C/yr} \qquad (7)$$

where C_t is the amount of carbon (CO_2 and CH_4) in tonnes in the year t, C_0 is the initial amount of organic carbon in soil and vegetation in tonnes, and t is the time in years.

This model calculates only GHG emissions from the decomposition of organic matter, not GHG emissions from turbines and spillway.

Paper [38] calculated the proportion of carbon emitted as CO_2 and CH_4, observing that CO_2 emissions are 73% of carbon emissions and CH_4 emissions are 27%. Carbon emissions can be converted to CO_2 and CH_4 by multiplying the values by 44/12 and 16/12, respectively. CH_4 emissions are converted to CO_2 equivalents by multiplying the global warming potential of CH_4 specified in IPCC's Fifth Assessment Report [33].

Ref. [39] presented, in addition to Equation (7), two other calculation equations of CH_4 emissions, from the organic matter cycle (OMC model) (Equation (8)) and downstream emissions (from turbines and spillways) (downstream emission model) (Equation (9)):

$$E_{OMCM} = 10\left(15.5S^{0.841} + 1.73S^{0.927} + 35.2A^{0.649}\right), \text{ g } CH_4/\text{yr} \qquad (8)$$

$$E_{DEM} = Q \times d \times c \times t, \text{ g } CH_4/\text{yr} \qquad (9)$$

where S is the reservoir surface area (m^2); Q is the waterflow (m^3/d); d is the effective emission assumed to be 50% to 80%; c is the CH_4 concentration in turbine/spillway water intakes (g/m^3); and t is the annual operation of spillways (1/4 year) and turbines (3/4 year) (days).

In paper [6], their own measurements of methane emissions from reservoirs were analyzed together with the emissions from 49 other reservoirs from temperate and boreal zones in order to find the best mathematical relations for estimating emissions according to the characteristics of the reservoirs. The following regression equations for bubbling, diffusive and storage emissions were proposed:

- CH_4 bubbling emission:

$$(CH_4)_{bub} = 10^{0.838+0.934\times\log(S)+0.881\times\log(TP)}, \text{ g } CH_4/yr \Big) \tag{10}$$

- CH_4 diffusive emission:

$$(CH_4)_{diff} = 10^{0.234+0.927\times\log(S)}, \text{ g } CH_4/yr \tag{11}$$

- CH_4 storage emission:

$$(CH_4)_{stor} = 10^{7.068+3.304\times\log(TP)-1.904\times\log(DOC)}, \text{ g } CH_4/yr \tag{12}$$

where S is the reservoir area (m^2), TP is the concentration of total phosphorous (μmol/L), and DOC is the concentration of dissolved organic carbon (mg C/L).

The International Energy Agency Hydropower Implementing Agreement on Hydropower Technologies and Programs (IEA Hydro) developed a framework, which describes the steps for data collection and analysis and the modeling tools to estimate the GHG emissions from a reservoir on the principles of net emissions defined by the IPCC. The estimation of GHG emissions from reservoirs is based on the analysis of the collected data and four elements of the new reservoir projects (flooded area, reservoir, upstream catchment area, reservoir outflow facilities, downstream river) [35].

The International Government Panel on Climate Change guidelines [2] recommend estimating the diffusion emission of CO_2 from reservoirs by using the following equation:

$$(CO_2)_{diff} = P \times E(CO_2)_{diff} \times S \times 10^{-3}, \text{ tonne } CO_2/yr \tag{13}$$

where $(CO_2)_{diff}$ is the total CO_2 emission from the reservoir, tonne CO_2/yr; P is the number of days without ice cover during a year, days/yr; and E(CO2)diff is the average daily diffusive emissions, kg CO2/ha/d (Table 3).

Table 3. CO_2 measured emissions for flooded land [2].

Climate	Diffusive Emission (Ice-Free Period) $E(CO_2)_{diff}$ (kg CO_2/ha/d) Range and Median Values
Boreal wet	(0.8–34.5) 11.8
Cold temperate, moist	(4.5–86.3) 15.2
Warm temperate, moist	(−10.3–57.5) 8.1
Warm temperate, dry	(−12.0–31.0) 5.2
Tropical, wet	(11.5–90.9) 44.9
Tropical, dry	(11.7–58.7) 39.1

To these emissions must be added the degassing emissions, which represent up to 30% of the total CO_2 emissions from reservoirs in a temperate moist region and less than 5% in cold temperate regions.

For a preliminary estimate of the total annual CH_4 emissions (reservoir surface emissions plus emissions originating from reservoir but emitted downstream of dam) from reservoirs older than 20 years, the 2019 Refinement to the 2006 IPCC Guidelines for National Greenhouse Gas Inventories, Volume 4: Agriculture, Forestry and Other Land Use

(Chapter 7: Wetlands) [36] recommend the following equation with the emission factors derived from the G-Res model given in Table 2:

$$(CH_4)_{tot} = (CH_4)_{res} + (CH_4)_{downstream} = \alpha \times EF_{CH_4} \times S + \alpha \times EF_{CH_4} \times S \times R_d, \text{ kg } CH_4/yr \quad (14)$$

where $(CH_4)_{res}$ is the annual reservoir surface emissions of CH_4 from reservoirs > 20 years old, kg CH_4/yr; $(CH_4)_{downstream}$ is the annual emissions of CH_4 emitted downstream of dam, kg CH_4/yr; EF_{CH4} is the emission factor for CH_4 emitted from the reservoir surface, kg CH_4/ha/yr (Table 4); R_d is a constant equal to the ratio of total downstream emission of CH_4 to the total flux of CH_4 from the reservoir surface, (median value of R_d is 0.09); α is the emission factor adjustment for the trophic state in the reservoir (it can be estimated from the trophic index (TI), total phosphorus (TP), total nitrogen (TN) and Secchi depth (SD), and mean annual chlorophyll-a concentration in the reservoir (Chl-a)) (Table 5).

Table 4. CH_4 Emission factors EF_{CH4} age > 20 [36].

Climate Zone	Average Values (kg CH_4/ha/yr)
Cool temperate	54.00
Warm temperate/dry	150.90
Warm temperate/moist	80.30

Table 5. Relationship between TI, Chl-a, TP, TN, SD, trophic class and trophic state adjustment factor [36].

TI	Chl-a (μg/L)	TP (μg/L)	TN (μg/L)	SD (m)	Trophic Class	Trophic State Adjustment Factor, α Range and Recommended Values
<30–40	0–2.6	0–12	-<350	>4	Oligotrophic	0.7 (0.7)
40–50	2.6–20	12–24	−350–650	2–4	Mesotrophic	0.7–5.3 (3)
50–70	20–56	24–96	650–1200	0.5–2	Eutrophic	5.3–14.5 (10)
70–100+	56->155	96->384	>1200	<0.5	Hypereutrophic	14.5–39.4 (25)

Paper [36] also recommends the use of the greenhouse gas reservoir (G-res) model as "currently the only easily and widely applicable model" that uses empirical relationships between environmental drivers and emissions to estimate reservoir GHG fluxes. The International Hydropower Association (IHA) in cooperation with the UNESCO Chair in Global Environmental Change has developed a detailed measurement guide for net GHG assessment [37] and the powerful and user-friendly G-res tool to assess GHG emissions [21]. Scientists from the University of Quebec at Montreal (UQAM), the Norwegian Foundation for Scientific and Industrial Research (SINTEF) and the Natural Resources Institute of Finland (LUKE) have also contributed to the development of the G-res tool [22]. The free online G-res tool (https://g-res.hydropower.org/ accessed on 5 May 2021) estimates the GHG emissions resulting from the impoundment of an existing or planned reservoir and emissions related to human activities and infrastructure. The modeling of GHG emissions (diffusive CO_2 flux, diffusive CH_4 flux, bubbling of CH_4 and CH_4 degassing) is based on the statistical analysis of gross emissions from 223 reservoirs in all climatic zones considering the following key governing variables: temperature, reservoir age, littoral area, solar radiance, phosphorous concentration in the reservoirs, soil carbon content, operating regime of reservoir, climate zone, land cover, reservoir area and soil type. The user of the G-res tool should prepare the input data before accessing the tool. The estimated net annual GHG emissions from a hydropower reservoir result from subtracting the pre-impoundment emissions and emissions from the reservoir due to unrelated anthropogenic sources (activities within the catchment as sources of nutrients and carbon flowing into the reservoir) from the post-impoundment emissions. The total annual GHG emissions are calculated by adding to the net annual emissions the emissions related to reservoir construction.

3. Case Study: The Stânca-Costești Multipurpose Reservoir

The Stânca-Costești (Figure 1) multipurpose reservoir is located on the middle course of the Prut River, on the border between Romania and the Republic of Moldova. It was built between 1973 and 1978. It has an area of 6000 ha at normal retention level (NRL) and a maximum volume of 1400 million m^3. In the Romanian Register of Large Dams, the Stânca-Costești reservoir appears on the 49th place in order of height and on the second place, according to the useful volume of the reservoir (1290 million m^3, after the Iron Gates I with 2100 million m^3). The length of the reservoir, at normal retention level (NRL 90.80 m), is 70 km, and the length at maximum level (Nmax 99.50 m) is 90 km. The surface of the reservoir at normal retention level (NRL), is of 6000 ha and at maximum level (Nmax) is of 9200 ha. The reservoir has a reserve provided for flood control of 550 million cubic meters, which ensures the removal from flood risk of 100,000 ha of agricultural land, irrigation of 140,000 ha (70,000 on each shore), ensuring the necessary flows for water supply downstream (10–16 m^3/s) and energy production with the help of two power plants with a hydro unit of 15 MW, at a flow rate of 2×65 m^3/s with an average annual energy supply of 2×65 GWh [40,41].

To build the reservoir, eight villages were relocated on both banks of the Prut River. The surface of the river basin of the Prut River in the Stânca-Costesti section is about 12,000 km^2, and the average multiannual flow is 81 m^3/s.

The dam, made of concrete and local materials, uses the favorable morphological situation due to the presence of calcareous reefs, which reduce the average width of the major riverbed from 3–4 km to 400 m.

The GHG emission from the multipurpose reservoir Stânca –Costești is estimated by using the above mentioned modeling methodologies and the powerful web-based tool G-res. The input data required by the G-res tool are given in Table 6, and the input data for the other models are the following: dissolved organic carbon concentration in reservoir water (DOC = 6.5 mg/L), maximum temperature (T_{max} = 30 °C); ice-free period (IFP = 319 days/yr); mean annual chlorophyll-a concentration in reservoir (Chl-a = 10 mg/L) [41] organic carbon in soil and vegetation (C_0 = 50 tonne C/ha) [42]; erosion rate (Er = 0.5 t/ha/yr) [43]; and concentration of total phosphorous in reservoir water(TP = 0.97 µmol/L).

To determine the average water volume of the Stânca-Costești reservoir, the daily records made by the "Romanian Waters" National Administration ("Prut-Bârlad" Water Basin Administration) in the period 2002–2018 were used.

Figure 1. *Cont.*

Figure 1. Stânca-Costești multipurpose reservoir. Reprinted with permission from ref. [44]. Copyright 2021 www.planiada.ro (accessed on 15 January 2021).

Table 6. Data on the multipurpose reservoir Stânca-Costești [45].

Catchment Information	Value	
Catchment area	12,000 km^2	
Population in the catchment	150,000	
Precipitation	380 mm/yr	
Catchment annual runoff	10 m^3/s [46]	
Community wastewater treatment	None	
Industrial wastewater treatment	None	
Land cover in the catchment area	Post-impoundment	Pre-impoundment
Croplands	64%	64%
Grassland	18%	19%
Forest	16%	16%
Water bodies	2%	1%
Reservoir Information		
Country	Romania	
Longitude of dam	27°14′00″ E	
Latitude of dam	47°51′30″ N	
Climate zone	Temperate	
Water uses	flood control, water supply (10 m^3/s), irrigation (0.8 km^3/yr), hydroenergy, aquaculture (2.9 m^3/s)	
Impoundment year	1978	
Power generation	30 MW	
Power connection	110 V	
Yearly electricity generation	130 GWh	
Reservoir area	59 km^2	
Reservoir volume (multiannual average)	1.4 km^3	
Water Level (m above sea level)	126 m	
Maximum depth	32 m	
Mean depth	23.33 m (Calculated by the G-res Tool)	
Littoral area	3.59% (Calculated by the G-res Tool)	
Water intake depth	28 m	
Soil carbon content under impounded area	0.8 kg C/m^2	
Annual wind speed at 10 m	6.6 m/s	
Water residence time	11.68 yr (Calculated by the G-res Tool)	
Annual discharge from the reservoir	3.8 m^3/s (Calculated by the G-res Tool)	
River Length before Impoundment (m)	70,000 m	
Reservoir mean global horizontal radiance	3.24 kWh/m^2/d	
Mean annual air temperature	13.3 °C	

Table 6. *Cont.*

Catchment Information	Value
Construction information	
Material excavated and/or used for construction	500,000 m^3
All concrete brought to site for the dam, tunnels, foundations	4,000,000 m^3
All steel brought to site for reinforcement, pipelines, mechanical and electrical equipment	5000 tonne

4. Results and Discussions

Using the models presented above and the G-res tool to evaluate the GHG emissions from the Stânca-Costești multipurpose reservoir, we obtained the results given in Table 7.

Table 7. Comparison of different approaches to estimate reservoir emissions.

Model		Emissions						CO$_{2eq}$ (GWP$_{100}$ for CH$_4$ = 34)		
		CO$_2$			CH$_4$					
		mg/m^2/d	kg/MWh	tonne/yr	mg/m^2/d	kg/MWh	tonne/yr	mg/m^2/d	kg/MWh	tonne/yr
Scherer 2016 [26]		327.56	155.88	-	28.57	5.93	-	1298.94	210.72	27,972.67
Barros 2011 [5]		414.35	-	-	1.15	-	-	453.65	73.59	9769.44
IPCC 2006 [2]		520	-	10,124.4	-	-	-	520	77.88	10,124.4
IPCC 2019 [36]		-	-	-	13.11	-	287.10	445.73	72.31	9761.40
Total		520	-	10,124.4	13.11	-	287.10	965.73	150.19	19,885.80
Bastviken 2004 [6]	Bubbling	-	-	-	2.46	-	53.88	83.65	13.57	1831.91
	Diffusion	-	-	-	1.27	-	27.82	43.19	7.01	945.86
	Storage	-	-	-	0.18	-	3.93	6.10	0.99	133.51
	Total	-	-	-	3.91	-	85.63	132.93	21.56	2911.27
Vilela 2017 [38]		-	-	1141.04	19.27	-	422.03	3497.3	567.34	15,490.06
Bergier 2007 [39]	OMCM	-	-	-	39.24	-	859.321	1334.11	218.42	29,216.91
	DEM	-	-	-	18.25	-	399.72	620.57	100.67	13,590.48
G-res tool [22]	Post-impoundment	205.48	-	4425.00	2.98	-	64.21	306.85	50.83	6608.00
	Pre-impoundment	216.44	-	4661.00	3.06	-	65.94	320.55	53.10	6903.00
	Unrelated Anthropogenic Source (UAS)	0.00	-	0.00	2.98	-	64.21	101.37	16.79	2183.00
	Net reservoir emission	−10.96	0.00	−236.00	−3.06	0.00	−65.94	−115.07	−19.06	−2478.00

There are only three models (Scherer; Barros, Villela) and the G-res tool that calculates both CO$_2$ and CH$_4$ emissions. The only model that calculates the net GHG emission (post-impoundment emission minus pre-impoundment emission minus emissions from unrelated anthropogenic sources) is the G-res tool. It is also the only model that allows estimating GHG emissions related to reservoir construction, and which calculates the emissions related to each use of the reservoir (in the case of multipurpose reservoirs).

Regarding the CO$_2$ emission, the IPCC 2006 model (based on measured emissions) gave the largest emission (520 mg CO$_2$/m^2/d), followed by the Barros model (which takes into account the reservoir age and dissolved organic carbon (414.35 mg CO$_2$/m^2/d)), the Scherer model (which takes into account the reservoir age and erosion rate (327.56 mg CO$_2$/m^2/d)), and the G-res tool (which gave the lowest emission (205.48 mg CO$_2$//m^2/d)) (post-impoundment) (Figure 2).

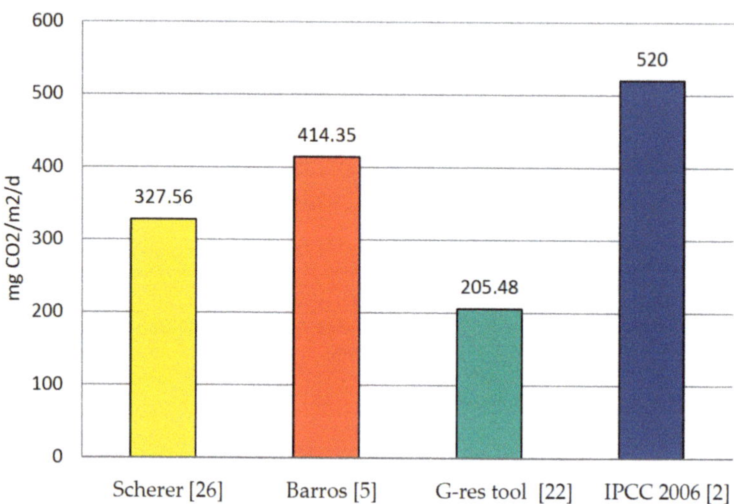

Figure 2. Estimated CO_2 emission from the Stânca-Costeşti multipurpose reservoir.

The values obtained for methane emission can be divided into two groups: a group of high values and a group of low values. The first group includes the Vilela plus Bergier model with 76.52 mg $CH_4/m^2/d$, the Scherer model with 28.57 mg $CH_4/m^2/d$ and the IPCC 2019 model with 13.11 mg $CH_4/m^2/d$. The second group includes the Bastviken model with 3.91 mg $CH_4/m^2/d$, the G-res tool with 2.42 mg $CH_4/m^2/d$ (post-impoundment) and the Barros model with 1.16 mg $CH_4/m^2/d$ (Figure 3).

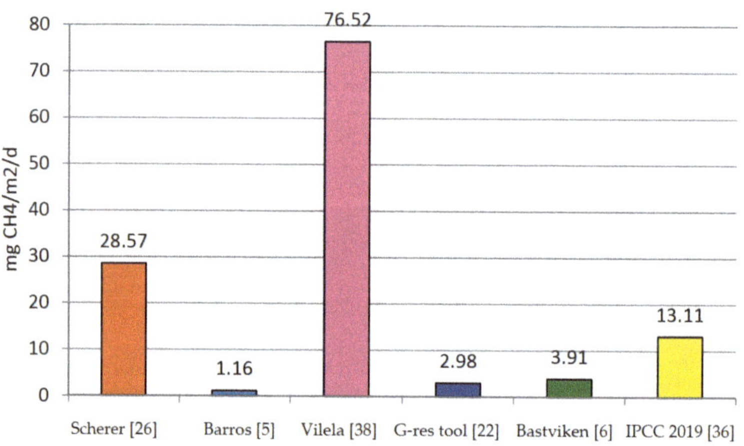

Figure 3. Estimated CH_4 emission from the Stânca-Costeşti multipurpose reservoir.

The G-res tool indicates that the reservoir has a negative net GHG emission of -819 mg $CO_{2eq}/m^2/d$, which means that the reservoir is a carbon sink, with a lower emission after impoundment (403 mg $CO_{2eq}/m^2/d$) than before impoundment (1222 mg $CO_{2eq}/m^2/d$).

Among the models that calculate the annual global GHG emission, the G-res tool gave the lowest emission (6608 tonne CO_{2eq}/yr) (post-impoundment), followed by the Barros model with 9769.44 tonne CO_{2eq}/yr and the Scherer model with the highest emission of 27,972.67 tonne CO_{2eq}/yr, more than four times higher than the lowest emission.

The calculation of GHG emission from a multipurpose reservoir as function of energy produced (expressed in kg CO_{2eq}/MWh) is not conclusive, as energy generation is one of multiple uses. For the studied multipurpose reservoir, given the reduced installed capacity of only 30 MW (and therefore the reduced amount of energy produced, 130 GWh/yr) the calculated amount of GHG emitted per unit of energy produced resulted in being two times higher than the IHA estimated emission of hydropower [47] (50.83 kg CO_{2eq}/MWh versus of 23 kg CO_{2eq}/MWh).

The G-res tool solves this shortcoming because it compares the GHG emissions obtained by using it with the emissions of reservoirs of the same type from the same climate zone.

The net reservoir footprint of -9 g CO_{2eq}/m^2/yr is placed at the base of the Gaussian distribution, which ranges from -90 to $+1960$ g CO_{2eq}/m^2/yr and has a peak near $+210$ g CO_{2eq}/m^2/yr. CH_4 diffusive emissions of $+43$ g CO_{2eq}/m^2/yr is very close to the value that gives the peak of the Gaussian distribution ($+40$ g CO_{2eq}/m^2/yr). CH_4 bubbling emissions of $+3$ g CO_{2eq}/m^2/yr is very close to the value that gives the peak of the Gaussian distribution (0 g CO_{2eq}/m^2/yr). CO_2 diffusive emissions of $+75$ g CO_{2eq}/m^2/yr are very close to the value that gives the peak of the Gaussian distribution ($+90$ g CO_{2eq}/m^2/yr). CH_4 degassing emissions could not be calculated because the reservoir has a deep water intake. However, its contribution to the overall GHG footprint is relatively small (mean: 14%, median: 4%) [23]. The allocated GHG emissions intensity of 4.1 g CO_{2eq}/kWh is very close to the value that gives the peak of the Gaussian distribution (0 g CO_{2eq}/kWh). The CH_4 post-impoundment emissions have the following distribution: 93% CH_4 diffusive emissions and 7% CH_4 bubbling emissions. The contribution of each use of the reservoir is as follows: 26.7% flood control; 5% fisheries; 15% irrigation; 26.7% water supply; and 26.7% hydroelectricity.

Of course, for a preliminary and rapid estimate of GHG emissions from a reservoir, simple models can be used that require only a few reservoir-specific factors. For a more accurate assessment, it is necessary to use a more complex and complete model, such as the G-res tool. The G-res tool offers the possibility to find out the potential impact that the creation of a reservoir can have and also to find out if measures are needed to reduce GHG emissions from a certain reservoir (regardless of the reservoir purpose), or to find out that the emissions of a reservoir may be the result of human activity that is not related to the reservoir creation. When the necessary time and tools are available, it is most desirable to perform long-term measurements to report the actual values of emissions from all sources related to reservoirs (drawdown area, reservoir surface, turbines, spillway and downstream river).

Taking into account the lifecycle GHG emission intensity from Table 8, if the energy produced by the Stânca-Costeşti reservoir is produced by a coal-fired power plant, additional GHG emissions of 103,610 tonnes CO_{2eq}/yr would be generated, and if it was produced by a natural gas-fired power plant, additional GHG emissions of 60,710 tonnes CO_{2eq}/yr would be generated.

Table 8. Lifecycle GHG emissions for different electricity generation technologies [47].

Technology	Lifecycle GHG Emission Intensity kg CO_{2eq}/MWh
Thermal–coal	820
Thermal–natural gas	490
Biomass	230
Solar–PV	48
Hydropower	23
Nuclear	12
Wind	12

5. Conclusions

Built to meet human needs, multipurpose reservoirs increase human well-being, but they cause changes in the water quality, ecosystem and flow regime of river networks. They are considered neutral in terms of GHG emissions, but they may become considerable sources of GHG depending, especially, on the climatic zone in which they are located and their uses. The creation of a water reservoir on a river leads to the generation of GHG, due to biogeochemical processes in the reservoir. The calculation of GHG emissions of the studied reservoir, which is placed in a temperate zone and has multiple uses of water, shows that they are lower than those of a lake (306.85 g $CO_{2eq}/m^2/yr$ versus 953.73 g $CO_{2eq}/m^2/yr$).

Knowing the GHG emissions from the reservoir is useful to accurately report the greenhouse gas (GHG) emissions. To calculate the CO_2 emission, four models were used; to calculate the CH_4 emission, six models were used. If the difference between the highest (520 mg $CO_2/m^2/d$) and the lowest CO_2 emission value (205.48 mg $CO_2/m^2/d$) is more than two-fold, the difference between the highest CH_4 emission (76.52 mg $CH_4/m^2/d$) and the lowest emission value (1.16 CH_4 mg $CH_4/m^2/d$) is much larger, by about 65 times.

Because not all the methodologies reviewed make an overall assessment of GHG emissions and because some are used only for hydropower reservoirs—except the G-res tool, which estimates the GHG emissions from the reservoir surface, drawdown, turbines and spillway—it is difficult to compare the results obtained by applying the methodologies to the multipurpose reservoir, Stânca-Costești.

In the absence of a standardized methodology for calculating GHG emissions from the reservoirs, the reviewed models can be used in correlation with the available data on reservoirs.

Author Contributions: Conceptualization, I.V.I. and A.E.; methodology, I.V.I.; software, I.V.I.; validation, I.V.I. and A.E.; formal analysis, I.V.I.; investigation, I.V.I. and A.E.; resources, I.V.I. and A.E.; data curation, I.V.I. and A.E.; writing—original draft preparation, I.V.I.; writing—review and editing, I.V.I. and A.E.; visualization, I.V.I. and A.E.; supervision, A.E.; project administration, A.E.; funding acquisition, A.E. All authors have read and agreed to the published version of the manuscript.

Funding: This work was supported by the project with code BSB 165 "HydroEcoNex", funded by the European Union through the Joint Operational Programme "Black Sea Basin 2014–2020". The content of this publication is the sole responsibility of the authors, and in no case should it be considered to reflect the views of the European Union. The APC is funded by "Dunarea de Jos" University of Galati, Romania, grant no. RF 3621/2021.

Institutional Review Board Statement: Not applicable.

Informed Consent Statement: Not applicable.

Data Availability Statement: The data that support the findings of this study are available from the authors upon reasonable request.

Acknowledgments: The authors are grateful to the "Romanian Waters" National Administration for providing, on request, construction and operation data of the reservoir, Stânca-Costești, and to the International Hydropower Association (IHA) for the permission to use the GHG reservoir (G-res) tool.

Conflicts of Interest: The authors declare no conflict of interest. The funders had no role in the design of the study; in the collection, analyses, or interpretation of data; in the writing of the manuscript, or in the decision to publish the results.

References

1. Levasseur, A.; Mercier-Blais, S.; Prairie, Y.T.; Tremblay, A.; Turpin, C. Improving the accuracy of electricity carbon footprint: Estimation of hydroelectric reservoir greenhouse gas emissions. *Renew. Sust. Energ. Rev.* **2021**, *136*, 110433. [CrossRef]
2. IPCC (Intergovernmental Panel on Climate Change). *2006 IPCC Guidelines for National Greenhouse Gas Inventories*; Eggleston, H.S., Buendia, L., Miwa, K., Ngara, T., Tanabe, K., Eds.; Institute for Global Environmental Strategies (IGES): Kanagawa, Japan, 2006.
3. World Bank. *Greenhouse Gases from Reservoirs Caused by Biogeochemical Processes*; World Bank: Washington, DC, USA, 2017. Available online: https://openknowledge.worldbank.org/handle/10986/29151 (accessed on 16 December 2020).
4. Almeida, R.M.; Paranaíba, J.R.; Barbosa, Í.; Sobek, S.; Kosten, S.; Linkhorst, A.; Mendonça, R.; Quadra, G.; Roland, F.; Barros, N. Carbon dioxide emission from drawdown areas of a Brazilian reservoir is linked to surrounding land cover. *Aquat. Sci.* **2019**, *81*, 68. [CrossRef]
5. Barros, N.; Cole, J.J.; Tranvik, L.J.; Prairie, Y.T.; Bastviken, D.; Huszar, V.L.M.; del Giorgio, P.; Roland, F. Carbon emission from hydroelectric reservoirs linked to reservoir age and latitude. *Nat. Geosci.* **2011**, *4*, 593–596. [CrossRef]
6. Bastviken, D.; Cole, J.; Pace, M.; Tranvik, L. Methane emissions from lakes: Dependence of lake characteristics, two regional assessments, and a global estimate. *Glob. Biogeochem. Cycles* **2004**, *18*, GB4009. [CrossRef]

7. Bastviken, D.; Tranvik, L.J.; Downing, J.A.; Crill, P.M.; Enrich-Prast, A. Freshwater methane emissions offset the continental carbon sink. *Science* **2011**, *331*, 50. [CrossRef]

8. Chen, H.; Wu, Y.Y.; Yuan, X.Z.; Gao, Y.S.; Wu, N.; Zhu, D. Methane emissions from newly created marshes in the drawdown area of the Three Gorges Reservoir. *J. Geophys. Res.* **2009**, *114*, D18301. [CrossRef]

9. Chen, Z.; Ye, X.; Huang, P. Estimating Carbon Dioxide (CO_2) Emissions from Reservoirs Using Artificial Neural Networks. *Water* **2018**, *10*, 26. [CrossRef]

10. Deemer, B.R.; Harrison, J.A.; Li, S.; Beaulieu, J.J.; Delsontro, T.; Barros, N.; Bezerra-Neto, J.F.; Powers, S.M.; Dos Santos, M.A.; Vonk, J.A. Greenhouse gas emissions from reservoir water surfaces: A new global synthesis. *Bioscience* **2016**, *66*, 949–964. [CrossRef]

11. Du, H.L.; Li, Z.; Guo, J.S. Carbon footprint of a large hydropower project in the upstream of the Yangtze: Following ISO14067. *Resour. Environ. Yangtze Basin* **2017**, *26*, 1102–1110.

12. Gallagher, J.; Styles, D.; McNabola, A.; Williams, A.P. Life cycle environmental balance and greenhouse gas mitigation potential of micro-hydropower energy recovery in the water industry. *J. Clean Prod.* **2015**, *99*, 152–159. [CrossRef]

13. Galy-Lacaux, C.; Delmas, R.; Kouadio, J.; Richard, S.; Gosse, P. 1999, Long-term Greenhouse Gas Emissions from Hydroelectric Reservoirs in Tropical Forest Regions. *Glob. Biogeochem. Cycles* **1999**, *13*, 503–517. [CrossRef]

14. Jiang, T.; Shen, Z.; Liu, Y.; Hou, Y. Carbon Footprint Assessment of Four Normal Size Hydropower Stations in China. *Sustainability* **2018**, *10*, 2018. [CrossRef]

15. Kadiyala, A.; Kommalapati, R.; Huque, Z. Evaluation of the Life Cycle Greenhouse Gas Emissions from Hydroelectricity Generation Systems. *Sustainability* **2016**, *8*, 539. [CrossRef]

16. Li, S.; Zhang, Q. Carbon emission from global hydroelectric reservoirs revisited. *Environ. Sci. Pollut. Res.* **2014**, *21*, 13636–13641. [CrossRef] [PubMed]

17. Li, X.; Gui, F.; Li, Q. Can Hydropower Still Be Considered a Clean Energy Source? Compelling Evidence from a Middle-Sized Hydropower Station in China. *Sustainability* **2019**, *11*, 4261. [CrossRef]

18. Mäkinen, K.; Khan, S. Policy considerations for greenhouse gas emissions from freshwater reservoirs. *Water Altern.* **2010**, *3*, 91–105.

19. Mosher, J.J.; Fortner, A.M.; Phillips, J.R.; Bevelhimer, M.S.; Stewart, A.J.; Troia, M.J. Spatial and Temporal Correlates of Greenhouse Gas Diffusion from a Hydropower Reservoir in the Southern United States. *Water* **2015**, *7*, 5910–5927. [CrossRef]

20. Prakash, R.; Bhat, I.K. Life cycle greenhouse gas emissions estimation for small hydropower schemes in India. *Energy* **2012**, *44*, 498–508.

21. Prairie, Y.T.; Alm, J.; Beaulieu, J.; Barros, N.; Battin, T.; Cole, J.; del Giorgio, P.; Del Sontro, T.; Guérin, F.; Harby, A.; et al. Greenhouse gas emissions from freshwater reservoirs: What does the atmosphere see? *Ecosystems* **2017**, *21*, 1058–1071. [CrossRef]

22. Prairie, Y.T.; Alm, J.; Harby, A.; Mercier-Blais, S.; Nahas, R. *The GHG Reservoir Tool (Gres) Technical documentation v2.1 (2019-08-21). UNESCO/IHA research project on the GHG status of freshwater reservoirs*; Joint publication of the UNESCO Chair in Global Environmental Change and the International Hydropower Association: London, UK, 2017. Available online: https://www.hydropower.org/publications/the-ghg-reservoir-tool-g-res-technical-documentation (accessed on 30 March 2021).

23. Prairie, Y.T.; Mercier-Blais, S.; Harrison, J.A.; Soued, C.; Del Giorgio, P.; Harby, A.; Alm, J.; Chanudet, V.; Nahas, R. A new modelling framework to assess biogenic GHG emissions from reservoirs: The G-res, tool. *Environ. Model Softw.* **2021**, *143*, 105117. [CrossRef]

24. Rosa, L.P.; dos Santos, M.A.; Matvienko, B.; Sikar, E. Hydroelectric reservoirs and global warming. In Proceedings of the RIO 02—World Climate and Energy Event, Rio de Janeiro, Brazil, 6–11 January 2002; pp. 123–129. Available online: http://www.rio12.com/rio02/proceedings/pdf/123_Rosa.pdf (accessed on 10 April 2021).

25. Rosa, L.P.; dos Santos, M.A.; Matvienko, B.; dos Santos, E.O.; Sikar, E. Greenhouse Gases Emissions by Hydroelectric Reservoirs in Tropical Regions. *Clim. Chang.* **2004**, *66*, 9–21. [CrossRef]

26. Scherer, L.; Pfister, S. Hydropower's Biogenic Carbon Footprint. *PLoS ONE* **2016**, *11*, e0161947. [CrossRef]

27. Song, C.; Gardner, K.H.; Klein, S.J.W.; Souza, S.P.; Mo, W. Cradle-to-grave greenhouse gas emissions from dams in the United States of America. *Renew. Sustain. Energ. Rev.* **2018**, *90*, 945–956. [CrossRef]

28. St. Louis, V.L.; Kelly, C.A.; Duchemin, É.; Rudd, J.W.M.; Rosenberg, D.M. Reservoir surfaces as sources of greenhouse gases to the atmosphere: A global estimate. *BioScience* **2000**, *50*, 766–775. [CrossRef]

29. Suwanit, W.; Gheewala, S.H. Life cycle assessment of mini-hydropower plants in Thailand. *Int. J. LCA* **2011**, *16*, 849–858. [CrossRef]

30. Zhang, Q.F.; Karney, B.; MacLean, H.L.; Feng, J.C. Life-cycle inventory of energy use and greenhouse gas emissions for two hydropower projects in China. *J. Infrastruct. Syst.* **2007**, *13*, 271–279. [CrossRef]

31. Zhang, S.R.; Pang, B.H.; Zhang, Z.L. Carbon footprint analysis of two different types of hydropower schemes: Comparing earth-rockfill dams and concrete gravity dams using hybrid life cycle assessment. *J. Clean Prod.* **2015**, *103*, 854–862. [CrossRef]

32. Schlömer, S.; Bruckner, T.; Fulton, L.; Hertwich, E.; McKinnon, A.; Perczyk, D.; Roy, J.; Schaeffer, R.; Sims, R.; Smith, P.; et al. Annex III: Technology-specific cost and performance parameters. In *Climate Change 2014: Mitigation of Climate Change. Contribution of Working Group III to the Fifth Assessment Report of the Intergovernmental Panel on Climate Change*; Edenhofer, O., Pichs-Madruga, R., Sokona, Y., Minx, J.C., Farahani, E., Kadner, S., Seyboth, K., Adler, A., Baum, I., Brunner, S., et al., Eds.; Cambridge University Press: New York, NY, USA, 2014.

33. Myhre, G.; Shindell, D.; Bréon, F.M.; Collins, W.; Fuglestvedt, J.; Huang, J.; Koch, D.; Lamarque, J.F.; Lee, D.; Mendoza, B.; et al. Anthropogenic and Natural Radiative Forcing. In *Climate Change 2013: The Physical Science Basis. Contribution of Working Group I to the Fifth Assessment Report of the Intergovernmental Panel on Climate Change*; Stocker, T.F., Qin, D., Plattner, G.-K., Tignor, M., Allen, S.K., Boschung, J., Nauels, A., Xia, Y., Bex, V., Midgley, P.M., Eds.; Cambridge University Press: Cambridge, UK; New York, NY, USA, 2013.
34. Li, Z.; Du, H.L.; Xiao, Y.; Guo, J.S. Carbon footprints of two large hydro-projects in China: Life-cycle assessment according to ISO/TS 14067. *Renew. Energy* **2017**, *114*, 534–546. [CrossRef]
35. IEA Hydropower. Annex XII: Guidelines for Quantitative Analysis of Net GHG Emissions from Reservoirs. 2012. Available online: https://www.ieahydro.org/publications/iea-hydro-reports (accessed on 23 November 2020).
36. Lovelock, C.E.; Evans, C.; Barros, N.; Prairie, Y.; Alm, J.; Bastviken, D.; Beaulieu, J.J.; Garneau, M.; Harby, A.; Harrison, J.; et al. *2019 Refinement to the 2006 IPCC Guidelines for National Greenhouse Gas Inventories, Volume 4: Agriculture, Forestry and Other Land Use (Chapter 7: Wetlands)*; IPCC: Kanagawa, Japan, 2019.
37. Goldenfum, J.A. (Ed.) *UNESCO/IHA GHG Measurement Guidelines for Freshwater Reservoirs*; International Hydropower Association (IHA): London, UK, 2010.
38. Vilela, T.; Reid, J. Improving hydropower choices via an online and open access tool. *PLoS ONE* **2017**, *12*, e0179393. [CrossRef]
39. Bergier, I.; Bambace, L.; Ramos, F.M. GHG life cycle analysis and novel opportunities arising from emerging technologies developed for tropical dams. In Proceedings of the UNESCO-IHP Workshop on the Greenhouse Gas Status of Freshwater Reservoirs, Foz do Iguaçu, Brazil, 4–5 October 2007. Available online: https://www.researchgate.net/publication/323073413_GHG_LIFE_CYCLE_ANALYSIS_AND_NOVEL_OPPORTUNITIES_ARISING_FROM_EMERGING_TECHNOLOGIES_DEVELOPED_FOR_TROPICAL_DAMS (accessed on 15 April 2021).
40. Corobov, R.; Ene, A.; Trombitsky, I.; Zubcov, E. The Prut River under Climate Change and Hydropower Impact. *Sustainability* **2021**, *13*, 66. [CrossRef]
41. Dumitran, G.E.; Vuta, L.I.; Popa, B.; Popa, F. Hydrological Variability Impact on Eutrophication in a Large Romanian Border Reservoir, Stânca–Costesti. *Water* **2020**, *12*, 3065. [CrossRef]
42. Zomer, R.J.; Bossio, D.A.; Sommer, R.; Verchot, L.V. Global Sequestration Potential of Increased Organic Carbon in Cropland Soils. *Sci. Rep.* **2017**, *7*, 15554. [CrossRef] [PubMed]
43. Panagos, P.; Borrelli, P.; Poesen, J.; Ballabio, C.; Lugato, E.; Meusburger, K.; Montanarella, L.; Alewell, C. The new assessment of soil loss by water erosion in Europe. *Environ. Sci. Policy* **2015**, *54*, 438–447. [CrossRef]
44. Available online: https://planiada.ro/destinatii/botosani/lacul-stanca-costesti-289 (accessed on 15 January 2021).
45. Available online: https://www.hidroconstructia.com/dyn/2pub/proiecte_det.php?id=114&pg=8 (accessed on 15 January 2021).
46. Fekete, B.M.; Vörösmarty, C.J.; Grabs, W. Global Composite Runoff Fields Based on Observed River Discharge and Simulated Water Balances. In *UNH/GRDC Composite Runoff Fields v1.0*; Complex Systems Research Center, University of New Hampshire: Durham, NH, USA; Global Runoff Data Center: Koblenz, Germany, 2000. Available online: http://www.grdc.sr.unh.edu (accessed on 1 December 2020).
47. International Hydropower Association (IHA). Carbon Emissions from Hydropower Reservoirs: Facts and Myths, 2021. Available online: https://www.hydropower.org/blog/carbon-emissions-from-hydropower-reservoirs-facts-and-myths (accessed on 12 April 2021).

 *sustainability*

Article

Residents' Willingness to Pay for a Carbon Tax

Ie Zheng Goh [1] and Nitanan Koshy Matthew [1,2,*]

[1] Department of Environment, Faculty of Forestry and Environment, Universiti Putra Malaysia, Serdang 43400 UPM, Selangor, Malaysia; nickey.goh@gmail.com

[2] Institute of Tropical Agriculture and Food Security (ITAFoS), Universiti Putra Malaysia, Serdang 43400 UPM, Selangor, Malaysia

* Correspondence: nitanankoshy@upm.edu.my

Abstract: Addressing environmental issues has been a significant challenge. Malaysia is one of the fastest-growing countries in terms of economic, social, and land use development but high in CO_2 emission rates. The introduction of a carbon tax is seen to reduce greenhouse gases emission (GHG), but the uncertain extent of implementation, based on economic theory, remains unknown. Hence, the current study's objectives are to assess residents' knowledge and attitude towards GHG. It is also to analyse the factors influencing residents' Willingness to Pay (WTP). Three hundred and eleven (311) residents from Klang were selected using convenience sampling. The result shows that most of the respondents were willing to pay and had medium knowledge and a high level of attitude towards GHG. Poisson regression analysis results showed that gender, age, income, education, number of households, and marital status variables significantly influenced the maximum WTP. Overall, the residents' WTP for a carbon tax was estimated at RM36.31 per year for open-ended (CVM): RM36.96 and double bound (CVM): RM35.65. A mechanism such as investment in green technology, eco-transportation, and green energy using the tax can be applied. This study is pivotal towards achieving SDG 13: Climate action.

Keywords: greenhouse gases emissions; willingness to pay; carbon tax; knowledge; attitude; contingent valuation method

Citation: Goh, I.Z.; Matthew, N.K. Residents' Willingness to Pay for a Carbon Tax. *Sustainability* **2021**, *13*, 10118. https://doi.org/10.3390/su131810118

Academic Editor: Baojie He

Received: 28 July 2021
Accepted: 25 August 2021
Published: 9 September 2021

1. Introduction

1.1. Introduction and Problem Statement

Addressing environmental issues has been a significant challenge for many countries striving for sustainable economic development [1]. Carbon dioxide is identified as one of the main components of greenhouse gases and can be used to produce diesel fuel [2,3]. These days, the use of fossil fuels has become a worldwide issue. Human activity in development has released large quantities of carbon dioxide (CO_2) and other greenhouse gas into the atmosphere [4].

Human activities, including burning fossil fuels, deforestation, land development, and electricity production contribute to climate change. In the end, this climate change will cause a greenhouse effect. Atmospheric greenhouse gases concentrations, such as CO_2, methane (CH_4), and nitrous oxide (N_2O) have been attributed to human activity since the 1750's [5,6]. Emissions of greenhouse gases have contributed significantly to air pollution and affected climate changes by increasing the atmosphere's temperature [7,8]. The release of CO_2 gas, resulting from fossil fuels while producing fuel worldwide, has also contributed to global warming [1].

The most significant increase in energy consumption and CO_2 emissions occurred in cities, especially where growing populations enjoyed higher living standards and material prosperity [9]. Increasing demand for energy resources also affects living standards through urbanisation and industrialisation [10]. This is because the increase in demand for energy resources, especially fossil fuels, will increase CO_2 emissions. As a response, about

40 national jurisdictions and more than 20 cities, states, and regions have implemented or planned an explicit carbon price, covering approximately $7GtCO_2e$, accounting for about 12% of global annual greenhouse gas emissions [8,11]. The number of carbon pricing tools implemented or planned has increased from US$20 to US$38 [12]. There has been concern that carbon pricing will damage industrial competitiveness. As such, most clear prices are still low, about less than US$10 per ton of carbon dioxide only, and there is no mechanism or plan to increase them [11]. Some countries also provided exemptions or special treatment for their most polluting energy-intensive industries, thereby limiting the effectiveness of the carbon price [11]. For example, the British Government pledged to reduce greenhouse gas emissions by at least 80% by 2050 [13]. In order to reduce the emissions, they introduced the Contracts for Difference (CFD), Renewable Obligation (RO), and Feed in Tariff (FIT) exemption schemes to all the industries [13].

In Southeast Asia, Malaysia is one of the fastest-growing countries in terms of economic, social, and land use development [14]. The CO_2 emission rate in Malaysia in 2018 measured 2210.6 ton, while in 2017 it was only 2123.3 ton. In 2016, the CO_2 emission rate measured 2044.1 ton [15]. Many sources of CO_2 have led to increased emissions in Malaysia [16]. Coal power plants function as one of Malaysia's major sources of CO_2 emissions [17]. Expanding tourism development will also increase CO_2 emissions in Malaysia [18]. This is because tourist arrivals in Malaysia will increase CO_2 emissions through transportation services [19]. Industrialisation in Malaysia can also create pollution in the environment, resulting in rising CO_2 emissions [20].

Malaysia's CO_2 emissions are mainly caused by electricity consumption, mobility, and municipal solid waste accumulated in landfills [21]. CO_2 emissions in Malaysia are associated with fossil fuels for the production of commodities and from the household sector's demand [22]. Not only that, but particulate matter 10 (PM_{10}) also exceeds the Malaysian air quality guideline in Petaling Jaya, Gombak, Kelang, Kajang and Kuala Lumpur. This is affecting human health already [23]. The primary contributor sources of PM_{10} in Malaysia include power generation, motor vehicles, and industries [24,25]. Usually, the elderly, children, patients with respiratory problems, heart disease, and allergy patients are the victims of the effects of particulate matter [26,27]. When air pollution rises to a dangerous level, the fatality rate will peak [28]. In order to raise standards and enact pollution control measures, local and national governments increasingly gather cost and benefit information about the level of pollution levels to support them in overcoming the issue [23]. Hence, a WTP study is helpful to estimate the economic benefits of air pollution reduction.

Economic valuation is defined as a measurement of the economic value of the benefit of conservation. It reveals a price for ecosystem services to provide information to decision-makers, and hence, facilitates quantification of the trade involved and help in the decision-making process [29]. Willingness to Pay (WTP) measures the maximum amount of money an individual is willing to pay to increase the quality of an item or service that can be experienced [30].

Few countries' citizens also faced issues about citizen acknowledgement about air pollution. In China, most of the public had little knowledge level concerning the impacts of air pollution on natural resources availability and people's health [31]. Attitudes towards air pollution is low, especially towards climate change and WTP for reducing air pollution [32]. In Germany, consumers lacked knowledge and information about air pollution and the voluntary carbon offset market [33]. The citizens there also experienced low or zero knowledge of voluntary carbon offsetting. Knowledge about carbon offset will influence the demand for voluntary carbon offsets [34]. The resident's or citizen's attitudes may lead to good intentions. However, it must be noted that certain elements such as social norms, lack of knowledge, change of behaviour, and education may act as a barrier in combating air pollution [35].

In Malaysia, many citizens did not acknowledge aspects of air pollution. Many Malaysians did not have any experience or ideas about carbon reduction programmes in combating air pollution [36]. Some Malaysians also did not know about the origin and source of air pollution [37]. They also lacked sufficient knowledge of environmental protection and conservation [38]. The typical citizen attitude towards air pollution is somewhat thoughtless. Only a few Malaysians have experience buying a carbon offset initiative [36]. Most of them were not interested in paying for air quality improvement because they lack environmental awareness [39]. Malaysian citizens also have a negative attitude towards public transportation as they preferred to drive their private vehicles [39]. Overall, knowledge about and attitudes towards air pollution will affect the WTP, which is particularly important [40]. The citizen's understanding of air pollution may influence their attitudes towards the effectiveness of those policies set by The Intergovernmental Panel on Climate Change. The citizen's practices and lifestyle could influence the emission of greenhouse gases [41].

There are scant studies done in Malaysia on the WTP for a carbon tax that involves the residents' greenhouse gas emissions reduction. This is because a large majority of studies focused on the developed country only. No study or research evidence has investigated the issue in Malaysia, or developing countries, on a carbon tax.

Therefore, Klang, Selangor was chosen as the research area due to the air pollution index of Malaysia. Klang holds the highest air pollution index among Kuala Selangor, Petaling Jaya, Shah Alam, and Banting [42]. The statistic by the Department of Statistics Malaysia stated that, throughout the year 2018, Klang holds the highest record in August, which is 227API. The overall result in Selangor is the status index of air quality in Klang, where 271 days is good, 91 days is moderate, and two days is unhealthy [43]. The cargo and container traffic in Port Klang, also known as the busiest port in Malaysia, has many imports and exports, and contributes to this situation [44].

The choice of the area was considered strategic because there are many residents in Klang. As the country grows, the population will also increase. In 2018, there were 1,025,000 people in Klang, of which 552,400 were male, and 472,700 were female, compared to 2017 and 2016, where the population had a total of 1,008,000 and 991,600 individuals, respectively [43]. The annual population growth rate in Klang was 1.6% in 2018, 1.7% in 2017, and 1.8% in 2016 [43]. However, of 1,025,000 individuals, 117,000 are non-citizens [43]. The rise in population will increase the emissions rate due to cascade effects serving people's needs [7]. Pollution, such as fossil fuel use, increases the number of buildings and cars, will cause a rise in greenhouse gas emissions [45].

Hence, this study focuses on estimating the WTP for a carbon tax in Malaysia using the double bound CVM and the open-ended CVM method for comparison purposes. The specific objective of the study is to assess Klang residents' knowledge and level of attitude towards greenhouse gases emission. It is also to analyse the factors influencing the residents' willingness to pay for a carbon tax generally and estimate their willingness to pay for a carbon tax in Klang specifically. It will guide future consideration in determining the carbon tax and hopefully help achieve sustainable development for future generations. Not only that, but this research will also help to achieve the Sustainable Development Goals (SDG) in Malaysia, principally SDG 13: Climate Action, which is mainly focused on reducing GHG emissions [8].

1.2. Literature Gap

Existing studies as shown in Table 1 include general studies on WTP for air quality improvements through greenhouse gases emissions reduction [31,32,39,45–51]. Some studies focused on specific aspects of transportation with a focus on vehicle owners, such as those by Brouwer, Brander, and Van Beukering [52]; Gupta [53]; Rotaris and Danielis [54]; Rizali et al. [30]; Schwirplies et al. [34] and airline services by Jou and Chen [55]; Shaari et al. [36]. Other aspects include companies' carbon emissions trading schemes [56]. Hence, in terms of a literature gap, following Zhang; Wang; Sun and Liu [57], only 12% of the total world publications on carbon tax were from 1991 to 2014. In addition, general studies found on WTP for greenhouse gas emissions reduction are slightly outdated and not getting much attention in ASEAN countries per se, including Malaysia. Therefore, there is a need to conduct a study that focuses on the residents' willingness to pay for a carbon tax to ensure the knowledge on this subject matter is updated from time to time.

Table 1. Summary of existing literature on carbon tax.

Author	Country	Type of Carbon Tax	Method of Study	Factors	WTP (Price per Unit)
Brouwer et al. (2008)	United Kingdom	Emission based	CVM (Open ended, Double bounded)	Nationality, Flying frequency, Awareness, Price ticket, Household income	€25 (RM122.02) per ton CO_2-eq
Carlsson et al. (2010)	Sweden, China, United States	Emission based	CVM (Open ended, Payment card)	Gender, Age, Household size, Education, Income, Religious, Political affiliation	2000 SEK (RM972.88) per year per household
Diederich and Goeschl (2011)	Germany	Emission based	CVM (Single bounded, Payment card)	Cash prize, Gender, Age, Number of children, Education, Personal benefit, Future benefit, Lifestyle, Carbon footprint	€6.30 (RM30.75) per ton of CO_2
Tsang and Burge (2011)	United Kingdom	Emission based	CVM (Iterative bidding)	Level of income, Social-economic background	Between £1.45 (RM7.97) and £2.97 (RM16.33) per year
Blasch (2013)	Switzerland	Emission based	CVM (Single bounded, Payment card)	Age, Gender, Academic level, Monthly gross income, Marital status, Knowledge of offsetting	78 CHF (RM349.35) per tCO_2
Duan et al. (2014)	China	Emission based	CVM (Open ended, Iterative bidding)	Gender, Annual income, Education, Political orientation, Member of environmental organisation, House ownership, Car ownership	CNY201.86 (RM124.29) per year or CNY16.82 (RM10.36) per month for each person
Jou and Chen (2015)	Taiwan	Emission based	CVM (Open ended, Single bounded)	Education level, Annual number of flights, Monthly income, Age, Gender	NT$39.05 (RM5.91) per passenger
Tolunay and Başsüllü (2015)	Turkey	Emission based	CVM (Open ended, Payment card)	Unplanned urbanisation, Residence, Age, Gender, Marital status, Occupation, Number of household members, Income per capita	US$23.52 (RM94.61) per consumer
Gupta (2016)	India	Emission based	CVM (Open ended, Single bounded)	Interest, Environmental activeness, Use of public transport, Quality of public transport, Age, Education, Family size, Individual income	Rs581.5 (RM32) per people
Bazrbachi et al. (2017)	Malaysia	Emission based	CVM (Single bounded)	Gender, Age, Efficiency of public transport, Education level, Health index, Income, Air pollution concern	RM4.99 per trip
Akhtar et al. (2017)	Pakistan	Emission based	CVM (Open ended, Single bounded)	Gender, Age, Education level, Marital status, Number of children, Number of households, Monthly income, Air quality area	US$9.86 (RM39.66) per month or US$118 (RM474.65) per year
Jones et al. (2017)	United States	Emission based	CVM (Open ended, Single bounded)	Age, Education, Gender, Ideology, Income, Attitudinal belief	US$3.66 (RM14.72) per year per household
Kotchen et al. (2017)	United States	Emission based	CVM (Single bounded, Double bounded)	Education, Gender, Household size, Income, Age	US$177 (RM711.98) annually
Rizali et al. (2017)	Indonesia	Emission based	Open ended CVM	Car ownership, Level of education, Car insurance availability	Rp 432.182,70 (RM1225.12) average per year
Schwirplies et al. (2017)	Germany	Emission based	CVM (Iterative bidding)	Level of contribution, Politics, Religious, Age, Gender, Number of children, Education level, Residents, Compensation scheme	€52 (RM253.80) or €53 (RM258.68) per tCO_2e

Table 1. *Cont.*

Author	Country	Type of Carbon Tax	Method of Study	Factors	WTP (Price per Unit)
Nastis and Mattas (2018)	Greece	Emission based	CVM (Open ended, Single bounded)	Age, Education level, Level of income, Household size, Gender	€81 (RM395.34) per household
Zhao et al. (2018)	China	Emission based	CVM (Open ended, Double bounded)	Types of company, Carbon market, Potential, Sector type, Company size, Experience	35 yuan (RM22.81) per tCO_2e
Rotaris and Danielis (2019)	Italy	Emission based	CVM (Single bounded, Double bounded)	Attitudes and belief, Environmental awareness, Political affiliation, Place of residents, Car ownership, Gender, Age, Education, Employment status, Income level	€101 (RM492.95) to €154 (RM751.63) per litre
Shaari et al. (2020)	Malaysia	Emission based	CVM (Open ended, Double bounded)	Bid price, Income, Gender, Age, Education, Job, Offset information, Occupation	RM86.00 per passengers

2. Materials and Methods

2.1. Research Location

The study was conducted in Klang, Selangor, as shown in Figure 1. Klang is part of Klang Valley, which is the primary economic zone in Malaysia [58]. The land area of Klang is 632 km^2 [59]. There are nine districts in Klang Valley which include Gombak, Hulu Langat, Hulu Selangor, Klang, Kuala Langat, Kuala Selangor, Petaling, Sabak Bernam and Sepang [60]. The state legislative area consists of Kota Anggerik, Batu Tiga, Kota Kemuning, Sungai Kandis, Sentosa, Pandamaran, Bandar Baru Klang, Pelabuhan Klang, Selat Klang, Sementa and Meru.

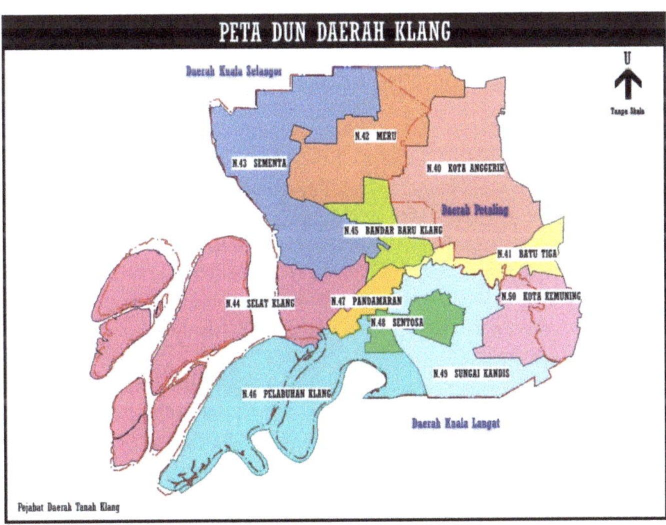

Figure 1. Geographical map of Klang, Selangor. Source: Klang Land and District Office [60].

2.2. Model Specification for Double Bound CVM and Open-Ended CVM

Table 2 shows the variables used in the analysis of WTP.

Double Bound CVM

$$WTP = \beta0 + \beta1 Price + \beta2 Environment\ attitude + \beta3 Gender + \beta4 Age +$$
$$\beta5\ Monthly\ gross\ income + \beta6 Education + \beta7 Household\ size +$$
$$\beta8 Marriage\ status + \mathcal{E}$$

Open-Ended CVM

$$MAX\ WTP = \beta0 + \beta1 Environment\ attitude + \beta2 Gender + \beta3 Age +$$
$$\beta4\ Monthly\ gross\ income + \beta5 Education + \beta6 Household\ size +$$
$$\beta7 Marriage\ status + \mathcal{E}$$

Table 2. Variables-variables used in the analysis are listed below.

	Dependent Variable with 1 If a Respondent Is Willing to Pay for the Amount Asked to Them, 0 Otherwise
	Maximum (MAX) WTP
Initial BID	Bid price levels set out in the CVM question (Dichotomous choice format) RM 5, RM 10, RM 15, RM 20
BID2	Follow-up the bid assigned
Environment attitude	Likert scale
Gender	1 for male, 2 for female
Age	Age of the respondent (years)
Monthly gross income	Income of the respondents (RM/month)
Education	1 for primary, 2 for secondary, 3 for diploma, 4, degree, 5 for master/PhD, 6 for others
Family size	Household size of the respondents (people)
Marital status	1 for single, 2 for married
$\mathcal{E}$	Random error

2.3. Research Design

2.3.1. Data Sources

This study uses primary data and secondary data. For secondary data involved the use of reading sources: libraries, texts, journals, magazines, and reports to assist in collecting data and information to facilitate the process of completing research. For example, the researcher obtained additional information from the Klang Town Planning Department that is unavailable in the written material in libraries or published on a website.

For all the objectives, primary data were collected. This research uses quantitative methods to obtain data. This method only involved the use of questionnaires. The questionnaire was distributed to the targeted respondents. Researchers distributed the questionnaires online using an online platform to the targeted respondents living in the vicinity of Klang, Selangor. Internet surveying is expected to obtain higher response rates and is more affordable than face to face or phone based surveying [61]. Online platforms such as Facebook, Messenger and WhatsApp were used to distribute the questionnaire. Besides that, the chain referral technique was used, where friends and family members who reside at Klang can fill and subsequently refer to their friends, colleagues, and neighbours. Email services were also used to email the questionnaire to the government and non-governmental offices to boost the response rate. Before that, a letter of application of distribution was sent to the authorities to approve distribution to make sure proper permissions were obtained and the flow of data collection was smooth. Finally, with all parties' cooperation, all the data were collected and analysed, supported, and presented.

2.3.2. Questionnaire Design and Structure

The questionnaires were provided in English. The questionnaires consist of closed and open-ended questions.

The questionnaire has five main sections. Section one covers the air quality in Klang, Selangor. Section two is the residents' knowledge of greenhouse gases emission. Section three involved the residents' attitude towards greenhouse gases reduction. Section four covers their willingness to pay. Section five includes respondent demographics.

In the questionnaire, both the Open-ended and Single Bound and Double Bound CVM elicitation techniques were used for the willingness to pay part. For the latter technique, the dichotomous choice format question was used. A double-bounded logit model is more efficient than the single-bounded as the value obtained is deemed to be more reliable about the respondent's willingness to pay [62]. For Double Bound CVM: four different sets were used in which the WTP price is different in terms of the starting bid price for CVM question which set A (RM5), set B (RM10), set C (RM15), and set D (RM20). Based on Bid 1: RM5 (Set A), if the option "yes" is chosen, the WTP amount will be increased by RM5 for each bid. If the option "no" is chosen, then the bid's lower amount will be presented as shown in Table 3. Next, for open-ended CVM, the respondents were asked about the maximum value they are willing to pay for air pollution as shown in Table 3. The respondents were given a scenario about air pollution as shown below:

Table 3. Willingness to Pay (WTP).

D1.Let us say a household must pay RM 5 a year for a carbon tax in Klang, Selangor. Are you willing to pay? ☐ **Yes (please proceed to question D2)** ☐ **No (please proceed to question D3)**
D2. Let us say a household must pay **RM 10 a year** for a carbon tax in Klang, Selangor. Are you willing to pay? ☐ **Yes (please proceed to question D4)** ☐ **No (please proceed to question D4)**
D3. Let us say a household must pay **RM 2.50 a year** for a carbon tax in Klang, Selangor. Are you willing to pay? ☐ **Yes (please proceed to question D4)** ☐ **No (please proceed to question D4)**
D4. What is the maximum amount that you are willing to pay for a carbon tax Klang, Selangor? **(please state)** • **Maximum payment is** RM per year.

Example of scenario:

Klang is well known for being the most polluted city in Selangor state, with the highest index of Air Pollution Index (an air quality measurement) almost every day. Klang also has the third-highest population in Selangor. The increase in population in Klang throughout the years also will result in further impacts on the environment, especially air pollution. With the money collected through a carbon tax, activities such as investment in new sustainable energy, technology (e.g., solar energy, wind energy, energy-efficient cars), awareness program, policy, and many more will be proposed to help in greenhouse gases reduction.

This study will help understand residents' willingness to pay for a carbon tax in Klang, Selangor.

Before answering the following question, think about:

• The amount of willingness to pay is based on the ability to pay once every year.

Based on the scenario above, please mark (√) how much you are willing to pay for a carbon tax in Klang, Selangor.

2.4. Validity and Reliability Analysis

Before distributing the questionnaire to respondents, the questionnaire was validated by five experts in the field.

Validity is an instrument in which an idea is precisely measured in quantitative research [63]. Validity is also an essential term in instrument development [64]. Validity only calculates what it wants to measure. There are three types of validity: criterion validity, content validity, and construct validity [63,64]. For this study, only content validity was used to minimise the potential error for the questionnaire.

Content validity refers to the extent to which the study instrument measures all aspects of the structure accurately [63]. Content validity can help improve the possibility of obtaining the effectiveness of the support structure at a later stage [65].

The researchers selected a total of five panels of experts in economics to measure the instrument. Each panel received a questionnaire to provide their comments and evaluate them. The personnel from the panels conducted a four-point scale questionnaire to rate them by. The point scale used was 1 = not relevant, 2 = somewhat relevant, 3 = relevant, 4 = very relevant. All these scales were used in all sections of the questionnaire for each question. Content Validity Index (CVI) and Aiken's V method was conducted to analyse the point filled by the panel to obtain the accurate result. However, Scale-level-CVI (S-CVI) was calculated using the number of items in the tool that received a "very relevant" rating [66]. It is recommended that scales with excellent content validity be 0.78 or higher for I-CVI while for S-CVI/UA and S-CVI/Ave for 0.8 and 0.9 or higher, respectively [67]. In addition, ideally, a value of 1.00 in I-CVI should be present if there are five or fewer judges, and in the case of six or more judges, I-CVI should not be less than 0.78 [65]. Furthermore, another researcher supported that an S-CVI/Ave value of more than 0.9 has excellent content validity [66]. It is recommended that a minimum S-CVI should be 0.8 for reflecting content validity [65]. Otherwise, the question will have to change or be removed to gain better validity.

Fifty items were identified in the questionnaire. The result shows the average validity for all sections using Aiken's formula gained 0.73 for all sections. However, S1 gained 0.74, followed by S2, S4, and S5, each gaining 0.73. Lastly, S3 gained 0.68 only. Overall, the result shows that content validity is slightly lower than the proposed 0.78 [65]. Therefore, an adjustment towards the questionnaire was made.

After validation, comments, and suggestions were shared, the researcher made the necessary correction. After correction, the final questionnaire was distributed to the respondents.

Next, for reliability, a pilot study was conducted with 40 respondents. For Section 2: knowledge, the reliability was Kuder-Richardson coefficient of reliability (K-R20). The result shows that the value obtained is 0.6347, which ranked strong (0.61–0.79), as shown in Table 4.

Table 4. Kuder-Richardson coefficient of reliability rank.

Reliability Coefficient	Level of Reliability
0.81 or more	Near complete agreement
0.61–0.80	Strong
0.41–0.60	Moderate
0.21–0.40	Fair
0.00–0.20	Poor agreement

Source: Kuder and Richardson [68].

Next, for section three (Residents' attitude towards greenhouse gas emissions), Cronbach's Alpha strength analysis tested the reliability. The result showed that the Cronbach's Alpha values obtained is 0.804 for only 12 questions in total, whereby the level of reliability is good. Therefore, after removing two questions, question number five and eight, the results of Cronbach's Alpha values turn out to be 0.860, suggesting that the level of reliability is very good, as shown in Table 5.

Table 5. Cronbach's Alpha.

Reliability Coefficient	Level of Reliability
0.90 or more	Very good
0.80–0.89	Good
0.60–0.79	Normal
0.40–0.59	Doubted
0.00–0.39	Rejected

Source: Faizal, Lee, Leow, Wei; Pallant [69,70].

2.5. Open-Ended CVM

Mean WTP can be easily affected by the assumed formation of the end of the distributions [71]. Mean WTP will be measured by confirming no negative amount from the respondents for a carbon tax using the equation proposed by Honu [72] as shown in Equation (1).

$$Mean\ WTP = \frac{1}{N}\sum_{i}^{N} WTP_i \tag{1}$$

Source: Honu [72].

2.6. Sampling Technique

The total population in 2018 in Klang, Selangor was 1,025,000 people, and 552,400 were male while 472,700 were female [15]. Thus, after computing using the formula by Yamane [73], as shown in Equation (2), the total number of respondents needed for this research was 400. The study respondents were selected using non-probability sampling, which is an easy sampling method to select respondents. The sampling procedure was taken using convenience sampling, where the respondent was easier to reach. The lack of a sample size frame is suitable for this type of sampling [74].

$$n = \frac{N}{1 + N(e)^2} \tag{2}$$

Source: Yamane [73].

$$n = \frac{1025000}{1 + 1025000(.05)^2}$$
$$= 399.844 \tag{3}$$
$$\approx 400$$

Nonetheless, the study could not obtain the required sample size, which was possible for 311 only. Although the study did not meet the requirement as proposed by Yamane [73], using the G* Power 3.1 application for sample size calculation, the minimum sample size required for a multiple regression analysis such as Poisson regression was estimated at only 89 samples based on (Effect size: f^2) = 0.15, α err prob: 0.05, Power (1-β err prob) = 0.95 and (6) number of predictors. Therefore, 311 respondents were sufficient for analytic purposes.

2.7. Data Analysis

The analysis used in this study was descriptive analysis and inferential statistical analysis. This research used percentage statistics, mean scores, standard deviations, frequency, crosstab, central tendency distribution, and standard deviation for descriptive analysis. The data were retrieved and analysed using descriptive analysis of mean scores using the Statistical Package for Social Sciences (SPSS) software version 25 to obtain the frequency and mean scores and then draw a conclusion to obtain the results. The mean score value was determined based on Lendal's [75] guidance, which interprets the mean score according to the mean score average set. The overall result was used to carry out the answer for objectives one and two. For inferential statistical analysis, this research used Poisson regression for open-ended CVM and logit regression for double bound CVM. The data was prepared and analysed using the STATA software version 15 to identify the coefficient and p-values by looking at the variables. Descriptive analysis was used to answer objective one for knowledge and attitude, covering percentage statistics, frequency, and mean scores. However, inferential statistics analysis was used to answer objectives two and three, involving a regression model. Both analyses were used to answer the objectives and research questions.

3. Results and Discussion

3.1. General Information on Respondent Demographic

Table 6 shows the respondents' demographic data. The result shows that female respondents were 57.6% (179) of the respondents, while male respondents were 42.4% (132). Female respondents are more sensitive and worry about health and environmental problems [36]. This is different from the result from Chang [41]; Diederich and Goeschl [51], whereby male respondents were more common than females, as females said they were not familiar with such a statement and are unwilling to participate in the survey. As for the age factor, the result shows that the highest category is age 18–25 years, which was 35.4% (110) of the respondents. However, the lowest category is the age higher than 56 years at 8% (25) of the respondents. The elderly are unwilling to pay for air quality improvement as it will not benefit them [23]. Age is also one of the variables determining the willingness to pay [36]. The result also shows that 79.1% (246) of the respondents have a degree in education. However, only 0.3% (1) of the respondents choose other qualifications in their qualification level. The higher the respondents' level of education, the higher the willingness to pay for carbon offset [36]. The result also shows that marital status shows few differences, in which marital status was 50.5% (157) of the respondents, and a single status was 49.5% (154) of the respondents. The marital status does not influence the willingness to pay to improve air quality [39].

Table 6. Respondent demographic.

Gender	Frequency ($n = 311$)	Percent
Male	132	42.4
Female	179	57.6
Age		
18–25	110	35.4
26–35	84	27
36–45	56	18
46–55	36	11.6
>56	25	8
Education level		
Primary	5	1.6
Secondary	7	2.3
Diploma	14	4.5
Degree	246	79.1
Master and PhD	38	12.2
Others	1	0.3
Marital status		
Single	154	49.5
Married	157	50.5
Number of individuals within a household (including you)		
1–3 people	59	19
4–6 people	217	69.8
>7 people	35	11.2

Table 6. *Cont.*

Gender	Frequency (*n* = 311)	Percent
Employment status		
Student	76	24.4
Self-employed	44	14.1
Government sector	69	22.2
Private sector	95	30.5
Retired	22	7.1
Others	5	1.6
Monthly gross household income (overall)		
B40 (<RM4360)	190	61.1
M40 (RM4361–RM9619)	75	24.1
T20 (>RM9619)	46	14.8
Monthly gross income		
<RM2000	92	29.6
RM2001–RM3000	59	19
RM3001–RM4000	46	14.8
RM4001–RM5000	31	10
>RM5001	83	26.7

Results also show that the number of households from four to six people has the highest percentage, which was 69.8% (217) of the respondents. In comparison, the number of households with more than seven people have the least respondents, at only 11.2% (35) of the respondents. The increasing number of adults will decrease the willingness to pay for improving air quality [39]. Approximately 30.5% (95) of respondents work in the private sector, while only 1.6% (5) work in the others sector. Most of the respondents who work in the professional sector are willing to pay more than the non-professional sector [23,36]. There are 61.1% (190) of the respondents whose monthly gross income is in the B40 category, while there were 14.8% (46) of the respondents whose monthly gross income is in the T20 category. A respondent with a higher income is willing to pay more since they can afford it even at a higher price [36]. People in a high-income category, have an illness, or are able to witness the depletion of air quality are more likely pay to improve the air quality [39]. As much as 29.6% (92) of the respondents earn a monthly gross income less than RM2000, while only 10% (31) of the respondents earn a monthly gross income from RM4001 to RM5000. This is supported by Fong et al. [7] that the higher the income, the higher the energy used, and emission produced.

3.2. General Information on Air Quality in Klang, Selangor

Table 7 shows a total of 311 respondents from the distribution of the questionnaire. Regarding satisfaction with air quality in Klang, results show that 69.5% (216) of the overall respondents were not satisfied with the air quality. In comparison, 30.5% (95) of the respondents were satisfied with the air quality. This is because Klang is currently undergoing an urbanisation process, which leads to an increase in population. The increase in population will eventually lead to the occurrence of many pollutions, including air pollution. This is supported by Fong et al.; Tolunay and Başsüllü and Chen [7,45,76] who argued that the rapid increase in population will increase the emissions rate.

Table 7. Air quality in Klang, Selangor.

Are you satisfied with the air quality in Klang?	Frequency ($n = 311$)	Percent
Yes	95	30.5
No	216	69.5
Are you concerned about the air pollution in the community where you live?		
Yes	253	81.4
No	58	18.6
How severe would you say is the air pollution in the community where you live?		
Low	49	15.8
Moderate	230	74
High	32	10.3
How would you feel about the quality of air pollution?		
Worried	246	79.1
Not worried	65	20.9
Who do you think should be primarily responsible for the reduction of air pollution?		
Government	11	3.5
Citizen	13	4.2
Industries	41	13.2
Non-Governmental Organisation	9	2.9
All the above	237	76.2
What is your most favourite way to obtain knowledge related to air pollution and related protective measures?		
Television	42	13.5
Internet	227	73
Books	7	2.3
Newspaper	9	2.9
Lecturer	4	1.3
Friends	21	6.8
Others	1	0.3
Are you aware of the greenhouse gases emission reduction measure?		
Yes	236	75.9
No	75	24.1
If you were responsible for designing a plan to address greenhouse gases emission reduction, which of the following technologies would you use? (Multiple responses possible)		
Solar energy	189	13.5
Energy-efficient appliances	120	73
Energy-efficient cars	123	2.3
Wind energy	82	2.9
Nuclear energy	26	1.3
Carbon capture and storage	53	6.8

Note: For the last questions, the respondent may choose more than one answer.

For the concern about air pollution in the respondents' community, results show that 81.4% (253) of respondents were concerned about the air pollution in the community they lived, while only 18.6% (58) respondents were not concerned. This is because air pollution can lead to various illnesses and diseases. This is supported by Gupta [53] that the increase in local air pollution can become a dilemma in health and welfare impacts.

About 74% (230) of the respondents rank their air pollution level in their community to be moderate. This is because the areas in which they reside are not exposed to anthropogenic activities. This is supported by Chang [41] that most of the respondents answered that man-made activities cause pollution. However, 15.8% (49) of respondents ranked their air pollution level in the community as low. This is because their community area might be considerably far from the industrial or town area. Air pollution can contribute to many adverse side effects, especially in human health, agriculture, and industrial production [39]. Only 10.3% (32) of respondents ranked their air pollution level in the community they live in as high. This is because they live near urban areas exposed to pollution, as the city's heart tends to cause air pollution resulting from more anthropogenic activities. This is supported by Rotaris and Danielis [54] that urban communities are more vulnerable to air pollution resulting from transportation and others.

There were 79.1% (246) of respondents worried about air pollution quality, while only 20.9% (65) were not worried about the quality of air pollution. Exposure to significant air pollution can lead to many dangerous diseases that contribute to various health problems. Thus, respondents are very aware of it. This is supported by Krupnick, Rowe, Lang; Cropper, Simon, Alberini, Arora, Sharma [77,78] who argued that the rise in air pollution levels will lift the public concern rate. Individuals that live in highly polluted areas are willing to pay more than the slightly polluted area [23].

Taking responsibility for the reduction in air pollution is essential in combating air pollution. A total of 76.2% (237) of respondents selected "all of the above" that everyone (government, citizen, industries, non-governmental organisation) should be primarily responsible for reducing air pollution. This is because air pollution can be solved with the cooperation of all parties. However, only 2.9% (9) of the respondents selected that non-governmental organisation should be responsible for reducing air pollution. This is because non-governmental organisations such as The Clean Air Forum Society of Malaysia (MYCAS) can help share more information effectively throughout the whole nation. Companies and the authorities should work together to control air pollution [36]. This is supported by Chang [41] that the central and local governments, industrial firms, non-governmental organisations, international organisations, and individuals and families should be responsible for air pollution. This is also supported by Fong et al. [7] that researchers and policymakers oversee reducing air pollution. Research by Akhtar et al. [39] shows that 65.5% of the respondents believe that every citizen should be responsible for pollution.

A total of 73% (227) respondents prefer to obtain knowledge related to air pollution and related protective measures by the Internet; nowadays, getting information online is much more efficient and affordable. Internet surveys tend to obtain higher response rates than face to face and phone surveys [61]. This is supported by Chang [41] that 70% of his respondents obtain awareness through mass media. However, only 0.3% (1) of respondents obtained such knowledge and measure through family members.

For awareness of greenhouse gas emission reduction measures, 75.9% (236) of respondents are aware of it, while only 24.1% (75) respondents are unaware of it. This is because the air pollution issue is prevalent globally. Thus, it became a hot topic in news coverage. This is supported by Ameyaw and Yao [79] that global environmental change is crucial for humans. Among those technologies provided to address greenhouse gas emission reductions, those 236 respondents choose solar energy to be their highest priority at 31.87% (189). This is because solar energy is clean and renewable. The majority of the respondents support the ideology that solar energy is one of the ways to reduce carbon from the atmosphere [61].

Secondly, an energy-efficient car was an option chosen by a total of 20.74% (123) of respondents. This is because energy-efficient cars primarily use electrical energy, which produces less pollution and is better for the environment than fossil fuel-based cars. Eco-friendly transportations can reduce air pollution significantly [7,80]. This is supported by Chang [41] that his respondents can help reduce pollution by bringing down daily transportation. Thirdly, energy-efficient appliances are chosen by 20.23% (120) of respondents. This is because energy-efficient appliances use lesser energy to operate. Thus, it can help reduce electricity bills. This is also supported by Chang [41] who showed that his respondents were willing to pay for green products. Wind energy was selected by 13.82% (82) of respondents supporting it. This is because wind energy can produce electricity that can minimise the use of burning fossil fuels. The local support for wind energy is highly supported as it will decrease the annual electricity costs [81]. Subsequently, carbon capture and storage only accumulate 8.93% (53) of respondent choice. This is because this strategy is too expensive in terms of capturing it. Curry [61] mentioned that a majority of the respondents did not know about carbon capture and storage before. Lastly, nuclear energy has the lowest percentage, with only 4.38% (26) of the respondents. This is because nuclear energy is too dangerous if it is not managed correctly and professionally. Therefore, it is firmly rejected by the respondents. This is supported by Curry [61] that the public is puzzled about using nuclear power plants to solve climate change issues.

3.3. General Information on Residents' Knowledge of Greenhouse Gases Emission

Table 8 shows the residents' knowledge of greenhouse gases emissions. From the result, we can identify that 98.4% (304) of respondents know that global warming is one of the issues most countries face, while only 0.3% (1) of the respondents are unsure about this. Fong et al. [7] also mentioned that global warming is one of the issues most countries face. The majority of the respondents acknowledged global warming and its consequence of increasing temperature [61]. This is also supported by Shah et al. [1] that the release of carbon dioxide gases contributed to global warming worldwide. Next, respondents also know that the costs for carbon sequestration service can ensure that future generations live healthily when most of the respondents, or 69.1% (215), answered yes. This is different from the research by Curry [61] that carbon sequestration terms are unfamiliar to the respondents. Therefore, most of them are unsure about it. However, only 5.5% (17) of respondents do not know about it. In addition, the result shows that 61.4% (191) of the respondents know that political changes will affect climate regulation, while only 13.8% (43) of the respondents are unsure about this statement. This is supported by Rotaris and Danielis [54] that political understanding is one factor affecting environmental policy. Besides that, 37.3% (116) of the respondents know about carbon offset, while 29.3% (91) of the respondents are unsure about carbon offset. This is not supported by Blasch [47] who showed only 17% of the respondents are aware of carbon offset. The same goes for research by Shaari et al. [36] that almost half of the respondents have no idea about carbon offset, and only a few got involved in it.

Furthermore, the result shows that 83.6% (260) of the respondents do not know that temperature has not increased globally. However, only 8% (25) of the respondents were unsure about this statement. This is different from research done by Harris Interactive [82], whereby 74% of the respondents believe that carbon dioxide and other harmful gases can contribute to global warming and finally lead to an increase in temperature. Plus, 83.9% (261) of the respondents know that the emission of the carbon monoxide causes air pollution, while only 4.5% (14) of the respondents are unsure of such a statement. Emission of carbon monoxide from transportation, industrial and power plants released into the atmosphere will result in air pollution [61].

Moreover, the result shows that 93.2% (290) of the respondents know that the emission of waste gases causes air pollution. However, only 1.3% (4) of the respondents do not know about that. Furthermore, 88.4% (275) of the respondents know that investing in energy-saving technology can aid in combating air pollution, while only 3.2% (10) of the respondents do not know about this statement.

Table 8. Residents' knowledge of greenhouse gases emission.

Item	Frequency (*n* = 311)			Ranking
	Yes	No	Do Not Know	
Knowledge 1	306 (98.4)	4 (1.3)	1 (0.3)	1
Knowledge 2	215 (69.1)	17 (5.5)	79 (25.4)	6
Knowledge 3	191 (61.4)	77 (24.8)	43 (13.8)	12
Knowledge 4	116 (37.3)	104 (33.4)	91 (29.3)	8
Knowledge 5	26 (8.4)	260 (83.6)	25 (8)	10
Knowledge 6	261 (83.9)	36 (11.6)	14 (4.5)	5
Knowledge 7	290 (93.2)	4 (1.3)	17 (5.5)	4
Knowledge 8	275 (88.4)	10 (3.2)	26 (8.4)	11
Knowledge 9	290 (93.2)	16 (5.1)	5 (1.6)	2
Knowledge 10	86 (27.7)	209 (67.2)	16 (5.1)	7
Knowledge 11	32 (10.3)	263 (84.6)	16 (5.1)	9
Knowledge 12	286 (92)	18 (5.8)	7 (2.3)	3

Knowledge 1: Global warming is one of the issues faced by most countries. Knowledge 2: Costs for carbon sequestration service can ensure that future generations live in a healthy manner. Knowledge 3: Political changes will affect climate regulation. Knowledge 4: I know about carbon offset. Knowledge 5: The temperature has not increased globally. Knowledge 6: Emission of carbon dioxide causes air pollution. Knowledge 7: Emission of waste gases causes air pollution. Knowledge 8: Investing in energy-saving technology can help in combating air pollution. Knowledge 9: Global climate change is already taking place. Knowledge 10: Global climate change is not happening now, but it will happen in the future. Knowledge 11: Global climate change will not occur at all. Knowledge 12: Humans have caused the temperature to increase. Note: The ranking ranges from 1 to 10, signifying participants' knowledge from most to least knowledgeable.

Results also show that 93.2% (290) of the respondents know that global climate change is already taking place. This is supported by research done by Rotaris and Danielis [54] that most of their samples believe that climate change has occurred. On the other hand, only 1.67% (5) of the respondents do not know about this statement. Besides that, we can identify that 67.2% (209) of the total respondents do not know that global climate change is not happening now. However, considering if it will happen in the foreseeable future, only 5.1% (16) of respondents do not know about this statement. In addition, the result shows that 84.6% (263) of the respondents do not know that global climate change will not occur at all, while only 5.1% (16) of the respondents are unsure about this statement. Lastly, humans have caused the temperature increase; the result shows that 92% (286) of the respondents know about it, while only 2.3% (7) of the respondents are not sure about this statement.

3.4. General Information on Residents' Attitude towards Greenhouse Gases Emission Reduction

Table 9 show residents' attitude towards greenhouse gas emission reduction. First, it clearly shows that every effort towards climate protection is effective because almost all the respondents strongly agree with this statement and obtain the highest percentage, which is 36.7% (114) of respondents. In comparison, only 1.9% (6) of the respondents strongly disagreed with this statement. Next, we can also notice that the residents would also contribute part of their income if they were sure that the money would be used to prevent atmospheric pollution with 36.7% (114) of the respondents agreeing with it. However, there are only 4.5% (14) of the respondent who strongly disagreed with the statement. Respondents would pay more if they were convinced that quick action would prevent pollution [61]. Besides that, for statements educating younger generations about environmental protection is important shows that 72.7% (226) of the respondents strongly agree with it (e.g., encourage carpool). However, only 0.3% (1) of the respondents disagreed with this statement. Promoting environmental education should be implemented during primary school age [36].

Table 9. Shows residents' attitude towards greenhouse gases emission reduction.

Item	Frequency					Mean	Ranking	Score Level
	1	**2**	**3**	**4**	**5**			
Attitude 1	6 (1.9)	16 (5.1)	74 (23.8)	101 (32.5)	114 (36.7)	3.9678	5	High
Attitude 2	14 (4.5)	23 (7.4)	91 (29.3)	114 (36.7)	69 (22.2)	3.6463	7	Medium
Attitude 3	1 (0.3)	0 (0)	8 (2.6)	76 (24.4)	226 (72.7)	4.6913	1	High
Attitude 4	2 (0.6)	9 (2.9)	77 (24.8)	80 (25.7)	143 (46)	4.1350	4	High
Attitude 5	25 (8)	43 (13.8)	128 (41.2)	66 (21.2)	49 (15.8)	3.2283	10	Medium
Attitude 6	9 (2.9)	43 (13.8)	112 (36)	78 (25.1)	69 (22.2)	3.4984	9	Medium
Attitude 7	11 (3.5)	14 (4.5)	89 (28.6)	108 (34.7)	89 (28.6)	3.8039	6	High
Attitude 8	6 (1.9)	51 (16.4)	90 (28.9)	87 (28)	77 (24.8)	3.5723	8	Medium
Attitude 9	0 (0)	3 (1)	44 (14.1)	112 (36)	152 (48.9)	4.3280	2	High
Attitude 10	3 (1)	3 (1)	44 (14.1)	131 (42.1)	130 (41.8)	4.2283	3	High

Attitude 1: Every single effort towards climate protection is effective. Attitude 2: I would contribute part of my income if I were certain that the money would be used to prevent atmospheric pollution. Attitude 3: Educating younger generations about the knowledge of environmental protection (ex. encourage carpool) is important. Attitude 4: Reduction in the use of air-conditioning can be made by me to improve the current atmospheric situation. Attitude 5: Protecting the environment should be given priority, even it might increase the unemployment rate. Attitude 6: I often cut back on driving a car to protect the environment. Attitude 7: Protecting the environment is necessary, even if it will slow down economic growth. Attitude 8: Government must bear the full cost of reducing air pollution. Attitude 9: Citizen is responsible for climate change. Attitude 10: I feel obligated to protect the climate.

Moreover, the result shows that there are 46% (143) of the respondents who know that reducing of the use of air-conditioners can be made with an improvement of the current atmospheric situation, while only 0.6% (2) of the respondents strongly disagreed with this statement. Reducing pollution is a more efficient way than energy conservation to prevent climate change [61]. Furthermore, we can also identify that 41.2% (128) of the respondents feel neutral about the statement that protecting the environment should be given priority even though it might increase the unemployment rate. However, only 8% (25) of the respondents strongly disagreed with this statement. This is supported by Fong et al. [7] who showed it is essential to regulate environmental quality, although it may reduce job opportunity, education, and quality of life. Enhancing the environment quality can offer more job opportunities [83].

Plus, we can know that 36% (112) of the respondents feel neutral about cutting back often on driving a car to protect the environment. Meanwhile, there are only 2.9% (9) of the respondents who strongly disagreed with this statement. This result is supported by Fong et al. [7] that reducing the transportation sector can protect the environment. In addition, the result shows that 34.7% (108) of the respondents agreed that protecting the environment is necessary, even it will slow economic growth, while only 3.5% (11) of the respondents strongly disagreed with that. Air pollution can result in economic loss. Therefore, the public put pressure on the authorities to approach the situation with action [39].

On the other hand, the result shows that 28.9% (90) of the respondents feel neutral about the government's statement that the government must bear the total cost of reducing air pollution. In comparison, only 1.9% (6) of the respondents strongly disagreed with this statement. Respondents believe that they should not bear the cost of reducing air pollution, but the authorities and companies of service should instead [36]. Authorities must put effort into controlling pollution [39].

Furthermore, the result shows that 48.9% (152) of the respondents strongly agreed that the citizens are responsible for climate change, whereas only 1% (3) of the respondents disagreed with such a statement. Ideally, passengers and the public must be well educated about climate change [36]. Lastly, 42.1% (131) of the respondents agreed that they feel obligated to protect the climate. However, only 1% (3) of the respondents strongly disagreed with this statement that they feel obligated to protect the climate. This is supported by Akhtar et al. [39] that most respondents worry about the air quality as it is crucial towards their health.

3.5. General Information on the Willingness to Pay Responses

3.5.1. Single Bound CVM and Double Bound CVM

The variables used from the single bound CVM and double bound CVM are gender, age, number of households, income, education, marital status, knowledge, attitude, and gross income to identify the p-value. However, the result shows that all the variables are insignificant, whereby the WTP estimations using the bid 1 and bid 2 variables is acceptable. Hence, to solve this situation, this study was conducted using the Poisson regression for open-ended CVM, where the dependent variables are the maximum WTP. Poisson regression can allow no random item in the variables, and the variance does not contain an error component [84]. Therefore, most of the p-value results show significance at level 1% and 10%.

Table 10 shows the Poisson regression residents' maximum WTP as the dependent variable. All the results below are based on the collection of 311 respondents from the questionnaire. The result shows that the variable gender is significant at a 1% level. A negative coefficient for gender means that the male is more willing to pay more than the female. This shows that male respondents are more willing to pay for environmental improvement. This result is inconsistent with Safian and Hamzah; Shaari et al. [36,80] that females are more willing to pay for environmental conservation. The result shows that variable age is also significant at the 1% level. A negative coefficient for age means that the younger respondents are more willing to pay for a carbon tax. This is supported by Safian and Hamzah; Lu and Shon [40,80] that younger respondents believed in paying for environmental protection. However, this is different from the research done by Shaari et al. [36] that the older respondents are more willing to pay for a carbon tax to ensure the future generations' environment is conserved. The result shows that variable income is also significant at the 1% level. A positive coefficient for income indicates that the higher the monthly gross income of the respondents, the more they are willing to pay for a carbon tax. This is supported by Fong et al. [7] that a passenger with high earning is willing to pay for carbon prevention to minimise emissions and pollution.

Table 10. Poisson regression.

Variables	Coef.	Std. Err.	z	$p > |z|$	[95% Conf. Interval]	
Gender	−0.0991372	0.0191298	−5.18	0.000 ***	−0.136631	−0.0616434
Age	−0.007413	0.0009105	−8.14	0.000 ***	−0.0091975	−0.0056285
Income	0.0000177	2.30×10^{-6}	7.68	0.000 ***	0.0000132	0.0000222
Education	0.1040986	0.0155919	6.68	0.000 ***	0.0735391	0.1346581
Number of households	−0.046041	0.0060985	−7.55	0.000 ***	−0.0579939	−0.0340881
Attitude	−0.0141071	0.0199381	−0.71	0.479	−0.0531851	0.0249709
Marital status	−0.0376537	0.0213806	−1.76	0.078 *	−0.0795588	0.0042514
_cons	3.840645	0.1167101	32.91	0.000 ***	3.611898	4.069393

Note: *** significant at 1% level of confidence; * Significant at 10% level of confidence.

The result shows that variable education is significant at a 1% level. A positive coefficient for education means that the higher the level of education obtained by the respondents, the more the willingness to pay for a carbon tax. This is because educated respondents know more aspects concerning environmental issues [36]. This is supported by Masud, Al-Amin, Akhtar, Kari, Afroz, Rahman MS, Rahman [85] that education level plays a role in determining the willingness to pay a respondent for a carbon tax. The result shows that a variable number of households is significant at the 1% level. A negative coefficient for the number of households means that when the number of households is lower, the more willing they are to pay a carbon tax. This is because it will cause a burden to a big family, especially low-income families. Akhtar et al. [39] proved that the increasing number of family members decreases the willingness to pay for improved air quality. The result also shows that the attitude of the respondents is not significant. The result shows that the marital status of the respondents is significant at the 10% level. A negative coefficient for marital status means that if the respondents are singles, they are more willing to pay a carbon tax. Tolunay and Başsüllü [45] proved that the singles would be willing to pay for air quality improvement than those who are married, divorced, or widowed.

3.5.2. Estimation of WTP for Open-Ended and Double Bound CVM

The result shows that the mean WTP (based on the open-ended CVM) for this research is RM36.97 per year, while double bound CVM is RM35.65 per year. Comparing with the double bound WTP result, this defines that the amount of WTP is higher by RM1.32 only. The proposed average value is RM36.31 per year. Parts of Malaysians based in Klang, Selangor, are willing to pay more money and contribute part of their income for environmental protection [86].

The mean WTP findings by Safian and Hamzah [80] show that Malaysian consumers are willing to pay about RM6.5 yearly towards environment protection, which is lower than our result. In addition, the mean WTP findings worth a total of US$3.66 (RM14.82) over 20 years are willing to pay by the respondents in Colorado River, United States, to reduce greenhouse gas emissions [50]. Compared to our study, the result is lower than our mean WTP. However, the mean WTP findings from a study by Shaari et al. [36] mentioned that passengers in Malaysia are willing to pay RM86 per trip for a carbon offset, which is higher than our result to reduce the emissions. Lastly, the mean WTP findings from Tolunay and Başsüllü [45] study is higher with a maximum of US$23.52 (RM95.26), with a willingness to pay by the community in Turkey to implement carbon sequestration services.

Overall, the result is acceptable. Therefore, the average maximum WTP for a carbon tax is RM36.31 per year.

3.5.3. Open-Ended CVM

The mean WTP can be calculated by dividing the sum of WTP, starting with WTP_i and ending with WTP_N, with the total number of items of WTP where 1 is the total amount of WTP, N is the total number of respondents, i is the exponential function.

$$Mean\ WTP = \tfrac{1}{311}(11497)$$
$$= 36.97$$

3.5.4. Double Bound CVM

Table 11 shows the double bound CVM. Maximum WTP is the dependent variable in this study. The result shows the maximum willingness to pay by the respondents towards carbon tax based on double-bound contingent valuation methods worth RM35.65 only. The government can levy as much as this amount per year if the carbon tax proposal is proposed.

Table 11. Double bound CVM.

	Coef.	Std. Err.	z	$p > |z|$	[95% Conf. Interval]	
Beta_cons	35.64819	3.2316	11.03	0.000 ***	29.31437	41.98201
Sigma_cons	21.98887	3.072689	7.16	0.000 ***	15.96651	28.01123

Note: *** significant at 1% level of confidence.

3.5.5. Reasons

Figure 2 shows the reasons why people are willing to pay. There are many reasons why people are willing to pay a carbon tax. Based on this research, the result shows that the main reason that people are willing to pay for a carbon tax is that they feel responsible for their contribution to climate change, which shows a total of 37%. This is supported by Shaari et al. [36] that a traveller is presumably willing to pay extra for their airfare to minimise emissions. The second reason that influences them is that they care about the environment in general, which accumulate 32.5%. The younger generations are more responsible for the environment [36]. The third reason, which obtained 14.1% of the respondents, was to avoid future natural disasters. People wish to preserve the forest to ensure that future generations are safe from future disasters [45,51]. The fourth reason, which has a percentage of 7.4% from the respondents, states that the environment has the right to be protected irrespective of the costs. The rest of the reasons, to reduce future economic damage costs and not willing to share their opinion, each obtained 6.8% and 2.3% from the respondents.

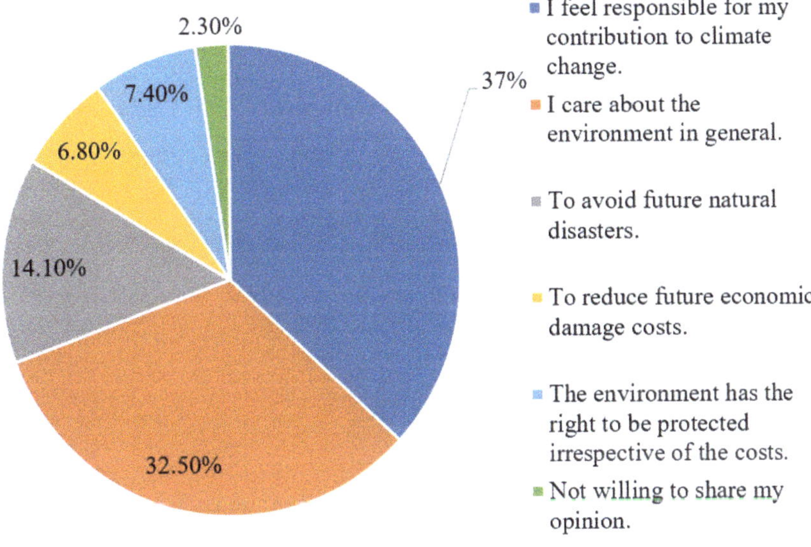

Figure 2. Reasons why people are willing to pay.

Figure 3 shows some reasons why people are not willing to pay. There are many reasons why people are not willing to pay a carbon tax. Based on this research, the result shows two main reasons why people are not willing to pay for a carbon tax, which each accumulates a total of 33.8%. First is their income being too low. This is supported by Chang [41] that the low-income level would become a burden for the respondents, especially when the prices are too high. The second is that they do not believe that such a program would have any real impact. This is supported by Rotaris and Danielis [54] that the failure of government programmes on environmental motive has reduced the public's confidence towards those programmes. However, this is not supported by Blasch [47] that the respondents appreciate the authority's effort on carbon offset. The third reason which obtains a 17.4% from the respondent is that they prefer to spend their money on other things. The fourth reason is that they are unwilling to share their opinions, which shows 7.1% of the respondents. The rest of the reason is that climate change does not affect them or their family, and they feel irresponsible for their contribution to climate change which each obtain a result of 4.2% and 3.9%. Some respondents think that emissions reduction is not their responsibility [36]. Women are more willing to pay and be responsible for their children to ensure their future [36]. This is supported by Chang [41], whereby an estimation of 30% of his respondents said that global warming is not their responsibility, and it is something they cannot control. Research by Shaari et al. [36] mentioned that respondents are also not willing to pay as airfare costs are already high.

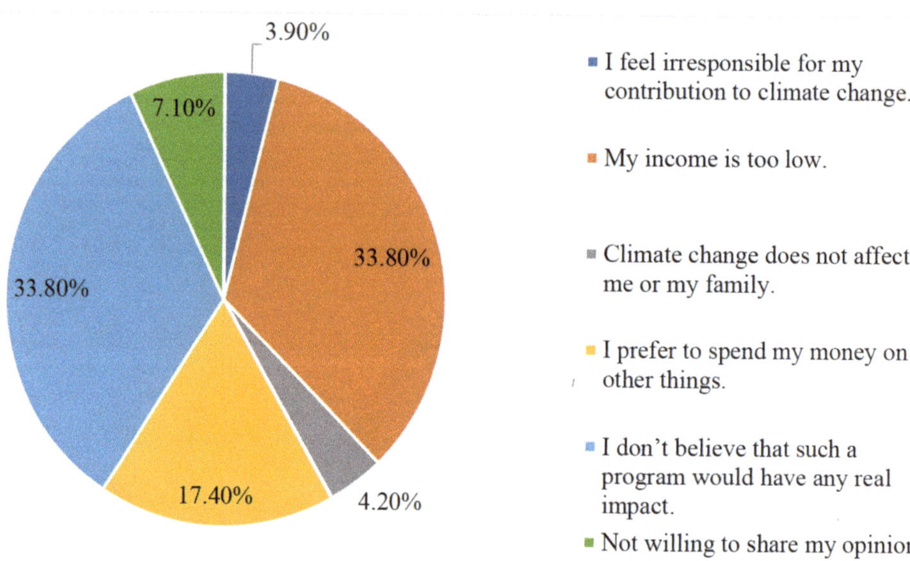

Figure 3. Reasons why people are not willing to pay.

Figure 4 shows the administrator deemed to be appropriate to collect the money obtained from this carbon tax. This research shows that the Ministry of Science, Technology and Innovation has the highest percentage, 61.4% of the respondents. The second administrator hat is appropriate to collect the money obtained from this carbon tax is the Inland Revenue Board Malaysia. This is the second highest which obtain a percentage of 21.2% from the respondents. Next, the result shows that 10.3% of the respondents think that the Ministry of Finance should collect the carbon tax. Followed by a total of 5.8% of the respondents think that other entities should collect the carbon tax. Lastly, the respondents believe that the Royal Malaysian Customs Department should collect it, which only obtains 1.3% of the total percentage.

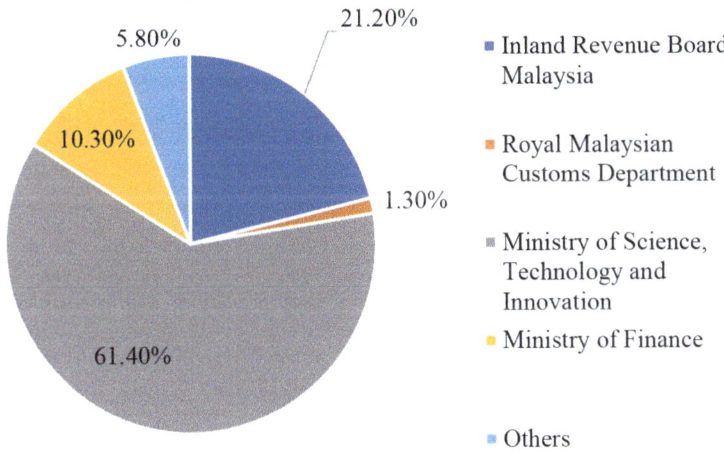

Figure 4. Administrator deemed appropriate to collect the money obtained from this carbon tax (ONE only).

4. Conclusions

This study assessed the knowledge and level of attitude towards greenhouse gases emissions and estimated the willingness to pay for a carbon tax. Most of the respondents were willing to pay because they feel responsible for their contribution to climate change. Most of the respondents are not willing to pay because their income is too low, and they do not believe that such a program would have any real impact. Most of the respondents are not satisfied with the air quality in Klang, and most of the respondents are also concerned and worried about the air pollution in their respective community areas. Moreover, respondents also rank their air pollution level in the community they live in as moderate only. Many of the respondents also think that all the above parties, including the government, citizens, industries, and non-governmental organisations, should be primarily responsible for reducing air pollution. They prefer to obtain their knowledge related to air pollution and related protective measure from the internet. Most of them are also aware of the measures concerning greenhouse gases emission reduction. Moreover, they supported the idea that solar energy is the perfect technology to address greenhouse gas emissions reduction. The result shows that the variables that significantly influenced the maximum WTP were gender, age, income, education, number of households at 1% level, and marital status at 10% level. Overall, the conclusion revealed that the average maximum of residents' WTP for a carbon tax is RM36.31 per year. The CVM was used to estimate the number of residents willing to pay for a carbon tax. The respondents also think that the Ministry of Science, Technology and Innovation should be the one who is appropriate to collect the money from this tax.

As a recommendation, the decision-makers should consider residents' preference on WTP for a carbon tax. This is to make sure that the amount is acceptable by the residents and worth the value. For example, the results derived from the residents' WTP amount from this research can be proposed as a start to the implementation process in Malaysia. Thus, it will be acceptable among Malaysians.

Other than that, the result suggests that governments can conduct awareness campaigns to expose residents to environmental issues and the importance of a carbon tax. By exposing the information about an environmental issue and the importance of carbon tax, residents will get involved in more environmental conservation activities that contribute to the environment more. For example, campaigns can be conducted using various mediums such as newspapers, social media, broadcasts, and talks to reach a diverse audience. Thus, it will be highly informative and accessible to all parties.

Another suggestion is that the current education curriculum can also be reformed. This is to instil and expose younger generations of students or citizens about the importance of the environment. Not only that but additionally, knowledge about the current environmental issues will also be led to their acceptance of carbon taxation and its importance. In the foreseeable future, they are more willing to accept the implementation of a carbon tax. For example, reforming the education curriculum at the primary school stage can provide greater insight and awareness during early age.

Lastly, the government should serve as a role model towards citizens about carbon emission reduction. With the amount of tax collected, subsidies should be allocated to green technologies to promote clean technology. This can help build a solid and promising potential in reducing carbon emission socially and, most importantly, serve as a trial before making it mandatory. For example, the government can reduce or set green technologies to a more affordable price to encourage more citizens to invest in it to ensure a better future generation and life.

In conclusion, this research is limited in scope because it was only conducted in residents around Klang, Selangor. Therefore, it is recommended that future studies should be implemented throughout the Klang Valley or within Malaysia. Thus, all various opinions can be collected. Not only that, but it also required more respondents to gain more accurate results for this study. It is suggested that a larger number can improve the validity and reliability of the research result.

Author Contributions: Conceptualisation, I.Z.G. and N.K.M.; Data curation, I.Z.G. and N.K.M.; Formal analysis, I.Z.G.; Investigation, I.Z.G. and N.K.M.; Methodology, I.Z.G.; Supervision, N.K.M.; Project administration, I.Z.G.; Resources, I.Z.G. and N.K.M.; Software, N.K.M.; Validation, N.K.M.; Writing—Original draft, I.Z.G.; Writing—Review and editing, I.Z.G. and N.K.M. All authors have read and agreed to the published version of the manuscript.

Funding: Funding for the publication fee was provided by the Research Management Centre, Universiti Putra Malaysia, Serdang.

Institutional Review Board Statement: Not applicable.

Informed Consent Statement: Not applicable.

Data Availability Statement: Not applicable.

Acknowledgments: The authors would gratefully thank the support provided by MASPUTRA, Faculty of Forestry and Environment, Universiti Putra Malaysia. The help from the five panels of experts who validated the research instrument is remembered and appreciated. Reviews by two anonymous peers greatly enhanced the final manuscript.

Conflicts of Interest: The authors declare no conflict of interest.

References

1. Shah, R.M.; Yunus, R.M.; Kadhum, A.A.H.; Yin, W.W. Kajian fotomangkin berasaskan grafin untuk penurunan karbon dioksida. *J. Kejuruter. SI* **2018**, *1*, 19–32.
2. Centi, G.; Perathoner, S. Opportunities and prospects in the chemical recycling of carbon dioxide to fuels. *Catal. Today* **2009**, *148*, 191–205. [CrossRef]
3. Jiang, Z.; Xiao, T.; Kuznetsov, V.L.; Edwards, P.P. Turning carbon dioxide into fuel. *Philos. Trans. R. Soc. A Math. Phys. Eng. Sci.* **2010**, *368*, 3343–3364. [CrossRef] [PubMed]
4. Idris, N.; Mahmud, M. Kajian jejak karbon di Kuala Lumpur. *E-Bangi J. Soc. Sci. Humanit.* **2017**, *12*, 165–182.
5. IPCC. Global Warming of 1.5 °C. 20 April 2020. Available online: https://www.ipcc.ch/sr15/ (accessed on 7 May 2020).
6. Liu, W.; Liu, T.; Li, Y.; Liu, M. Recycling carbon tax under different energy efficiency improvements: A CGE analysis of China. *Sustainability* **2021**, *13*, 4804. [CrossRef]
7. Fong, W.K.; Matsumoto, H.; Ho, C.S.; Lun, Y.F. Energy consumption and carbon dioxide emission considerations in the urban planning process in Malaysia. *PLanning Malays.* **2008**, *6*, 30.
8. Ghazouani, A.; Xia, W.; Ben Jebli, M.; Shahzad, U. Exploring the role of carbon taxation policies on CO_2 emissions: Contextual evidence from tax implementation and non-implementation European countries. *Sustainability* **2020**, *12*, 8680. [CrossRef]

9.   IGES. *Urban Energy Use and Greenhouse Gas Emission in Asian Mega Cities: Policies for a Sustainable Future*; Kanagawa; Institute of Global Environmental Strategies: Hayama, Japan, 2004.

10.   Chik, N.A.; Rahim, K.A.; Radam, A.; Shamsudin, M.N. CO_2 emissions induced by household's lifestyle in Malaysia. *Int. J. Bus. Soc.* **2013**, *14*, 344–357.

11.   Rydge, J. Implementing effective carbon pricing. *New Clim. Econ.* **2015**, 1–24. Available online: http://2015.newclimateeconomy.report/wpcontent/uploads/2015/10/Implementing-Effective-Carbon-Pricing.pdf (accessed on 19 May 2020).

12.   The World Bank. State and Trends of Carbon Pricing 2015. 11 March 2015. Available online: http://data.worldbank.org/indicator (accessed on 21 May 2020).

13.   Council of the European Union. Energy Intensive Industries. 2017. Volume 2019. Available online: https://www.consilium.europa.eu/register/nl/content/out/?&DOC_LANCD=EN&typ=ENTRY&i=ADV&DOC_ID=ST-8263-2017-INIT (accessed on 13 May 2020).

14.   Awang, A.; Ali, S.S.S.; Razman, M.R. Tren penggunaan tenaga elektrik dan pembebasan gas rumah hijau di Malaysia. *Asian J. Environ. Hist. Herit.* **2019**, *3*, 39–45.

15.   Department of Statistic Malaysia. Air Pollutant Emission, 2011–2018. 2019. Available online: https://newss.statistics.gov.my/ (accessed on 16 May 2020).

16.   Knoema. Malaysia—Total CO_2 Emissions from Fossil Fuels. 2020. Available online: https://knoema.com/atlas/Malaysia/topics/Environment/CO2-Emissions-from-Fossil-fuel/CO2-emissions-from-fossil-fuels (accessed on 13 May 2020).

17.   OECD. *OECD Economic Surveys Malaysia July 2019*; OECD Publications: Paris, France, 2019.

18.   Ling, C.C.; Leong, C.Z.; Chuen, G.P.; Vern, L.S.; Hong, L.Z. Carbon Dioxide (CO_2) Emissions and GDP by Sectors: A study of Five Selected ASEAN Countries. 2017. Available online: http://eprints.utar.edu.my/2687/1/fyp-BF-1206547-CZL.pdf (accessed on 14 May 2020).

19.   Solarin, S.A. Tourist arrivals and macroeconomic determinants of CO_2 emissions in Malaysia. *Anatolia Int. J. Tour. Hosp. Res.* **2014**, *25*, 228–241.

20.   Begum, R.A.; Sohag, K.; Abdullah, S.M.S.; Jaafar, M. CO_2 emissions, energy consumption, economic and population growth in Malaysia. *Renew. Sustain. Energy Rev.* **2015**, *41*, 594–601. [CrossRef]

21.   Khoo, E. Malaysia Continue Efforts to Reduce Carbon Footprint. 19 September 2019. Available online: https://www.theedgemarkets.com/article/malaysia-continues-efforts-reduce-carbon-footprint (accessed on 18 May 2020).

22.   Othman, J.; Yahoo, M. Reducing CO_2 emissions in Malaysia: Do carbon taxes work? *Pros. PERKEM* **2014**, *9*, 17–19.

23.   Masahina, S.; Afroz, R.; Duasa, J.; Mohamed, N. A framework to estimate the willingness to pay of household for air quality improvement: A case study in Klang Valley, Malaysia. *OIDA Int. J. Sustain. Dev.* **2012**, *4*, 11–16.

24.   Department of Environment. Air Quality. 2020. Available online: https://www.doe.gov.my/portalv1/en/awam/maklumat-alam-sekitar/kualiti-udara (accessed on 20 May 2020).

25.   Noor, N.M.; Yahaya, A.S.; Ramli, N.A.; Luca, F.A.; Al Bakri Abdullah, M.M.; Sandu, A.V. Variation of air pollutant (particulate matter—PM10) in Peninsular Malaysia: Study in the southwest coast of Peninsular Malaysia. *Rev. De Chim.* **2015**, *66*, 1443–1447.

26.   MOH. Laporan. 2020. Available online: https://www2.moh.gov.my/index.php/pages/view/971?mid=55 (accessed on 22 May 2020).

27.   WHO. *World Report on Violence and Health*; World Health Organization: Geneva, Switzerland, 2002.

28.   Peng, X.Z.; Tian, W.H. Study on willingness to pay about air pollution on economic loss in Shanghai. *World Econ. Forum* **2003**, *3*, 32–44.

29.   Salcone, J.; Brander, L.; Seidl, A. *Economic Valuation of Marine and Coastal Ecosystem Services in the Pacific Guidance Manual*; Fernandes, L., Ed.; Report to the MACBIO Project; GIZ, IUCN, SPREP: Suva, Fiji, 2016; Available online: https://www.researchgate.net/publication/319244012_Economic_valuation_of_marine_and_coastal_ecosystem_services_in_the_Pacific_Guidance_manual (accessed on 23 May 2020).

30.   Rizali, R.; Sa'roni, C.; Sopiana, Y.; Muzdalifah, M. Estimasi keinginan membayar (willingness to pay) terhadap udara bersih untuk penentuan pajak emisi (survei terhadap pelanggan bengkel uji emisi di Kota Banjarmasin). *At-Taradhi J. Studi Ekon.* **2017**, *8*, 65. [CrossRef]

31.   Duan, H.X.; Lü, Y.L.; Li, Y. Chinese public's willingness to pay for CO_2 emissions reductions: A case study from four provinces/cities. *Adv. Clim. Chang. Res.* **2014**, *5*, 100–110. [CrossRef]

32.   Carlsson, F.; Kataria, M.; Krupnick, A.; Lampi, E.; Löfgren, Å.; Qin, P.; Chung, S.; Sterner, T. Paying for mitigation: A multiple country study. *Land Econ.* **2012**, *88*, 326–340. [CrossRef]

33.   Dütschke, E.; Schleich, J.; Schwirplies, C.; Ziegler, A. *German Consumers' Willingness to Pay for Carbon Emission Reductions: An Empirical Analysis of Context Dependence and Provider Participation*; EEA: Mannheim, Germany, 2015.

34.   Schwirplies, C.; Dütschke, E.; Schleich, J.; Ziegler, A. *Consumers' Willingness to Offset Their CO_2 Emissions from Traveling*; Working Paper Sustainability and Innovation; EconStor: Kiel, Germany, 2017.

35.   Eluwa, S.E.; Siong, H.C. Willingness to engage in energy conservation and CO_2 emissions reduction: An empirical investigation. *IOP Conf. Ser. Earth Environ. Sci.* **2014**, *18*. [CrossRef]

36.   Shaari, N.F.; Abdul-Rahim, A.S.; Afandi, S.H.M. Are Malaysian airline passengers willing to pay to offset carbon emissions? *Environ. Sci. Pollut. Res.* **2020**, *27*, 24242–24252. [CrossRef]

37.   Fatihah, S.N.; Rahim, A.A. The willingness to pay of air travel passengers to offset their carbon dioxide (CO_2) emissions: A Putrajaya resident case study. *J. Tour. Hosp. Environ. Manag.* **2017**, *2*, 15.

38.   Bazrbachi, A.; Sidique, S.F.; Shamsudin, M.N.; Radam, A.; Kaffashi, S.; Adam, S.U. Willingness to pay to improve air quality: A study of private vehicle owners in Klang Valley, Malaysia. *J. Clean. Prod.* **2017**, *148*, 73–83. [CrossRef]

39.   Akhtar, S.; Saleem, W.; Nadeem, V.M.; Shahid, I.; Ikram, A. Assessment of willingness to pay for improved air quality using contingent valuation method. *Glob. J. Environ. Sci. Manag.* **2017**, *3*, 279–286. [CrossRef]

40. Lu, J.L.; Shon, Z.Y. Exploring airline passengers' willingness to pay for carbon offsets. *Transp. Res. Part D Transp. Env.* **2012**, *17*, 124–128. [CrossRef]

41. Chang, G. Rural residents' understanding and willingness to pay higher prices for mitigation against global warming in China. *Int. J. Clim. Chang. Strateg. Manag.* **2018**, *10*, 711–728. [CrossRef]

42. APIMS. Air Pollutant Index of Malaysia. 2020. Available online: http://apims.doe.gov.my/public_v2/api_table.html (accessed on 27 May 2020).

43. Department of Statistic Malaysia. *Malaysia Statistical Handbook 2019*; Pusat Pentadbiran Kerajaan Persekutuan; Department of Statistic Malaysia: Putrajaya, Malaysia, 2019.

44. Ministry of Transport Malaysia. 2019. Available online: https://www.mot.gov.my/en/browse-by-topic/reports-and-stats (accessed on 16 May 2020).

45. Tolunay, A.; Başsüllü, Ç. Willingness to pay for carbon sequestration and co-benefits of forests in Turkey. *Sustainability* **2015**, *7*, 3311–3337. [CrossRef]

46. Nastis, S.A.; Mattas, K. Income elasticity of willingness-to-pay for a carbon tax in Greece. *Int. J. Glob. Warm.* **2018**, *14*, 510–524. [CrossRef]

47. Blasch, J. Consumers' valuation of voluntary carbon offsets—A choice experiment in Switzerland. *J. Environ. Econ. Policy* **2013**, 1–27. Available online: https://www.unine.ch/files/live/sites/irene/files/shared/documents/SSES/Blasch.pdf (accessed on 23 June 2020).

48. Tsang, F.; Burge, P. *Paying for Carbon Emissions Reduction*; Rand Corporation: Santa Monica, CA, USA, 2011.

49. Kotchen, M.J.; Turk, Z.M.; Leiserowitz, A.A. Public willingness to pay for a US carbon tax and preferences for spending the revenue. *Environ. Res. Lett.* **2017**, *12*. [CrossRef]

50. Jones, B.A.; Ripberger, J.; Jenkins-Smith, H.; Silva, C. Estimating willingness to pay for greenhouse gas emission reductions provided by hydropower using the contingent valuation method. *Energy Policy* **2017**, *111*, 362–370. [CrossRef]

51. Diederich, J.; Goeschl, T. Willingness to pay for individual greenhouse gas emissions reductions: Evidence from a large field experiment. *Environ. Resour. Econ.* **2011**, *57*, 517.

52. Brouwer, R.; Brander, L.; Van Beukering, P. "A convenient truth": Air travel passengers' willingness to pay to offset their CO_2 emissions. *Clim. Chang.* **2008**, *90*, 299–313. [CrossRef]

53. Gupta, M. Willingness to pay for carbon tax: A study of Indian road passenger transport. *Transp. Policy* **2016**, *45*, 46–54. [CrossRef]

54. Rotaris, L.; Danielis, R. The willingness to pay for a carbon tax in Italy. *Transp. Res. Part D Transp. Environ.* **2019**, *67*, 659–673. [CrossRef]

55. Jou, R.C.; Chen, T.Y. Willingness to pay of air passengers for carbon-offset. *Sustainability* **2015**, *7*, 3071–3085. [CrossRef]

56. Zhao, Y.; Wang, C.; Sun, Y.; Liu, X. Factors influencing companies' willingness to pay for carbon emissions: Emission trading schemes in China. *Energy Econ.* **2018**, *75*, 357–367. [CrossRef]

57. Zhang, K.; Wang, Q.; Liang, Q.; Chen, H. A bibliometric analysis of research on carbon tax from 1989 to 2014. *Sust. Energy Rev.* **2016**, *58*, 297–310. [CrossRef]

58. Abdullah, A.M.; Abu Samah, M.A.; Jun, T.Y. An overview of the air pollution trend in Klang Valley, Malaysia. *Open Environ. Sci.* **2012**, *6*, 13–19. [CrossRef]

59. Department of Survey and Mapping Malaysia. Station: Pelabuhan Klang, Selangor. 2019. Available online: https://www.jupem.gov.my/page/station-pelabuhan-klang-selangor-1 (accessed on 1 July 2020).

60. Klang Land and District Office Peta DUN Daerah Klang. 10 September 2019; 2021. Available online: https://www.selangor.gov.my/klang.php/pages/view/98 (accessed on 1 July 2020).

61. Curry, T.E. Public Awareness of Carbon Capture and Storage: A Survey of Attitudes toward Climate Change Mitigation. 2004. Available online: http://sequestration.mit.edu/pdf/Tom_Curry_Thesis_June2004.pdf (accessed on 9 July 2020).

62. Hanemann, M.; Loomis, J.; Kanninen, B. Statistical efficiency of double-bounded dichotomous choice contingent valuation. *Am. J. Agric. Econ.* **1991**, *73*, 1255–1263. [CrossRef]

63. Heale, R.; Twycross, A. Validity and reliability in quantitative studies. *Evid.-Based Nurs.* **2015**, *18*, 66–67. [CrossRef]

64. Retnawati, H. Proving content validity of self-regulated learning scale (The comparison of Aiken index and expanded Gregory index). *Router A J. Cult. Stud.* **2016**, *22*, 123–146. Available online: http://routerjcs.nctu.edu.tw/router/word/11542212017.pdf (accessed on 12 July 2020). [CrossRef]

65. Shrotryia, V.K.; Dhanda, U. Content validity of assessment instrument for employee engagement. *SAGE Open* **2019**, *9*. [CrossRef]

66. Rodrigues, I.B.; Adachi, J.D.; Beattie, K.A.; MacDermid, J.C. Development and validation of a new tool to measure the facilitators, barriers and preferences to exercise in people with osteoporosis. *BMC Musculoskelet. Disord.* **2017**, *18*, 1–9. [CrossRef] [PubMed]

67. Shi, J.; Mo, X.; Sun, Z. Content validity index in scale development. *J. Cent. South Univ. Med Sci.* **2012**, *37*, 152–155. [CrossRef]

68. Kuder, G.F.; Richardson, M.W. The theory of the estimation of test reliability. *Psychometrika* **1937**, *2*, 151–160. [CrossRef]

69. Faizal, M.; Lee, N.; Leow, A.; Wei, T. Kesahan dan kebolehpercayaan instrumen penilaian kendiri pembelajaran geometri tingkatan satu (Learning form one geometry: Validity and reliability of a self-evaluation instrument). *Malays. J. Learn. Instr.* **2017**, *14*, 211–265.

70. Pallant, J. *SPSS Survival Manual: A Step by Step Guide to Data Analysis Using SPSS*, 4th ed.; Open University Press: Buckingham, London, UK, 2005; Volume 181.

71. Li, H.; Jenkins-Smith, H.C.; Silva, C.L.; Berrens, R.P.; Herron, K.G. Public support for reducing US reliance on fossil fuels: Investigating household willingness-to-pay for energy research and development. *Ecol. Econ.* **2009**, *68*, 731–742. [CrossRef]

72. Honu, B. Contingent Valuation Method for General Practitioners: A Cookbook Approach. *Inst. Dev. Stud.* **2007**, *11*, 83–96.

73. Yamane, T. *Statistics, An Introductory Analysis*; Harper & Row: New York, NY, USA, 1970.

74. Yeo, H.Y.; Shafie, A.A. The acceptance and willingness to pay (WTP) for hypothetical dengue vaccine in Penang, Malaysia: A contingent valuation study. *Cost Eff. Resour. Alloc.* **2018**, *16*, 1–10. [CrossRef] [PubMed]
75. Lendal, K. *Management by Menu London*; Wiley and Son Inc.: Hoboken, NJ, USA, 1997.
76. Chen, S. The Urbanisation Impacts on the Policy Effects of the Carbon Tax in China. *Sustainability* **2021**, *13*, 6749. [CrossRef]
77. Krupnick, A.J.; Rowe, R.D.; Lang, C.M. Transportation and air pollution: The environmental damages. In *The Full Costs and Benefits of Transportation*; Springer: Berlin/Heidelberg, Germany, 1997; Volume 1, pp. 337–369.
78. Cropper, M.N.; Simon, N.B.; Alberini, A.; Arora, S.; Sharma, P.K. The health benefits of air pollution control in Delhi. *Am. J. Agric. Econ.* **1997**, *79*, 1625–1629. [CrossRef]
79. Ameyaw, B.; Yao, L. Analysing the impact of GDP on CO_2 emissions and forecasting Africa's total CO_2 emissions with non-assumption driven bidirectional long short-term memory. *Sustainability* **2018**, *10*, 3110. [CrossRef]
80. Safian, S.S.S.; Hamzah, H.Z. Consumers' willingness to pay (WTP) of special tax for non-green vehicles towards environmental performance in Malaysia. *J. Tour. Hosp. Environ. Manag.* **2019**, 66–77. [CrossRef]
81. Deutch, J.; Moniz, E.; Ansolabehere, S.; Driscoll, M.; Gray, P.E.; Holdren, J.P.; Joskow, P.L.; Lester, R.K.; Todreas, N.E. *The Future of Nuclear Power: An Interdisciplinary MIT Study*; MIT: Cambridge, MA, USA, 2003.
82. Harris Interactive. Majorities Continue to Believe in Global Warming and Support Kyoto Treaty Harris Interactive. 2002. Available online: http://www.harrisinteractive.com/harris_poll/printerfriend/index.asp?PID=335 (accessed on 26 October 2020).
83. Baranzini, A.; Goldemberg, J.; Speck, S. A future for carbon taxes. *Ecol. Econ.* **2000**, *32*, 395–412. [CrossRef]
84. Khoiriyah, S.; Suam Toro, M.J. Faktor-Faktor Yang Mempengaruhi Kesediaan Membeli Produk Hijau. *J. Bisnis Manaj.* **2014**, *1*, 63–76.
85. Masud, M.M.; Al-Amin, A.Q.; Akhtar, R.; Kari, F.; Afroz, R.; Rahman, M.S.; Rahman, M. Valuing climate protection by offsetting carbon emissions: Rethinking environmental governance. *J. Clean. Prod.* **2015**, *89*, 41–49. [CrossRef]
86. Chin, Y.S.J.; De Pretto, L.; Thuppil, V.; Ashfold, M.J. Public awareness and support for environmental protection-A focus on air pollution in peninsular Malaysia. *PLoS ONE* **2019**, *14*, e0212206. [CrossRef] [PubMed]

 *sustainability*

Article

How Would We Cycle Today If We Had the Weather of Tomorrow? An Analysis of the Impact of Climate Change on Bicycle Traffic

Anton Galich [1,*], Simon Nieland [1,*], Barbara Lenz [1] and Jan Blechschmidt [2]

[1] Institute of Transport Research, German Aerospace Center, Rudower Chaussee 7, 12489 Berlin, Germany; barbara.lenz@dlr.de

[2] INWT Statistics GmbH, Hauptstrasse 8, 10827 Berlin, Germany; j.blechschmidt@yahoo.de

* Correspondence: anton.galich@dlr.de (A.G.); simon.nieland@dlr.de (S.N.); Tel.: +49-30-6705-5507 (A.G.); +49-30-670-559-109 (S.N.)

Abstract: Bicycle usage is significantly affected by weather conditions. Climate change is, therefore, expected to have an impact on the volume of bicycle traffic, which is an important factor in the planning and design of bicycle infrastructures. To predict bicycle traffic in a changed climate in the city of Berlin, this paper compares a traditional statistical approach to three machine learning models. For this purpose, a cross-validation procedure is developed that evaluates model performance on the basis of prediction accuracy. XGBoost showed the best performance and is used for the prediction of bicycle counts. Our results indicate that we can expect an overall annual increase in bicycle traffic of 1–4% in the city of Berlin due to the changes in local weather conditions caused by global climate change. The biggest changes are expected to occur in the winter season with increases of 11–14% due to rising temperatures and only slight increases in precipitation.

Keywords: bicycle traffic; time series; weather; climate change; machine learning

Citation: Galich, A.; Nieland, S.; Lenz, B.; Blechschmidt, J. How Would We Cycle Today If We Had the Weather of Tomorrow? An Analysis of the Impact of Climate Change on Bicycle Traffic. *Sustainability* **2021**, *13*, 10254. https://doi.org/10.3390/su131810254

Academic Editors: Baojie He, Ayyoob Sharifi, Chi Feng and Jun Yang

Received: 12 August 2021
Accepted: 10 September 2021
Published: 14 September 2021

Publisher's Note: MDPI stays neutral with regard to jurisdictional claims in published maps and institutional affiliations.

1. Introduction

Cities around the world are facing challenges due to climate change and are increasingly required to formulate and implement adaptation strategies for changed climate conditions [1]. In transportation planning, this is particularly important for trips that are greatly affected by weather conditions, such as cycling trips. Many cities promote cycling as a mode of transport as it does not emit carbon or any other harmful air pollutants and contributes to a healthier lifestyle by increasing the amount of daily physical activity [2]. Therefore, planning future bicycle infrastructure will benefit from understanding the variations in bicycle usage brought about by the characteristics and the extent of climate change.

However, the scholarly literature on the impacts of climate change on bicycle usage is rather thin. Indeed, it is well understood how cultural, infrastructural, economic, and sociodemographic factors affect bicycle usage [3–9]. Furthermore, many studies looked into the impact of weather conditions on cycling rates and found significant effects of temperature [10–14], precipitation [9,15–17], wind speed [18–20], snow [21,22], and humidity [6]. Rising temperatures generally tend to increase bicycle traffic, while adverse weather conditions such as precipitation, snow, and high humidity usually have a negative impact on bicycle usage. Ref. [23] provides a more comprehensive review of the weather conditions whose impact on bicycle traffic was analysed.

In contrast, only two studies explicitly address the potential impact of climate change on bicycle usage with different approaches. Ref. [24] compared people's mobility behaviour in years that weather-wise resemble the conditions that are expected to be brought about by climate change to that in years which resemble average seasonal weather conditions at the

213

moment in order to highlight the differences. Ref. [25] trained a negative binomial model on bicycle count station data and current weather conditions and applied it to predict future cycling rates on the basis of future weather conditions generated by climate models.

Against this background, this study aims at enriching the scholarly literature by using machine learning prediction algorithms and data from regional climate models to analyse the potential impact of climate change on bicycle usage. For this purpose, we focus on the changes in air temperature, precipitation, and wind speed as predicted by regional climate models, while keeping constant all other relevant factors such as infrastructure, economic development, etc. In this way, it is analysed how we would cycle today if we had the weather of tomorrow.

Hence, we follow the conceptual framework put forward by [25] but attempt to improve its accuracy by using machine learning models, which are known for their excellent prediction accuracy [26,27]. We first train four different time series models (seasonal autoregressive integrated moving average with exogenous factors (SARIMAX), Facebook Prophet (Prophet), XGBoost, long short-term memory neural network (LSTM)) on local weather and bicycle count data for the city of Berlin. Hence, we compare the traditional statistical approach SARIMAX to three are machine learning models suitable for analysing time series data. In a second step, we select the model with the highest prediction accuracy (XGBoost) and apply it to predict bicycle traffic based on the altered weather conditions that we expect in the future due to climate change.

The study is organised as follows. Section 2 presents the database and highlights necessary pre-processing and data fusion. Section 3 introduces the modelling approach. Section 4 presents the results, which are further discussed in Section 5, before conclusions for practitioners are drawn in Section 6.

2. Materials

Bicycle counts, weather, and climate data constitute the primary database for our study, which is described in more detail in the following two subsections. First, we explain how the data from the bicycle count stations in Berlin were prepared and aggregated to the daily level for further analysis. Thereafter, we provide a brief overview of the sources of weather and climate data used in this paper. In particular, we explain how the projections of future weather conditions for regional climate models are used in combination with the local weather conditions measured in Berlin in 2017, 2018, and 2019 to generate three scenarios of future weather conditions that we expect to be brought about by climate change.

2.1. Bicycle Count Data

Bicycles constitute an important means of transport in Berlin and account for 15% of the modal split of all trips conducted on a typical day [28]. For a more precise picture of its bicycle traffic, the city of Berlin started to install automatic bicycle count stations in 2012 [29]. Altogether, 26 bicycle count devices were installed throughout the city by 2020. Each device automatically detects the bicycles passing by and reports the aggregated number of cyclists per hour.

After data preparation and the exclusion of count stations with too many missing values (see Appendix A for more details), 17 bicycle count devices remained. Figure 1 shows the locations of the remaining bicycle count devices.

The count stations cover the whole area of the city including central, northern, eastern, southern, and western parts. Furthermore, most of the count stations are located in areas in which large shares of the population of Berlin live (see also Figure A2 in the Appendix A). Therefore, we assume that they represent the entire volume of bicycle traffic in Berlin quite well so that daily, weekly, monthly, seasonal, or yearly variations in the actual volume of bicycle traffic are also present in our data. Our rather conservative approach to data preparation gave rise to a data set with consistent patterns across the different count stations.

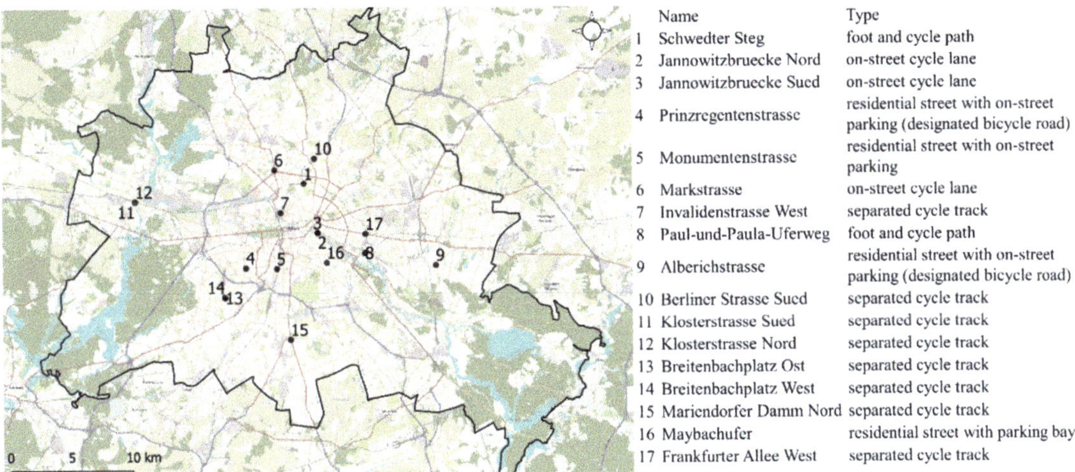

Figure 1. Locations of bicycle count stations in Berlin and corresponding infrastructure type.

After data preparation, various statistical tests were performed to check the stationarity of the time series and to analyse whether the bicycle counts at the different count stations follow similar seasonal patterns and react to weather conditions in a similar manner. In particular, we looked at the mean and variance of the different count stations, computed correlation matrices and correlation coefficients, and conducted the Augmented Dickey–Fuller Test and the Kwiatkowski–Philipps–Schmidt–Shin Test. The details of this procedure are described in Appendix A.

All of these analyses have shown that our data can be regarded as stationary. In general, stationarity is an important concept in time series analysis and denotes that the statistical properties of a time series such as the mean and the variance do not change over time. Stationarity also implies that there is no trend in the time series data such as a linear growth trend, for example, which leads to increasing values in the data from year to year. In our case, stationarity implies among other things, that the volume of bicycle traffic measured at each count station reacts in a very similar way to changes in weather conditions over the entire period of time covered by our data.

The bicycle counts from these remaining devices were aggregated to a daily level for two reasons. First, we are interested in predicting the impact of climate change on the total volume of bicycle traffic in Berlin and not on the number of bicycle counts at individual stations. Second, the climate change data are generally provided on that level. There are procedures to narrow the data down to the hourly level but these would increase the uncertainty considerably.

Our data show a mean of 34,452 bicycle rides per day with a standard deviation of 479. The maximum number of bicycle rides was recorded on 21 June 2017 at 71,367, while the minimum was detected on 8 January 2017 at 2864. The total number of bicycle counts per year varied from 11.7 million in 2017 to 13.1 million in 2018 and 13.0 million in 2019. Figure A1 in the Appendix A provides a graphical illustration of our data with plots of the daily bicycle counts at each station as well as the aggregated number of counts over all stations.

2.2. Using Outputs from Regional Climate Models to Generate Future Weather Scenarios

In order to predict the impact of climate change on the volume of bicycle traffic, we generate changed local weather conditions for the year 2050 according to the three different Representative Concentration Pathways of the IPCC: RCP2.6, RCP4.5, and RCP8.5. RCPs are trajectories of greenhouse gas concentration in the atmosphere describing different climatic futures adapted by the IPCC.

These emission scenarios for climate change should not be seen as forecasts or predictions of future climate conditions but rather as expert judgements of plausible future emissions based on socioeconomical, environmental, and technological developments incorporated in integrated assessment models. Since the future evolution of anthropogenic factors cannot be known in advance, the possible effects are presented in different scenarios describing several possible emission pathways [30].

RCP2.6 (meaning the radiative forcing is 2.6 Watts per square metre (W/m^2) in 2100) shows a peak in greenhouse emission concentration around 2040 and then declines, RCP4.6 stabilises at ~4.5 (W/m^2) after 2100, and RCP8.5 increases constantly even after 2100. Therefore, RCP8.5 can be regarded as a "business as usual" scenario in which no climate protection actions are realised, even after 2100. RCP2.6 represents a "moderate" emission scenario and RCP2.6 can be seen as a "climate protection" scenario. For more detailed information on RCPs and their societal and environmental implications, see [30].

This study uses outputs of a regional climate model (domain: EUR-11, driving model name: MPI-ESM-LR, realisation: r1i1p1, frequency: day) developed and calculated by the EURO-CORDEX initiative, which aims at downscaling global climate projections to a regional scale (12 km ground resolution) for the European continent [31–33]. In fact, the EURO-CORDEX initiative ran many different simulations with various regional climate models downscaling the output of 10 different global climate models. In order to decrease uncertainty and to avoid biases due to the features of a particular model, this study uses the output of one of those models whose simulations of maximum air temperature, sum of precipitation, and mean wind speed most often fell into the range of median values of the outputs of the different models for the three RCPs. To produce realistic climatic scenarios, the characteristics of their weather conditions need to be consistent with the actual weather in Berlin. Realistic refers to yearly distributions of weather conditions in terms of their mean, variance, and range as well as the number of extreme weather events such as hot days ($\geq$30 °C) or days with heavy rain ($\geq$10 mm). To do this, we adapt the output of regional climate models that predict the expected climatic changes to local weather conditions that were measured in Berlin in 2017, 2018, and 2019.

Therefore, we carry out the following data preparation steps: First, we link daily measurements from a local weather station to the bicycle counts to generate the training basis for the forecast model (see Section 3.2). Second, we calculate the absolute changes in weather conditions between historical and forecast periods from a regional climate model to extract the expected impact of climate change. Third, we adapt these expected changes to the weather conditions measured in Berlin in 2017, 2018, and 2019 in order to generate realistic local weather conditions for 2050. The details of this procedure are described in Appendix B.

The results of the data processing are synthetic distributions of the weather variables for each day of the year, reflecting the expected changes in weather from the future climate scenarios of RCP2.6, RCP4.5, and RCP8.5. These synthetic distributions will be fed into our time series model to predict the volume of bicycle traffic in 2020 under the changed weather conditions that we expect climate change to bring about. Figure 2 illustrates the results of our adaptation procedure for the RCP8.5 scenario and outlines the synthetic future values generated in relation to the weather station's average daily maximum air temperature, sum of precipitation, and mean wind speed values in 2017, 2018, and 2019.

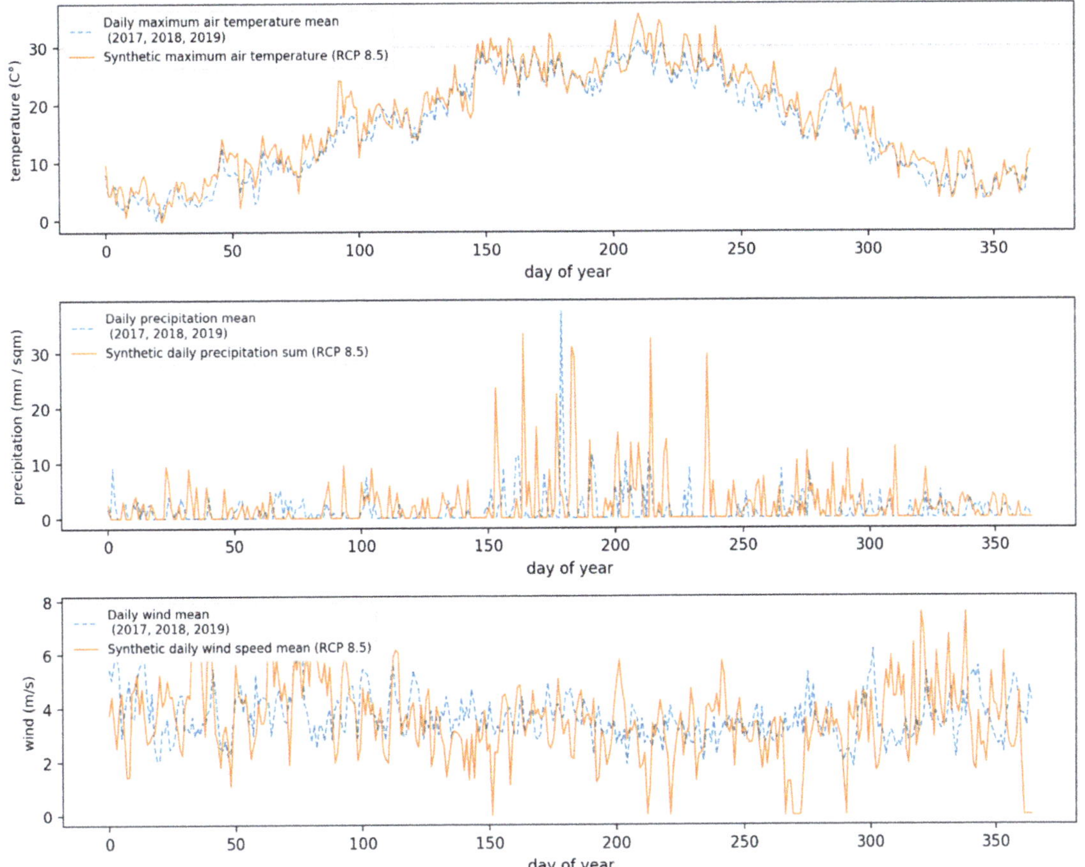

Figure 2. Yearly distributions of the synthetically generated future weather conditions and averaged weather station data from 2017, 2018, and 2019. The dashed blue line shows the reference value of the respective weather variable (top: maximum air temperature averaged over 2017, 2018, and 2019; centre: absolute precipitation averaged over 2017, 2018, and 2019; bottom: absolute wind speed averaged over 2017, 2018, and 2019). The orange line illustrates the yearly distributions of the synthetically generated weather variables based on the data from the RCP8.5 scenario. The RCP2.6 and RCP4.5 scenarios are excluded from the figure for illustrative purposes.

As can be seen, our adaptation procedure resulted in realistic yearly distributions of the three weather variables: maximum air temperature, sum of precipitation, and mean wind speed (see also the Section 4 for a more quantitative comparison). Realistic in this case means two things. First, the data meet the expected changes to be brought about by the different climate change scenarios in the number of extreme weather events as defined by the German Weather Service [34,35], such as hot days ($\geq$30 °C) or days with heavy rain ($\geq$10 mm) (see [33]). Second, the mean, the variation, and the range of the daily values of the three weather variables also reflect the expected changes to be brought about by the different climate change scenarios [33]. Hence, our adaptation procedure resulted in three different yearly distributions for 2020 that realistically portray the mean changes in local weather conditions in Berlin that can be expected until 2050 on the basis of RCP2.6, RCP4.5, and RCP8.5.

3. Methods

In this section, we explain our methodological framework in more detail. For this purpose, we first illustrate the general research design of this study. Subsequently, the four time series models that were chosen for comparison are described before we illustrate the cross-validation procedure developed for model selection. Finally, it is outlined what independent variables were used (see Table 1) and how the different models were tuned on the basis of the cross-validation procedure.

3.1. Research Design

Figure 3 illustrates the research design of this study including model selection, data preparation, and prediction.

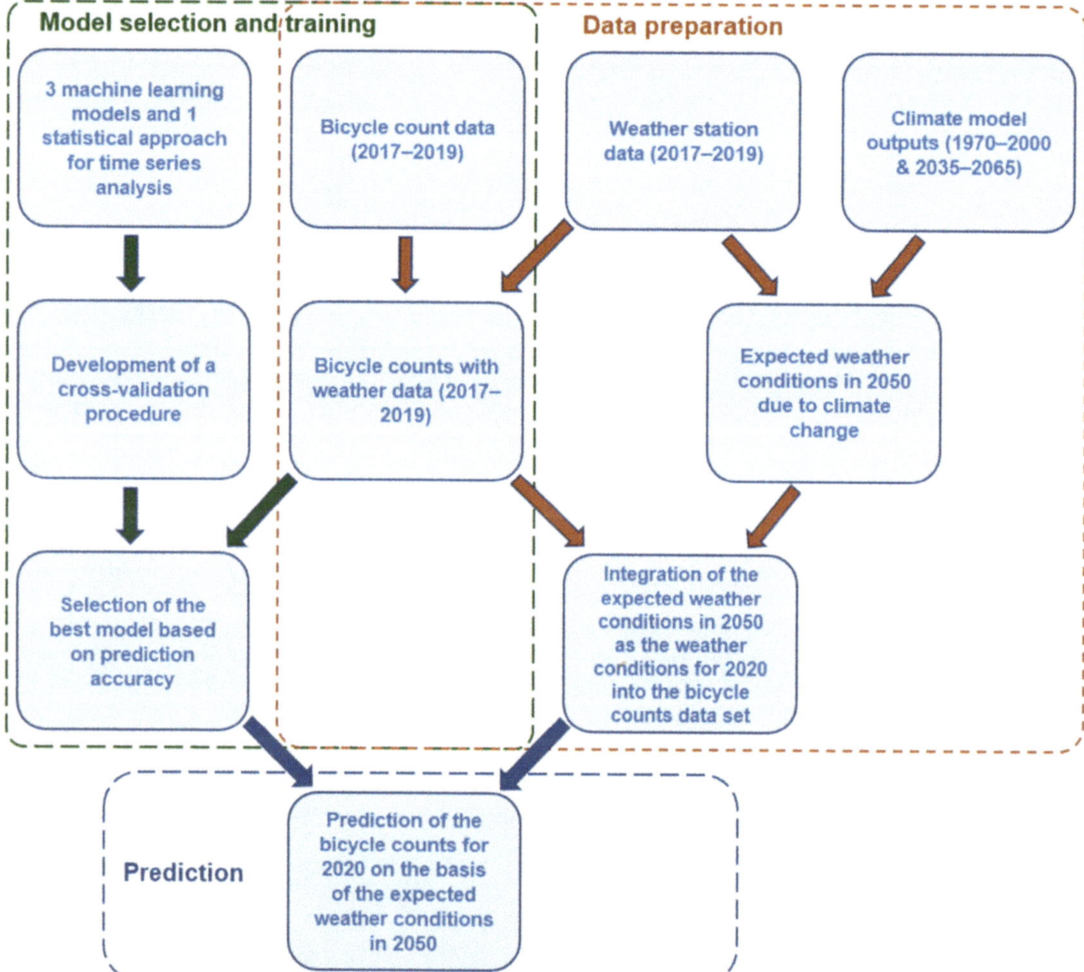

Figure 3. Research design. Green arrows indicate processes that refer to model selection and training, while red arrows denote data preparation steps and blue arrows refer to prediction processes.

We chose to compare three machine learning models to one traditional statistical approach for time series analysis which can be seen as a sort of benchmark. For this comparison, a cross-validation procedure was developed based on the bicycle count data of 2017–2019 which were enriched with the measured weather conditions for the same period. The cross-validation procedure was used to tune the four models and to select the best model on the basis of prediction accuracy.

The trained algorithm is then used to predict bicycle counts for a future year using synthetically generated weather conditions. As we have measured weather data and bicycle counts for 2017, 2018, and 2019, we predict the bicycle traffic volume of 2020 under the changed climatic conditions of 2050 considering the three climate change scenarios RCP2.6, RCP4.5, and RCP8.5 in order to generate a consistent time series. This means that we use the different models to predict bicycle volume in 2020 in combination with weather data for 2050.

3.2. Forecast Models

In the following, we describe the four models that we chose to compare in this study. All four methods were selected due to their ability to model time series data. As the objective is to find the best method for the prediction of bicycle counts, prediction accuracy was used as the central benchmark in the cross-validation procedure. This suggests the use of machine learning models instead of traditional statistical approaches, as the former are directly optimised on increasing the prediction accuracy, while the latter are optimised on improving model fit [36]. Thus, three machine learning models and one traditional statistical approach are compared with regard to their prediction accuracy in this paper.

Seasonal autoregressive integrated moving average with exogenous factors (SARI-MAX) constitutes the classic statistical approach for modelling time series data with seasonality [37]. SARIMAX models are parametric approaches based on the concepts of autoregression and moving averages [37]. As such, they allow a relatively straightforward interpretation of the relationships between the independent and the dependent variables and constitute a popular tool for analyses of time series data which aim at clarifying these relationships [6].

Prophet is a generalised additive model developed by Facebook for time series analyses [38]. It is a non-parametric approach in which the dependent variable is estimated by an addition of functions of the trend in the data, the seasonality, and the holidays as well as an error term [38]. While the trend function uses either a logistic or a linear growth model to capture the trend in the data accurately, the seasonal function relies on Fourier series to model periodic seasonality [38]. The holiday function incorporates the changes induced by specific, reoccurring events such as public holiday by assigning each holiday a parameter corresponding to its impact on the dependent variable [38]. These features make prophet models more flexible in terms of detecting complex relationship patterns in the analysis of time series data than SARIMAX approaches, while preserving some degree of the interpretability.

XGBoost constitutes a regression tree-based ensemble algorithm with a gradient boosting framework [39]. Hence, XGBoost is used for fitting multiple regression trees, each based on a random subsample of the data, which are pooled into a so-called ensemble. The boosting principle denotes that each tree is built on previous trees with particular emphasis on the mis-modelled values so that the overall prediction performance improves gradually. Thus, each tree constitutes a separate function and the final prediction output of the dependent variable is the sum of these functions [39]. These mechanisms increase the predictive power of XGBoost, yet they also decrease its interpretability. While approaches to measure the importance of the independent variables in predicting the dependent one have been developed, these interpretations rely solely on the data used and do not make any assumptions on general distributions of the variables beyond the data set [40].

Long short-term memory is an artificial recurrent neural network architecture which is well suited for predictions based on time series data due to its reliance on sequences rather than single values [41,42]. As all neural networks in general, lstm is a highly flexible model able to adapt to diverse and complex patterns in different kinds of data ranging from tabular data (as in our case) to audio or visual data [41,42]. Yet, this flexibility comes at the price of decreased interpretability. Indeed, in spite of increasing efforts to open up the black box of neural networks [43], it still remains difficult to reveal the detected patterns in the data analysed in an understandable manner.

3.3. Cross-Validation

For selection of the best model, we developed a cross-validation procedure based on the concepts of simulated historical forecasts and rolling windows. This means that first, we had to specify a cut-off date. Then, we performed the first training on the data from the first date the data provide up to this cut-off date, and we made predictions for a specified time period of the remaining data after the cut-off date, called the forecast horizon. During consecutive training, the cut-off date was shifted forward in time by half the number of days of the forecast horizon and we repeated training based on the increased training data set before making predictions for the new forecast horizon, which extended into the future from the new cut-off date.

The development of this cross-validation procedure included the testing of many different specifications. It was, for instance, also tried to shift the cut-off date forward day by day, by a week, a month, a quarter of the year, etc. In fact, often the results in terms of the prediction accuracy did not differ that much between different cross-validation approaches. However, the cross-validation procedure presented in this paper is the one that generally led to the best results, even if not by large margins.

To provide an example of the cross-validation procedure developed: If 1 July 2017 was chosen as the first cut-off date and the forecast horizon was specified by 30 days, then the first training was performed on data from 1 January 2017 to 1 July 2017 and predictions were made for the first 30 consecutive days after 1 July 2017. For the next training, the cut-off date was shifted by half the number of days of the forecast horizon to 16 July 2017, the algorithm was trained on data from 1 January 2017 to this new cut-off date, and predictions were made for the first 30 consecutive days after 16 July 2017. This process was repeated until the shift of the forecast horizon extended beyond the last date of our data set (31 December 2019).

The final step in our cross-validation procedure was to calculate the mean absolute percentage error. This was realised by first calculating the mean absolute percentage error for each specific day of the predictions in consecutive order, then taking the mean of the mean absolute percentage error for the same days of the different prediction rounds, and finally taking the mean of the mean absolute percentage error for each day of the forecast horizon. In the example above, this means first calculating the mean absolute percentage error for the first day of the forecast horizon for the different predictions, then taking the mean of these calculations, repeating the same procedure for each other day of the forecast horizon, and finally taking the mean of the mean absolute percentage errors for each day of the forecast horizon.

3.4. Model Tuning

Based on our cross-validation procedure, we tuned the four different methods with the ambition of minimising the mean absolute percentage error for a forecast horizon of 365 days because our objective was to use the best model for the prediction year of 2020. All models were initially tested with the same set of features, including the numerical variables maximal air temperature, sum of precipitation, and mean wind speed as well as the categorical variables day of the week, month, and year, and the dummy variables public holiday and school holiday. Table 1 provides an overview of the independent and dependent variables:

Table 1. Independent and dependent variables.

Name	Data Type	Unit	Temporal Resolution
Independent variables			
Maximal air temperature	Numeric	° Celsius	Daily (maximum)
Sum of precipitation	Numeric	Millimetres	Daily (sum)
Mean wind speed	Numeric	Metre per second	Daily (mean)
Day of the week	Categorical	-	Daily
Month	Categorical	-	Monthly
Year	Categorical	-	Yearly
Public holiday	Dummy	-	Daily
School holiday	Dummy	-	Daily
Dependent variable			
Bicycle counts	Numeric	-	Daily

As single days constitute our temporal level of analysis, each feature comprises 1095 observations that together cover the three years 2017, 2018, and 2019. As there is no linear trend in the bicycle counts in Berlin (see Section 2.1), we do not have to apply differencing transformations to our data. Consequently, it also means that our predictions for the year 2020 will have no linear trend that would predict a rise or decline in the bicycle counts, irrespective of the weather.

We applied the auto.arima function from the forecast package in R to evaluate the SARIMAX method. As the auto.arima function automatically detects the best combinations of p (autoregression order), d (difference order), and q (moving average order) values as well as the seasonal components P (seasonal autoregression order), D (seasonal difference order), Q (seasonal moving average order), and m (the number of time steps for a single seasonal period) based on the Bayesian Information Criterion (BIC), we did not perform any further manual parameter tuning for the SARIMAX method. The best model had the configuration of SARIMAX (3, 1, 1) (0, 1, 0 (365)) and a BIC of 15,826.

Prophet was developed with the specific intention of providing a tool that works quite well out of the box. Thus, using its implementation in the prophet package in R, we allowed the model to automatically detect the yearly, weekly, and daily seasonalities in our data set (with the default Fourier order of 10). We also tested the model by manually adding monthly seasonalities with a Fourier order of 12 and chose the additive instead of the multiplicative seasonal effect because we did not observe any changes in the strength of the seasonal patterns over the three years in our data set. We also chose the linear growth model because in general, it still seems more appropriate for the development of bicycle counts than the logistic growth model, although we did not detect a linear trend in the three years of our data.

For XGBoost, implementation in the library named likewise in Python was used. We chose the "dart" booster and linear regression as the objective. Since XGBoost was not developed to automatically detect specific patterns in time series data, categorical variables for the day of the week, month, and year were added to the training data set as well as the first lag of our bicycle count data.

For training of the LSTM, based on the keras and tensorflow packages in R, variables such as day of the week, month, and year were transformed into binary subvectors using one-hot-encoding. Together with the numerical features (air temperature, precipitation, and wind speed), these one-hot-encoded variables were then scaled based on the minimum and maximum of the training data to a range of -1 to 1. In a manual trial and error approach, different configurations with one and two hidden layers; a batch size of one, five, and 73; SGD, Adam, and Adamax as optimisers; dropout rates of 0.05, 0.1, 0.2, 0.3, and 0.5; varying numbers of units in the first and the second hidden layer were tested with 50, 100, 500, 1000, and 2000 epochs. All algorithms were tested with and without weather information.

4. Results

This section first describes the results of our model comparison and selection. Thereafter, the prediction results of the XGBoost model are illustrated for the bicycle traffic in 2020 under the synthetically generated weather conditions.

4.1. Model Performance, Comparison and Selection

We used our cross-validation procedure to perform an exhaustive grid search for Prophet and XGBoost to find the optimal configuration of hyperparameters. For Prophet, this resulted in 5 for seasonality.prior.scale, 0.05 for changepoint.prior.scale, and 20 for holidays.prior.scale. The optimal values for XGBoost were 0.01 for eta, 0 for gamma, 15 for max_depth, 5 for min_child_weight, 0.5 for subsample, and 1 for colsample_bytree. The manual trial and error approach applied for the LSTM resulted in two hidden layers with 100 and 50 units, dropout rates of 0.2, a batch size of 5, the Adam optimiser, the mean absolute error as loss function, the hyperbolic tangent activation function, and 50 epochs yielding the lowest mean absolute percentage error.

Figure 4 shows the results of our cross-validation procedure for the best model specifications of all four methods for six different forecast horizons.

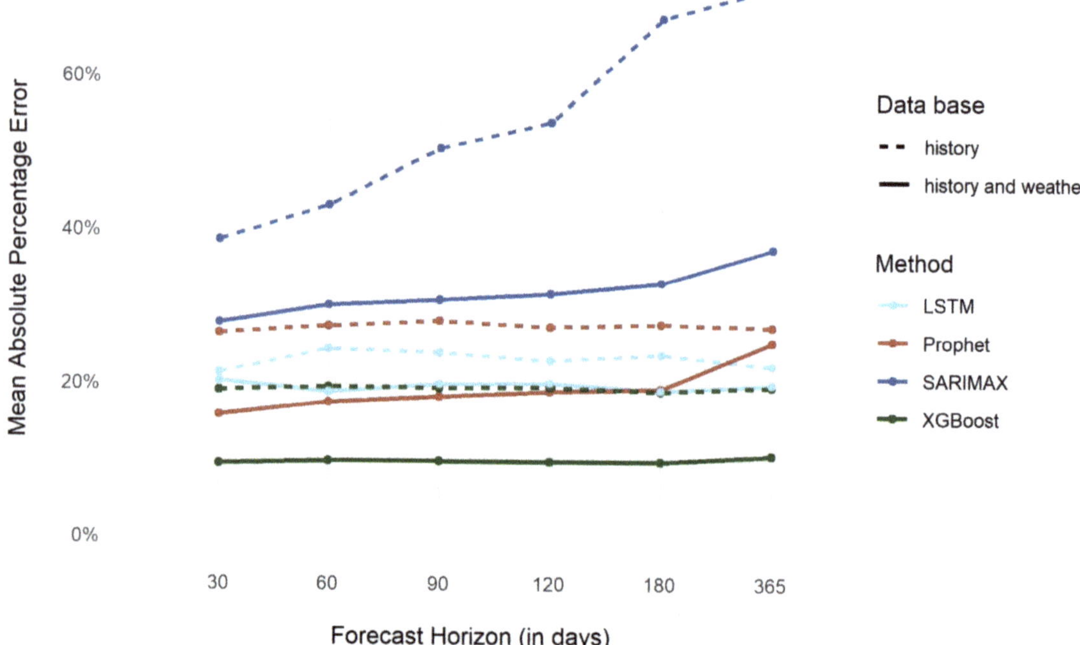

Figure 4. Cross-validation results for different forecast horizons. The colours represent the different methods compared. The dashed lines show the mean absolute percentage error when only the bicycle counts were used for prediction. The solid line illustrates the mean absolute percentage error for the predictions based on bicycle counts and weather variables.

The prediction accuracy of all four methods improves considerably if not only the bicycle count data are used for training but also the weather station measurements on daily maximum air temperature, sum of precipitation, and mean wind speed. This illustrates once more the importance of including weather conditions when modelling bicycle traffic. In particular, the parametric SARIMAX approach improves considerably when the weather conditions are included as further independent variables in the model. Furthermore, it can be seen that the prediction error increases for all methods if the forecast horizon is extended to 365 days.

Our XGBoost configuration, trained on the bicycle counts and the weather station measurements, clearly yields the best prediction accuracy for all six forecast horizons. Its mean absolute percentage error is 11% for the forecast horizon of 365 days, while the errors for LSTM, Prophet, and SARIMAX are 20%, 25%, and 38% respectively. XGBoost was therefore chosen for prediction of the future bicycle counts based on the synthetically generated future weather conditions derived from global and regional climate science models.

Unfortunately, the relatively high prediction accuracy of machine learning approaches comes at the disadvantage of a more difficult interpretation of the results. Indeed, it is rather difficult to illustrate why one model performs better than the other. We assume, however, that the structure of our data in terms of size and heterogeneity is not complex enough to fully exploit the advantages of neural network approaches such as LSTM. These constitute generally the most flexible machine learning methods being able to adapt relatively well to complex non-linear relationships in the data. Yet, our data only comprised a very limited number of variables and cases by machine learning standards.

In contrast, XGBoost is based on regression-trees which in our case appears to better capture the underlying relationships in our data. This might be due to the fact that the relationship between temperature and bicycle traffic is of a rather linear nature. However, the relatively bad performance of the SARIMAX approach indicates that there are revenant non-linear relationships in our data that are better captured by non-parametric machine learnings methods than by traditional parametric statistical approaches.

4.2. Predictions

Figure 5 illustrates the actual bicycle count data of 2017, 2018, and 2019 and the predicted bicycle counts per day for the year 2020 on the basis of the weather conditions of the three different RCPs in 2050.

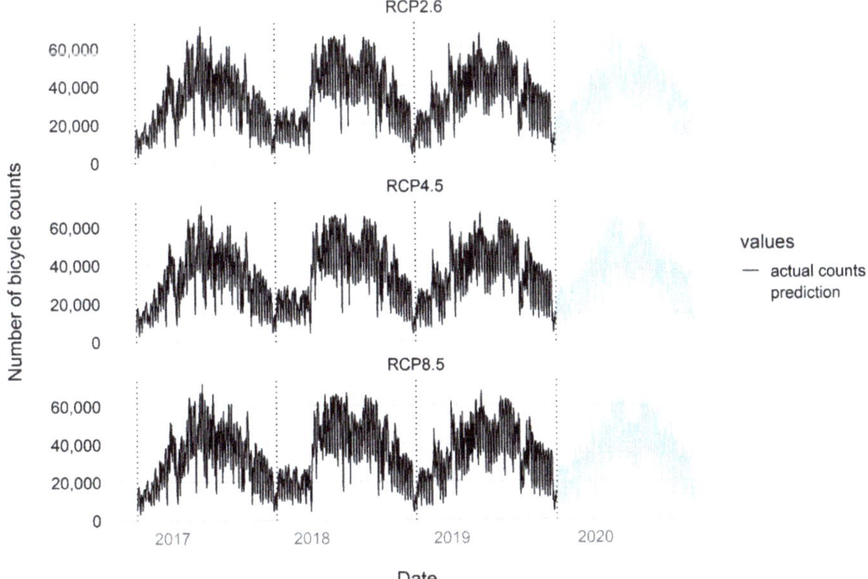

Figure 5. Actual daily bicycle counts and future predictions. The black line illustrates the actual bicycle counts detected in 2017–2019. Therefore, these values are the same ones for all three RCPs. The light blue line shows the prediction of the bicycle counts which different among the RCPs due to different weather conditions.

Our model seems to capture the general seasonal patterns quite well and to some extent also accounts for the expectable daily variations. Table 2 summarises the synthetically generated future weather conditions for 2020 and the predicted bicycle traffic for the three representative concentration pathways compared to the weather station measurements and the actual bicycle count data of 2017, 2018, and 2019.

Table 2. Weather station measurements, synthetically generated future weather conditions, actual bicycle counts, and prediction results.

Season	2017	2018	2019	2020 (RCP2.6)	2020 (RCP4.5)	2020 (RCP8.5)
Maximal daily air temperature in °Celsius (mean)						
Spring	15.35	16.38	15.21	15.94	16.34	16.98
Summer	24.18	27.19	27.30	26.43	27.65	27.92
Autumn	14.22	16.11	15.17	16.18	15.92	17.19
Winter	4.19	5.17	6.65	5.74	5.97	6.52
Daily precipitation in mm (sum)						
Spring	89.8	103.5	91	120.92	116.36	129.18
Summer	400.4	97	198.6	393.33	349.76	401.68
Autumn	191.4	55.7	140.1	194.54	246.66	208.18
Winter	114.8	119.7	94.8	137.18	113.84	125.99
Total	796.4	375.9	524.5	845.86	825.61	865.01
Daily wind speed in m/s (mean)						
Spring	3.85	3.88	4.23	3.95	3.90	3.92
Summer	3.35	3.47	3.26	3.23	3.17	3.24
Autumn	3.62	3.39	3.22	3.41	3.32	3.35
Winter	3.91	3.86	3.87	3.76	3.71	3.70
Specific weather events (number of days)						
Hot days * ($\geq$30 °C)	8	31	26	16	24	26
Dry days *	171	240	207	213	227	209
Days with heavy rain *	22	7	13	20	17	19
Days with moderate or strong winds *	40	34	37	37	37	34
Bicycle counts (sum)						
Spring	3,189,554	3,369,275	3,301,122	3,215,209	3,317,420	3,340,132
Summer	4,014,512	4,487,628	4,386,066	4,281,880	4,317,930	4,377,109
Autumn	3,004,748	3,487,274	3,268,441	3,285,980	3,222,224	3,396,897
Winter	1,450,916	1,755,762	2,009,325	1,931,819	1,974,888	1,982,382
Total	11,659,730	13,099,939	12,964,954	12,714,888	12,832,461	13,096,522

* Hot days refers to days with a maximum air temperature of at least 30 °C. Dry days are days with less than 0.1 mm precipitation. Days with heavy rain refers to days with a precipitation of at least 10 mm. Days with moderate or strong winds refers to days with a mean wind speed of at least 5.5 m/s (Beaufort 4).

In a nutshell, the results indicate that climate change will contribute to an overall rise in bicycle traffic in Berlin if all other relevant factors are held constant. More specifically, we expect an increase in bicycle traffic compared to the mean total of 12,574,874 for the three years 2017, 2018, and 2019 for all representative concentration pathways (RCP2.6: +140,013; RCP4.5: +257,587; RCP8.5: +521,647). This corresponds to an overall expected relative increase of 1.1% for RCP2.6, 2.1% for RCP4.5, and 4.1% for RCP8.5. This general increase, however, varies significantly between the seasons and is slowed down partly by the expected increase in precipitation.

We expect the highest rise in bicycle traffic in the winter season. Compared to the mean bicycle traffic for the winters of 2017, 2018, and 2019, our model predicts an increase of 11.1% in RCP2.6, 13.6% in RCP4.5, and 14.0% in RCP8.5. The main reason for this is the expected increase in air temperature in winter. Its positive impact on bicycle traffic outweighs the negative impact of the increase in precipitation that is also expected. In addition, comparison of the future predictions with the actual bicycle counts highlights the very high bicycle traffic in the exceptionally warm and dry winter months of 2019. Only the winter conditions in RCP8.5 lead to similar bicycle traffic, while the bicycle traffic predicted for a typical winter season in RCP2.6 and RCP4.5 remains below the level of 2019.

This large increase in bicycle traffic expected during winter outweighs all the changes in bicycle traffic predicted for the other seasons together. In a typical spring season in RCP2.6, we expect a decrease in bicycle traffic by 2.2% due to relatively high precipitation. In RCP4.5, the average maximum air temperature rises a bit, while precipitation is a bit lower than in RCP2.6, so an increase of 0.9% in bicycle traffic is predicted. In a typical spring season in RCP8.5, bicycle traffic is predicted to increase by 1.6% as the effects of higher air temperatures outweigh the impact of the increasing precipitation.

In fact, the relative changes in air temperature and precipitation due to climate change are the most important factors affecting the increase or decrease in bicycle traffic predicted for the different seasons. In summer and autumn in RCP2.6 and RCP4.5, the average maximum daily air temperature is expected to rise a bit but the expected relative increase in precipitation is even higher with the result that the predicted overall changes in bicycle traffic remain at a low level. For a typical summer season, our model predicts a decrease of 0.3% in RCP2.6 and an increase of 0.5% in RCP4.5, while in autumn, bicycle traffic is predicted to increase by 1.0% in RCP2.6 and to decrease by 1.0% in RCP4.5. In contrast, in RCP8.5, bicycle traffic is predicted to increase by 1.9% in summer and by 4.4% in autumn, reflecting, above all, the higher increases in the average maximum daily air temperature in comparison to RCP2.6 and RCP4.5.

5. Discussion

Our results show that climate change will lead to an overall increase in annual bicycle traffic in Berlin of between 1% and 4%. During winter, in particular, bicycle traffic might increase by 11–14% due to higher air temperatures and only a relatively low increase in precipitation. Although increases in air temperature are also expected in the other seasons, their positive effect on bicycle traffic is offset by relatively high increases in precipitation, which can even lead to a decrease in predicted bicycle traffic for some seasons, considering the representative concentration pathways 2.6 and 4.5.

This overall positive effect of climate change on bicycle usage in Berlin corresponds to the findings of [24] for the region of Randstad in the Netherlands. In addition, [24] also found that climate change might lead to a higher bicycle usage in winter due to milder temperatures and only slightly more precipitation. However, the results of [24] for the summer season show a decrease in bicycle traffic due to more intense precipitation expected to be brought about by climate change. The findings of our study also differ a bit from those of Wadud [25], produced for bicycle traffic in London. First, our expected annual increase with 1–4% is larger than his of 0.5% [25]. Second, he predicts the largest seasonal increase of 2.5% for the season of summer, while we expect the largest increases to occur in future winters [25].

The differences in our results and the other two studies could be based on the different historical weather data for Berlin, London, and Randstad, on different changes in the local weather expected to be brought about by climate change due to geographical differences, or on different reactions of cyclists to weather conditions in the three locations. Thus, the comparison of the results should be treated with care and rather illustrates the importance of further research on the impact of weather conditions on mobility behaviour in different climate zones.

However, given the fact that nearly all studies worldwide observe similar effects of weather patterns on cycling rates [23], it can be assumed that cities with similar climatic conditions (humid continental climate with dry winters and warm summers [44]) as Berlin and similar expected changes due to climate change can also expect similar impacts on cycling rates. This counts especially for European cities with a continental climate such as Prague, Warsaw, Vienna, Bratislava, Budapest, Kiev, etc.

Furthermore, it should be kept in mind that the annual changes in bicycle traffic might also be affected by population growth. Based on the data of the Office for Statistics for the region of Berlin and Brandenburg [45], we calculated annual growth rates of 1.08%, 0.87%, and 0.68% for the years 2017–2019. These are modest growth rates but other cities might have a more dynamic population growth, in which case, this factor should definitely be addressed in the analysis and also in the comparison with the results of this study.

Finally, the reliability of the results of this study also depends on the outputs of the climate change models. As already illustrated, regional and global climate models rely on various assumptions about ecological, economic, social, and technological developments and thus naturally come with a lot of uncertainties. To account for this uncertainty, various regional climate models based on the output of different global climate models simulating three different scenarios were run and compared with each other. This constituted the state-of-the-art procedure in climate science at the moment when this study was conducted.

6. Conclusions

In this paper, we introduce machine learning methods for the prediction of future bicycle traffic based on bicycle count, weather station, and regional climate model data. Our results have shown a higher prediction accuracy of machine learnings methods in comparison to a traditional statistical approach for time series modelling. This should be considered by future studies that are more interested in predicting accurate results based on well-known relationships between dependent and independent variables than on exploring the nature of these relationships.

Furthermore, in contrast to model fit measures such as the Akaike information criterion, which is often used for model selection in traditional statistical approaches, but which cannot be directly calculated for non-parametric approaches, we developed a cross-validation procedure with the mean absolute percentage error as the central benchmark for model selection. This increases the comparability of the performance of our methodological framework to future studies as the mean absolute percentage error can easily be calculated for both parametric and non-parametric approaches.

Our results show that bicycle traffic in Berlin will most likely increase due to the effects of climate change, if all other factors remain constant. City planners should consider these findings since they need to prepare infrastructures that are suitable for changes in demand and allow for increased requirements for more sustainable mobility. In particular, the expected increase in cycling in the darker winter months contributes to a higher traffic load on bicycle lanes throughout the year, providing an additional argument for the further extension of street lighting and bicycle infrastructure in general. This might also increase public and political acceptance of the need to redistribute public space for bicycle usage.

In addition to the specific results, our research also highlights the importance of including weather conditions in any analysis of mobility behaviour in general. The prediction accuracy of all four methods compared improved considerably if not only bicycle count data were used for training but also the weather information regarding maximum daily air temperature, sum of precipitation, and mean wind speed. This not only illustrates the importance of weather conditions for bicycle traffic but also paves the way for further research which might investigate what other modes of transport benefit or suffer from increases or decreases in bicycle traffic due to climate change, and to what extent different groups of people, in terms of age, gender, etc., adjust their mobility behaviour to weather conditions.

However, the inclusion of weather variables has also illustrated the limitations of our approach. Our findings on the potential impact of climate change on bicycle traffic in Berlin are hardly comparable to the results of similar studies on other locations due to different present weather conditions and differing impacts of climate change. Therefore, more studies in different geographic and climatic regions are needed to better understand the impacts that climate change might have on mobility behaviour in different parts of the world.

Author Contributions: A.G.: Conceptualisation, Methodology, Software, Validation, Data Curation, Writing—Original Draft Preparation. S.N.: Conceptualisation, Methodology, Software, Data Curation, Writing—Original Draft Preparation. B.L.: Conceptualisation, Writing—Review and Editing. J.B.: Methodology, Formal Analysis, Writing—Review and Editing. All authors have read and agreed to the published version of the manuscript.

Funding: This study was produced within the scope of the Helmholtz Association's mobility subproject for the Climate Initiative.

Data Availability Statement: The bicycle count data used in this paper are publicly accessible via the homepage of the company Eco-counter (https://www.eco-public.com/ParcPublic/?id=4728 (accessed on 21 February 2021)). The weather station data can be retrieved in the open data portal of the German Weather Service (https://www.dwd.de/DE/klimaumwelt/cdc/cdc_node.html (accessed on 20 November 2020)). The data outputs of the regional climate models of the German Climate Service Centre (GERICS) used for this study can generally be accessed via the Euro-Cordex project (https://euro-cordex.net/060378/index.php.en (accessed on 25 November 2020)). However, at the moment of the submission of this paper, the concrete data used have not yet been published.

Acknowledgments: We would like to express our thanks to Thomas Remke from the German Climate Service Centre (GERICS) who not only provided the data for the representative concentration pathways for us but also gave us advice on how to adjust these data to our weather station measurements for the region of Berlin. We would also like to thank Michael Hardinghaus for carefully reading the manuscript and giving valuable advice on all parts of this paper.

Conflicts of Interest: The authors declare that they have no known competing financial interests or personal relationships that could have appeared to influence the work reported in this paper.

Appendix A

As the count devices were installed at different points in time and as some of them were occasionally out of operation or showed unreasonable results, some data preparation was needed to produce a consistent data set for our analysis. As our objective was to predict the volume of bicycle traffic for each day of an entire year, we decided to train our algorithm on data available for entire years. This left us with data for the years 2017, 2018, and 2019, as around half of the stations went into operation in 2016. Four count devices were excluded from further analysis because they showed missing values over several days, weeks, or months in a row.

Missing values at the remaining stations were imputed by taking the mean of the previous and following hour. If missing values occurred at 11 pm on 31 December 2019, the value of the previous hour was used. Days with zero bicycle counts were regarded as extremely unlikely under regular conditions and thus taken as an indicator for irregularities. Three stations showed consecutive days with zero bicycle counts over several months. Two other stations showed extremely low or extremely high values over several months. In both cases, these time periods were deemed too long for data imputation and thus, all five stations were excluded from further analysis.

Figure A1 illustrates the sum of the bicycle counts per day per stations for the remaining 17 devices after data preparation.

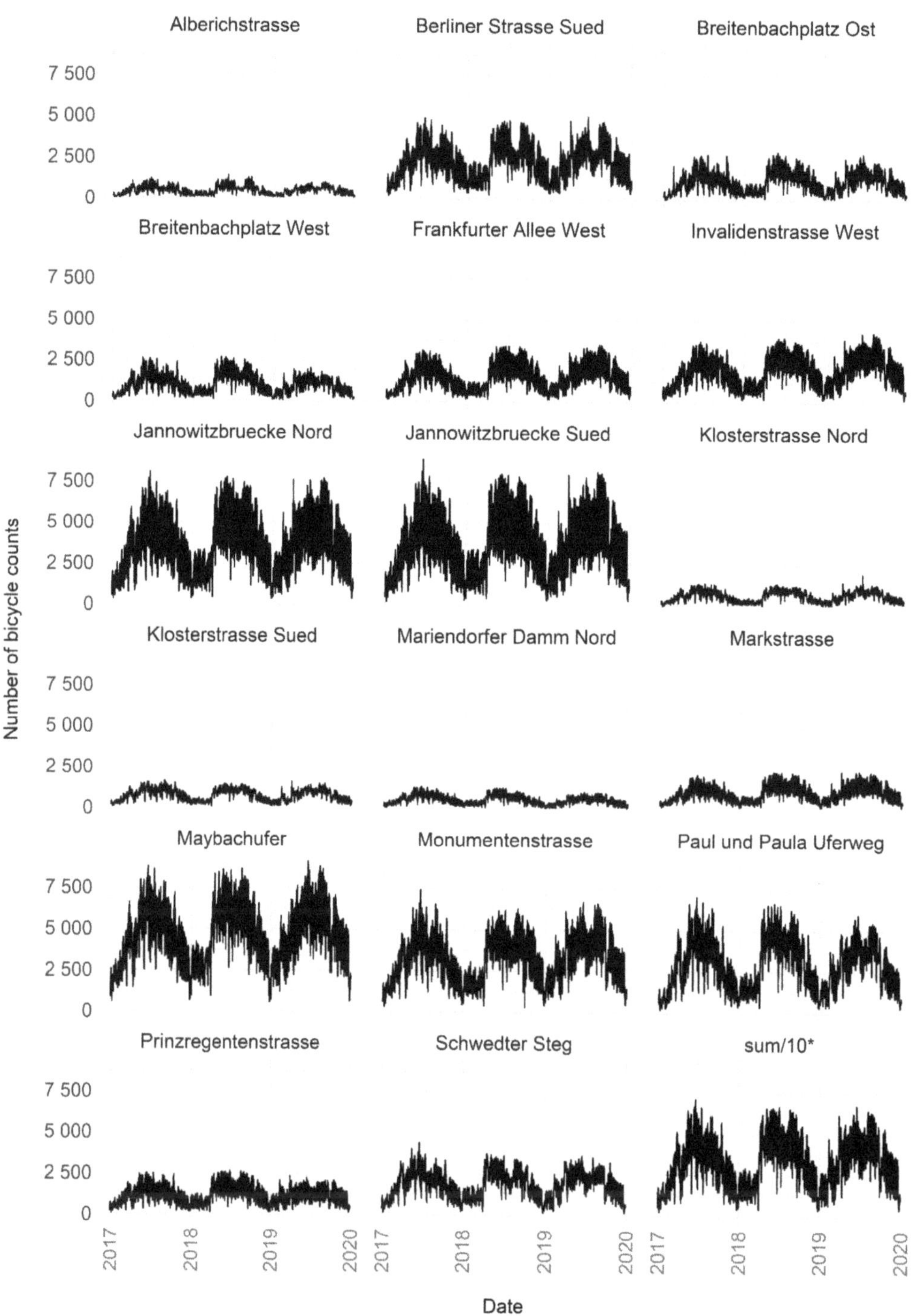

Figure A1. Bicycle counts for each count station per day from 2017 to 2020. * The aggregated number of all stations on the bottom right of the figure was divided by ten in order to fit on the same scale as the counts from the individual stations.

We first analysed whether the bicycle counts at the different count stations follow similar seasonal patterns and react to weather conditions in a similar manner. For this purpose, various statistical tests were conducted to check the stationarity of the time series of the different count stations. As can be seen in Figure A1, each station shows the same general seasonal trend with peaks in summer and the lowest number of bicycle counts during the winter months and has a relatively constant mean and variance per year. In addition, the Augmented Dickey–Fuller Tests and Kwiatkowski–Philipps–Schmidt–Shin Tests performed had p-values smaller than 0.05, allowing us to reject the null hypotheses of a unit root or trend-stationarity being present. Therefore, the time series of the different count stations can be regarded as stationary. This allows us to investigate the correlation between the bicycle counts per day at the different stations directly without the need of detrending the data first.

In fact, the lowest value in the correlation matrix of all count stations is the Pearson correlation coefficient of 0.87 between the count stations of Alberichstrasse and Frankfurter Allee West. For the combinations of the large majority of all count stations, the correlation coefficient has a value higher than 0.9. Consequently, the seasonal patterns of the individual count stations are also visible in the bicycle counts aggregated over all stations, shown in the bottom right corner of Figure 2.

The count station data are not just highly correlated between the individual stations but also the counts of the different stations show a very similar correlation to the weather variables taken from the weather station Berlin-Tempelhof of the German Weather Service. Pearson's correlation coefficient between the maximum daily temperature and the daily bicycle count of each station falls into the range of 0.60 to 0.82. Moreover, the correlation between the bicycle counts of the individual stations per day and the sum of precipitation per day and the mean wind speed per day fall into the rather narrow ranges of -0.08 to -0.12 and -0.21 to -0.26. This shows that the volume of bicycle traffic measured at each count station reacts in a very similar way to changes in weather conditions.

We also inspected whether the locations of the remaining count stations after data preparation correspond to the population distribution in Berlin:

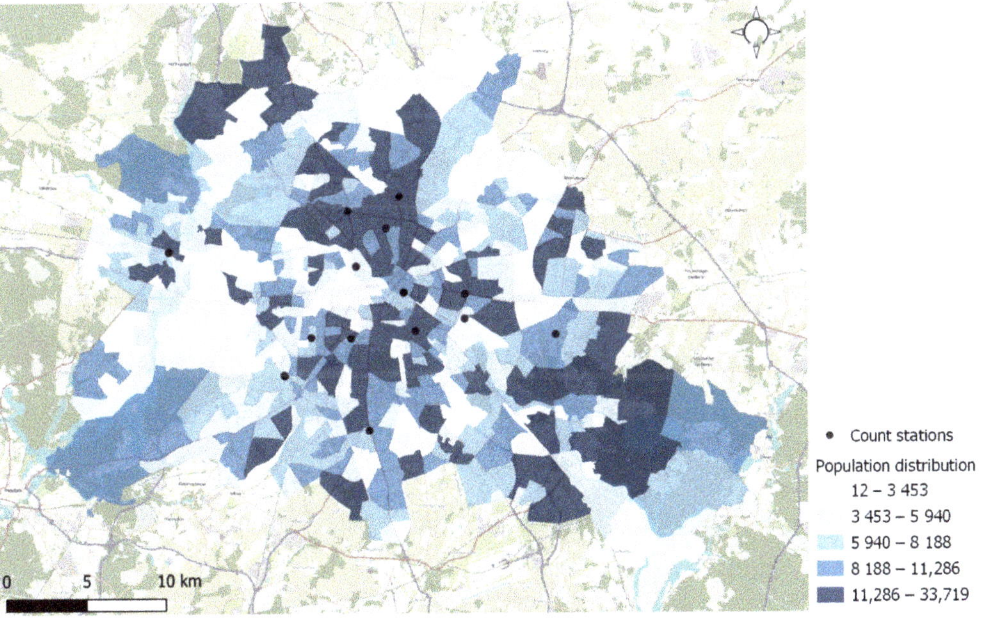

Figure A2. Count station locations and population distribution.

As can be seen in Figure A2, most of the count stations are located in areas where also large shares of the population reside. Only some areas with relatively many inhabitants in the north and in the east and southeast are not covered or not covered well.

Appendix B

Two types of weather and climate data were used in this paper: daily measurements from a nearby weather station (further called weather station measurements) and weather projections of a regional climate model for the time periods 1970–2000 (further called historical data) and 2035–2065 (further called forecast data). All data sources include the three weather variables of maximum air temperature, sum of precipitation, and mean wind speed for each day of the year (see Figure A2).

The weather station measurements are taken from the Berlin-Tempelhof station for 2017, 2018, and 2019. The station is operated by the German Weather Service (DWD) and provides data on an hourly and daily basis. For this study, we use the daily measurements for each day of the three years considered. These data are linked to the daily bicycle count data by their date to provide the basis for training the time series models.

To make reasonable assumptions about future changes in the weather, this study applies outputs from the regional climate model (REMO) [46] (domain: EUR-11, driving model name: MPI-ESM-LR, realisation: r1i1p1, frequency: day) for the historical data and the forecast data, considering the three different Representative Concentration Pathways (RCP): RCP2.6, RCP4.5, and RCP8.5.

The data of regional climate models are available in approximately 12.5 × 12.5 km ground resolution. In order to reflect the local uncertainties of the climate models, a mask of 6 × 6 pixels (appr. 75 × 75 km) is placed around Berlin. Finally, we use the mean of the resulting data.

Figure A3 illustrates the need to adapt the outputs of the regional climate model to the local weather conditions. This is performed by comparing the yearly distribution of the historical and forecast results from the regional climate model with the distribution of the measured values from the weather station in Berlin in 2018. Generally, the daily maximum air temperature from the regional climate model shows higher values than the measured weather data in 2018 since they illustrate the highest values of a 30-year interval instead of one single year. However, the absolute changes from historical to forecast data appear realistic, as they lead to an overall annual increase of 1.6 °C in the average maximum daily air temperature, which is consistent with the output of recent findings for Germany (Brasseur et al., 2017).

Therefore, in the case of air temperature, we can simply add the expected absolute changes from the climate models to the averaged weather station data to create realistic synthetic future temperature conditions for 2050 (see Figure A3). In contrast to air temperature, however, wind speed and precipitation are less continuous over time and rainy days and storms are mostly discrete events. This leads to an unwanted effect when averaging the historical and forecast data: the values are smoothed. Just adding the differences means there would be no cases of heavy rain and no cases without rain in the data set, leading to a bias in the variance in respect of single-year data. Hence, a more sophisticated procedure is required to adapt the expected changes in precipitation and wind speed from the climate models to the weather station data.

For this reason, the synthetic future precipitation values are generated using a variance adjustment procedure to harmonise measured and forecast yearly distributions. It can be assumed, however, that there are different effects on the variation of precipitation based on the time of year. Findings from the regional climate simulations in Germany (Pfeifer et al., 2015), for instance, suggest that summer precipitation is expected to decrease, while winter precipitation is expected to increase. Therefore, the measured, historical, and forecast data were first split into the meteorological seasons of winter (1 December to 28 February), spring (1 March to 30 May), summer (1 June to 30 August) and autumn (1 September to 30 November).

For each season, the absolute differences between the precipitation sums of the historical and forecast data are calculated. Here, we use the season with the highest variance because extreme events are expected to increase in the future and we want to adapt to the most "extreme" season we have. The expected absolute differences in precipitation are then added to the sum of the corresponding season in the measured weather station data to generate an expected future precipitation sum for the respective RCP. Then, we calculate the proportion of precipitation for each day in relation to the sum of precipitation for the respective season and multiply the result by the expected future precipitation sum. The outcome of this process is the expected daily precipitation with the distribution of the respective measured season.

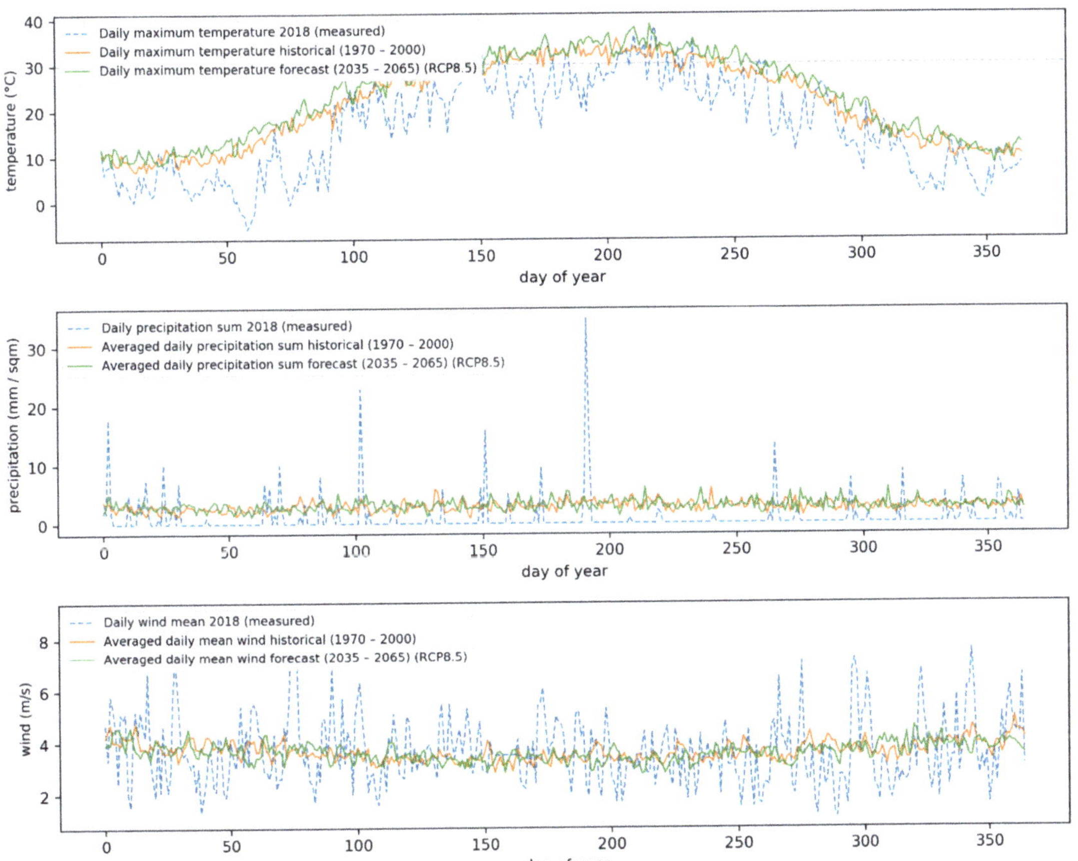

Figure A3. Yearly distribution of the historical (orange) and forecast (green) weather variables based on the regional climate model and the measured weather variables for 2018 (blue dashed) from a local weather station. Forecast values are based on RCP.

Next, the variance of this synthetically generated future precipitation data is adjusted to meet the probability distribution of the year with the highest variance in the respective season (winter and spring of 2018 and summer and autumn of 2017) in order to achieve a realistic (but highly variable) distribution of precipitation. The variance adjustment is realised by performing the following steps. First, the synthetically generated precipitation values are centred by subtracting the yearly mean of each value. Second, the centred values are multiplied by the square root of the variance of the measured data divided by the variance of the forecast data. Third, the data are brought out of centre again by adding the yearly mean to each value. Since the results include negative values, the data are split into positive and negative values. The latter are iteratively eliminated by setting negative values to zero and subsequently adding the proportion of the sum of all negative values in relation to the sum of positive values to each positive value as a fourth step. This step is performed until all negative values are eliminated.

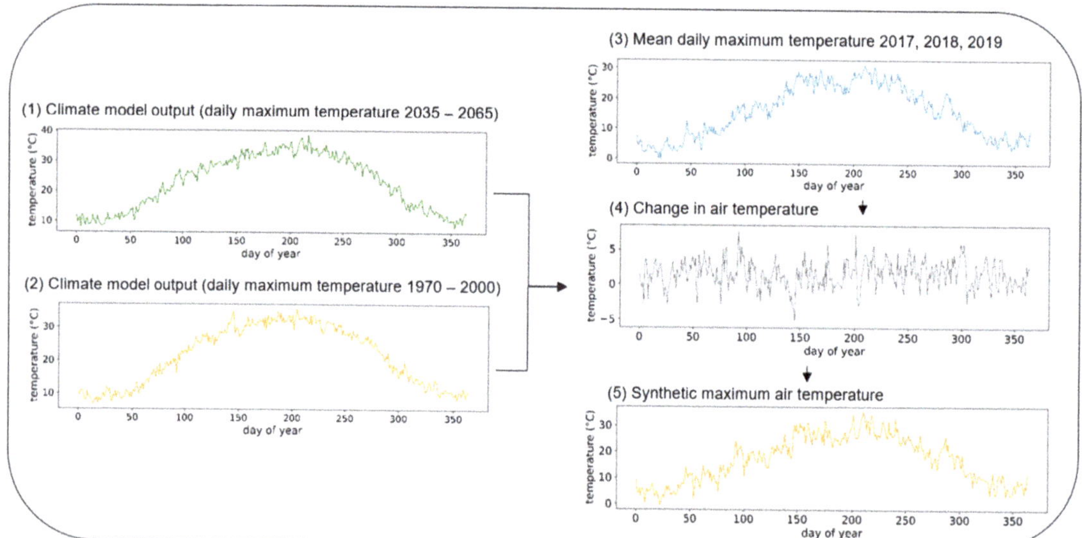

Figure A4. Process of generating daily synthetic temperature values based on the climate models. All line graphs represent annual air temperature curves in °C. The process comprises the derivation of daily air temperature changes (**4**) between historical (**2**) and forecasted (**1**) data and generation of future temperature air data (**5**) by calculating the sum of temperature changes (**4**) and averaged measured values (**3**).

Wind speed values are generated analogously to precipitation, but since there were very few negative values in the result (1 in RCP2.6 and RCP4.5, 2 in RCP8.5), these were set straight to zero.

Although the synthetically generated weather data for 2050 are based on the outputs of a single forecast model instead of an ensemble of different climate models, the results of the climate data preparation are consistent with the findings of regional climate models in Germany [33]. For the Berlin region, climate change leads to a general increase in maximum air temperature throughout the year with precipitation sums rising but with a trend towards more extreme events and therefore fewer wet days in summer.

References

1. Araos, M.; Berrang-Ford, L.; Ford, J.; Austin, S.E.; Biesbroek, R.; Lesnikowski, A. Climate change adaptation planning in large cities: A systematic global assessment. *Environ. Sci. Policy* **2016**, *66*, 375–382. [CrossRef]
2. Raser, E.; Gaupp-Berghausen, M.; Dons, E.; Anaya-Boig, E.; Avila-Palencia, I.; Brand, C.; Castro, A.; Clark, A.; Eriksson, U.; Götschi, T.; et al. European cyclists' travel behavior: Differences and similarities between seven European (PASTA) cities. *J. Transp. Health* **2018**, *9*, 244–252. [CrossRef]
3. Aldred, R.; Jungnickel, K. Why culture matters for transport policy: The case of cycling in the UK. *J. Transp. Geogr.* **2014**, *34*, 78–87. [CrossRef]
4. Dill, J.; Carr, T. Bicycle Commuting and Facilities in Major U.S. Cities: If You Build Them, Commuters Will Use Them. *Transp. Res. Rec. J. Transp. Res. Board* **2003**, *1828*, 116–123. [CrossRef]
5. Mertens, L.; Compernolle, S.; Deforche, B.; Mackenbach, J.D.; Lakerveld, J.; Brug, J.; Roda, C.; Feuillet, T.; Oppert, J.-M.; Glonti, K.; et al. Built environmental correlates of cycling for transport across Europe. *Health Place* **2017**, *44*, 35–42. [CrossRef]

6.   Nosal, T.; Miranda-Moreno, L.F. The effect of weather on the use of North American bicycle facilities: A multi-city analysis using automatic counts. *Transp. Res. Part A Policy Pract.* **2014**, *66*, 213–225. [CrossRef]
7.   Pucher, J.; Buehler, R. Making Cycling Irresistible: Lessons from The Netherlands, Denmark and Germany. *Transp. Rev.* **2008**, *28*, 495–528. [CrossRef]
8.   Pucher, J.; Komanoff, C.; Schimek, P. Bicycling renaissance in North America?: Recent trends and alternative policies to promote bicycling. *Transp. Res. Part A Policy Pract.* **1999**, *33*, 625–654. [CrossRef]
9.   Thomas, T.; Jaarsma, R.; Tutert, B. Exploring temporal fluctuations of daily cycling demand on Dutch cycle paths: The influence of weather on cycling. *Transportation* **2012**, *40*, 1–22. [CrossRef]
10.   Böcker, L.; Uteng, T.P.; Liu, C.; Dijst, M. Weather and daily mobility in international perspective: A cross-comparison of Dutch, Norwegian and Swedish city regions. *Transp. Res. Part D Transp. Environ.* **2019**, *77*, 491–505. [CrossRef]
11.   Gebhart, K.; Noland, R.B. The impact of weather conditions on bikeshare trips in Washington, DC. *Transportation* **2014**, *41*, 1205–1225. [CrossRef]
12.   Liu, C.; Susilo, Y.O.; Karlström, A. The influence of weather characteristics variability on individual's travel mode choice in different seasons and regions in Sweden. *Transp. Policy* **2015**, *41*, 147–158. [CrossRef]
13.   Liu, C.; Susilo, Y.O.; Karlström, A. Investigating the impacts of weather variability on individual's daily activity–travel patterns: A comparison between commuters and non-commuters in Sweden. *Transp. Res. Part A Policy Pract.* **2015**, *82*, 47–64. [CrossRef]
14.   Sabir, M. Weather and Travel Behaviour. Ph.D. Thesis, Free University of Amsterdam, Amsterdam, The Netherlands, 2011.
15.   Flynn, B.S.; Dana, G.S.; Sears, J.; Aultman-Hall, L. Weather factor impacts on commuting to work by bicycle. *Prev. Med.* **2012**, *54*, 122–124. [CrossRef] [PubMed]
16.   Parkin, J.; Wardman, M.; Page, M. Estimation of the determinants of bicycle mode share for the journey to work using census data. *Transportation* **2008**, *35*, 93–109. [CrossRef]
17.   Phung, J.; Rose, G. Temporal variations in usage of Melbourne's bike paths. In Proceedings of the 30th Australasian Transport Research Forum, Melbourne, VIC, Australia, 25–27 September 2007.
18.   Heinen, E.; Maat, K.; Van Wee, B. Day-to-Day Choice to Commute or Not by Bicycle. *Transp. Res. Rec. J. Transp. Res. Board* **2011**, *2230*, 9–18. [CrossRef]
19.   Helbich, M.; Böcker, L.; Dijst, M. Geographic heterogeneity in cycling under various weather conditions: Evidence from Greater Rotterdam. *J. Transp. Geogr.* **2014**, *38*, 38–47. [CrossRef]
20.   Saneinejad, S.; Roorda, M.J.; Kennedy, C. Modelling the impact of weather conditions on active transportation travel behaviour. *Transp. Res. Part D Transp. Environ.* **2012**, *17*, 129–137. [CrossRef]
21.   Creemers, L.; Wets, G.; Cools, M. Meteorological variation in daily travel behaviour: Evidence from revealed preference data from the Netherlands. *Theor. Appl. Clim.* **2015**, *120*, 183–194. [CrossRef]
22.   Winters, M.; Buehler, R.; Götschi, T. Policies to Promote Active Travel: Evidence from Reviews of the Literature. *Curr. Environ. Health Rep.* **2017**, *4*, 278–285. [CrossRef]
23.   Liu, C.; Susilo, Y.O.; Karlström, A. Weather variability and travel behavior—What we know and what we do not know. *Transp. Rev.* **2017**, *37*, 715–741. [CrossRef]
24.   Böcker, L.; Prillwitz, J.; Dijst, M. Climate change impacts on mode choices and travelled distances: A comparison of present with 2050 weather conditions for the Randstad Holland. *J. Transp. Geogr.* **2013**, *28*, 176–185. [CrossRef]
25.   Wadud, Z. Cycling in a changed climate. *J. Transp. Geogr.* **2014**, *35*, 12–20. [CrossRef]
26.   Lv, Y.; Duan, Y.; Kang, W.; Li, Z.; Wang, F.-Y. Traffic Flow Prediction With Big Data: A Deep Learning Approach. *IEEE Trans. Intell. Transp. Syst.* **2014**, *16*, 865–873. [CrossRef]
27.   Zhao, Z.; Chen, W.; Wu, X.; Chen, P.C.Y.; Liu, J. LSTM network: A deep learning approach for short-term traffic forecast. *IET Intell. Transp. Syst.* **2017**, *11*, 68–75. [CrossRef]
28.   Nobis, C.; Kuhnimhof, T. *Mobilität in Deutschland–Tabellarische Grundauswertung, in Studie von Infas, DLR, IVT und Infas 360 im Auftrag des Bundesministers für Verkehr und Digitale Infrastruktur*; (FE-Nr. 70.904/15); Bundesministerium für Verkehr und Digitale Infrastruktur: Bonn, Germany, 2018.
29.   Senatsverwaltung für Umwelt, Verkehr und Klimaschutz Berlin. Karte: Zählung der Radfahrer. Available online: https://www.berlin.de/senuvk/verkehr/lenkung/vlb/de/karte.shtml (accessed on 3 April 2020).
30.   Moss, R.H.; Edmonds, J.A.; Hibbard, K.A.; Manning, M.R.; Rose, S.K.; van Vuuren, D.P.; Carter, T.R.; Emori, S.; Kainuma, M.; Kram, T.; et al. The next generation of scenarios for climate change research and assessment. *Nat. Cell Biol.* **2010**, *463*, 747–756. [CrossRef]
31.   Jacob, D.; Petersen, J.; Eggert, B.; Alias, A.; Christensen, O.B.; Bouwer, L.M.; Braun, A.; Colette, A.; Déqué, M.; Georgievski, G.; et al. EURO-CORDEX: New high-resolution climate change projections for European impact research. *Reg. Environ. Chang.* **2014**, *14*, 563–578. [CrossRef]
32.   Giorgi, F.; Jones, C.; Asrar, G.R. Addressing Climate Information Needs at the Regional Level: The CORDEX Framework. *WMO Bull.* **2009**, *58*, 175–183.
33.   Pfeifer, S.; Bülow, K.; Gobiet, A.; Hänsler, A.; Mudelsee, M.; Otto, J.; Rechid, D.; Teichmann, C.; Jacob, D. Robustness of Ensemble Climate Projections Analyzed with Climate Signal Maps: Seasonal and Extreme Precipitation for Germany. *Atmosphere* **2015**, *6*, 677–698. [CrossRef]

34. German Weather Service. Heißer Tag. 2021. Available online: https://www.dwd.de/DE/service/lexikon/Functions/glossar.html?lv2=101094&lv3=101162 (accessed on 6 September 2021).
35. German Weather Service. Starkregen. 2021. Available online: https://www.dwd.de/DE/service/lexikon/begriffe/S/Starkregen.html (accessed on 6 September 2021).
36. Breiman, L. Statistical Modeling: The Two Cultures (with comments and a rejoinder by the author). *Stat. Sci.* **2001**, *16*, 199–231. [CrossRef]
37. Hyndman, R.J.; Athanasopoulos, G. *Forecasting: Principles and Practice*, 2nd ed.; OTexts: Melbourne, VIC, Australia, 2018.
38. Taylor, S.J.; Letham, B. Forecasting at Scale. *PeerJ Prepr.* **2017**, *5*, 37–45. [CrossRef]
39. Chen, T.; Guestrin, C. XGBoost: A Scalable Tree Boosting System. In Proceedings of the 22nd ACM SIGKDD International Conference on Knowledge Discovery and Data Mining, San Francisco, CA, USA, 13–17 August 2016; pp. 785–794. [CrossRef]
40. Hastie, T.; Tibshirani, R.; Friedman, J. *The Elements of Statistical Learning: Data Mining, Inference, and Prediction*, 2nd ed.; Springer Science & Business Media: New York, NY, USA, 2009.
41. Aggarwal, C.C. *Neural Networks and Deep Learning. A Textbook*; Springer: Cham, Switzerland, 2018.
42. Skansi, S. *Introduction to Deep Learning: From Logical Calculus to Artificial Intelligence*; Springer: Cham, Switzerland, 2018.
43. Fan, F.-L.; Xiong, J.; Li, M.; Wang, G. On Interpretability of Artificial Neural Networks: A Survey. *IEEE Trans. Radiat. Plasma Med. Sci.* **2021**. [CrossRef]
44. Köppen, W.P. *Handbuch der Klimatologie: In Fünf Bänden*, Allgemeine Klimalehre by W. Borchardt, Das geographische System der Klimate, 1st ed.; Köppen, W., Geiger, R., Eds.; Verlag von Gebrüder Borntraeger: Berlin, Germany, 1936.
45. Office for Statistics Berlin-Brandenburg. Bevölkerungsstand. 2021. Available online: https://www.statistik-berlin-brandenburg.de/bevoelkerung/demografie/bevoelkerungsstand (accessed on 6 September 2021).
46. Jacob, D. A note to the simulation of the annual and inter-annual variability of the water budget over the Baltic Sea drainage basin. *Theor. Appl. Clim.* **2001**, *77*, 61–73. [CrossRef]

sustainability

Article

Integrating Political Science into Climate Modeling: An Example of Internalizing the Costs of Climate-Induced Violence in the Optimal Management of the Climate

Shiran Victoria Shen [1,2]

1   Hoover Institution, Stanford University, Stanford, CA 94305-6003, USA; svshen@stanford.edu
2   Woodrow Wilson Department of Politics, University of Virginia, Charlottesville, VA 22904, USA

Abstract: Extant modeling of the climate has largely left out political science; that needs to change. This paper provides an example of how a critical political concept—human security—can be accounted for in climate modeling. Scientific evidence points to an active link between climate change and the incidence of interpersonal and inter-group violence. This paper puts forth a new method to internalize the costs of climate-induced violence in the optimal management of the climate. Using the established MERGE integrated assessment model, this paper finds that based on the median estimates of the climate–violence relationship, such internalization can roughly double the optimal carbon price—the carbon price at which the net social benefit of carbon emissions would be maximized—consistently over time in most sensitivity scenarios. Sub-Saharan Africa is estimated to be the biggest beneficiary of such internalization in terms of avoided damages related to climate-induced violence as a percentage of the regional GDP, avoiding up to a 27 percent loss of GDP by 2200 under high-end estimates. That is significant for many African countries that have been suffering from underdevelopment and violence. The approach of this paper is a first for the climate modeling community, indicating directions for future modeling that could further integrate relevant political science considerations. This paper takes empirical findings that climate change mitigation can reduce violence-related damages to the next step toward understanding required to reach optimal policy decisions.

Keywords: carbon externality; climate management; climate impact; violence; avoided damages; integrated assessment modeling

Citation: Shen, S.V. Integrating Political Science into Climate Modeling: An Example of Internalizing the Costs of Climate-Induced Violence in the Optimal Management of the Climate. *Sustainability* **2021**, *13*, 10587. https://doi.org/10.3390/su131910587

Academic Editor: Baojie He

Received: 4 August 2021
Accepted: 14 September 2021
Published: 24 September 2021

Publisher's Note: MDPI stays neutral with regard to jurisdictional claims in published maps and institutional affiliations.

1. Introduction

Identifying the optimal ways to tackle climate change, which poses one of the biggest threats to human survival and well-being, is more crucial and timelier than ever. To provide important information to policymakers, natural scientists and economists have developed what are known as "integrated assessment models" (IAMs), which combine different strands of knowledge to illuminate how human development, societal choices, and the natural world affect each other in complex systems [1,2]. Works using IAMs have been recognized with two Nobel Prizes, awarded to the Intergovernmental Panel on Climate Change (IPCC) in 2007 and William Nordhaus in 2018, respectively. Extant modeling work has integrated economic, technological, and biophysical processes that produce greenhouse gas (GHG) emissions, but political science—despite its high relevance—has mostly been left out of the picture. This is not entirely surprising because political phenomena can be very challenging to model.

Making reasonable assumptions about human behaviors can provide essential insights into alternative future climates. A critical step towards achieving that goal is to conceptualize and quantify political science ideas, such as power, violence, and legitimacy, in a way that they can be incorporated into existing models. These concepts are consequential yet unincorporated or under-incorporated in existing modeling efforts. Violence is destructive

235

to social order and economic growth [3]. The annual total cost of violence is estimated to be USD 9.4 trillion, which is equivalent to 11 percent of the world's GDP [4–6]. While violence has shaped social and economic development tremendously across human history, most extant economic models have failed to take violence into account, and the impact of security on prosperity remains mostly understudied [3] (It is important to differentiate between "carbon price" and "optimal carbon price." A carbon price is a cost imposed on carbon emissions to discourage polluters from emitting, which usually takes the form of a carbon tax or a requirement to purchase emissions permits. The optimal carbon price is the carbon price at which the net social benefit of carbon emissions would be maximized). Some recently published econometric studies are paving the way, and this study showcases how a critical political concept—human security—can and must be integrated into policy-relevant climate modeling.

Some scientific studies have established an active link between climate change and the incidence of interpersonal and inter-group conflicts. Hotter temperatures lead to higher levels of aggression and violence [7–10]. Increasing temperatures can also indirectly contribute to more violence by decreasing agricultural production, industrial outputs, and political stability [11]. After surveying 60 of the most rigorous quantitative studies of the relationship between temperature change and the incidence of violence, Hsiang, Burke, and Miguel identify strong evidence that climatic events change the frequency or intensity of human violence across substantial spatial and temporal scales [12].

In that spirit, this paper utilizes a new method to internalize the costs of climate-induced violence in the established MERGE integrated assessment model and evaluate how such internalization affects the optimal carbon price returned by the model, along with associated projections of temperature and damages under climate policy (It is important to differentiate between "carbon price" and "optimal carbon price." A carbon price is a cost imposed on carbon emissions to discourage polluters from emitting, which usually takes the form of a carbon tax or a requirement to purchase emissions permits. The optimal carbon price is the carbon price at which the net social benefit of carbon emissions would be maximized). It is based on my working paper that previously appeared in a World Bank report [13]. Deploying recent econometric findings on the costs of different types of violence for different global regions, this paper finds that internalizing the damages from climate-induced violence can roughly double the carbon externality that is priced by the model, and this relationship holds across time and different specifications regarding climate sensitivity, GDP growth rate, and the catastrophic temperature. This relationship can be sensitive when (1) the willingness to pay (WTP) to avoid nonmarket damages (e.g., damages related to mortality, health, quality of life) is low, (2) the nonmarket damages are excluded, or (3) the magnitude of climate-violence damage is at a high boundary of the uncertainty range in empirical studies. Under the assumption that the WTP to avoid nonmarket damages equates to 1 percent of regional income, the avoided damages from climate-induced violence in sub-Saharan Africa is modeled to reach about 0.5 percent of the region's GDP in 2050, 2 percent in 2100, and almost 4 percent in 2200. When the magnitude of climate damage reaches the high-end, the avoided damages from climate-induced violence in sub-Saharan Africa are projected to reach close to 2 percent in 2050, 10 percent in 2100, and 30 percent in 2200 in terms of the region's GDP. This exercise shows that socially contingent damages, such as human violence, can and must be integrated into policy-relevant climate models. Thus, there is vast space for political scientists to integrate their insights and make significant contributions to the climate modeling enterprise.

The rest of the paper is organized as follows. Section 2 provides an overview of the existing literature on the externality of climate-induced violence and the motivation behind this modeling exercise. Section 3 describes the methods and procedures. Section 4 presents modeling results from scenarios under different assumptions. Section 5 concludes with policy implications and future research directions.

2. The Externality of Climate-Induced Violence

Scientific evidence suggests that climate change contributes to a more violent society. On an interpersonal scale, higher temperature has demonstrably led to increased rates of a variety of personal violence [14], including violent crimes [15] and domestic violence [16]. On a collective or intergroup level, empirical evidence points to an active link between climate change and conflicts in both sub-Saharan Africa [17–21] and elsewhere [22]. The causal mechanisms behind the effects of weather shocks on conflicts are sundry: crop yields [20], economic growth [17], government revenue [23], and migration [24].

A model that assesses carbon externality trades off the benefits associated with carbon emissions, largely stemming from associated energy use, with the costs, which can be categorized into market and nonmarket damages. Market damages refer to damages for marketed goods and services, such as property losses due to increased flood risks and declines in agricultural production due to temperature increases. Nonmarket damages include mortality, health, quality of life, as well as effects on environmental goods and services, habitats and ecosystems, and biodiversity. The discount rate captures the temporal dimension of paying at present to avoid future climate damages. To study this tradeoff, researchers have employed integrated assessment models (IAMs), one of whose many functions is to simulate a "causal chain" where carbon emissions lead to climate change and finally to climate damages [25].

Greater unpredictability of temperature resulting from climate change will likely lead to sustained increases in violence. Nevertheless, there does not yet exist any study that considers the cost of climate change-induced violence in computing the optimal carbon price systematically. Existing studies that assess the impact of climate change on mortality and morbidity effects focus on thermal stress, ozone exposure, diarrhea, labor productivity loss, and disease—but not violence [26]. To the author's knowledge, there is only one study that considers one of the many forms of climate change-induced violence—violent crime—in the damage function, but the other types of climate-induced violence damages remain unaccounted [27] (Another study that is broadly related to the climate–conflict relationship focuses instead on a positive sociopolitical feedback loop [28]. Using the Dynamic Integrated Climate-Economy (DICE) model, it illustrates that climate-induced violence makes it harder to sign environmental treaties, including global climate treaties, which, in turn, jeopardizes climate change mitigation further. However, the study does not explicitly model the costs of violence, but it provides further evidence that the estimates of carbon externality reached in this paper are lower bounds).

Thanks to finely disaggregated and newly available data on the costs of a wide range and types of violence for different regions of the world, this paper seeks to internalize the costs of climate-induced violence in the IAM's damage function to calculate the optimal carbon price. This paper then examines how the newly priced carbon externality impacts future temperature and the avoided damages from climate-induced violence for different regions of the world. To assess the effect of uncertainties inherent to projecting future carbon externalities, models are run for multiple sets of scenarios that differ in climate sensitivity, GDP growth rate, the WTP to avoid nonmarket climate damages, the inclusion/exclusion of nonmarket damages, the magnitude of the climate–violence relationship, and the catastrophic temperature—the temperature at which the entire regional product is wiped out.

It shall be noted that the relationship between climate change and the incidence of interpersonal and intergroup violence is under ongoing debate. Including a wider range of studies and deploying clearer methodological standards, the more recent reviews have concluded a significant and robust relationship between climate change and conflict across a wide range of spatial and temporal scales [12,29]. In the meantime, some question the criteria for case inclusion and the appropriateness of meta-analysis as a method to quantify climate-induced violence in these studies [30]. The debate continued in subsequent papers [31,32].

However, it is still a worthwhile exercise to calculate the optimal carbon price by incorporating the damage costs of a plausible relationship between climate and violence. The purpose of this exercise is not to provide precise predictions or propose a new carbon pricing policy. Rather, this study seeks to understand how any increased risk of violence associated with climate change may affect the tradeoff in choices relating to carbon emissions. It assesses model outputs under different assumptions about climate damages.

3. Methods

3.1. Definition and Data

This paper follows the World Health Organization (WHO), which defines violence as "the intentional use of physical force or power, threatened or actual, against oneself, another person, or against a group or community, that either results in or has a high likelihood of resulting in injury, death, psychological harm, maldevelopment or deprivation" [33]. The WHO records deadly violence in three general categories: self-directed violence, collective violence, and interpersonal violence. Here, the latter two types of violence are in focus, which correspond with the data on the costs of violence published in recent years by James Fearon and Anke Hoeffler [4–6]. Collective violence refers to violence perpetrated by organized groups, such as states, rebel organizations, terrorists, street mobs, or criminal organizations. Interpersonal violence refers to violence committed by an individual, which, depending on the relationship between the perpetrator and the victim, can be further broken down into intimate partner violence and child abuse.

This paper uses the cost estimations by Hoeffler [6], who relied on unit cost estimates for various types of violence (e.g., homicides, assaults, and rapes) in the United States by McCollister et al. [34] to estimate the cost of violence for the year 2013. Hoeffler extended the cost estimations to other countries by multiplying the US cost of homicide by the ratio of a country's GDP to US GDP and then used data on violent events by global region to develop regional cost estimates of violence. There are assumptions associated with both the approach taken by Hoeffler and that by McCollister et al. McCollister et al. combined several approaches to pricing crimes, such as the cost of illness, contingent valuation, and jury compensation methods, and all of these approaches assume that the impacts of crimes are fully understood (The cost of illness approach tries to quantify the tangible costs of crime on outcomes of interest based on the best available information, which usually comes from self-reports from the victims and assigned prices. Similarly, the jury award approach seeks to assess the total social cost of crime by employing actual compensation from civil personal injury cases. Contingent valuation involves surveys to gauge respondents' willingness to pay to avoid different crimes, which, in theory, should provide a measure of both tangible and intangible costs). Hoeffler's approach also comes with nontrivial assumptions. First, applying social costs of homicide to calculate the costs of deaths from collective violence equates two drastically different types of death, which generate different social losses. Second, using the ratio of GDPs to map US-based estimates into other countries, especially non-high-income countries, ignores fundamental differences across geographies such as life expectancy, which would affect the social loss of homicidal and non-homicidal offenses. Violence is often of greater concern to and considered "worse" for richer countries than lower-income ones.

Despite these nuances, the estimates made by Hoeffler and Fearon are arguably the most comprehensive at the time of this writing. There is also a long tradition of using simple benefit transfer in the IAM literature to develop first-order approximations [35,36]. Specifically, IAMs often require transferring estimates from developed countries with greater data availability to developing countries with less availability of data.

With that in mind, it is worth noting that the cost estimates of violence likely lie on the lower-bound of the real costs. As detailed by McCollister et al. [34], which estimates most tangible and intangible losses, excludes such costs as psychological injury and additional costs associated with sexual violence (e.g., sexually transmitted infections, pregnancy, suicide, and substance abuse) [34]. This exclusion is transported into Hoeffler's estimates.

Furthermore, in Hoeffler's calculation, self-directed violence costs are excluded and so are some relevant costs in cases where data is unavailable. These include the cost of war injuries, widespread destruction of infrastructure in a war, and economic and security concerns resulting from war, in addition to the costs of nonfatal domestic violence against women and children, violence perpetrated by women against their male partners, and violence amongst homosexual couples. In addition to the missing items specified by Hoeffler, scholarly classics in the social sciences enlighten us that one critical example of valuing the invaluable or pricing the priceless involves measuring the decline or loss of social capital. First put forth by Alexis de Tocqueville, the concept of social capital was developed further by Robert Putnam [37,38]. Social capital refers to "features of social organization such as networks, norms, and social trust that facilitate coordination and cooperation for mutual benefit" [38]. The benefit of social capital is complicated to quantify because measures such as the number and membership size of associations cannot precisely fathom the frequency and quality of association. Violence provides a negative drag on social capital. Hence, any empirical attempt to quantify the costs of violence will lack most intangible costs, and existing estimates are best understood as appropriate underestimates.

I group the various costs into market damages (e.g., GDP losses from war, economic costs of medical care, criminal justice system, and lost income) and nonmarket damages (e.g., deaths from war, fear) as percentages of GDP for seven regions of the world in Table 1. (It is plausible that war deaths entail both nonmarket and market losses. Losses from deaths from war are grouped into nonmarket damages to stay consistent with Hoeffler's method of using intangible costs. Hoeffler calculated the costs of the lives lost in civil wars by "multiplying the number of fatalities by the cost of homicide," whose "intangible cost . . . is assumed to be $8.44 million, and this value is inflated to 2013 prices and scaled by the relative GDP ratio to approximate the cost of homicide across different countries" [34]).

Table 1. Costs of collective and interpersonal violence as percentages of GDP for different global regions [6], by market versus nonmarket damages classified by the author (The percentage figures are reported to three decimal places so that those for East Asia and Pacific have significant digits. The non-market cost of violence as a percentage of regional GDP is much higher in the Middle East and North Africa because the region had a much higher figure for deaths from civil war, as shown in Table 1 in Anke Hoeffler's paper [6]).

Region	Collective Violence		Interpersonal Violence	
	Market	Nonmarket	Market	Nonmarket
East Asia and Pacific	0.007	0.003	0.119	9.151
Europe and Central Asia	0.963	0.017	0.426	10.394
Latin America and Caribbean	0.494	0.046	0.698	18.512
The Middle East and North Africa	0.877	0.603	0.206	27.424
South Asia	0.249	0.011	0.078	20.502
Sub-Saharan Africa	0.595	0.035	0.166	37.114
High Income	0.000	0.000	0.899	5.371

3.2. The Relationship between Climate Change and the Incidence of Violence

In their meta-analysis of the 60 most quantitatively rigorous studies of 45 different conflict datasets, Hsiang, Burke, and Miguel collected findings across a wide range of conflict outcomes that spanned from 10,000 BCE to the present day and across all major regions of the world [12]. They identify that the median effect of a one-standard-deviation increase from normal temperature (i.e., 0.5 °C) induces a 14 percent rise in the frequency of intergroup conflict and a 4 percent increase in the incidence of interpersonal violence globally (While the median effect on intergroup conflict is higher than that on interpersonal violence, the base number of incidents of interpersonal violence is substantially higher; in other words, a small percentage rise can entail a massive increase in total incidents). Based on these figures, the rate of the incidence of climate-induced violence at time t, $v(t)$, can be expressed as a function of temperature increase at time t from the pre-industrial level,

$atp(t)$, and as a multiple of the rate of climate-induced violence in the pre-industrial period (In the version of MERGE used in this paper, $atp(t)$ does not include annual variability. This is an area where future research can improve). Suppose that the rate for the pre-industrial period is "1," the median rate of intergroup violence, $v(group, t)$, and the median rate of interpersonal violence, $v(person, t)$, can be expressed as the following (While some studies identify climate change to have an approximately linear effect on conflicts [39], Hsiang, Burke, and Miguel elaborate that reported linear relationships should be interpreted as local linearizations of a global relationship that is nonlinear and possibly curved [12]. Future extensions of this study can evaluate carbon externality under the assumption of a linear climate-conflict relationship):

$$v(group,\ t) = 1.14^{\frac{atp(t)}{0.5}} \tag{1}$$

$$v(person,\ t) = 1.04^{\frac{atp(t)}{0.5}} \tag{2}$$

In order to factor in the uncertainty represented in Hsiang et al.'s estimates, I would expect to run the model to optimize over the uncertainty in the climate–violence response. This feature is not yet built-in for the version of the model used in this paper. Nevertheless, this paper will try to account for uncertainty by replacing the median estimates in Equations (1) and (2) with high- and low-bound estimates of the climate–violence response in Section 4.6.

3.3. Internalization of the Costs of Climate-Induced Violence into the Damage Function of the IAM

The IAM chosen for this research is Model for Evaluating Regional and Global Effects (MERGE) of GHG reduction policies, an intertemporal general equilibrium model that optimizes discounted utility. It was initially developed at Stanford University and has led to a significant amount of scholarship. It has a relatively detailed climate module, allowing for sufficient flexibility for an alternative view on a wide range of contentious issues, including damages from climate change. The version of the model used in this paper is the same version used in the Stanford Energy Modeling Forum Study (EMF 27) [40], which modified the original version [41] by incorporating more features, refining the global regions, and updating some data. More details on the EMF 27 MERGE model can be found in the online appendix of Blanford et al. (2014) [40]. The EMF 27 global regions include Canada–Australia–New Zealand (Other OECD), China, the Greater European Union, Group 3, India, Japan, the Rest of Asia, the Rest of the World, and the United States (The global regions in the original MERGE model are the USA, other OECD countries (Western Europe, Japan, Canada, Australia, and New Zealand), FSU (the former USSR), China, and the ROW (rest of the world) [41]). Canada–Australia–New Zealand, the Greater European Union, Japan, and the United States are high-income and thus are assigned costs of violence values based on Hoeffler's estimates for the "High Income" region. The climate damages already represented in the model do not include those from violence, so it does not appear necessary to remove any portion of the damage functions for this paper [41].

The modeling process operates in either the benefit-cost mode, which considers climate damages and GHG mitigation costs, or the cost-effective mode, which finds the least-cost emissions mitigation pathway to satisfy a climate-related constraint, such as a limit on concentrations or temperature rise. This study opts for the benefit-cost mode because it seeks the socially optimal price of carbon, given the assumptions, needed to internalize the externalities associated with climate change and maximize the net benefit to society. It does so by invoking the damage module, where damages are calculated from temperature, which influences current production. The objective function below represents the discounted utility of consumption in a given global region after allowing for the disutility of climate change or damages from climate change:

$$Discounted\ utility = \sum_{t=1}^{T} U(c(t))(1+\rho)^{-t} \tag{3}$$

where U stands for the single-period level of utility or social well-being; $c(t)$ is the flow of consumption at time t; ρ represents the rate of time preference for utility.

The carbon price is calculated endogenously in the optimization process. An increase in temperature contributes to climate damage, resulting in a higher carbon price. A higher carbon price then exerts pressure on the producers, forcing them to reduce emissions, which leads to a slowed temperature increase. The loop continues until the algorithm finds the optimal carbon price, which is the one that will yield the highest discounted utility.

The market and nonmarket damages are modified separately for the climate module. The abstract representation of damages is introduced and discussed in Manne, Mendelsohn, and Richels (1995), which is a high-level representation designed to focus on the core tradeoffs. Building on Manne, Mendelsohn, and Richels (1995), the function for market damage in global region i at time t, $MD(i, t)$, can be expressed as:

$$MD(i, t) = \alpha(i) \times gdp(i,t) \times \frac{atp(t)}{\delta} \tag{4}$$

where $\alpha(i)$ is the default market damage factor in given region i, which is represented by the proportion loss in GDP; $gdp(i,t)$ is the GDP in a given region i at a given time t; $atp(t)$ is actual temperature increase from the pre-industrial period at a given time t; δ is the reference temperature for market damage coefficients, which by default equals 2 °C.

To account for the market damages of climate-induced violence, the new market damage factor $\gamma(i)$ is created in the following way:

$$\gamma(i) = \alpha(i) + \beta(i) \tag{5}$$

where $\beta(i)$ refers to the market damage factor from climate-induced violence, both collective and interpersonal. The new market damage is expressed as:

$$MDnew(i, t) = \gamma(i) \times gdp(i,t) \times \frac{atp(t)}{\delta} \tag{6}$$

The nonmarket damages of climate-induced violence are accounted for similarly. The economic loss factor (ELF), which is a component of the nonmarket damage function, hinges on two parameters, $catt$ and hsx [42], and is calculated as follows:

$$ELF(i,t) = [1 - (\frac{reftemp}{catt})^{2^{\,hsx(i,t)}}]$$

$$= [1 - NMD(refwtp) \times \left(\frac{atp(t)}{NMD(reftemp)}\right)^{2^{\,hsx}}] \tag{7}$$

where NMD stands for non-market damage; $refwtp$ is the reference willingness to pay as a fraction of consumption, with a default value of 0.04 in MERGE; $reftemp$ is the reference temperature rise relative to the pre-industrial level for non-market damages, with a default value of 2 °C; $catt$ is a catastrophic temperature parameter chosen in such a way that the entire regional product is wiped out, which is 10 °C by default; hsx is the hockey-stick parameter, representing the quadratic loss due to temperature increase, whose default value is 1. The ELF represents the fraction of consumption that remains available after accounting for nonmarket damages for conventional uses.

The modified ELF function, $ELFnew(i)$, would subtract from the original ELF, $ELF(i)$, the nonmarket damage factor relating to the nonmarket portion of climate-induced violence damages, $\zeta(i)$. Subtraction rather than addition applies here because ELF represents the remaining consumption for society. $ELFnew(i)$ is calculated as follows:

$$ELFnew(i) = ELF(i) - \zeta(i) \tag{8}$$

The $\beta(i)$ and $\zeta(i)$ are calculated by combining the relationship between climate change and the incidence of violence [12] and the costs of collective and interpersonal violence as fractions of GDP for different regions of the world [6]. For given region i, the $\beta(i)$ and $\zeta(i)$ can be expressed based on Equations (1) and (2) as:

$$\beta(i) = \beta(group, i) \times v(group,\ t) + \beta(person,\ i) \times v(person,\ t) \tag{9}$$

$$\zeta(i) = \zeta(group, i) \times v(group,\ t) + \zeta(person, i) \times v(person,\ t) \tag{10}$$

where $\beta(group, i)$ refers to the part of market damage factor from climate-induced intergroup violence and $\beta(person,\ i)$ is the part of market damage factor from climate-induced interpersonal violence. Similar designations apply for $\zeta(i)$.

3.4. Avoided Damage from Internalization

To assess the normative significance of a slowed temperature increase, I calculate the avoided damage from climate-induced violence as a percentage of regional GDP for region i at time t, $\frac{\Delta Damage(i,t)}{gdp(i,t)}$. The avoided damage is the difference in climate damages as a percentage of regional GDP between the Violence scenario and the Business as Usual (BAU) scenario, where the former internalizes the costs of climate-induced violence while the latter does not (For each global region, the MERGE model uses an exogenous trajectory for reference economic growth. For more information, please refer to the online appendix of Blanford et al. (2014)). For a given region i at time t, $\frac{\Delta Damage(i,t)}{gdp(i,t)}$ can be expressed as:

$$\frac{\Delta Damage(i, t)}{gdp(i, t)} = \frac{DamageViolence(i, t) - DamageNoViolence(i, t)}{gdp(i, t)} \tag{11}$$

4. Updating the Carbon Externality and Its Effects under Different Scenarios

This section presents changes in model outcomes when the social costs of climate-induced violence are internalized vis-à-vis when they are not. For each scenario, this paper seeks to compare BAU and Violence scenarios to calculate the avoided damages. This paper presents results under the default assumptions first and then performs five sets of sensitivity analyses: climate sensitivity, GDP growth rate, the WTP to avoid nonmarket climate damages, the inclusion/exclusion of nonmarket damages, and the assumption about the catastrophic temperature at which the WTP at infinite income reaches 100 percent.

While the built-in horizon for MERGE is 2200, most of my reporting of values will be more near-term—until 2050—though projected values are shown up until 2200. Scholars and practitioners may disagree among themselves as to the most appropriate time horizon within which to interpret values. Some believe that near-term values are likely of more interest to policymakers because the further we look into the future, the more pronounced the model uncertainty (i.e., uncertainty regarding the representation of the climate) and the scenario uncertainty (i.e., uncertainty regarding people and their actions). On the other hand, policymakers in the United States, for instance, use a 300-year time span. Uncertainty is not a reason to look at a shorter time horizon, as uncertainty should lead policymakers to take more aggressive climate action via risk premiums and option value. In that light, the 200-year time horizon in MERGE would be short unless the discount rate is relatively high. Either way, it is worth noting that the doubling of the optimal carbon price in the Violence scenario vis-à-vis the BAU scenario holds across time between 2020 and 2200.

4.1. Reference Scenario

I begin with projections for the reference scenario, where all variables other than the market and nonmarket damage factors are left to their default values. The model prices the externality of carbon for all regions of the world. Since the model output carbon prices are un-normalized and the rise in carbon price over time is largely driven by economic growth, it is important to focus on relatively how much larger endogenizing the costs of climate-induced violence would increase the optimal carbon price. Hence, I calculate the ratio of carbon prices in the Violence scenario and the BAU scenario. Figure 1 suggests that internalization consistently raises the optimal carbon price to 1.8 times (i.e., a near doubling) during the 2020–2200 period.

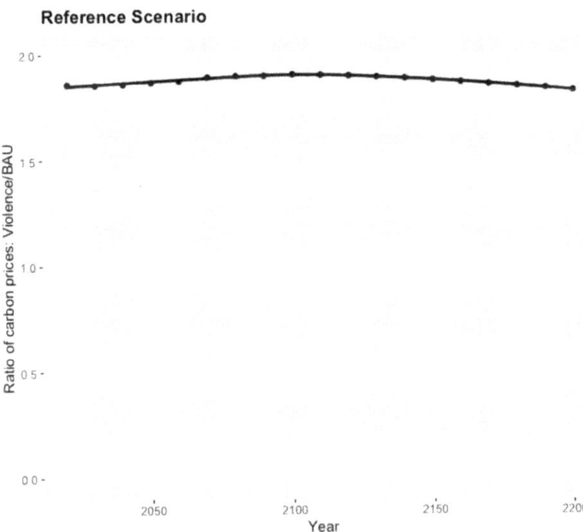

Figure 1. Projections of the ratio of optimal carbon prices in the Violence vs. BAU scenarios.

With the cost of violence internalized, the optimal path of the model shows a lower speed of temperature increase. The projected temperature increases for both the BAU and the Violence scenarios are 1.16 °C in 2020; the values diverge to 1.75 °C and 1.73 °C in 2050, respectively (Figure 2). Based on Equations (1) and (2), the trajectories for temperature give rise to the projected trends for the rates of collective and interpersonal violence, represented as multiples of that in the pre-industrial period (i.e., the pre-industrial rate is "1").

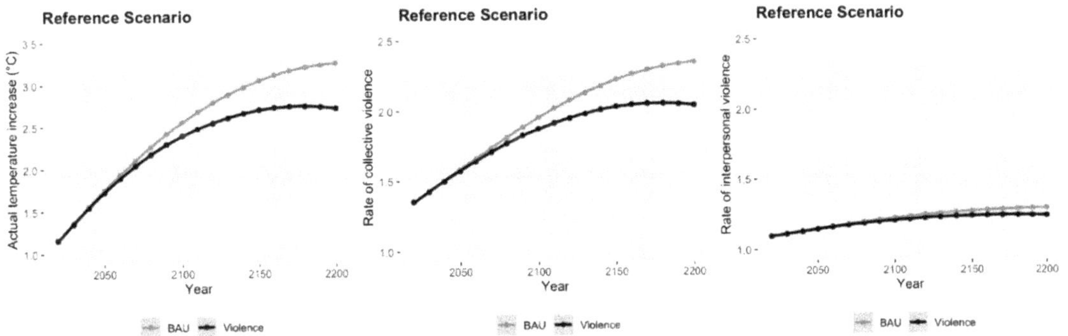

Figure 2. Projections of temperature increase and the rates of collective and interpersonal violence using the pre-industrial levels as the baseline.

To assess the normative significance of a slowed temperature increase, I calculate the avoided damages from climate-induced violence as a percentage of regional GDP (Figure 3). The most prominent beneficiary is sub-Saharan Africa, which is projected to avert 0.09 percent of GDP loss from climate-induced violence in 2050 when the model assesses outputs after internalizing the costs of climate-induced violence. This figure of averted loss is about 14 percent of the region's current incurred costs from collective violence, estimated by Hoeffler, to be at 0.63 percent of the regional GDP [6]. Sub-Saharan Africa is followed by the Middle East and North Africa, India, Eastern Europe and Central Asia, and China. High-income countries, while the lowest on the list, are estimated to be still able to avoid 0.01 percent of GDP worth of loss from climate-induced violence in 2050 with the new carbon pricing in place.

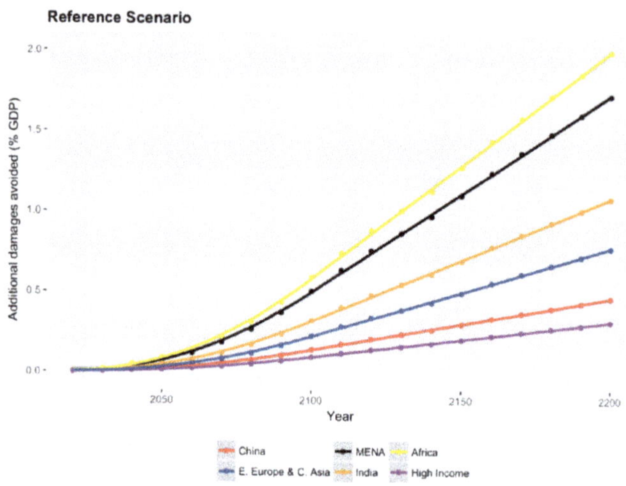

Figure 3. Avoided damages from climate-induced violence as a percentage of regional GDP.

4.2. Sensitivity Analysis: Climate Sensitivity

The climate sensitivity, which is the equilibrium global mean surface temperature change following a doubling of atmospheric CO_2 concentration, is 3.5 °C at default in MERGE (The value of climate sensitivity in MERGE is comparable to those in other IAMs, e.g., FUND at 3 °C, PAGE at 2.54 °C, DICE-2010 at 3.2 °C, and DICE-2013 at 2.9 °C, which are all clustered around the IPCC's Fourth Assessment Report's modal estimate of 3 °C [43]). The upper- and lower-bound values of climate sensitivity also explored are 1.5 °C and 6 °C. As shown in Figure 4, the relationship of a near doubling of the optimal carbon price due to the internalization of climate-induced violence costs holds across time in all three scenarios.

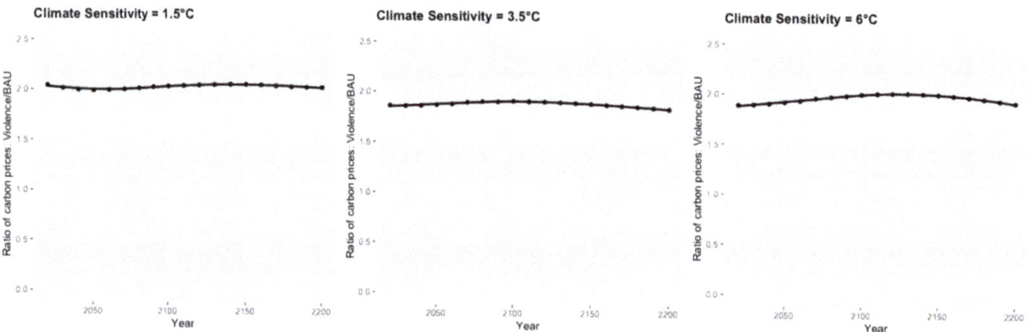

Figure 4. Projections of the ratio of optimal carbon prices in the Violence vs. BAU scenarios, under different assumptions about climate sensitivity.

Under different climate sensitivity specifications, the projection of temperature increases vis-à-vis preindustrial levels in 2050 can reach 1.68 °C and 1.66 °C when the climate sensitivity is high and 1.21 °C and 1.14 °C when the climate sensitivity is low (Figure 5).

With higher climate sensitivity, the projected rates of climate-induced violence during each time period are consistently higher. Under which assumption about climate sensitivity does endogenizing the costs of climate-induced violence yield the most reduced rates of violence? It depends on the time horizon. The rates for collective and interpersonal violence are most reduced at 0.03 and 0.01 in 2050 when the climate sensitivity is low, 0.08 and 0.02 in 2100 when the climate sensitivity is at default, and 0.54 and 0.07 in 2200 when the climate sensitivity is high.

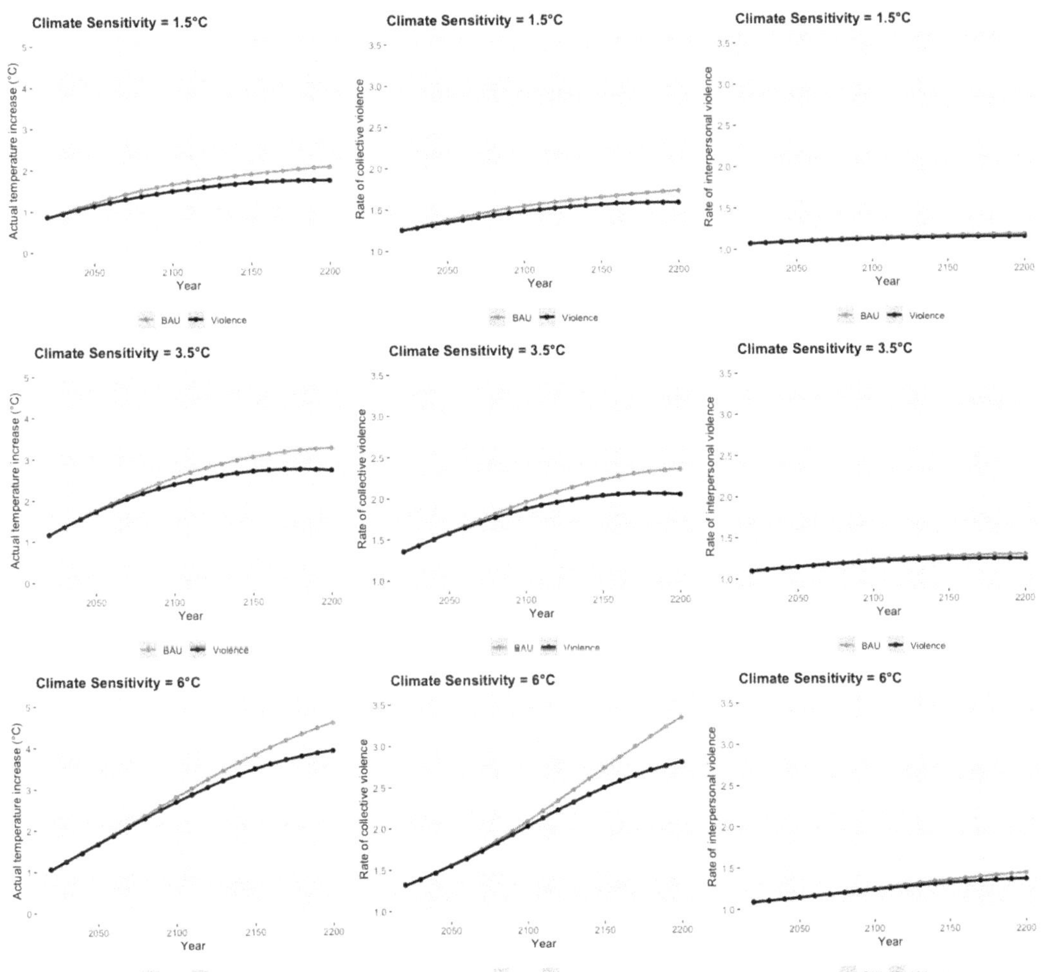

Figure 5. Projections of temperature increase and the rates of collective and interpersonal violence, under different assumptions about climate sensitivity.

Based on the trajectories of temperature and the rates of violence, I calculate the avoided damages from climate-induced violence as percentages of regional GDP under different climate sensitivity scenarios. Figure 6 suggests that in the long run, the higher the climate sensitivity, the higher the avoided damages from climate-induced violence as a percentage of regional GDP. When the climate sensitivity equals 1.5 °C, sub-Saharan Africa is projected to prevent 0.24 percent of GDP loss due to climate-induced violence in 2050. Under the assumption of a highly sensitive climate (6 °C), the percentage for Africa is estimated to reach 0.05 percent in 2050. At the other end of the spectrum, the avoided damages from climate-induced violence will be worth 0.04 percent of GDP when the climate is the least sensitive and 0.01 percent of GDP when the climate is the most sensitive in high-income countries.

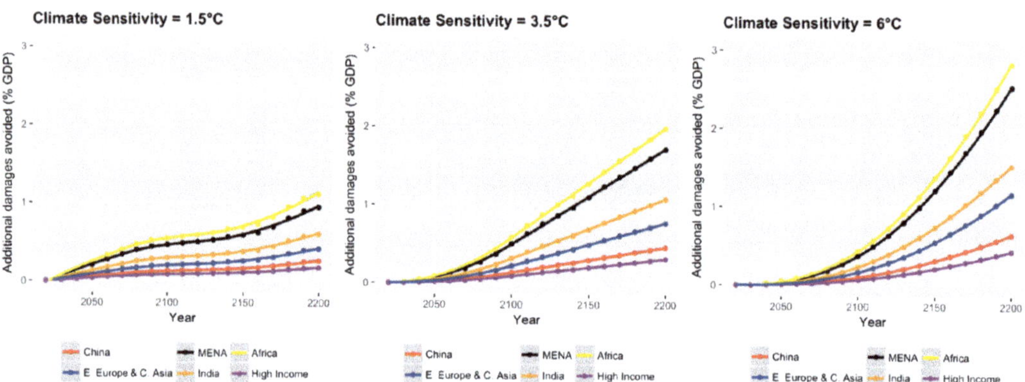

Figure 6. Projections of avoided damages from climate-induced violence as percentages of regional GDP, under different assumptions about climate sensitivity.

4.3. Sensitivity Analysis: GDP Growth Rate

The second dimension along which I seek to test the sensitivity of the results is the GDP growth rate. The default GDP growth rates in MERGE are increased by 1 percent and 2 percent. Figure 7 shows that accounting for climate-induced violence costs nearly doubles the pricing of carbon externalities, a relationship that remains robust across time and growth rate scenarios.

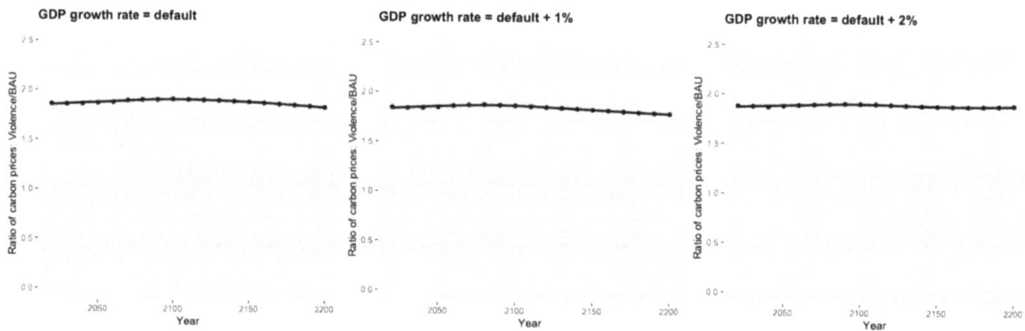

Figure 7. Projections of the ratio of optimal carbon prices in the Violence vs. BAU scenarios, under different assumptions about GDP growth rate.

The higher the GDP growth rate, the higher the optimal carbon prices (even after normalization), which lowers the projected temperature increases. When the default GDP growth rate is incremented by 2 percent, the expected temperature increases are 1.72 °C and 1.71 °C in the BAU and the Violence scenarios in 2050, respectively (Figure 8). When the GDP growth rate is at default or is incremented by 1 percent, the rate of collective violence is reduced by 0.01 in the Violence scenario vis-à-vis the BAU scenario in 2050.

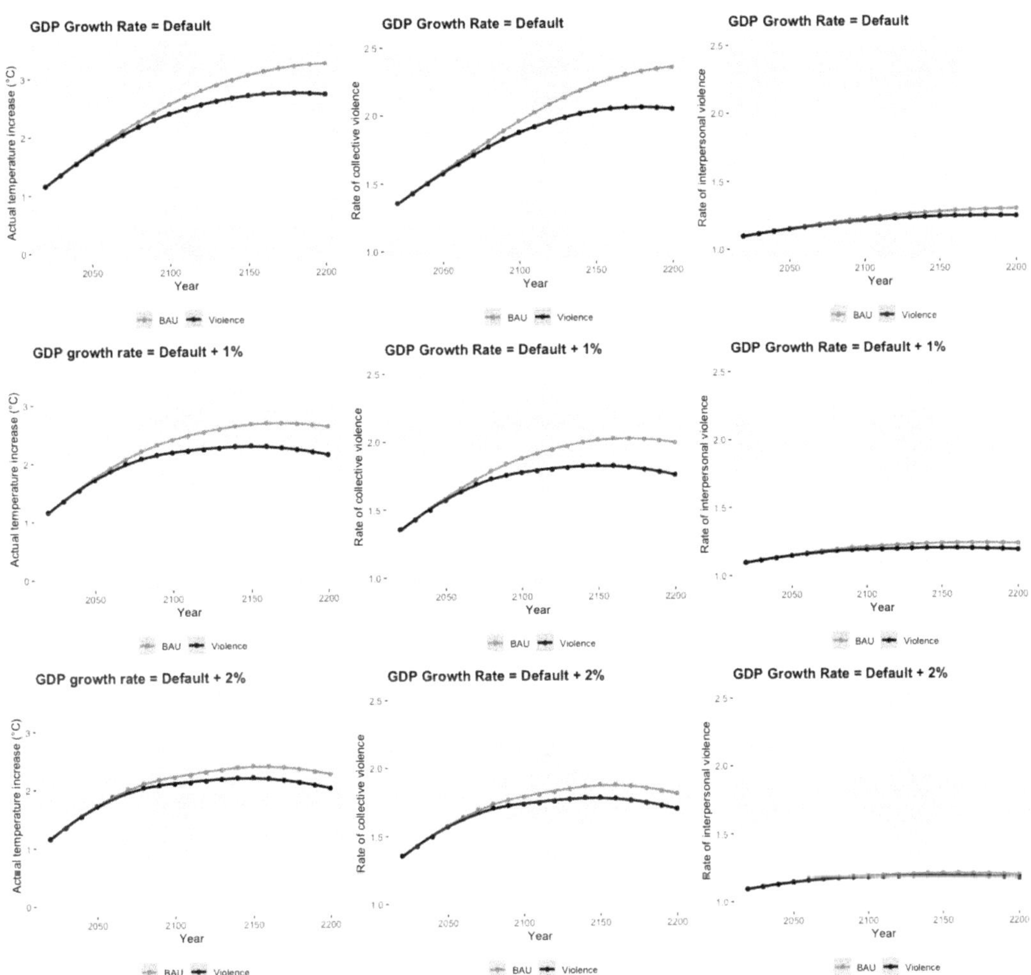

Figure 8. Projections of temperature increase and the rates of collective and interpersonal violence, under different assumptions about GDP growth rate.

The avoided damages from climate-induced violence as percentages of regional GDP are projected to decrease as the GDP growth rate increases (Figure 9). It is also worth noting that when the GDP growth rate is incremented by 2 percent, the avoided costs are negative in 2020, 2030, and 2040. That means that during those time periods internalizing climate-induced violence in generating the optimal carbon price would cause more damages from climate-induced violence. Nevertheless, the payoffs become evident in the longer run.

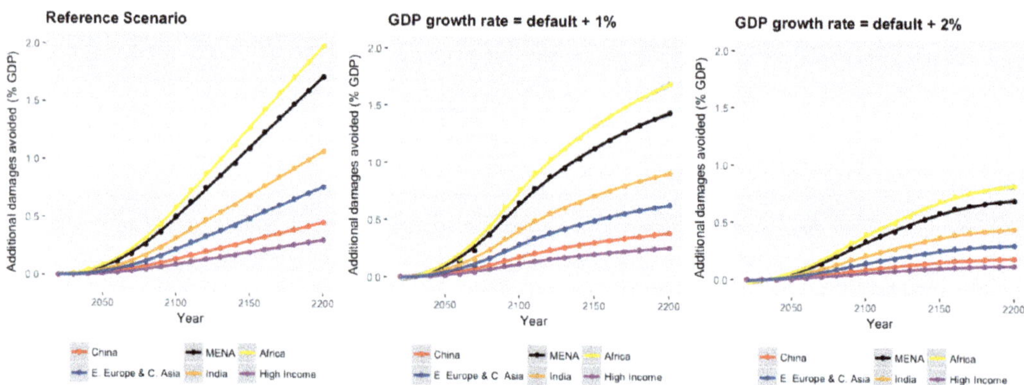

Figure 9. Projections of avoided damages from climate-induced violence as percentages of regional GDP, under different assumptions about GDP growth rate.

4.4. Sensitivity Analysis: WTP for Nonmarket Damages

The third dimension along which I seek to test the sensitivity of the results is the WTP to avoid nonmarket climate damages. For nonmarket damages, MERGE is built based on the highly speculative assumption that the expected losses would increase quadratically with the rise in temperature [42]. Changing the WTP, which is in itself uncertain, for nonmarket damages can influence how much nonmarket damages are factored into the calculation of optimal carbon prices. Furthermore, the default WTP is 4 percent in MERGE, meaning that residents of all regions are willing to devote 4 percent of their regional income to avoid nonmarket climate damages associated with 2 °C of warming. Since the figure could be lower in less-developed regions of the world, I change the WTP to avoid nonmarket damages to 1 percent and 2 percent for the sensitivity analysis.

As shown in Figure 10, incorporating the costs of climate-induced violence raises the projection of optimal carbon prices to about twice as much as before under the 2 percent of regional GDP assumption and a bit above three times as much under the 1 percent assumption. Hence, changes in pricing the carbon externality in the Violence scenario vis-à-vis BAU could be sensitive to the assumption about WTP to avoid nonmarket damages.

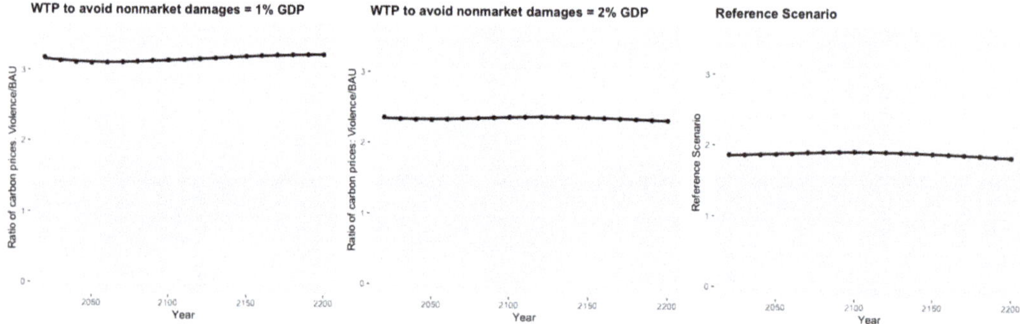

Figure 10. Projections of the ratio of optimal carbon prices in the Violence vs. BAU scenarios, under different assumptions about the WTP for nonmarket damages.

The projected temperature increases correspond with the fact that the higher the WTP, the more people care about avoiding climate change, which lowers the temperature increase (Figure 11). Under a WTP of 1 percent of regional income, the projected temperature increases are estimated to be 1.89 °C and 1.76 °C in 2050, and the rates of collective and interpersonal violence are projected to reduce by 0.06 and 0.01, respectively, in the Violence scenario vis-à-vis the BAU scenario.

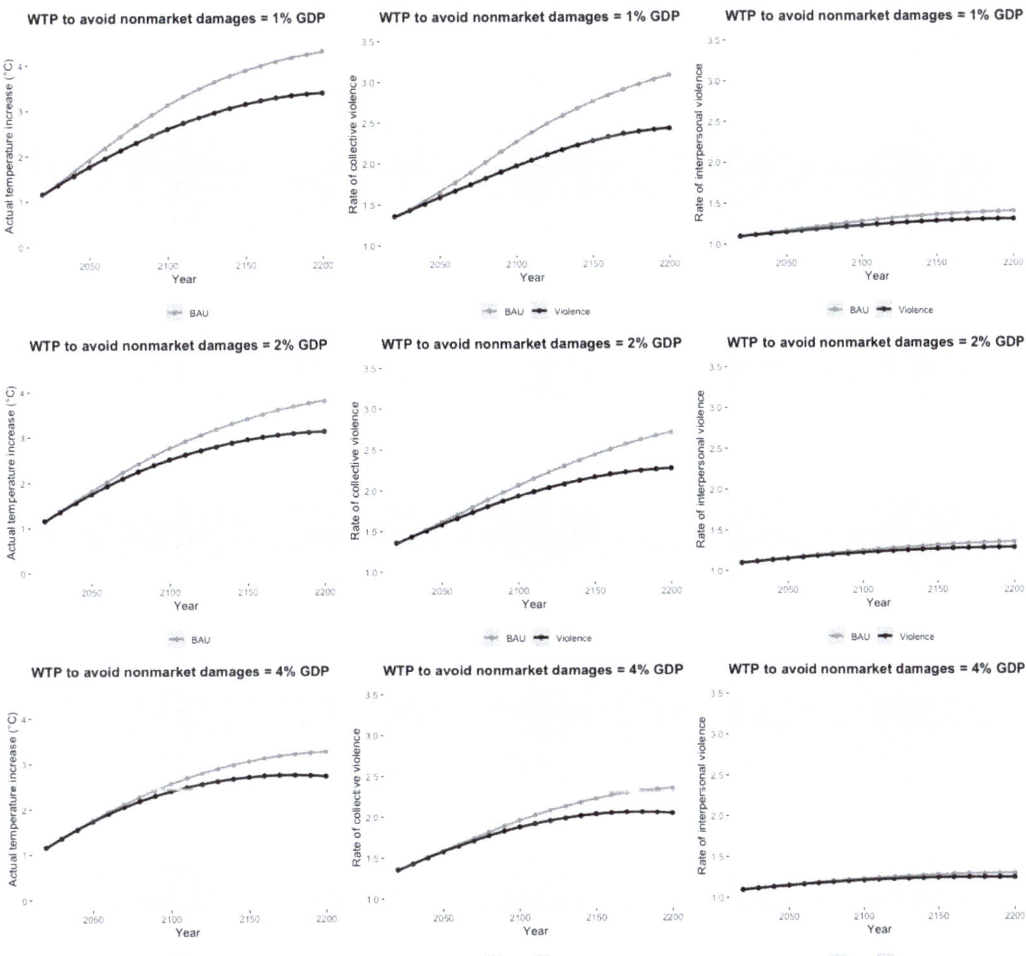

Figure 11. Projections of temperature increase and the rates of collective and interpersonal violence, under different assumptions about the WTP for nonmarket damages.

The lower the WTP to avoid nonmarket damages, the higher the avoided damages from climate-induced violence (Figure 12). This pattern is intuitive because as people care less about climate damages from sources other than climate-induced violence, climate-induced violence matters relatively more to people. At a WTP of 1 percent of regional income, which is perhaps the closest to reality based on existing empirical works, the avoided damages from climate-induced violence in sub-Saharan Africa are estimated to reach 0.46 percent of the region's GDP in 2050.

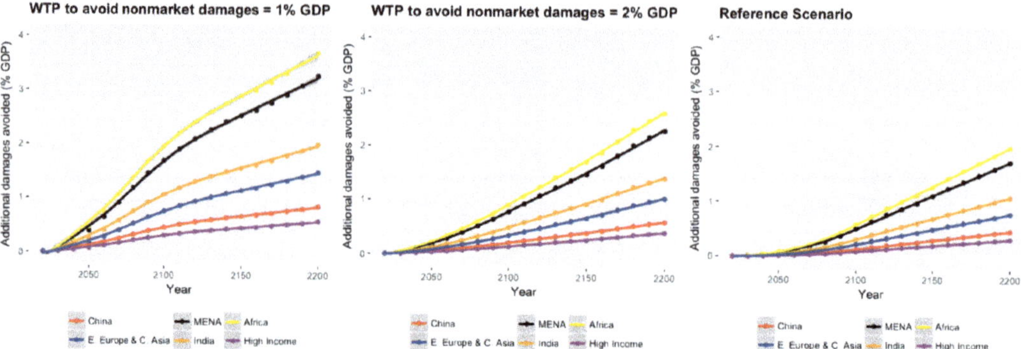

Figure 12. Projections of avoided damages from climate-induced violence as percentages of regional GDP, under different assumptions about the WTP for nonmarket damages.

4.5. Sensitivity Analysis: Inclusion/Exclusion of Nonmarket Damages

I examine how the projections of the optimal carbon price change when only the market damages from climate-induced violence are considered. This exercise is worthwhile because MERGE assumes that the expected losses would increase quadratically with the rise in temperature, which is highly speculative and prone to producing very high optimal carbon prices when the temperature is high. In light of this, the reference/default scenarios are rerun to exclude nonmarket damages. Between the Violence and BAU scenarios, the Violence scenario yields optimal carbon prices about twice the magnitude of those in the BAU scenario in the next few decades (2.04 times in 2020 and 2.11 times in 2050); the magnitude is projected to increase gradually to above three times by 2200 (Figure 13).

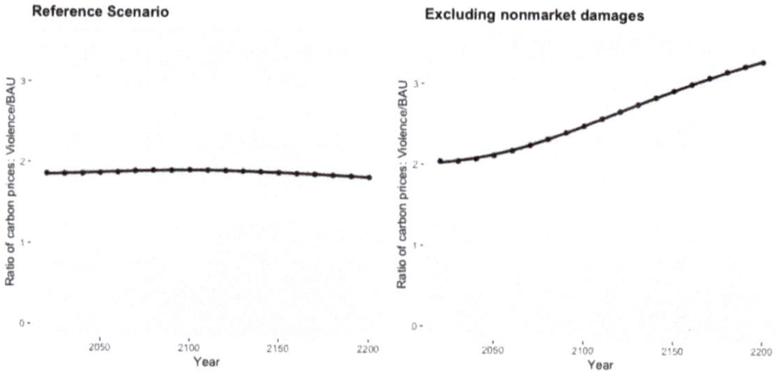

Figure 13. Projections of the ratio of optimal carbon prices in the Violence vs. BAU scenarios, with or without the inclusion of nonmarket damages.

Intuitively, lower optimal carbon prices give rise to higher projected temperatures and higher rates of climate-induced violence (Figure 14). The avoided damages from climate-induced violence as percentages of regional GDP are consistently lower with lower carbon prices in place (Figure 15).

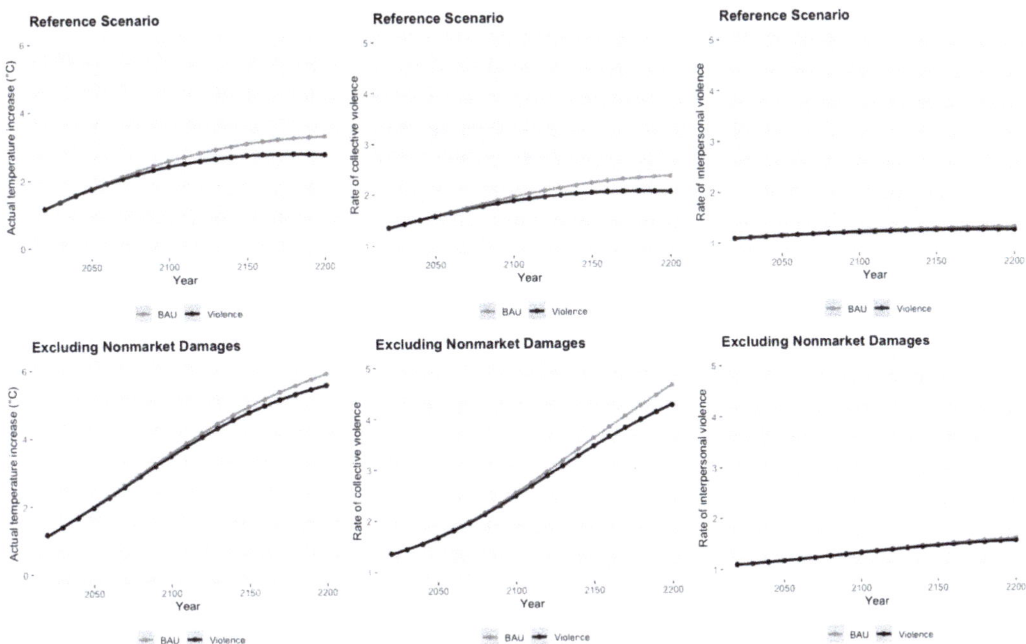

Figure 14. Projections of temperature increase and the rates of collective and interpersonal violence, with or without the inclusion of nonmarket damages.

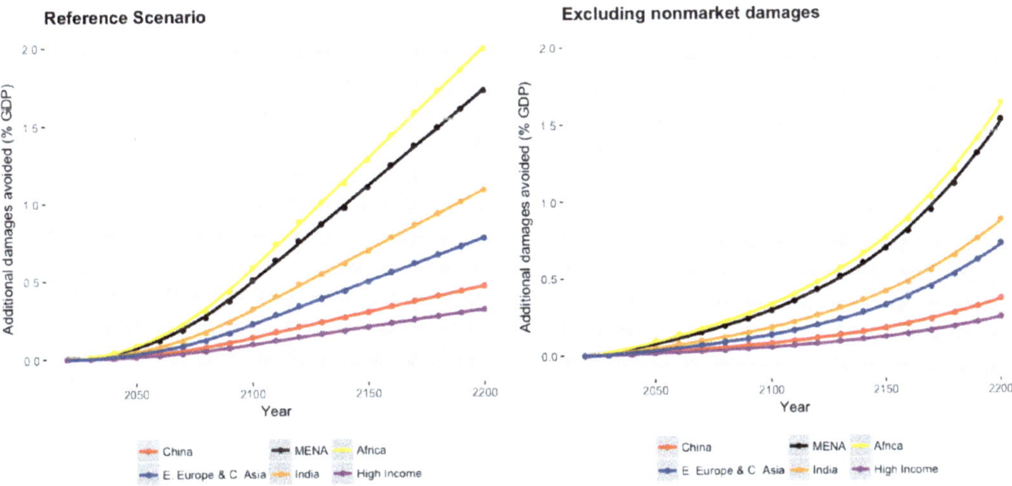

Figure 15. Projections of avoided damages from climate-induced violence as percentages of regional GDP, with or without the inclusion of nonmarket damages.

4.6. Accounting for Uncertainty: The Magnitude of the Climate–Violence Response Relationship

As foreshadowed in Section 3.2, this paper seeks to factor in the uncertainty surrounding the estimates in Hsiang et al. (2013). Specifically, this paper reassesses the results by replacing the median estimates of the increase in the incidence of climate-induced violence with high- and low-bound estimates. The calculations refer to the Supplementary Tables S2 and S3 in Hsiang et al. (2013), which provide their original estimated effects from each study on interpersonal violence and intergroup violence, respectively. For estimates on interpersonal violence, calculations take the second-highest (16) and the second-lowest (2), among 11, estimates. For estimates on intergroup conflict, calculations take the fourth-highest (42) and the fourth-lowest (6), among 21, estimates. The original Equations (1) and (2) are updated to (12) and (13), respectively, for high-bound estimates of the incidences of climate-induced violence and expressed as follows:

$$v(group,\ t) = 1.42^{\frac{atp(t)}{0.5}} \tag{12}$$

$$v(person,\ t) = 1.16^{\frac{atp(t)}{0.5}} \tag{13}$$

Similarly, the low-bound estimates of the incidences of climate-induced violence are expressed in Equations (14) and (15) as:

$$v(group,\ t) = 1.06^{\frac{atp(t)}{0.5}} \tag{14}$$

$$v(person,\ t) = 1.02^{\frac{atp(t)}{0.5}} \tag{15}$$

As shown in Figure 16, when the damage magnitude is low-bound, the Violence scenario yields optimal carbon prices nearly twice the size of those in the BAU scenario. Having a high-bound damage magnitude raises the optimal carbon prices to between six and eight times those in the BAU scenario, suggesting that the main finding is sensitive to the value of the high-bound damage magnitude.

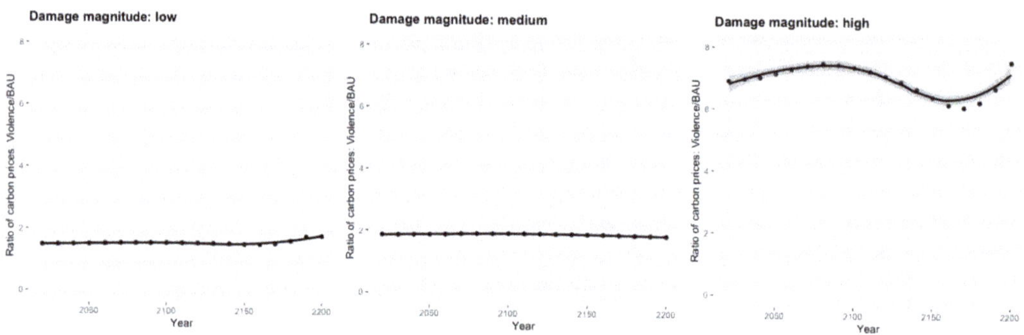

Figure 16. Projections of the ratio of optimal carbon prices in the Violence vs. BAU scenarios, under different assumptions about the magnitude of the climate-violence response relationship.

With the highest optimal carbon price across three scenarios, the high-damage-magnitude scenario is projected to have the least rise in temperature and, by extension, rates of climate-induced violence (Figure 17). As a result, such a scenario is also projected to have the highest avoided damages from climate-induced violence as percentages of regional GDP (Figure 18). Of particular note, sub-Saharan Africa is estimated to avoid damages related to climate-induced violence that is worth 1 percent of the regional GDP in 2050, 9 percent in 2100, and 27 percent in 2200.

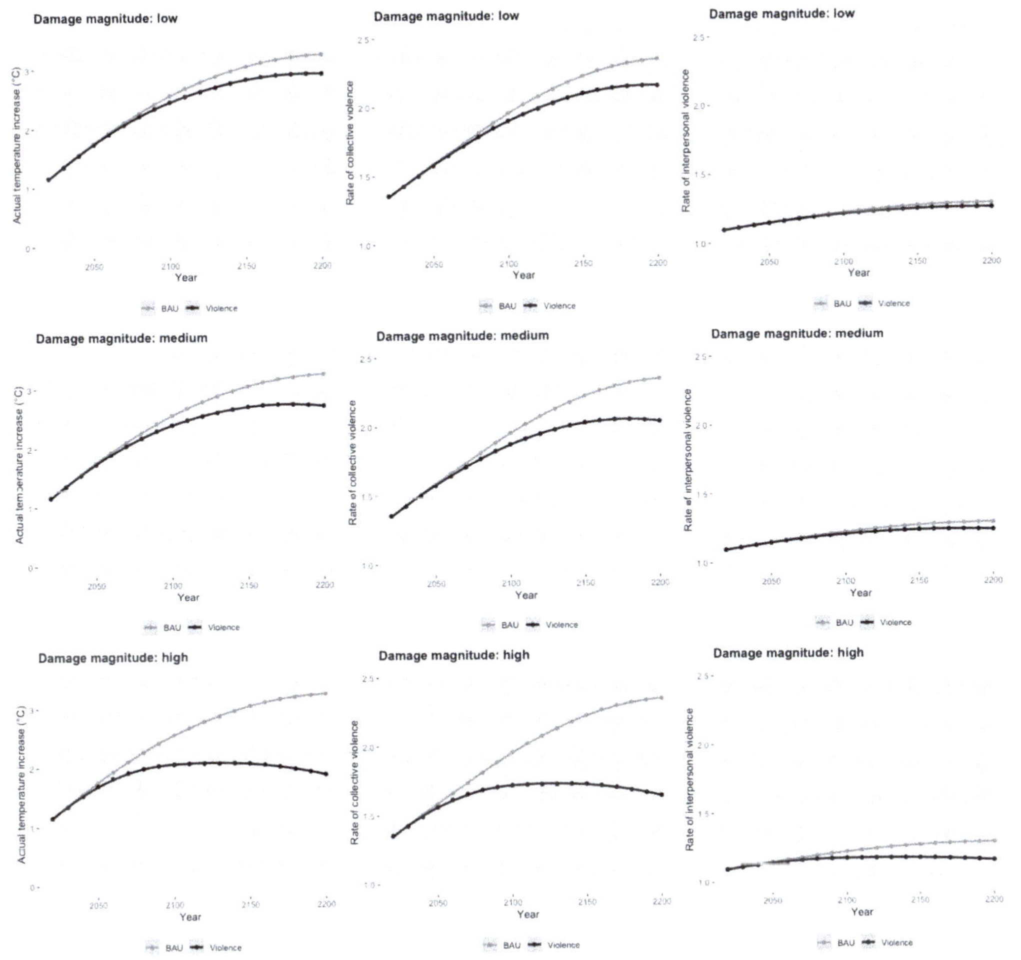

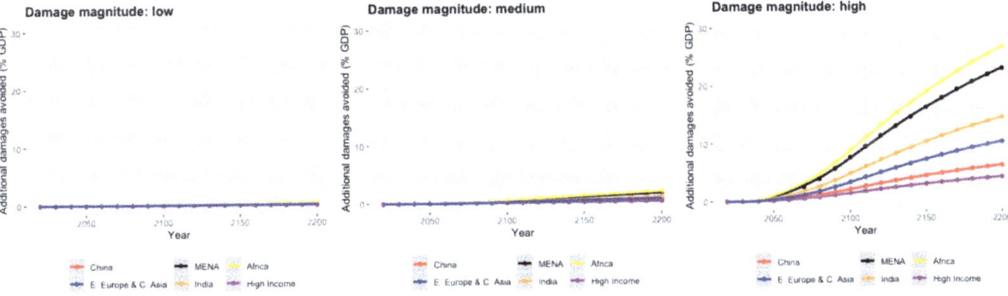

Figure 17. Projections of temperature increase and the rates of collective and interpersonal violence, under different assumptions about the magnitude of the climate-violence response relationship.

Figure 18. Projections of avoided damages from climate-induced violence as percentages of regional GDP, under different assumptions about the magnitude of the climate-violence response relationship.

4.7. Sensitivity Analysis: Catastrophic Temperature

Last but not least, a sensitivity analysis is performed on the value of the catastrophic temperature parameter. The economic loss factor (*ELF*), which is part of the nonmarket damage function, represents the fraction of consumption left for conventional use by households and government. In MERGE, the default WTP to avoid nonmarket climate damages associated with 2 °C of warming is 4 percent, meaning that $ELF(reftemp)$ has a default value of 0.96. *reftemp* is the reference temperature rise relative to the pre-industrial level for nonmarket damages with a default value of 2 °C and is calculated as follows:

$$ELF(reftemp) = [1 - (\frac{reftemp}{catt})^2]^{hsx} \tag{16}$$

This *ELF* depends on the values of two parameters: *catt* and *hsx* [42]. Here, *catt* is a catastrophic temperature parameter chosen in such a way that the entire regional product is wiped out; in other words, it represents the catastrophic temperature at which the WTP at infinite income reaches 100 percent. Its default value in MERGE is 10 °C. In addition, *hsx*, the hockey-stick parameter that represents the degree of loss due to temperature increase. A default value of 1 assumes that the loss is quadratic in terms of the temperature rise, and this value is assumed for high-income countries and is highly speculative [42]. If the amount of loss as a result of one unit of temperature rise is less than the quadratic, the value of *hsx* should be less than 1, and the value for *catt* should be lower than the default. After a new value for *catt* is set at 8.5 °C, leaving all other parameters unchanged, the analysis is re-executed.

With a lower assumed catastrophic temperature, the optimal carbon price projections for the *catt* = 8.5 °C scenario is very similar to those for the default scenario until 2170 (Figure 19). The near doubling of the optimal carbon prices in the Violence scenario vis-à-vis the BAU scenario persists between 2020 and 2180 when *catt* = 8.5 °C. After 2180, the default scenario, where the catastrophic temperature is assumed to be higher, the ratio of the optimal carbon prices is projected to decline in the *catt* = 8.5 °C scenario.

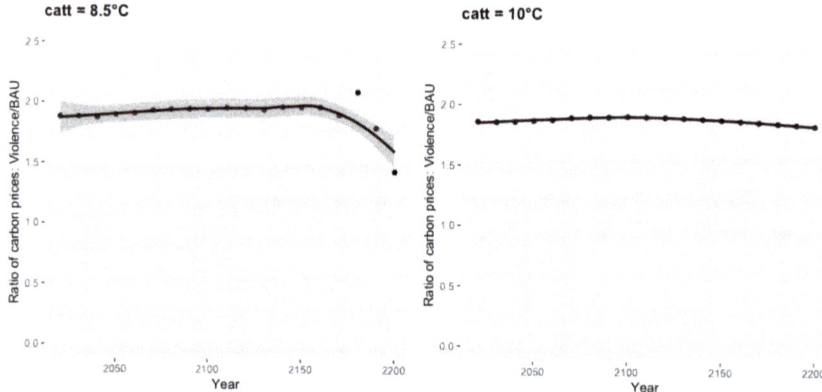

Figure 19. Projections of the ratio of optimal carbon prices in the Violence vs. BAU scenarios, under different assumptions about the catastrophic temperature.

A similar trend is observed for the projected temperatures, and by extension, the rates of collective and interpersonal violence and the avoided damages from climate-induced violence (Figures 20 and 21). Changing the assumption about the catastrophic temperature does not seem to affect the projections until 2170, with a horizon towards 2200. After 2170, the default scenario yields lower projected temperate and rates of violence and higher avoided damages as percentages of regional GDP thanks to higher optimal carbon prices.

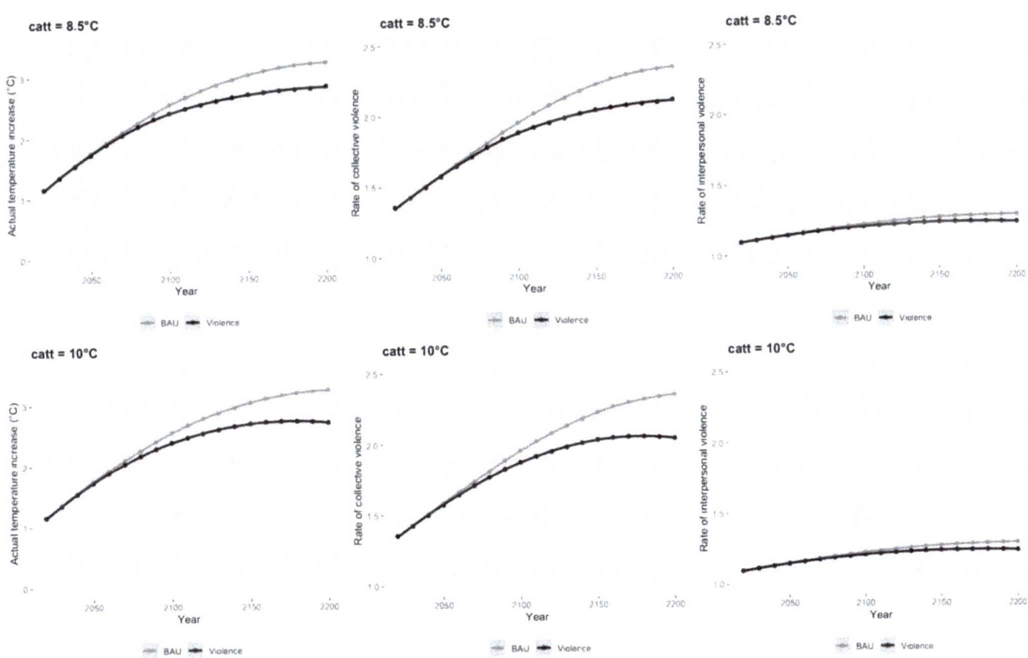

Figure 20. Projections of temperature increase and the rates of collective and interpersonal violence under different assumptions about the catastrophic temperature, under different assumptions about the catastrophic temperature.

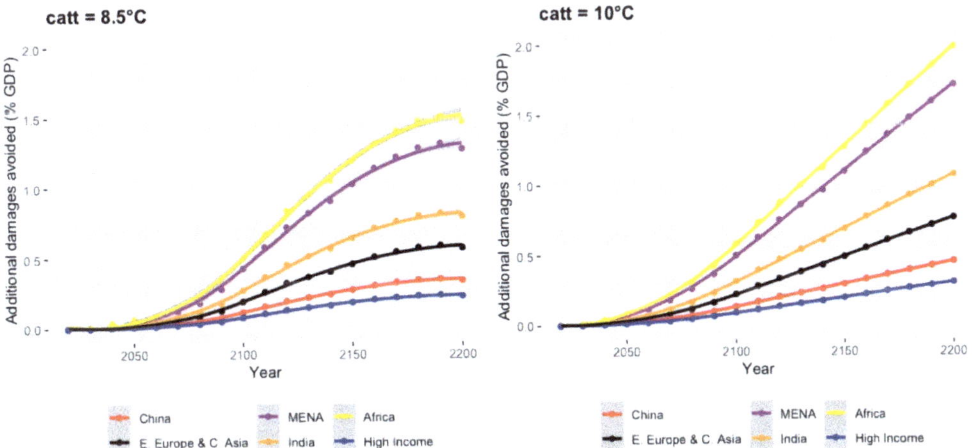

Figure 21. Projections of avoided damages from climate-induced violence as percentages of regional GDP under different assumptions about the catastrophic temperature, under different assumptions about the catastrophic temperature.

5. Conclusions and Policy Implications

This study puts forth a new way of modeling carbon externalities by endogenizing a critical yet previously missing aspect of climate impacts—violence. Utilizing recently published estimates of the costs of violence for different global regions [6] and the median estimates of the incidence of violence per one standard deviation change in temperature [12], this study derives the costs of climate-induced violence. By internalizing such costs in MERGE, which is a methodological contribution of the paper, the results indicate that the optimal carbon price roughly doubles across a range of scenarios with different assumptions regarding climate sensitivity, GDP growth rate, and the catastrophic temperature. This relationship can be sensitive when the WTP to avoid nonmarket damages is low, when nonmarket damages are excluded, or when the climate–violence damage magnitude is at a high-bound of the uncertainty range represented in Hsiang et al.'s estimates.

However, there are two known biases. First, the costs of violence are plausibly low-bound estimates, downwardly biasing the estimates of carbon externality. Second, since the climate–violence relationship coefficient in Hsiang, Burke, and Miguel (2013) is mainly based on local temperature deviations, which are likely higher than the global temperature variation, the estimates of carbon externality could be biased upward [12]. To mitigate these biases, future research can utilize improved estimates of the costs of violence, which would be based on the country-level estimate of the value of a statistical life (VSL), projected country-level weather trends, and country- or region-level climate–violence relationships, when such estimates become available. (Furthermore, since Hsiang, Burke, and Miguel (2013) provide one approximation of the "true" functional form of the relationship between climate change and conflict, future research can utilize other functional forms in the modeling exercise (e.g., linear). Moreover, researchers can conduct similar analyses using other IAMs and perform inter-model comparisons of results. Last but not least, future research can try to factor in the effects of projected adaption).

For the modeling community, the take-home message is that climate-induced violence may have a material impact on the results and should be considered in any model trading off the costs and benefits of greenhouse gas emissions. For the policy community, the approach of this paper bears normative significance by incorporating climate-induced violence costs into the broader tradeoff between the costs and benefits of carbon emissions across the global economy. Based on the median estimate of the effect of climate change on violence and under the assumption that the WTP to avoid nonmarket damages equates 1 percent of regional income, the avoided damages from climate-induced violence in sub-Saharan Africa is estimated to reach 0.5 percent of the region's GDP in 2050, 2 percent in 2100, and almost 4 percent in 2200. When the magnitude of climate damage is near high-bound, the avoided damages from climate-induced violence in sub-Saharan Africa are projected to reach 1 percent of the region's GDP in 2050, 9 percent in 2100, and 27 percent in 2200. These figures are very significant for a region that has long suffered from underdevelopment and violence and thus deserve policy attention.

More broadly, this exercise shows that socially contingent damages, such as violence, can and must be integrated into policy-relevant climate models. Given the significant role that politics play in shaping climate-related policies and trajectories, political scientists who have not yet been actively involved in the development of IAMs have great potential of making significant contributions. Beyond the climate–violence example presented in this paper, another promising area is the integration of political constraints into IAMs, possibly in the form of a new module.

Funding: This research did not receive nor required funding.

Institutional Review Board Statement: Not applicable.

Informed Consent Statement: Not applicable.

Data Availability Statement: Data sharing not applicable.

Acknowledgments: I thank Raphael Calel, James Merrick, John Weyant, and audiences at Stanford University, the World Bank's 2019 World's First International Research Conference on Carbon Pricing, and the 2020 Integrated Assessment Modeling Consortium Annual Meeting for providing feedback on previous drafts of this paper; James Merrick for helping with MERGE; and Angus Binnie and Rand Perry for proofreading the draft. Any remaining errors are my own.

Conflicts of Interest: The author declares no conflict of interest.

References

1. Matson, P.; Clark, W.; Andersson, K. *Pursuing Sustainability: A Guide to the Science and Practice*; Princeton University Press: Princeton, NJ, USA, 2016.
2. Weyant, J. Some Contributions of Integrated Assessment Models of Global Climate Change. *Rev. Environ. Econ. Policy* **2017**, *11*, 115–137. [CrossRef]
3. North, D.; Wallis, J.; Weingast, B. *Violence and Social Orders: A Conceptual Framework for Interpreting Recorded Human History*; Cambridge University Press: Cambridge, UK, 2009.
4. IEP. *The Economic Cost of Violence Containment: A Comprehensive Assessment of the Global Cost of Violence*; Institute for Economics & Peace: Sydney, Australia, 2014.
5. Fearon, J.; Hoeffler, A. *Beyond Civil War: The Costs of Interpersonal Violence and the Next Round of MDGs*; Cambridge University Press: Cambridge, UK, 2014.
6. Hoeffler, A. What Are the Costs of Violence? *Politics Philos. Econ.* **2017**, *16*, 422–445. [CrossRef]
7. Anderson, C.A. Temperature and aggression: Ubiquitous effects of heat on occurrence of human violence. *Psychol. Bull.* **1989**, *106*, 74–96. [CrossRef]
8. Anderson, C.A.; Anderson, K.B.; Dorr, N.; DeNeve, K.M.; Flanagan, M. Temperature and aggression. *Adv. Exp. Soc. Psychol.* **2000**, *32*, 63–133.
9. Kenrick, D.T.; MacFarlane, S.W. Ambient Temperature and Horn Honking: A Field Study of the Heat/Aggression Relationship. *Environ. Behav.* **1986**, *18*, 179–191. [CrossRef]
10. Larrick, R.P.; Timmerman, T.A.; Carton, A.M.; Abrevaya, J. Temper, Temperature, and Temptation: Heat-Related Retaliation in Baseball. *Psychol. Sci.* **2011**, *22*, 423–428. [CrossRef]
11. Dell, M.; Jones, B.F.; Olken, B.A. Temperature shocks and economic growth: Evidence from the last half century. *Am. Econ. J. Macroecon.* **2012**, *4*, 66–95. [CrossRef]
12. Hsiang, S.; Burke, M.; Miguel, E. Quantifying The Influence of Climate On Human Conflict. *Science* **2013**, *341*, 1235367. [CrossRef]
13. Shen, S.V. Pricing Carbon to Contain Violence. In *The First International Research Conference on Carbon Pricing*; World Bank: Washington, DC, USA, 2019; pp. 331–348.
14. Ranson, M. Crime, weather, and climate change. *J. Environ. Econ. Manag.* **2014**, *67*, 274–302. [CrossRef]
15. Jacob, B.; Lefgren, L.; Moretti, E. The Dynamics of Criminal Behavior: Evidence from Weather Shocks. *J. Hum. Resour.* **2007**, *42*, 489–527. [CrossRef]
16. Auliciems, A.; DiBartolo, L. Domestic Violence in a subtropical environment: Police calls and weather in Brisbane. *Int. J. Biometeorol.* **1995**, *39*, 34–39. [CrossRef]
17. Miguel, E.; Satyanath, S.; Sergenti, E. Economic Shocks and Civil Conflict: An Instrumental Variables Approach. *J. Polit. Econ.* **2004**, *112*, 725–753. [CrossRef]
18. Burke, M.; Miguel, E.; Satyanath, S.; Dykema, J.; Lobell, D. Warming increases risk of civil war in Africa. *Proc. Natl. Acad. Sci. USA* **2009**, *106*, 20670–20674. [CrossRef]
19. Burke, M.; Miguel, E.; Satyanath, S.; Dykema, J.A.; Lobell, D.B. Climate robustly linked to African civil war. *Proc. Natl. Acad. Sci. USA* **2010**, *107*, E185. [CrossRef]
20. Harari, M.; La Ferrara, E. Conflict, Climate and Cells: A disaggregated analysis. *Rev. Econ. Stat.* **2018**, *100*, 594–608. [CrossRef]
21. O'Loughlin, J.; Linke, A.M.; Witmer, F.D.W. Effects of temperature and precipitation variability on the risk of violence in sub-Saharan Africa, 1980–2012. *Proc. Natl. Acad. Sci. USA* **2014**, *111*, 16712–16717. [CrossRef]
22. Hsiang, S.; Meng, K.C.; Cane, M.A. Civil conflicts are associated with the global climate. *Nature* **2011**, *476*, 438–441. [CrossRef]
23. Shilling, A.K. *Climate Change and Conflict: Identifying the Mechanisms*; Stanford University: Stanford, CA, USA, 2012.
24. Bhavnani, R.R.; Lacina, B. The Effects of Weather-Induced Migration on Sons of the Soil Riots in India. *World Politics* **2015**, *67*, 760–794. [CrossRef]
25. Rose, S.; Diaz, D.; Blanford, G. Understanding the Social Cost of Carbon: A Model Diagnostic and Inter-Comparison Study. *Clim. Chang. Econ.* **2017**, *8*, 1750009. [CrossRef]
26. Hurd, B.; Jorgenson, D.W.; Goettle, R.J.; Smith, J.U.S. *Market Consequences of Global Climate Change*; Center for Climate and Energy Solutions: Arlington, VA, USA, 2004.
27. Hsiang, S. Estimating economic damage from climate change in the United States Downloaded from. *Science* **2017**, *356*, 1362–1369. [CrossRef] [PubMed]
28. Howard, P.; Livermore, M.A. Sociopolitical Feedbacks and Climate Change. *Harvard Environ. Law Rev.* **2019**, *43*, 119–174.
29. Burke, M.; Hsiang, S.; Miguel, E. Climate and Conflict. *Annu. Rev. Econ.* **2015**, *7*, 577–617. [CrossRef]
30. Buhaug, H. One effect to rule them all? A comment on climate and conflict. *Clim. Chang.* **2014**, *127*, 391–397. [CrossRef]
31. Hsiang, S.; Burke, M.; Miguel, E. Reconciling climate-conflict meta-analysis: Reply to Buhaug et al. *Clim. Chang.* **2014**, *127*, 399–405. [CrossRef]
32. Buhaug, H.; Nordkvelle, J. *Climate and Conflict: A Comment on Hsiang et al.'s Reply to Buhaug et al.*; University of Maryland: College Park, MD, USA, 2014; Volume 127, pp. 399–405.
33. WHO. *World Report on Violence and Health*; Peace Research Institute Oslo: Oslo, Norway, 2002.
34. McCollister, K.; French, M.; Fang, H. The Cost of Crime to Society: New Crime-Specific Estimates for Policy and Program Evaluation. *Drug Alcohol Depend.* **2010**, *108*, 98–109. [CrossRef] [PubMed]

35. Nordhaus, W.D. *Managing the Global Commons: The Economics of Climate Change*; The MIT Press: Cambridge, MA, USA, 1994.
36. Nordhaus, W.D.; Yang, Z. A Regional Dynamic General-Equilibrium Model of Alternative Climate-Change Strategies. *Am. Econ. Rev.* **1996**, *86*, 741–765.
37. Putnam, R. *Making Democracy Work: Civic Traditions in Modern Italy*; Princeton University Press: Princeton, NJ, USA, 1993.
38. Putnam, R. *Bowling Alone: The Collapse and Revival of American Community*; Simon & Schuster Paperbacks: New York City, NY, USA, 2000.
39. Carleton, T.; Hsiang, S.; Burke, M. Conflict in a changing climate. *Eur. Phys. J.* **2016**, *225*, 489–511. [CrossRef]
40. Blanford, G.; Merrick, J.; Richels, R.; Rose, S. Trade-offs between mitigation costs and temperature change. *Clim. Chang.* **2014**, *123*, 527–541. [CrossRef]
41. Manne, A.S.; Richels, R.G. MERGE: An Integrated Assessment Model for Global Climate Change. In *Energy and Environment*; Loulou, R., Waaub, J.-P., Zaccour, G., Eds.; Springer: Berlin/Heidelberg, Germany, 2005; pp. 175–189.
42. Manne, A.; Mendelsohn, R.; Richels, R. MERGE: A model for evaluating regional and global effects of GHG reduction policies. *Energy Policy* **1995**, *23*, 17–34. [CrossRef]
43. Calel, R.; Stainforth, D. On the Physics of Three Integrated Assessment Models. *Bull. Am. Meteorol. Soc.* **2017**, *98*, 1199–1216. [CrossRef]

 *sustainability*

Article

Assessing Energy Poverty in Urban Regions of Mexico: The Role of Thermal Comfort and Bioclimatic Context

Karla G. Cedano , Tiare Robles-Bonilla *, Oscar S. Santillán and Manuel Martínez

Instituto de Energías Renovables, Universidad Nacional Autónoma de Mexico, Temixco 04510, Mexico; kcedano@ier.unam.mx (K.G.C.); oss@ier.unam.mx (O.S.S.); mmf@ier.unam.mx (M.M.)

* Correspondence: trb@ier.unam.mx; Tel.: +52-17352069012

Abstract: The increase of energy access to households has been a global priority. By 2018, 89.59% of the world population had access to electricity, while 97.26% of the persons living in urban areas (The Mexican Government reports it at 99.99%) had access. We must now move beyond access to electricity and address energy poverty in urban spaces. A household is energy poor when their inhabitants are incapable of securing proper domestic energy services. Several different methodologies were developed to measure energy poverty. The Multidimensional Energy Poverty Index (MEPI) by Nussbaumer has been successfully used in Africa and in Latin-America. The MEPI considers five dimensions: cooking, lighting, household appliances, entertainment/education and communication. We developed a Multidimensional Energy Deprivation Index (MEDI), based on MEPI. Thermal comfort has been included as sixth dimension, by considering the temperature of the region where the household is located. We found important differences between MEPI and MEDI for Mexico at the national level (urban-MEPI at 0.028 vs. 0.071 urban-MEDI, which implies a higher degree of energy poverty). Also, differences between geopolitical and bioclimatic regions were found. Having better ways to assess energy poverty in the urban context is a key factor to develop effective public policies that might alleviate it.

Keywords: energy poverty; bio-climatic region; MEPI; thermal comfort; urban

Citation: Cedano, K.G.; Robles-Bonilla, T.; Santillán, O.S.; Martínez, M. Assessing Energy Poverty in Urban Regions of Mexico: The Role of Thermal Comfort and Bioclimatic Context. *Sustainability* **2021**, *13*, 10646. https://doi.org/10.3390/su131910646

Academic Editors: Baojie He, Ayyoob Sharifi, Chi Feng and Jun Yang

Received: 1 September 2021
Accepted: 23 September 2021
Published: 25 September 2021

Publisher's Note: MDPI stays neutral with regard to jurisdictional claims in published maps and institutional affiliations.

1. Introduction

Energy poverty (EP) is a growing and urgent issue currently affecting many people in the world. EP may manifest in several ways and at different levels (which make it hard to measure and compare), and despite its importance as a social phenomenon, there is not a universal definition at the moment [1]. Nevertheless, as a general approximation of the subject, it can be said that EP is the inability to access adequate, affordable, reliable, high-quality, safe, and environmentally benign energy services to support economic and human development [2].

EP is of particular interest in cities. Currently, around half of the world's population lives in cities, and for the year 2050, 70% of the world inhabitants is predicted to live in urban settlements [3]. Cities account for between 60% and 80% of energy consumption and near 75% of carbon emissions, despite the fact that hey only occupy 3% of the Earth's surface [3]. The United Nations (UN) acknowledges the importance of cities in sustainable development, and for the accomplishment of Sustainable Development Goal (SDG) 11: For sustainable cities and communities, addressing urban energy poverty is imperative.

EP incidence can be found within a wide range of populations: from local groups to different national contexts, which highlights the relevance of assessing its spatial distribution [4]. For the urban areas, dwellings have a higher probability of being able to access modern energy services (especially electricity and liquefied petroleum gas), and for this reason, the lack of access of energy services might reflect poverty and unaffordability [5]. Roberts et al. (2015) undertook a study in the UK context, and they found that the experience of fuel poverty in urban areas lasts longer on average, and that fuel poverty also has a higher probability of persistence [6]; whilst Robinson et al. (2018) found that there is a higher prevalence of fuel poverty in urban areas [4]. In México, García Ochoa & Graizbord Ed (2016) evaluated the deprivation of energy services while distinguishing between rural and urban settlements, where they also discussed the importance of the climate when assessing EP [7].

Positive Energy Districts (PED) emerge as a strategy to decarbonise urban spaces in Europe. Bruck et al (2021) define a PED as a community that generates a surplus of energy with renewable sources in comparison with the energy it consumes from the grid on an annual basis, and that keeps its CO2 emissions balance at zero. They show that with actual electricity demand patterns PEDs are feasible and even pose a cheaper alternative against the import of energy when favourable climatic conditions are met. They also show a decrease in energy consumption. Thus, both issues imply an incentive in the householders" economy [8].

The dwellings or buildings that might be conceptualized as a PED have to actively manage their energy consumption and the energy flow between them and the national grid [9].

For the development of PEDs, the urban energy end-use must be very low, and this requires high energy efficiency, especially in the building, industry, and transport sectors [10]. An important feature of the PEDs is that achieving an annual surplus of net energy is not their only goal, they also must adequately handle such surpluses to prevent additional stress on the electrical grid [10]. For the accomplishment of the later objective, PEDs must offer alternatives for increasing the self-consumption and consumption on site, technologies for storage (both for short and long term) and provide energy flexibility by means of smart grids [11].

Hearn & Castaño-Rosa highlight that PEDs are a solution that might alleviate energy poverty (EP). Even though some places enforce public policies to promote the use of PV systems or the decrease of social tariffs as measures to alleviate, they are not enough. This is the main reason why the concept of PEDs gains interest as a better strategy, since households experience a decrease in their energy expenses and increase their energy behaviour [12].

The objective of the present work is to generate knowledge regarding the Mexican incidence of energy poverty in urban areas. The paper seeks to answer the research question: What is MEDI's relevance for energy poverty evaluation and its alleviation through public policy design in Mexican urban populations? The originality of the paper lies in two main issues: first, the introduction of the Multidimensional Energy Deprivation Index (MEDI), an indicator based in Nussbaumer's Multidimensional Energy Poverty Index (MEPI) that includes thermal comfort, considering both the bioclimatic region and the temperature in the measurement of energy poverty. Second, this is the first study that assesses energy poverty in urban areas in the Mexican context. The results attempt to act as an information source for the elaboration of public policies with the aim to address energy poverty in Mexico.

The article is presented as follows: in the Section "Materials and Methods", both MEPI and MEDI are described, as well as the data set used for the analysis. The results and the comparison for both indexes are presented in "Results", along with the identification of energy deprivation in terms of the different Mexican bioclimatic regions and some examples of specific urban areas. Finally, in "Discussion", the key results and the more important findings are discussed.

2. Materials and Methods

2.1. MEPI Description

The Multidimensional Energy Deprivation Index (MEDI) is based on Nussbaumer's Multidimensional Energy Poverty Index (MEPI). MEPI is an indicator that captures a set of energy deprivations that affects people by means of five dimensions and six indicators that represent basic energy services, as shown in Table 1. The people living in a particular dwelling are in an energy poverty condition if the combination of deprivations faced exceeds a predefined threshold [13].

Table 1. MEPI information [13].

Dimension	Indicator (Weight)	Variable	Deprivation Limit (Poor If …)
Cooking	Modern cooking fuel (0.2)	Type of cooking fuel	Use any fuel besides electricity, LPG, kerosene, natural gas or biogas
	Indoor pollution (0.2)	Food cooked on stove or open fire (no hood/chimney) if using any fuel beside electricity, LPG, natural gas or biogas	True
Lighting	Electricity access (0.2)	Has access to electricity	False
Services provided by means of household appliances	Household appliance ownership (0.13)	Has a fridge	False
Entertainment/ education	Entertainment/education appliance ownership (0.13)	Has a radio or television	False
Communication	Telecommunication means (0.13)	Has a phone land line or a mobile phone	False

MEPI measures energy poverty on d variables across a population of n individuals. The matrix $Y = [y_{ij}]$ represents the states matrix $n \times d$ for i persons through j variables. $y_{ij} > 0$ indicates the state of individual i on variable j. A weighting vector w is composed of w_j elements corresponding to the weight that is applied to variable j. It is defined by:

$$\sum_{j=1}^{d} W_j = 1 \tag{1}$$

The deprivation threshold z_j on variable j is set; then, all individuals with deprivations on any variable are detected. Subsequently, it is defined the deprivation matrix $g = [g_{ij}]$ where each element g_{ij} is determined by:

$$g_{ij} = \{w_j, \ y_{ij} < z_j \ 0, \ y_{ij} \geq z_j\} \tag{2}$$

In a MEPI calculation, elements on the states matrix are non-numerical, and for that reason the threshold is defined as a set of conditions to be fulfilled. Later, a column vector c of deprivations counts is built, where the *ith* entry indicates the sum of deprivations that i person is facing, where:

$$c_i = \sum_{j=1}^{d} g_{ij} \tag{3}$$

The dwellings on the energy poverty condition are identified with the definition of a limit $k > 0$, which is applied to the column vector c, a dwelling is considered to be one of energy poverty if its weighted deprivation count c_i exceeds k. The censored vector of deprivation count is represented by $c(k)$, which is different to c for it counts zero deprivations to the persons that are not identified on multidimensional energy poverty.

$$c_i(k) = \{0,\ c_i \leq k\ c_i,\ c_i > k\} \tag{4}$$

Headcount ratio H represents the proportion of the population considered as energy poor, and is calculated with $H = q/n$, where q is the number of persons on energy poverty $(c_i > k)$, and n, the total number of the sample. H indicates the incidence of multidimensional energy poverty. The average of the censored weighted deprivation count $c_i(k)$ represents the intensity of multidimensional energy poverty, and is calculated by:

$$A = \sum_{i=1}^{n} \frac{C_i(k)}{q} \tag{5}$$

MEPI captures information regarding incidence and intensity of energy poverty, and is defined as MEPI = $H \times A$. When calculating H and A, the number of persons in each dwelling are included.

2.2. MEDI Description

Robles & Cedano (2021) state the need to consider bioclimatic regions to better understand energy poverty (EP). This is in regard to the importance that thermal comfort has in the proper assessment of EP. They create a new index adding not only the dimension of thermal comfort, but also a climate analysis that includes the regional climatic conditions to properly assess the need of appliances to achieve it. This index, called the Multidimensional Energy Deprivation Index (MEDI), considers the extreme temperature ranges that take place in Mexico and the importance of being deprived of appliances to mitigate the effects of extreme weather [14]. We use annual average extreme temperatures considering that a 30 °C average implies thermal discomfort, whether the standard deviation is large or small. In other words, a small standard deviation can translate to having minimum variations around 30 °C, which makes livelihood or work-related activities uncomfortable without the aid of cooling appliances. And on the other hand, in those cases where the standard deviation is larger, we can infer that the temperature can rise to even higher values, thus increasing the need of cooling appliances to achieve thermal comfort.

MEDI methodology allows for the incorporation of new dimensions and variables, and an alternative dimension was added, as shown in Table 2.

Table 2. Thermal comfort dimension [14].

Dimension	Indicator	Variable	Deprivation Cut-Off (Poor If . . .)
Thermal comfort	Thermal comfort access	Has an air conditioner or heating	False

Thermal comfort is a very important topic when assessing energy poverty, since providing it is considered an energy service needed to achieve wellbeing. Its relevance can be shown in several case studies; for example, in the summer of 2003, more than 70,000 deaths caused by extreme heat in Europe were reported [15]. Without any doubt, accessing thermal comfort can decrease the number of deaths related to very low temperatures or to dehydration when very high temperatures occur, not only in Europe, where most of the recent events are being reported, but also in Latin America. Thermal comfort has been thoroughly studied since 1959 by Crowden, with studies analysing its effects on human health in households. We found 16,976 papers on the Web of Science regarding thermal comfort up to August 2021; however only 85 of those papers address energy poverty also. These 85 papers focus on mortality, vulnerability, deprivation, and extreme temperature, among other topics as can be found in Table 3 (Data from Minero bibliométrico Avanzado [16]).

Table 3. Keywords in 85 papers about thermal comfort and energy poverty.

Consumption	Low-Income Households
Health	Climate
Mortality	Vulnerability
Impact	Energy justice
Energy deprivation	Extreme temperature
Mitigation	Older people
Excess winter deaths	Efficiency

Population increases in urban areas and a building's inadequate design and construction that affects thermal comfort affects quality of life in urban spaces. Karakounos et al [17], Gaitani et al [18], and Pontes et al [19] consider climatic regions and bioclimatic design to alleviate the adverse effects of extreme temperature and climate change. This situation can be alleviated by retrofitting buildings with materials that have better behaviour against temperature increases. They also suggest adding green spaces to decrease outdoor temperature and, in turn, decrease energy consumption for cooling. However, there are some cases where adjacent buildings increase temperature, thus adding layers of complexity to alleviate the lack of thermal comfort and increasing the need for cooling appliances. Calixto & Huelz highlight the importance of climate considerations in building design, since this is directly related with energy consumption, especially in countries in the Global South, like Mexico, where climate diversity is high. In this region, bioclimatic diversity has to be considered in the design and construction of buildings to attain thermal comfort and energy savings [20].

Robles & Cedano (2021) modified in a proportional way the weights that Nussbaumer et al defined in their methodology, while including the dimension of thermal comfort. The first approach was to adjust the weights for the five original dimensions proportionally, so they were not significantly altered when adding thermal comfort [14]. The weights were as shown in Table 4.

Table 4. Proportional weights of original MEPI when adding thermal comfort [14].

Dimension	Indicator	Variable	Weight
Cooking	Modern cooking fuel	Type of cooking fuel	0.18
	Indoor pollution	Food cooked on stove or open fire	0.18
Lighting	Electricity Access	Has Access to electricity	0.18
Services provide by means of household appliances	Household appliance ownership	Has a fridge	0.115
Entertainment/education	Entertainment/education appliance ownership	Has a radio or television	0.115
Communication	Telecommunication means	Has a phone land line or a mobile phone	0.115
Thermal comfort	Thermal comfort access	Has an air conditioner or heating	0.115

Also, new weights were applied to all the dimensions as shown in Table 5. To establish MEDI's weights, experts on the subject were consulted. They agreed on considering cooking as essential, followed by electricity and then appliances, since until this point, health and safety were considered, as well as which appliances were useful for food preservation and to prevent indoor pollution. Later, thermal comfort was considered due to its importance as an energy service. Finally, entertainment/education and communication were included [14].

Table 5. MEDI information [14].

Dimension	Indicator	Variable	Weight
Cooking	Modern cooking fuel	Type of cooking fuel	0.13
	Indoor pollution	Food cooked on stove or open fire	0.13
Lighting	Electricity Access	Has Access to electricity	0.24
Services provide by means of household appliances	Household appliance ownership	Has a fridge	0.21
Entertainment/education	Entertainment/education appliance ownership	Has a radio or television	0.08
Communication	Telecommunication means	Has a phone land line or a mobile phone	0.07
Thermal comfort	Thermal comfort access	Has an air conditioner or heating	0.14

2.3. Databases Used

The data used for the evaluation was obtained from the National Survey of Income and Expenditures in Households (Encuesta Nacional de Ingresos y Gastos en los Hogares, ENIGH, in Spanish). This survey is applied every two years by the National Institute of Statistics and Geography (Instituto Nacional de Estadística y Geografía, INEGI). We used the data of the survey applying secondary data analysis [21]. The ENIGH database includes information associated with three levels: dwelling, household, and the household 's members; and the results can be extensive for the whole population, with a confidence interval of 90%. The analyzed information encompassed 73,405 dwellings in which 256,989 persons reside [22].

In Section 2.2 it is mentioned that Robles & Cedano use the temperature of the bioclimatic region as conditional for calculating MEDI, and it is taken from the National Water Comission (Comisión Nacional del Agua, CONAGUA) dataset. In its web page, the information includes the national monthly average maximum or minimum perspective temperature in a three-month period. However, the authors requested CONAGUA information regarding one year's daily temperatures by municipality. CONAGUA delivered information about the year daily minimum, average and maximum temperature available. Temperature is expressed in degrees Celsius [23].

3. Results

3.1. Identification of Energy Deprivation by Bioclimatic Region

Two cases were analyzed for the identification of energy deprivation in relation to the bioclimatic region. In the first one, the extreme temperatures in the bioclimatic region were used, meaning that the minimum temperature among the municipalities of the region was set as t_{min} and the maximum temperature among the municipalities of the region was set as t_{max}. The second case considers only the average temperature, leaving the extreme temperatures out of the analysis. The two cases together show two completely different results: when using extreme temperatures 99% of the people show deprivations regarding thermal comfort, while just one bioclimatic region presents the need to use thermal comfort appliances when using the average temperature, resulting in a low incidence in this dimension.

To better understand the incidence of deprivations within the bioclimatic regions, the indicator´s weights were adjusted and the calculation was made considering the variables instead of the dimensions. A weight of 1, as opposed to the previously allocated weights, was used for the calculation of each variable alone, setting the weights of the other variables to zero. As an example, to know the incidence of the access to electricity variable, its weight was set in 1 while the weights of all the other variables was set to zero.

For the first case, where the extreme temperatures are used, all of the bioclimatic regions present the lack of heating or air conditioning as the most emerging deprivation affecting 99% of the people. Warm weather has presented temperatures below 10 °C (in part due to climate change), which have caused them to face heating deprivation. Dry tempered, tempered and humid semi-cold weather also requires heating. Humid tempered, dry semi-cold and semi-cold weathers show a wider deprivation of air conditioning in relation to heating, while both are at concerning levels. Figure 1 shows humid warm bioclimate and highlights the lack of heating. Figure 2 shows humid tempered weather, where both deprivation of heating and air conditioning are at a similar level.

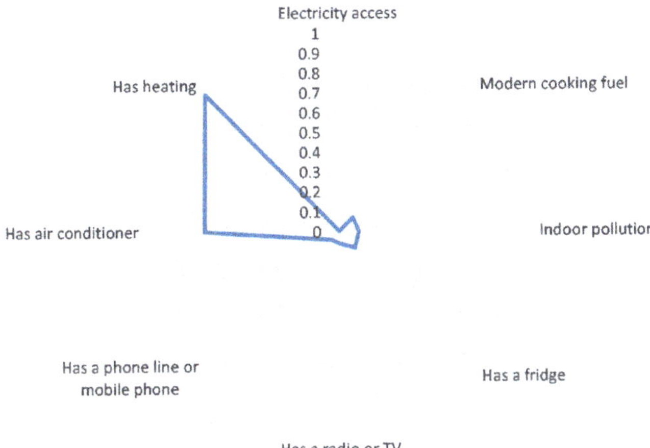

Figure 1. Humid warm weather.

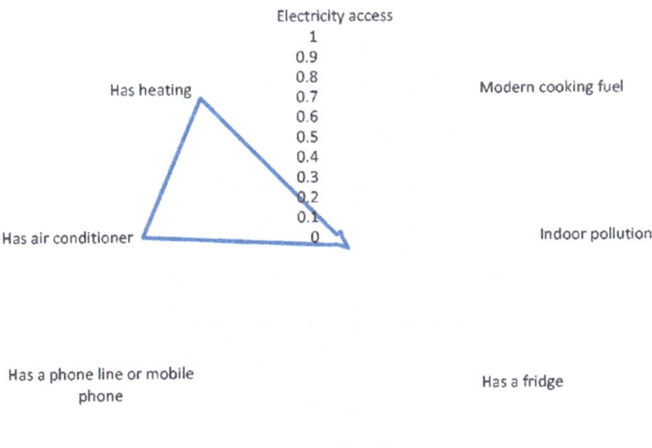

Figure 2. Humid tempered weather.

For the second case, where the average temperatures were used, only the semi-humid warm region needed air-conditioning appliances. However, it only had an incidence of 0.36. The larger deprivation presented in most of the regions is the lack of refrigerator. Only in the semi-humid warm and humid warm was the larger deprivation was the type of cooking fuel used. Figure 3 shows the semi-humid warm weather where this can be seen. The incidence for the air conditioning, while existing, cannot be seen due to the low value it poses in relation to the resolution of the figure. Figure 4 presents tempered weather, highlighting that the refrigerator is the deprivation with the larger incidence.

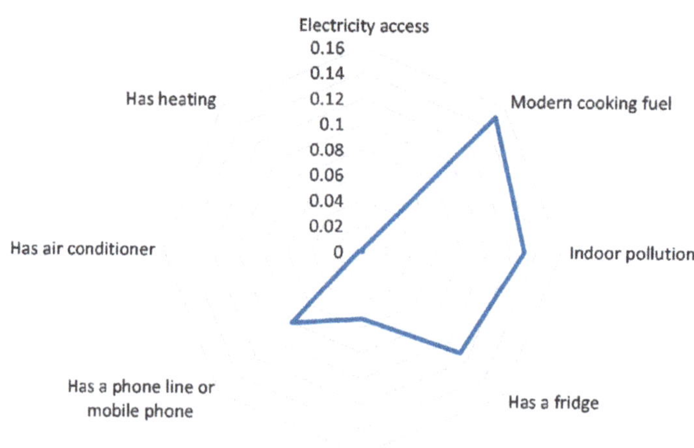

Figure 3. Semi-humid warm weather.

This analysis shows that the first case is more sensible to the weather variations and allows a more accurate assessment of thermal comfort conditions in each location. In this regard, we have a better approximation to the energy needed by a household to achieve thermal comfort. The following MEPI, MEPI comfort and MEDI will be calculated considering the annual average maximum and minimum temperatures.

3.2. Specific Urban Areas

A specific capital city was selected in each bioclimatic region as part of the bioclimatic analysis. The aim is to observe the differences on energy poverty measurements when using MEPI, MEPI with thermal comfort and MEDI. In all the cities, the level of the index regarding energy poverty is larger when using MEPI with thermal comfort instead of MEPI (except in Xalapa where is the same), and is larger when using MEDI instead of MEPI with thermal comfort, highlighting that MEDI might show better resolution when identifying people on energy poverty, since the weights for each dimension and variable were chosen by Mexican experts on energy access and poverty. Table 6 shows the differences between both MEPIs and MEDI as well as the maximum, minimum and average temperatures in each city.

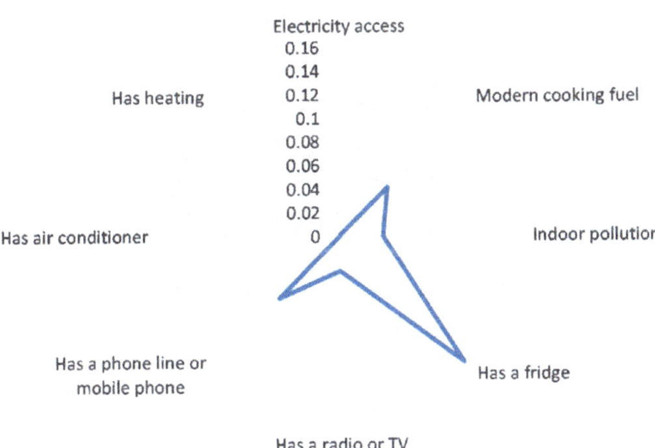

Figure 4. Tempered weather.

Table 6. MEPI and MEDI of capital cities in each bioclimatic region.

Region	Capital City	MEPI	MEPI Comfort	MEDI	T_{min}	T_{max}	T_{aver}
Humid warm	Campeche	0.024	0.060	0.070	6.0	44.0	27.0
Dry warm	Monterrey	0.000	0.033	0.039	−8.0	44.0	23.1
Extreme dry warm	Hermosillo	0.013	0.029	0.036	−7.5	49.0	25.5
Semi-humid warm	Ciudad Victoria	0.006	0.031	0.040	−4.0	44.0	22.7
Semi-cold	Toluca	0.007	0.014	0.045	−4.0	26.0	14.2
Humid semi-cold	Xalapa	0.050	0.050	0.060	6.0	34.0	19.0
Dry semi-cold	Pachuca	0.003	0.049	0.059	−2.7	31.0	16.0
Temperate	Guadalajara	0.000	0.018	0.022	4.0	37.5	21.0
Humid temperate	Tepic	0.004	0.042	0.049	8.2	35.0	23.0
Dry temperate	Oaxaca	0.015	0.080	0.094	8.0	39.5	23.0

When using average maximum and minimum temperatures, all the capital cities need heating and, excluding Toluca, all the cities need air conditioning. On the other hand, when using annual average temperatures, none of the selected cities would need heating or air conditioning. Another important finding is that the second and the third largest cities in the country, Guadalajara and Monterrey, have a MEPI of zero; that is, there seems to be no energy poverty in those cities; however, when using both MEPI with thermal comfort or MEDI, the two cities show households with energy poverty. This is due to the lack of appliances to attain thermal comfort.

4. Discussion

The analysis of energy poverty in cities becomes more relevant due to the high share of the population that lives in urban areas and, knowing the share is going to get significantly higher in the future, both in the world and in Mexico. The large number of people living together in mega urban areas might foster the lack of access to energy services. The alleviation of energy poverty is a challenge that must be addressed with context-related measures. One very interesting approach arises from fostering Positive Energy Districts. In this regard, we know that "In 2018 Europe, 39.3% of the population lived in the cities, 31.6% lived in towns and suburbs" [24], that is, more than 70% of Europe's population is concentrated in urban spaces. Also, according to [25], more than 40% of Europe's residential buildings were built before 1960, and 50% between 1960 and 1990. Considering that buildings account for 40% of energy consumption, 36% of CO2 emissions and 55% of electricity consumption, renovation of those dwellings is critical. Then for Europe, the reduction of energy consumption and CO2 emissions must be urgently addressed by city planners, developers, and operators by considering a systemic approach. A proposal to analyse the refurbishment of cities at district level has been considered as critical, and even the objective to obtain positive energy districts has been put forward by Yvann Nzengue and co-authors [26].

In 2013, Mexico announced a new approach to housing and urban policy, calling for a more explicit qualitative focus on housing and the urban environment. Nevertheless, it was more a quantitative push for formal housing, which came with quantitative costs: inefficient development patterns resulting in a hollowing out of city centres and the third-highest rate of urban sprawl in the Organization for Economic Cooperation and Development (OECD); increasing motorization rates; a significant share of vacant housing, with one-seventh of the housing stock uninhabited in 2010; housing developments with inadequate access to public transport and basic urban services; and social segregation [27].

The OECD [28] considered that Mexico´s moderate growth over the past two decades has been supported by oil wealth, working age population growth, and open trade and investment policies. Despite this growth, Mexico has not converged towards higher living standards and the gap in GDP per capita with the OECD average and the United States has not narrowed. Informal work remains high, encompassing nearly 60% of formal jobs and about a quarter of GDP. Inequality and poverty declined only moderately, and large gaps prevail between regions while poverty disproportionately affects the indigenous population.

Andres Manuel Lopez Obrador, President of Mexico, announced in one of his daily briefings that his government established as a priority area the recovery of homes in both low-income and housing areas to revive the construction sector and the consumption of building materials [29]. Furthermore, Roman Meyer, Mexican Minister for Agriculture, Territorial and Urban Development, explained in a briefing that of the 34 million households in the country, 9.4 million have underdevelopment issues, and that 79% of those are due to precarious building materials. He also revealed that there are about 650 thousand abandoned dwellings, of which 175 thousand will be recovered by the government, and the federal government will also provide all services to their urban surroundings. The programs will be financed directly to the inhabitants to promote self-construction or by individual contracting of local building workers. Therefore, in the short-term context, one can surmise that the establishment of federal energy programs for household development will be a difficult task.

Nevertheless, according to the Mexican National Council of Population (CONAPO, in Spanish) the Mexican population was 128 million as of 2020 and is projected to be 148 million by 2050 [30]. If accurate, if a 90 to 10 percentage ratio is taken for urban to rural distribution and a density of four people per house is considered, at least four million dwellings with pertinent urban infrastructure must be built. This gives the opportunity to establish programs for energy savings using renewable energies and developing positive energy districts.

5. Conclusions

Few papers address the evaluation of energy poverty in urban spaces and take into consideration thermal comfort and the climatic features of the evaluated regions. To have a better understanding of how energy poverty presents, and which are the factors that cause it, is a first step in alleviating the situations that many people in the world, especially in cities, are facing. There is no doubt that thermal comfort is a critical matter when analysing energy consumption. From heating devices in winter, to cooling devices in summer, energy consumption is greatly affected by these appliances. Its importance has been addressed by several studies [14,15,31]. We highlight its importance in urban dwellings not only because of the actual and forecasted population concentration in cities, but also considering possible district-focused strategies to alleviate it.

One of the main contributions of this paper is the application of MEDI, based on the structure of Nussbaumer et al.'s MEPI by adding the dimension of thermal comfort and using the bioclimatic regions for its evaluation. MEDI and MEPI are both useful for the evaluation of energy poverty; however, the results indicate that the multidimensional index is 2.5-fold when using MEDI instead of MEPI. This last part is evidence of the much larger incidence of energy poverty when considering thermal comfort. These findings are useful for policy makers when addressing energy poverty, both in rural and urban areas.

One of the most difficult challenges to properly assess thermal comfort lies in the regional differences on climate and the individual perception as well as its ability to adapt to it. Our research is one of the first attempts to include annual average temperatures (maximum, mean and minimum), as a parameter to consider while evaluating a household vulnerability in regard with thermal comfort. We chose 10 Mexican main cities, each in a different bioclimatic region, to show the difference that considering thermal comfort poses while measuring energy poverty. To do this, we used two different thresholds. The first only took into consideration annual average mean temperatures, while the second had a double-sided criterion, taking, for one side, the annual average minimum temperature, and for the other side, the annual average maximum temperature. The latter was a more accurate measure, since it considers a wider range of temperatures.

Assessing EP by considering thermal comfort in urban spaces is key to better understand the regional distribution of households classified by their energy accessibility and consumption. Since the calculation for EP can be done by localities or districts, mapping different EP conditions can help urban planning and policy making to foster locally the emergence of PEDs where EP is low, and connecting these PEDs with districts that exhibit high levels of EP. Since non-EP districts have the economic possibility to invest in renewable infrastructure, local governments can promote the development of PEDs. This energy solidarity model can be applied to urban and periurban zones, complementing energy needs of those in energy poverty.

Author Contributions: Conceptualization, K.G.C., T.R.-B.; Data curation, T.R.-B.; Formal analysis, K.G.C., T.R.-B., O.S.S. and M.M.; Investigation, K.G.C., T.R.-B., O.S.S. and M.M.; Methodology, K.G.C., T.R.-B. and O.S.S.; Project administration, K.G.C. and M.M.; Software, T.R.-B.; Supervision, K.G.C. and M.M.; Validation, K.G.C., O.S.S. and M.M.; Visualization, T.R.-B.; Roles/Writing–original draft, K.G.C., T.R.-B., O.S.S. and M.M.; Writing–review & editing, O.S.S., K.G.C. and M.M. All authors have read and agreed to the published version of the manuscript.

Funding: The authors want to gratefully to IER-UNAM internal project: Social Demand of energy for the financial support.

Institutional Review Board Statement: Not applicable.

Informed Consent Statement: Not applicable.

Acknowledgments: The authors O.S.S. and T.R.-B. want to gratefully acknowledge the financial support from the Mexican Consejo Nacional de Ciencia y Tecnología (CONACyT) awarded under their PhD. grant number 495905 and 794456, respectively.

Conflicts of Interest: The authors declare no conflict of interest.

References

1. Kyprianou, I.; Serghides, D.K.; Varo, A.; Gouveia, J.P.; Kopeva, D.; Murauskaite, L. Energy poverty policies and measures in 5 EU countries: A comparative study. *Energy Build.* **2019**, *196*, 46–60. [CrossRef]
2. United Nations Development Programme (2000). World Energy Assessment. Energy and the Challenge of Sustainability. Available online: https://www.undp.org/sites/g/files/zskgke326/files/publications/World%20Energy%20Assessment-2000.pdf (accessed on 1 May 2021).
3. United Nations. Goal 11: Make Cities Inclusive, Safe, Resilient and Sustainable. Available online: https://www.un.org/sustainabledevelopment/cities/. (accessed on 1 May 2021).
4. Robinson, C.; Bouzarovski, S.; Lindley, S. 'Getting the measure of fuel poverty': The geography of fuel poverty indicators in England. *Energy Res. Soc. Sci.* **2018**, *36*, 79–93. [CrossRef]
5. Nathan, H.S.K.; Hari, L. Towards a new approach in measuring energy poverty: Household level analysis of urban India. *Energy Policy* **2020**, *140*, 111397. [CrossRef]
6. Roberts, D.; Vera-Toscano, E.; Phimister, E. Fuel poverty in the UK: Is there a difference between rural and urban areas? *Energy Policy* **2015**, *87*, 216–223. [CrossRef]
7. García-Ochoa, R.; Graizbord-Ed, B. Privation of energy services in Mexican households: An alternative measure of energy poverty. *Energy Res. Soc. Sci.* **2016**, *18*, 36–49. [CrossRef]
8. Bruck, A.; Diaz Ruano, S.; Auer, H. A Critical Perspective on Positive Energy Districts in Climatically Favoured Regions: An Open-Source Modelling Approach Disclosing Implications and Possibilities. *Energies* **2021**, *14*, 4864. [CrossRef]
9. Rehman, H.; Reda, F.; Paiho, S.; Hasan, A. Towards positive energy communities at high latitudes. *Energy Convers. Manag.* **2019**, *196*, 175–195. [CrossRef]
10. Frontiers. Positive Energy Districts: Transforming Urban Areas into High Efficiency Districts with Local Renewable Generation and Storage. Available online: https://www.frontiersin.org/research-topics/14941/positive-energy-districts-transforming-urban-areas-into-high-efficiency-districts-with-local-renewab#overview. (accessed on 1 May 2021).
11. Hedman, Å.; Rehman, H.U.; Gabaldón, A.; Bisello, A.; Albert-Seifried, V.; Zhang, X.; Guarino, F.; Grynning, S.; Eicker, U.; Neumann, H.-M.; et al. IEA EBC Annex83 Positive Energy Districts. *Buildings* **2021**, *11*, 130. [CrossRef]
12. Hearn, A.X.; Castaño-Rosa, R. Towards a Just Energy Transition, Barriers and Opportunities for Positive Energy District Creation in Spain. *Sustainability* **2021**, *13*, 8698. [CrossRef]
13. Nussbaumer, P.; Bazilian, M.; Modi, V. Measuring energy poverty: Focusing on what matters. *Renew. Sustain. Energy Rev.* **2012**, *16*, 231–243. [CrossRef]
14. Robles-Bonilla, T.; Cedano, K.G. Addressing Thermal Comfort in Regional Energy Poverty Assessment with Nussbaumer's MEPI. *Sustainability* **2021**, *13*, 352. [CrossRef]
15. Thomson, H.; Simcock, N.; Bouzarovski, S.; Petrova, S. Energy & Buildings Energy poverty and indoor cooling: An overlooked issue in Europe. *Energy Build.* **2019**, *196*, 21–29. [CrossRef]
16. Kostoff, R.N.; del Río, J.A.; Cortés, H.D.; Smith, C.; Smith, A.; Wagner, C.; Karypis, G.; Malpohl, G.; Tshiteya, R. Clustering methodologies for identifying country core competencies. *J. Inf. Sci.* **2007**, *33*, 21–40. [CrossRef]
17. Karakounos, I.; Dimoude, A.; Zoras, S. The influence of bioclimatic urban redevelopment on outdoor thermal comfort. *Energy Build.* **2018**, *158*, 1266–1274. [CrossRef]
18. Gaitani, N.; Mihalakakou, G.; Santamouris, M. On the use of bioclimatic architecture principles in order to improve thermal comfort conditions in outdoor spaces. *Build. Environ.* **2007**, *42*, 317–324. [CrossRef]
19. Pontes, R.H.; Najjar, M.K.; Hammad, A.W.A.; Vazquez, E.; Haddad, A. Adapting the Olgyay bioclimatic chart to assess local thermal comfort levels in urban regions. *Clean Techn. Environ. Policy* **2021**, *23*, 1–15. [CrossRef]
20. Calixto Aguirre, V.I.; Huelsz, G. Consumo de energía en edificios en México. *Legado Arquit. Diseño* **2018**, *24*, 40–47.
21. INEGI (National Institute of Statistic and Geography). National Survey of Household Income and Expenditure (ENIGH). 2018 New Series, Microdata. Available online: https://www.inegi.org.mx/programas/enigh/nc/2018/#Microdatos (accessed on 1 May 2018).
22. INEGI (National Institute of Statistic and Geography). National Survey of Household Income and Expenditure (ENIGH). 2018 New Series, Documentation. Available online: https://www.inegi.org.mx/programas/enigh/nc/2018/ (accessed on 1 May 2018).
23. CONAGUA. National Meteorological Service. Available online: https://smn.conagua.gob.mx/es/climatologia/pronostico-climatico/temperatura-form, (accessed on 20 September 2019).
24. Eurostat. Urban and rural living in the EU. Available online: https://ec.europa.eu/eurostat/web/products-eurostat-news/-/EDN-20200207-1 (accessed on 19 December 2020).
25. Encore. Old European Building Stock. Available online: http://encorebim.eu/en/news/third-new (accessed on 19 December 2020).
26. Nzengue, Y.; Boishamon, A.; Laffont-Eloire, K. Planning City Refurbishment: An Exploratory Study at District Scale. Available online: http://www.sinfonia-smartcities.eu/contents/knowledgecenterfiles/planning-city-refurbishment--an-exploratory-study-at-district-scale.pdf (accessed on 10 November 2020).

27. OECD. OECD Urban Policy Reviews: Mexico 2015: Transforming Urban Policy and Housing Finance, OECD Urban Policy Reviews, OECD Publishing: Paris, France 2015. Available online: https://read.oecd-ilibrary.org/urban-rural-and-regional-development/oecd-urban-policy-reviews-mexico-2015_9789264227293-en#page1 (accessed on 20 September 2020).
28. OECD. OECD Economic Surveys: Mexico 2019. Available online: https://www.oecd.org/economy/surveys/Mexico-2019-OECD-economic-survey-overview.pdf (accessed on 2 December 2020).
29. Andrés Manuel López Obrador. Programas de Vivienda Generan Empleos y Reactivan la Economía. Conferencia Presidente AMLO [Video]. Available online: https://youtu.be/12BlYyebe8c (accessed on 3 December 2020).
30. CONAPO. Demographic Indicators of Mexico from 1950 to 2050. Available online: http://www.conapo.gob.mx/work/models/CONAPO/Mapa_Ind_Dem18/index_2.html (accessed on 20 September 2019).
31. Ormandy, D.; Ezratty, V. Health and thermal comfort: From WHO guidance to housing strategies. *Energy Policy* **2012**, *49*, 116–121. [CrossRef]

 *sustainability*

Article

Assessment on the Use of Meteorological and Social Media Information for Forest Fire Detection and Prediction in Riau, Indonesia

Anni Arumsari Fitriany [1,2,*], **Piotr J. Flatau** [3], **Khoirunurrofik Khoirunurrofik** [1] and **Nelly Florida Riama** [2]

[1] Faculty of Economics and Business, Universitas Indonesia, Depok 16424, Indonesia; khoirunurrofik@ui.ac.id
[2] Agency for Meteorology, Climatology, and Geophysics of the Republic of Indonesia, Jakarta 10720, Indonesia; nelly.florida@bmkg.go.id
[3] Scripps Institution of Oceanography, University of California San Diego, La Jolla, CA 92093, USA; pflatau@ucsd.edu
* Correspondence: anni.arumsari@bmkg.go.id

Abstract: In this study, tweets related to fires in Riau, Sumatra, were identified using carefully selected keywords for the 2014–2019 timeframe. The TAGGS algorithm was applied, which allows for geoparsing based on the user's nationality and hometown and on direct referrals to specific locations such as name of province or name of city in the message itself. Online newspapers covering Riau were analyzed for the year 2019 to provide additional information about the reasons why fires occurred and other factors, such as impact on people's health, animal mortality related to ecosystem disruption, visibility, decrease in air quality and limitations in the government firefighting response. Correlation analysis between meteorological information, Twitter activity and satellite-derived hotspots was conducted. The existing approaches that BMKG and other Indonesian agencies use to detect fire activity are reviewed and a novel approach for early fire detection is proposed based on the crowdsourcing of tweets. The policy implications of these results suggest that crowdsourced data can be included in the fire management system in Indonesia to support early fire detection and fire disaster mitigation efforts.

Keywords: forest fires; meteorology; newspaper; policy; Tweeter; Indonesia

Citation: Fitriany, A.A.; Flatau, P.J.; Khoirunurrofik, K.; Riama, N.F. Assessment on the Use of Meteorological and Social Media Information for Forest Fire Detection and Prediction in Riau, Indonesia. *Sustainability* 2021, *13*, 11188. https://doi.org/10.3390/su132011188

Academic Editor: Baojie He

Received: 15 August 2021
Accepted: 23 September 2021
Published: 11 October 2021

Publisher's Note: MDPI stays neutral with regard to jurisdictional claims in published maps and institutional affiliations.

1. Introduction

There are large forested areas in Indonesia. Based on 2019 data from the Ministry of Environment and Forestry of the Republic of Indonesia, total forested area covers about 94.1 million hectares, representing almost half of Indonesia's total land-covered area. Though Indonesia benefits from its forestry resources, forest fires also negatively impact aspects of life in Indonesia, in terms of both the economy and health. Tacconi [1] states that the economic cost associated with fires stems not only from the fires themselves, but also from associated smoke and haze, the loss of timber and the loss of opportunity to plant crops. Equally damaging are the costs of firefighting and the loss of property and life. By the same token, the indirect benefits of forested land, such as flood protection and biodiversity, are diminished. Further, smoke impacts health, tourism, transport and industrial production.

Indonesia's land and forest fires are strongly affected by peat soil and land management practices which are often caused by human activities [2], mainly illegal agricultural activities using slash-and-burn techniques to convert forests and peatlands into palm oil plantation fields [3]. This leads to a high frequency of fire activity in such regions, particularly during dry seasons [4]. The fires often produce haze in the peatland areas. There are several types of warnings which can be issued depending on haze severity, such as reduction of outdoor activities or evacuation of the affected areas due to haze-related health issues [5]. According to Ulya et al. [6], forest fires cost public and private companies

273

and the government approximately IDR 77.4 million per hectare of burned area. Therefore, urgent policy action in Indonesia requires the improvement of forest fire early detection and warning systems. Rogers and Tsirkunov [7] discuss costs and benefits of such early warning systems. It was shown in one case study in Europe that several hundred lives per year can be saved by employing hydro-meteorological information as part of an early warning system [8]. Prediction and early detection of forest fires are important in order to prevent fires from spreading more intensely. The Indonesian Agency Assessment and Application of Technology (BPPT), in collaboration with the Agency for Meteorology, Climatology, and Geophysics (BMKG), and the Ministry of Environment and Forestry (KLHK) has been operating the Fire Danger Rating System (FDRS) since 1999. Though the FDRS does gauge the probability of fire occurrence, a more direct system would provide valuable information.

This study explores how monitoring large-scale meteorological phenomena, such as El Niño/Southern Oscillation (ENSO) and Indian Ocean Dipole (IOD), as well as data collected from social media, such as Twitter, can contribute to early detection of and thus early warnings of forest fires in Indonesia. We also examine newspaper articles to gain additional information related to "sentiment analysis" related to fires. This will provide guidance and insight for policy makers as they formulate new practices to mitigate forest fires.

This paper consists of a literature review, a detailed discussion of methodology, an analysis of our results and a conclusion section.

2. Literature Review

2.1. Use of Social Media in Early Detection of and Warning of Disaster Events

Several studies have used information derived from social media and newspapers to study public policy issues, including disaster events. People use Twitter as a way to share their concerns about important societal issues [9]. A study by Biswas [10] shows how news media contribute to the process of policy making by identifying important issues. The use of social media in political communication was studied by Stieglitz and Dang-Xuan [11] who attempted to develop a toolset for collecting, storing, monitoring, analyzing and summarizing user-generated and politically relevant content from social media to be used by political institutions. Sinnenberg [12] has shown how Twitter can be used in health-related studies. Power, et al. [13] state that social media can be a valuable channel of communication. However, its adoption as a source of information to enhance public awareness is still rare [14] because it is not easy to frame social media content in proper context, process large volumes of information nor to trust the messages [15]. Twitter has been used for detecting disasters such as earthquakes [16–18] and floods [19]. Power et al. [20] applied tweets to develop a system for identifying fires in Australia.

Several papers study land and forest fires in Indonesia, including their meteorological and climate indicators [21–23]. Kibanov, et al. [4] reports on social media mining as the source of information related to peatland fires and haze disaster management. Forsyth [24] studied transboundary haze issues in Indonesia, Singapore and Malaysia. Panjaitan, et al. [25] studied forest fire policies in Indonesia, specifically the role of central and local governments and the moderating effects of good governance. Related studies deal with "sentiment analysis". Mustaqim et al. [26] studied public feelings (the "sentiment analysis") about the Indonesian government's response to handling forest fires in 2019 by using a semi-automated labeling and classification scheme. Mustaquim et al. state that sentiment analysis provides a way for businesspeople and academic institutions to better understand community thinking. A recent review of Twitter sentiment analysis can be found in Carvalho and Plastino [27].

2.2. Indonesian Agencies' Fire-Related Warnings

The Indonesian government has developed several ways to prevent and mitigate the occurrence of land and forest fires. The KLHK offers a web dashboard for monitoring forest fires called SiPongi. In conjunction with the Geospatial Information Agency, the KLHK provides a forest fire vulnerability map, using physical factors such as land cover, land topography, weather and climatology, as well as anthropogenic factors such as human behavior and local socio-culture. The National Disaster Management Agency (BNPB) provides a geospatial dashboard for forest fire mitigation and evacuation. The BMKG provides the Fire Danger Rating System (FDRS), which was developed in collaboration with the Agency for the Assessment and Application of Technology (BPPT) and the Ministry of Environment and Forestry (KLHK). It provides information about the Fine Fuel Moisture Code, Duff Moisture Code, Drought Code, Build Up Index, Initial Spread Index and Fire Weather Index, as well as smoke distribution imagery and hotspot distribution. The Fine Fuel Moisture Code and the Fire Weather Index information are based on weather analysis.

2.3. Social-Media Based Fire Detection

In Indonesia, Twitter is used by more than 75% of all active Internet users. In 2012, Indonesia had the largest global tweet/user ratio in the world [28]. Twitter messages can be merged with weather analysis, an idea which was recently exploited to predict floods in Indonesia [29]. Social media can be an effective communication method, though it remains difficult to analyze a large amount of social media content and determine its appropriateness to fire prediction. Twitter can be used for disaster management during earthquakes [16–18], as many people send messages about such events. Power et al. [20] applied a similar methodology to fire detection; they identified near-real-time tweets that describe fire occurrences and employed a text classifier to refine the results to obtain actual fire occurrences. In a study conducted by Aditya et al. [30], crowdsourcing information was used to validate satellite observations of hotspots and haze. They indicate that their Geocrowd app has the potential to help validate hotspots and to contribute to haze mitigation and environmental protection. They mention that the availability of mobile network signal in remote areas was one of the main obstacles to effective use of the Geocrowd app. They suggested that locally stored reports on the users' mobile devices should be sent when network signal returns.

Power, et al. [23] posits that social media could act as an effective canal of communication. On the other hand, Anderson [14] states that the use of social media has not been widely adopted. There are several difficulties in adopting social media, such as analyzing a large amount of available information and having trust in its content [15]. However, there have been successful studies on the detection of earthquake disasters using social media [16–18].

2.4. Meteorological Factors

While some researchers [31–33] mention that the key factor for the occurrence of fires are human activities which clear the land to be used for plantation and agriculture, severe drought conditions induced by the dry phase of climate oscillations can worsen land and forest fires. Several studies indicate that forest fires in Indonesia correlate with dynamical weather conditions and long-term climate indices representing climate oscillations such as El Niño/Southern Oscillation (ENSO) and Indian Ocean Dipole (IOD). In Indonesia, dry conditions are associated with the El Niño phase of ENSO and positive phase of IOD. El Niño events can be classified into two main categories or types based on the location of the maximum sea surface anomaly (SSTA). They are the Eastern Pacific (EP or canonical) El Niño with the maximum SSTA observed in the Eastern Pacific and the Central Pacific (CP or Moldoki) types (Ashok et al. [34]; Yu et al. [35]; Yeh et al. [36]) with the maximum SSTA located in the central Pacific. Studies by Ashok & Yamagata [37], Yu et al. [35], Yu & Kim [38], Wang & Wang [39], and Zhang et al. [40] showed that the different types of El Niño control Walker circulation location and intensity, modifying global circulation and precipitation patterns.

Numerous studies have shown that El Niño events cause intensive fires in Indonesia (C. C. Chen et al. [41]; Fanin & Werf [42]; Field et al. [43]; Wooster et al. [44]; Yin et al. [45]). A study by Wooster et al. [44] indicates that El Niño, by its influence on precipitation, has a great influence on fire activity magnitude in Kalimantan (Borneo). Their results also showed that forecasting El Niño conditions provides ways of estimating the extent and magnitude of fire events several months in advance. Pan et al. [23] recently showed that different types of El Niño can influence the occurrence of Indonesian fires each in a slightly different way; however, both El Niño types are associated with increased droughts and more severe fires (Field et al. [43]). According to Lee and McPhaden [46], in recent years, the El Niño CP type has been appearing more frequently.

Another factor that can modify the Indonesian precipitation pattern, and therefore influence fires, is the Indian Ocean Dipole (IOD). According to Saji et al. [47], IOD develops in the tropical Indian ocean (IO) and is characterized by the reversal of the SST gradient between the eastern and western equatorial Indian ocean. During the positive IOD event, the SST in the western IO is higher than in the eastern IO, modifying the zonal circulation and suppressing precipitation in Indonesia. IOD is usually initiated in a boreal summer, reaches its maximum height in autumn, and quickly falls off in winter. Interactions of the IOD and El Niño create variations in precipitation patterns for the different types of El Niño (Wang & Wang [39]; Zhang et al. [40]). Earlier studies of fires in Indonesia (Fanin and Werf [42]; Field et al. [43]) showed that the phase of Indian Ocean Dipole (IOD) might modulate fires, particularly when occurring in conjunction with the El Niño phase that also contributes to fires. Recently, Pan et al. [23], using the fire data from 1979 to 2016, showed that taking into consideration the presence of different El Niño and IOD regimes advanced understanding of the role of climate variability in Indonesian fire activity. They classified 12 El Niño events from 1979 to 2016 into Eastern Pacific (EP) and Central Pacific (CP) types (four and eight El Niño events, respectively) and analyzed observational datasets of sea surface temperature, precipitation, drought code, carbon emission associated with biomass burning, optical depth of aerosol and visibility. The results showed that in weather conditions associated with El Niño of the EP type, Indonesian droughts and fires occurred more intensely and were more prolonged, and the carbon amount emitted almost doubled when compared with the weather of the El Niño CP type. However, the IOD effect was evident when the El Niño CP events were further separated according to the phase of IOD. Weakly positive or negative phases of the IOD were associated with less intense burning during the fire season, consistent with increased precipitation during these IOD states.

3. Materials and Methods

In order to develop an initial understanding of the feasibility of using social media to augment fire monitoring in Indonesia and to provide supplementary tools for policy makers, we decided to perform a case study to understand the availability of fire-related tweets, newspapers reports and weather-related large-scale factors and their correlation with satellite-derived products. To this end, we focused our study on the Riau Province of Sumatra Island, Indonesia.

We choose Riau Province because it is situated in one of the most vulnerable locations in Indonesia for the occurrence of land and forest fires. Its location, which is close to neighboring countries such as Singapore and Malaysia, means that fires and associated smoke create a problem both in Indonesia itself as well as causing transboundary haze transport to the afore-mentioned countries.

Hotspot data from the MODIS satellite sensor on Terra and Aqua satellites from 2015 to 2019 is used to examine occurrences of forest fires in the Riau area. The ENSO and IOD events are identified using their respective indices. Newspaper data related to forest fires in Riau from 1 January 2019 to 31 December 2019 were examined, as well as five years of Twitter data from 2015 to 2019. This is described in more detail below.

3.1. Satellite Data Collection, Construction and Analysis

The satellite hotspot data from the MODIS satellite sensor on Terra and Aqua satellites and SNPP satellite were used to define the occurrences of land and forest fires from 2015 to 2019. The data was collected from a LAPAN site [16]. The MODIS (Moderate-Resolution Imaging Spectroradiometer) both on Terra and Aqua are based on channels 21 and 22

(3.92–3.98 μm) with spatial resolution of 1 km. The S-NPP (Suomi National Polar Orbiting Partnership) satellite data is based on the VIIRS sensor and uses channel 14 between 3.55 and 3.93 μm with spatial resolution of 375 m. These satellites observe Indonesia twice daily on their polar orbit. The LAPAN system allows for retrieval of data for a specific province or latitude and longitude box. One can also define a specific time period, confidence level and satellite observing system. The data is in the form of CSV files with latitude, longitude, relative humidity (%), name of satellite system (Aqua, Terra, SNPP) and date. Discussion of satellite data analysis is beyond the scope of this study, but in short, satellite inversion techniques retrieve the surface temperature by observing incoming radiance in the near-infrared spectrum region.

3.2. Meteorological Data Collection, Construction and Analysis

The ENSO and IOD indices data were used to identify El Niño and positive IOD events and confirm whether there is a connection between meteorological conditions and forest fires in Indonesia. Data for a 2015–2019 period were used. The Niño 3.4 index data and Dipole Mode Index (DMI) were used to identify ENSO and IOD phases, respectively. Both datasets are obtained from the Climate Prediction Center (CPC) of the National Weather Services, National Oceanic and Atmospheric Administration (NOAA). According to the CPC climate guide, the Niño 3.4 index reflects the SST anomaly averaged for the (5 N–5 S, 170 W–120 W) region. The Niño 3.4 index typically uses a 5-month running mean, and El Niño or La Niña events are defined as conditions when the Niño 3.4 SSTs exceed $+/- 0.4$ C for a period of six months or more. The meteorological data sets are quality controlled by the Climate Prediction Center (CPC) of the National Weather Services. No further quality control of the data was performed. DMI is defined as a difference between the SSTA in the western (10 S–10 N, 50 E–70 E) and eastern (10 S–0 N, 90 E–110 E) Indian Ocean.

3.3. Newspaper Data Collection, Construction and Analysis

The data was collected by searching fire-related keywords and reading online newspaper articles. We tried to estimate the duration, location, latitude and longitude (based on city mentioned in an article). We also collected the URL of the online newspaper and its name. We used Antara News for the Riau region and other online accessible sources of information. A qualitative content analysis related to fires from newspapers was conducted. Information from online newspapers was collected from the 1 January 2019 to 31 December 2019. The newspaper analysis also gave us information about the reason why fires occurred and other information such as impact on people's health, animal mortality related to ecosystem disruption, visibility and decrease in air quality, as well as limitations related to the government firefighting response. The purpose of newspaper analysis was not quantitative but was necessary to better understand issues related to policy making which would have been otherwise difficult to grasp just from numerical information about the number of hotspots and the number of tweets.

3.4. Twitter Data Collection, Construction and Analysis

Twitter messages form a mini blog, with a single tweet containing no more than 280 characters. Such messages were also used to confirm whether there is a connection between Twitter and the land and forest fire events from 2015–2019 in Riau, Indonesia. The limitation of the number of characters used on Twitter forces messages to be short and precise. Twitter messages provide some metadata with every tweet which may include location of tweet, author's name and the time when the tweet was posted. However, gathering and analyzing Twitter messages may be challenging because of data access and metadata uncertainties. There are several ways of obtaining historical tweets. For example, one can use Twitter API; however, for standard users there is only access to the previous 7 days. There are also some paid datasets, such as GNIP. There are some Twitter datasets available, but none addressing fires in Indonesia. Therefore, we used scrapping script to obtain the Twitter data.

Using the GetOldTweets-python, we were able to search twitter for the 5-year period between 2015 and 2019 for all of Indonesia using (separately) the following keywords: forest fire: kebakaran hutan, land fire: kebakaran lahan, smoke: asap, no rain: tidak ada hujan, fire: api, drought: kekeringan.

In the 5-year period, we located about 1,250,000 tweets which satisfied our search criteria. In contrast, only 50,000 of them contained the keyword "kekeringan" (draught). For fire events, we did not include "no-rain" and "drought" keywords; these were used to understand the number of tweets related to drought. The "smoke" keyword was used separately to understand possible "smoke" hazard related to fire events. Therefore, "fire" keywords dominated by far the total amount of retrieved tweets.

To analyze tweets, we needed to know date and place as well as number of relevant tweets. Dates are provided with individual tweets. Determining location is the most difficult task. Due to privacy concerns, tweets do not include GPS latitude and longitude position (even though users can provide such data) which would have been the simplest way of obtaining location. Unfortunately, less than 1% of all tweets contain such information. This significantly lowers the number of tweets available for analysis. To overcome this limitation, one can analyze tweet text and its metadata to provide an estimated event location. The geoparsing algorithm TAGGS of de Bruijn [19] was used to determine geographical locations. The TAGGS algorithm allows geoparsing based on the user's nationality (for example Bahasa language) and their hometown. One can also discover direct referrals to specific locations. For example, it can be the name of the province or the name of a city in the message itself. Such an estimation method may result in a certain number of messages being assigned to an inaccurate location, but since we rely on a large number of tweets to identify fire events, we obtained a large signal that minimizes noise from potentially inaccurate geoparsing location.

4. Results and Discussion

4.1. The Analysis of Meteorological and Climatological Conditions and Drought Effects

The Niño 3.4 and the indices of the IOD from the NOAA's CPC in the 2015–2019 timeframe are shown in Figure 1. We can see that positive Niño 3.4 and IOD occurred in April–December 2015, March–July 2017, July–December 2018 and May–December 2019, thus providing insights into climatological forcings in Indonesia for that period. It appears that the conditions preferable for increase in Indonesian fires were observed in 2015/2016 (El Niño) and 2019 (El Niño and positive IOD).

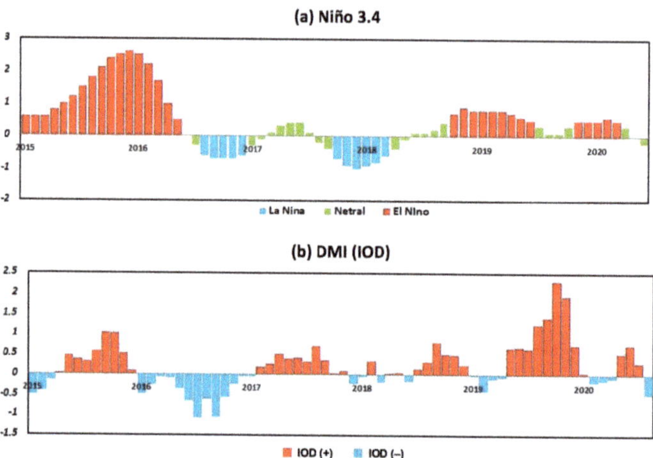

Figure 1. The Niño 3.4 and DMI (IOD) indices during periods of drought for 2015 to 2019.

4.2. The Analysis of the Newspaper Data

We analyzed the content of 31 articles on 2019 fire events as reported in the online Riau Antara Newspaper (https://riau.antaranews.com, accessed in 10 January 2020). From these articles one can deduce that the 2019 forest fires season started on 31 December 2018. The fires were mostly burning peatlands with some areas covered by shrub vegetation, palm oil trees and rubber plantation vegetation. In early 2019, some newspapers reported that forest fires occurred due to the dry season without rainfall and with strong winds and from land clearing for plantations and grass burning by people.

On 11 January 2019, newspapers reported that the burning area had reached up to 267.5 hectares and that local disaster management in Riau considered themselves on "standby" (ready for action) status related to forest and land fires in Riau, but to issue a final decision, they needed to coordinate with various institutions such as the Indonesian Agency for Meteorology, Climatology, and Geophysics (BMKG) and the Riau Provincial Government. Subsequently, on 18 February 2019, "standby" status related to land and forest fires had been declared for Bengkalis Regency and Dumai City but it was not yet determined at the level of the entire Riau province. It was also mentioned that "standby" status could be issued at the provincial level if two or more regencies or cities determined that such status was warranted. It took more than a month for "standby" status for forest and land fires in the Province of Riau to be declared. By that time, approximately 842 hectares in Riau had burned. It was also written that once the Riau Provincial Government determined "standby" status regarding land and forest fires, they could immediately coordinate with the National Disaster Management Agency (BNPB) for support and assistance, such as asking for additional information, requesting helicopters or other resources.

Some newspapers also reported difficulties in the field during efforts to extinguish fires. Locations were sometimes difficult to reach because of difficult terrain, no access to roads, the necessity of using small boats through mangrove forests or access by foot only. On occasion, winds were so strong that fires spread too quickly for people to react.

In summary, newspapers do provide insight into problems that occur in the field during firefighting and show how the government responds while dealing with many regulations—the result being that this bottom-up approach creates a delayed response. However, using newspapers has the constraint of a lack of information on the exact location of fires. Further, newspaper information is often delayed, and fire start times are not provided.

4.3. Analysis of Twitter Data

Twitter data on land and forest fires in Riau from 2015 to 2019 was compared with the occurrences of fires from MODIS satellite images for the same period (Figure 2). These plots, comparing the number of tweets and the number of hotspots, show consistency between them. For example, when the number of hotspots in 2015, 2016 and 2019 is large, the number of tweets is also high.

4.4. Empirical Analysis of Social Media Response

Hotspots vs. tweet scatterplots for 2015–2019 (Figure 3) show correlation of about 0.04 in 2015, 0.22 in 2016, 0.001 in 2017, 0.12 in 2018, and 0.089 in 2019, and the correlation between hotspot and Twitter data (not shown) for the whole year ranges from 0.04 to 0.39. This may be because of insufficient data in some months where there are no tweets or hotspots. However, for the dry season between August to October when drought conditions occur in Riau, there is better correlation, between 0.21 and 0.47 (Table 1). When hotspot occurrences are high as in 2015 and 2019, the correlation is lower compared to years when the hotspot occurrences are smaller. This may indicate that people are more concerned with fires during the dry season.

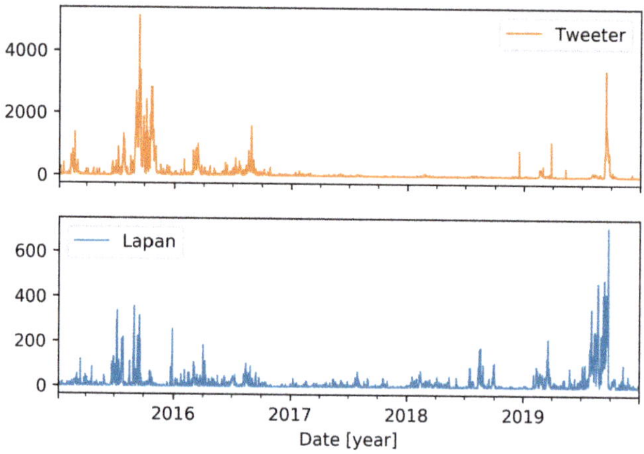

Figure 2. Number of tweets and hotspot occurrence from MODIS satellite images in Riau, 2015–2019.

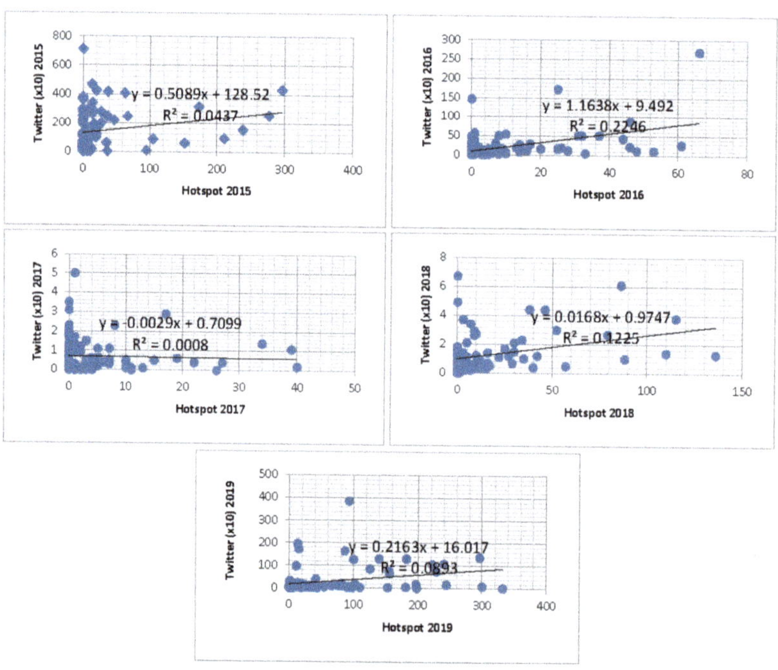

Figure 3. Annual trend of the scatterplot between hotspots and tweets from 2015 to 2019.

We grouped tweets and hotspots into monthly groups and calculated correlation between them for all the years. During 2015, 2016 and 2019, such correlation is high: 0.71, 0.85 and 0.84 respectively (Table 2 and Figure 4). Therefore, when monthly hotspot occurrences are high, such as in 2015, 2016 and 2019, the correlation is also high comparing to years when the hotspot occurrences are smaller and this monthly correlation for each year is also reflected in the total burnt data for each year.

Table 1. Yearly correlation between the number of hotspots and the number of tweets in Riau from 2015 to 2019.

Year	Tweet–Hotspot Correlation (January–December)	Tweet–Hotspot Correlation (August–October)
2015	0.2714	0.2091
2016	0.3847	0.4739
2017	0.0535	0.2091
2018	0.0421	0.4739
2019	0.3873	0.2060

Source: Authors' calculation.

Table 2. Hotspot–Twitter (monthly total) correlation and total burnt area in Riau, 2015–2019.

Year	Monthly Hotspot–Monthly Twitter Correlation	Total Burnt Area (in 100,000)
2015	0.7094	1.8381
2016	0.8567	0.8522
2017	0.3330	0.0687
2018	0.1094	0.3724
2019	0.8435	0.9055

Source: Authors' calculation.

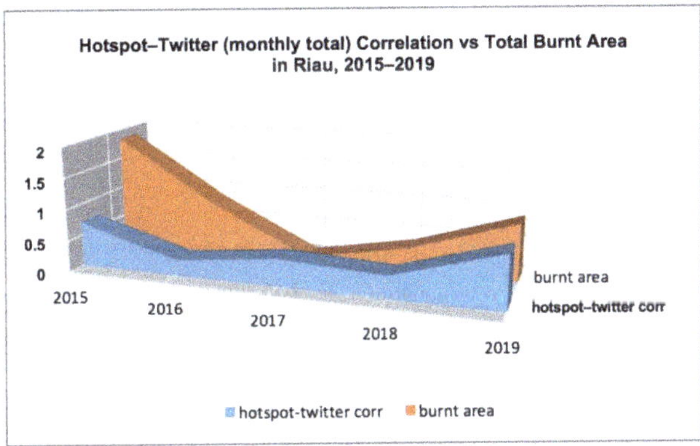

Figure 4. Total hotspot Twitter correlation and total burnt area in Riau, 2015–2019.

Social media information from newspapers and Twitter data provide important information that can support land and forest fire management decisions. For example, one can derive such information as fire location, time of fire occurrence, impact and risk faced by the community, health issues related to smoke, causes of fires and problems experienced in the field by firefighters.

The content analysis of the newspapers show that the requirements needed to establish "standby status" for land and forest fires at the provincial level, which results in support from the central government, are often too onerous. This leads to worsening conditions, causing the fire to spread and become more difficult to contain. The bottom-up approach would be fine for normal weather and climate conditions. However, a top-down approach is strongly required during extreme conditions of weather and climate.

The Twitter analysis shows that total monthly tweets and total monthly hotspots are well correlated, and this provides beneficial information for usage of Twitter data. Newspapers and Twitter information associated with forest fires show that they can provide additional near-real-time information on forest fire detection that is beneficial for forest fire management policy makers in deciding on approaches to fire prevention and mitigation.

Number of tweets, as it changes, might serve as an alert for decision makers to consider in order to determine appropriate steps to address problems on the ground.

Clearly there is no 1:1 correspondence between Twitter data and satellite hotspots. There can be several reasons for this: (1) Tweets can be sent before satellite hotspots are observed, particularly when fires are just starting, and satellite detection algorithms are not sensitive enough. On the other hand, satellites observe remote areas where mobile networks needed for tweeting do not have enough coverage, or where there are few people. In this case, hotspots will be recorded before tweets appear. (2) Another reason is the tweet "memory" effect in which tweets grow over time and "explode" when the intensity of the event is large. There may be also a long decay of tweets after a particularly intense fire event. In other words, tweets are a "social phenomenon". People discuss the event and tweet about subsequent health issues and displacement. This is why we observe long "Twitter memory". (3) There also may be a discrepancy because some events lead to tweets which are not from the locality where the fire occurred. For example, relatives in Jakarta may tweet about particularly large fires in Riau.

More research such as artificial intelligence Natural Language Processing (NLP) could be used to analyze fire-related Twitter messages and their discrepancies with hotspot data. For example, in the context of Twitter and floods, de Bruijin et al. [48] adopted the cased multilingual version of BERT to categorize tweets into two groups: the group which is associated with the ongoing event ("relevant") and the group which is not associated with the ongoing event ("irrelevant"). BERT processes the correlations between the words and sub-words in a text using the Natural Language Processing (NLP) model which is based on deep learning. However, such an analysis is beyond the scope of the current work. One intriguing angle would be the development of semi-automatic sentiment analysis as suggested recently by Mustaqim et al. [26], but one which would also provide satellite and meteorological data.

Even though tweets are "social media", our research illustrates how objectively derived information from hotspots correlate with tweets, thus providing a glimpse of how they can be used in future events for better policy making. For example, large events are well captured by a strong, sharp rise in tweet volume. We also observed a "memory effect" in which tweets are more "sentiment"-related, but these may nevertheless be useful for policy makers because even after the fires were extinguished, they were still being discussed in villages and cities.

5. Conclusions

We follow here the methodology of Saura, et al. [49] and Ribeiro-Navarrete, et al. [50] while summarizing the results. Social media information about fires, combined with data on forest and land fires, such as satellite-derived hotspots, wind velocity and speed, and rainfall and moisture content provide an additional source of information which can be used for near-real time forecasting and policy decision making. Newspapers do provide insight into problems that occur in the field during firefighting and how the government responds while faced with a complex regulatory framework. This often results in a bottom-up approach which can create delayed action. We also discuss that large-scale weather and oceanic state indices such as Indian Ocean Dipole and El Niño are well correlated with forest fire occurrences, thereby providing an additional estimation variable for forest fire detection. This work is relevant to the Riau province land and forest fires which are linked to the presence of peat soil and certain land management practices. We note that such practices lead to this region's high frequency of fire activity mostly during the dry seasons. It is observed that the remoteness of these areas is a factor in attempting to develop alternative detection tools for warning, such as tracking of Twitter messages. The government of Indonesia has tried many approaches to prevent and mitigate the occurrence of forest and land fires in Indonesia. In this study, existing approaches taken by the BMKG and other Indonesian agencies to estimate fire activity are reviewed and a novel approach based on crowdsourcing of Twitter messages is proposed.

5.1. Managerial Implication

Our results suggest the following: (1) A top-down approach is required for the extreme weather and climate conditions related to land and forest fires. (2) Since the dynamic condition of weather and climate phenomena is strongly correlated to fire intensity in Indonesia, the government needs to include these factors in its Fire Rating Danger System and forest fires vulnerability maps through the identification of the El Niño type and phase of IOD. In the future, even shorter time scale events such as Convectively Coupled Equatorial Waves and Madden–Julian Oscillations should be included. (3) Newspaper and Twitter data could also be included in the fire management system to support early detection of and timely warnings about fires.

5.2. Practical/Social Implications

Our findings provide some alternative ways for policy makers within the Indonesian government to prevent and mitigate the occurrence of land and forest fires.

5.3. Limitations and Future Research

This study can be extended in several ways. The development of a land and forest fire disaster threats index could be enhanced by adding information about weather and climate conditions. Of particular importance would be the forecasting of short term weather events such as Convectively Coupled Kelvin Waves and tropical disturbances such as Mesoscale Convective Systems and information about sudden onsets of droughts [51] which may contribute to sudden fire onsets. Incentives to provide Twitter information could be developed; enhanced mobile signal networking capabilities in remote areas would increase the viability of our proposed approach. "Public sentiment" analysis can be performed as to how people feel about government response and actions as concerns land and forest fire mitigation. We addressed this issue only in a cursory way in this study, but such sentiment analysis can be used to extend the methodology developed here.

Author Contributions: Conceptualization, A.A.F. and P.J.F.; methodology, A.A.F., P.J.F. and K.K.; software, P.J.F.; validation, A.A.F. and P.J.F.; investigation, A.A.F.; writing—original draft preparation, A.A.F.; writing—review and editing, A.A.F., P.J.F., K.K. and N.F.R.; visualization, A.A.F. and P.J.F.; supervision, K.K., project administration, K.K. and N.F.R. All authors have read and agreed to the published version of the manuscript.

Funding: P.J.F. has been supported in part by NSF grant 1724741 "Equatorial Line Observations (ELO) field campaign during the International Years of Maritime Continent" program managed by Eric T. DeWeaver. The Agency for Meteorology, Climatology, and Geophysics of the Republic of Indonesia (BMKG) contribution supported the Years of Maritime Continent ELO project.

Institutional Review Board Statement: Not applicable.

Informed Consent Statement: Not applicable.

Data Availability Statement: The satellite related to this paper can be downloaded from LAPAN site http://modis-catalog.lapan.go.id/monitoring/ (accessed 24 September 2021). El Niño and Indian Ocean Dipole (IOD) Indices are available from https://psl.noaa.gov/gcos_wgsp/Timeseries/DMI/ (accessed 24 September 2021). All the newspaper data were collected by searching online open newspaper sources. The tweet datasets analyzed and generated during the current study are not publicly available due to Twitter's privacy policy but are available from the corresponding author upon a reasonable request in line with the policy. The TAGGS code is publicly available on GitHub (https://github.com/jensdebruijn/TAGGS, accessed 23 September 2021). The Twittercode can be found on the github site: https://github.com/Jefferson-Henrique/GetOldTweets-python, accessed 23 September 2021.

Acknowledgments: The satellite hotspot data from MODIS satellite sensor on Terra and Aqua satellites and SNPP were obtained from LAPAN MODIS monitoring site. The Niño 3,4 index data were taken from the Climate Prediction Center (CPC) of the National Weather Services, National Oceanic and Atmospheric Administration (NOAA). The authors acknowledge the help of Maria Flatau with discussion of large-scale indices. Michał Łabuz processed the TAGGS algorithm and collected Twitter data. Katarzyna Barabasz and Beata Latos helped with analysis of newspaper data, and Dawid Gacek helped with data plotting. Tracy Stevens and Dariusz Baranowski helped to improve the manuscript.

Conflicts of Interest: The authors declare no conflict of interest.

References

1. Tacconi, L. Fires in Indonesia: Causes. *Cost Policy Implic. Bogor CIFOR* **2003**. [CrossRef]
2. Kim, D.; Sexton, J.O.; Townshend, J.R. Accelerated deforestation in the humid tropics from the 1990s to the 2000s. *Geophys. Res. Lett.* **2015**, *42*, 3495–3501. [CrossRef]
3. Holmgren, P. Fire and haze in Riau: Looking beyond the hotspots. *Cent. Int. For. Res. Bogor Indones.* 2013. Available online: https://www2.cifor.org/forestsasia/fire-haze-riau-looking-beyond-hotspots/index.html (accessed on 10 January 2020).
4. Kibanov, M.; Stumme, G.; Amin, I.; Lee, J.G. Mining social media to inform peatland fire and haze disaster management. *Soc. Netw. Anal. Min.* **2017**, *7*, 1–19. [CrossRef]
5. Frankenberg, E.; McKee, D.; Thomas, D. Health consequences of forest fires in Indonesia. *Demography* **2005**, *42*, 109–129. [CrossRef]
6. Ulya, N.A.; Yunardy, S. Analisis dampak kebakaran hutan di Indonesia terhadap distribusi pendapatan masyarakat. *J. Penelit. Sos. Dan Ekon. Kehutan.* **2006**, *3*, 133–146. [CrossRef]
7. Rogers, D.; Tsirkunov, V. Costs and benefits of early warning systems. *Global Assessment Report*, 1 January 2011.
8. Hallegatte, S. A cost effective solution to reduce disaster losses in developing countries: Hydro-meteorological services, early warning, and evacuation. *World Bank Policy Res. Work. Pap.* 2012. Available online: https://openknowledge.worldbank.org/handle/10986/9359 (accessed on 10 January 2020).
9. Kalogeropoulos, A.; Negredo, S.; Picone, I.; Nielsen, R.K. Who shares and comments on news?: A cross-national comparative analysis of online and social media participation. *Soc. Media Soc.* **2017**, *3*, 2056305117735754. [CrossRef]
10. Biswas, M. An exploratory research: A comparative analysis of mainstream and ethnic media coverage of social policy issues in the economic stimulus plan debate. *J. Comp. Soc. Welf.* **2010**, *26*, 13–26. [CrossRef]
11. Stieglitz, S.; Dang-Xuan, L. Social media and political communication: A social media analytics framework. *Soc. Netw. Anal. Min.* **2013**, *3*, 1277–1291. [CrossRef]
12. Sinnenberg, L.; Buttenheim, A.M.; Padrez, K.; Mancheno, C.; Ungar, L.; Merchant, R.M. Twitter as a tool for health research: A systematic review. *Am. J. Public Health* **2017**, *107*, e1–e8. [CrossRef]
13. Power, R.; Robinson, B.; Colton, J.; Cameron, M.A. A Case Study for Monitoring Fires with Twitter. In Proceedings of the ISCRAM, Kristiansand, Norway, 24–27 May 2015.
14. Anderson, M. Integrating social media into traditional emergency management command and control structures: The square peg into the round hole. In Proceedings of the Disaster and Emergency Management Conference Proceedings, Brisbane, Australia, 16–18 April 2012; pp. 18–34.
15. Lindsay, B.R. *Social Media and Disasters: Current Uses, Future Options, and Policy Considerations*; UNT Library: Denton, TX, USA, 2011.
16. Sakaki, T.; Okazaki, M.; Matsuo, Y. Tweet analysis for real-time event detection and earthquake reporting system development. *IEEE Trans. Knowl. Data Eng.* **2012**, *25*, 919–931. [CrossRef]
17. Robinson, B.; Power, R.; Cameron, M. A sensitive twitter earthquake detector. In Proceedings of the 22nd International Conference on World Wide Web, Rio de Janeiro, Brazil, 13–17 May 2013; pp. 999–1002.
18. Avvenuti, M.; Cresci, S.; Marchetti, A.; Meletti, C.; Tesconi, M. Ears (earthquake alert and report system) a real time decision support system for earthquake crisis management. In Proceedings of the 20th ACM SIGKDD International Conference on Knowledge Discovery and Data Mining, New York, NY, USA, 24–27 August 2014; pp. 1749–1758.
19. de Bruijn, J.A.; de Moel, H.; Jongman, B.; Wagemaker, J.; Aerts, J.C. TAGGS: Grouping tweets to improve global geoparsing for disaster response. *J. Geovisualization Spat. Anal.* **2018**, *2*, 2. [CrossRef]
20. Power, R.; Robinson, B.; Ratcliffe, D. Finding fires with twitter. In Proceedings of the Australasian Language Technology Association Workshop 2013 (ALTA 2013), Brisbane, Australia, 4–6 December 2013; pp. 80–89.
21. Reid, J.; Xian, P.; Hyer, E.; Flatau, M.; Ramirez, E.; Turk, F.; Sampson, C.; Zhang, C.; Fukada, E.; Maloney, E. Multi-scale meteorological conceptual analysis of observed active fire hotspot activity and smoke optical depth in the Maritime Continent. *Atmos. Chem. Phys.* **2012**, *12*, 2117–2147. [CrossRef]
22. Shawki, D.; Field, R.D.; Tippett, M.K.; Saharjo, B.H.; Albar, I.; Atmoko, D.; Voulgarakis, A. Long-lead prediction of the 2015 fire and haze episode in Indonesia. *Geophys. Res. Lett.* **2017**, *44*, 9996–10005. [CrossRef]
23. Pan, X.; Chin, M.; Ichoku, C.M.; Field, R.D. Connecting Indonesian fires and drought with the type of El Niño and phase of the Indian Ocean dipole during 1979–2016. *J. Geophys. Res. Atmos.* **2018**, *123*, 7974–7988. [CrossRef]
24. Forsyth, T. Public concerns about transboundary haze: A comparison of Indonesia, Singapore, and Malaysia. *Glob. Environ. Chang.* **2014**, *25*, 76–86. [CrossRef]
25. Panjaitan, R.B.; Sumartono, S.; Sarwono, S.; Saleh, C. The role of central government and local government and the moderating effect of good governance on forest fire policy in Indonesia. *Benchmarking Int. J.* **2019**, *26*, 147–159. [CrossRef]
26. Mustaqim, T.; Umam, K.; Muslim, M. *Twitter Text Mining for Sentiment Analysis on Government's Response to Forest Fires with Vader Lexicon Polarity Detection and K-Nearest Neighbor Algorithm*; IOP Publishing: Bristol, UK, 2020; Volume 1567, p. 032024.

27. Carvalho, J.; Plastino, A. On the evaluation and combination of state-of-the-art features in Twitter sentiment analysis. *Artif. Intell. Rev.* **2021**, *54*, 1887–1936. [CrossRef]
28. Carley, K.M.; Malik, M.M.; Kowalchuck, M.; Pfeffer, J.; Landwehr, P. Twitter Usage in Indonesia 2015. Available online: https://ssrn.com/abstract=2720332 (accessed on 23 September 2021).
29. Baranowski, D.B.; Flatau, M.K.; Flatau, P.J.; Karnawati, D.; Barabasz, K.; Labuz, M.; Latos, B.; Schmidt, J.M.; Paski, J.A. Social-media and newspaper reports reveal large-scale meteorological drivers of floods on Sumatra. *Nat. Commun.* **2020**, *11*, 1–10. [CrossRef] [PubMed]
30. Aditya, T.; Laksono, D.; Izzahuddin, N. Crowdsourced hotspot validation and data visualisation for location-based haze mitigation. *J. Locat. Based Serv.* **2019**, *13*, 239–269. [CrossRef]
31. Goldammer, J.G. History of equatorial vegetation fires and fire research in Southeast Asia before the 1997–98 episode: A reconstruction of creeping environmental changes. *Mitig. Adapt. Strateg. Glob. Chang.* **2007**, *12*, 13–32. [CrossRef]
32. Goldammer, J.G.; Seibert, B. The impact of droughts and forest fires on tropical lowland rain forest of East Kalimantan. In *Fire in the Tropical Biota*; Springer: Berlin/Heidelberg, Germany, 1990; pp. 11–31.
33. Simorangkir, D. Fire use: Is it really the cheaper land preparation method for large-scale plantations? *Mitig. Adapt. Strateg. Glob. Chang.* **2007**, *12*, 147–164. [CrossRef]
34. Ashok, K.; Behera, S.K.; Rao, S.A.; Weng, H.; Yamagata, T. El Niño Modoki and its possible teleconnection. *J. Geophys. Res. Oceans* **2007**, *112*. [CrossRef]
35. Yu, J.-Y.; Kao, H.-Y.; Lee, T.; Kim, S.T. Subsurface ocean temperature indices for Central-Pacific and Eastern-Pacific types of El Niño and La Niña events. *Theor. Appl. Climatol.* **2011**, *103*, 337–344. [CrossRef]
36. Yeh, S.-W.; Kug, J.-S.; Dewitte, B.; Kwon, M.-H.; Kirtman, B.P.; Jin, F.-F. El Niño in a changing climate. *Nature* **2009**, *461*, 511–514. [CrossRef]
37. Ashok, K.; Yamagata, T. The El Niño with a difference. *Nature* **2009**, *461*, 481–484. [CrossRef] [PubMed]
38. Yu, J.; Kim, S.T. Identifying the types of major El Niño events since 1870. *Int. J. Climatol.* **2013**, *33*, 2105–2112. [CrossRef]
39. Wang, X.; Wang, C. Different impacts of various El Niño events on the Indian Ocean Dipole. *Clim. Dyn.* **2014**, *42*, 991–1005. [CrossRef]
40. Zhang, W.; Wang, Y.; Jin, F.; Stuecker, M.F.; Turner, A.G. Impact of different El Niño types on the El Niño/IOD relationship. *Geophys. Res. Lett.* **2015**, *42*, 8570–8576. [CrossRef]
41. Chen, Y.; Morton, D.C.; Andela, N.; Van Der Werf, G.R.; Giglio, L.; Randerson, J.T. A pan-tropical cascade of fire driven by El Niño/Southern Oscillation. *Nat. Clim. Chang.* **2017**, *7*, 906–911. [CrossRef]
42. Fanin, T.; Werf, G.R. Precipitation–fire linkages in Indonesia (1997–2015). *Biogeosciences* **2017**, *14*, 3995–4008. [CrossRef]
43. Field, R.D.; Van Der Werf, G.R.; Fanin, T.; Fetzer, E.J.; Fuller, R.; Jethva, H.; Levy, R.; Livesey, N.J.; Luo, M.; Torres, O. Indonesian fire activity and smoke pollution in 2015 show persistent nonlinear sensitivity to El Niño-induced drought. *Proc. Natl. Acad. Sci. USA* **2016**, *113*, 9204–9209. [CrossRef] [PubMed]
44. Wooster, M.; Perry, G.; Zoumas, A. Fire, drought and El Niño relationships on Borneo (Southeast Asia) in the pre-MODIS era (1980–2000). *Biogeosciences* **2012**, *9*, 317–340. [CrossRef]
45. Yin, Y.; Ciais, P.; Chevallier, F.; Van der Werf, G.R.; Fanin, T.; Broquet, G.; Boesch, H.; Cozic, A.; Hauglustaine, D.; Szopa, S. Variability of fire carbon emissions in equatorial Asia and its nonlinear sensitivity to El Niño. *Geophys. Res. Lett.* **2016**, *43*, 10–472. [CrossRef]
46. Lee, T.; McPhaden, M.J. Increasing intensity of El Niño in the central-equatorial Pacific. *Geophys. Res. Lett.* **2010**, *37*. [CrossRef]
47. Saji, N.; Goswami, B.N.; Vinayachandran, P.; Yamagata, T. A dipole mode in the tropical Indian Ocean. *Nature* **1999**, *401*, 360–363. [CrossRef]
48. de Bruijn, J.A.; de Moel, H.; Jongman, B.; de Ruiter, M.C.; Wagemaker, J.; Aerts, J.C. A global database of historic and real-time flood events based on social media. *Sci. Data* **2019**, *6*, 1–12. [CrossRef]
49. Saura, J.R.; Palacios-Marqués, D.; Iturricha-Fernández, A. Ethical design in social media: Assessing the main performance measurements of user online behavior modification. *J. Bus. Res.* **2021**, *129*, 271–281. [CrossRef]
50. Ribeiro-Navarrete, S.; Saura, J.R.; Palacios-Marqués, D. Towards a new era of mass data collection: Assessing pandemic surveillance technologies to preserve user privacy. *Technol. Forecast. Soc. Chang.* **2021**, *167*, 120681. [CrossRef]
51. Pendergrass, A.G.; Meehl, G.A.; Pulwarty, R.; Hobbins, M.; Hoell, A.; AghaKouchak, A.; Bonfils, C.J.; Gallant, A.J.; Hoerling, M.; Hoffmann, D. Flash droughts present a new challenge for subseasonal-to-seasonal prediction. *Nat. Clim. Chang.* **2020**, *10*, 191–199. [CrossRef]

MDPI

St. Alban-Anlage 66

4052 Basel

Switzerland

Tel. +41 61 683 77 34

Fax +41 61 302 89 18

www.mdpi.com

Sustainability Editorial Office

E-mail: sustainability@mdpi.com

www.mdpi.com/journal/sustainability